D1748075

Edited by
Jean-Marie Dubois and
Esther Belin-Ferré

Complex Metallic Alloys

Related Titles

Bennet, D.W.

Understanding Single-Crystal X-Ray Crystallography

2010
Hardcover
ISBN: 978-3-527-32677-8

Jackson, K.A.

Kinetic Processes

Crystal Growth, Diffusion, and Phase Transitions in Materials

2010
Hardcover
ISBN: 978-3-527-32736-2

Kumar, C. S. S. R. (ed.)

Mixed Metal Nanomaterials

2009
Hardcover
ISBN: 978-3-527-32153-7

Kumar, C. S. S. R. (ed.)

Nanostructured Oxides

2009
Hardcover
ISBN: 978-3-527-32152-0

Kumar, C. S. S. R. (ed.)

Metallic Nanomaterials

2009
Hardcover
ISBN: 978-3-527-32151-3

Zehetbauer, M.J., Zhu, Y.T. (eds.)

Bulk Nanostructured Materials

2009
Hardcover
ISBN: 978-3-527-31524-6

Herlach, D.M., Kirchheim, R. (eds.)

Phase Transformations in Multicomponent Melts

2008
Hardcover
ISBN: 978-3-527-31994-7

Hirsch, J., Skrotzki, B., Gottstein, G. (eds.)

Aluminium Alloys

Their Physical and Mechanical Properties

2008
Hardcover
ISBN: 978-3-527-32367-8

Jackson, S.D., Hargreaves, J.S.J. (eds.)

Metal Oxide Catalysis

2008
Hardcover
ISBN: 978-3-527-31815-5

Pfeiler, W. (ed.)

Alloy Physics

A Comprehensive Reference

2007
Hardcover
ISBN: 978-3-527-31321-1

Foreword

The field of *Complex Metallic Alloys* can, although its roots reach back to the nineteen twenties, be considered one of the most recent research areas in modern materials science. The term *Complex Metallic Alloys* denotes a broad family of binary or multinary compounds consisting of either metallic elements or mixtures of metals to which metalloids, rare earth elements or chalcogenides are added. Their crystal structure is based on extraordinary large unit cells containing typically some ten to some hundred atoms. Cases with more than a thousand atoms per unit cell are also known. To understand how nature can organise such a high number of atoms into a highly ordered crystalline lattice presents a challenge to science as well as a chance to obtain new insights into the properties of condensed-matter systems. On the other hand the unusual structure gives rise to unusual physical properties with a potential for new technological applications.

In a pioneering paper published in 1923 Linus Pauling described for the first time the structure of an intermetallic compound [1]. He reported on an X-ray diffraction study of $NaCd_2$. In spite of the apparently "harmless" stoichiometry the diffraction patterns of this crystal were so complicated, however, that it was not then possible to assign indices to many of the diffraction spots. Only more than thirty years later, in 1955, Pauling was able to publish a model of the structure [2] based on a cubic space group $Fd\bar{3}m$. Pauling's unit cell has an edge length of about 3 nm and contains 384 sodium and 768 cadmium atoms, making a total of 1152. For metallic materials this is a very large number. Apart from the cases of elementary metals with 2, 2, and, 4 atoms per unit cell for the body-centred cubic, hexagonal close packed and face-centred cubic structure, respectively, we have, for example, 4 atoms for the γ'-phase, Ni_3Al, 16 atoms for Fe_3Al and the Heusler alloy $AlCu_2Mn$, and 52 atoms per unit cell in the γ-brass phase, Cu_5Zn_8. For these new structures the term "giant unit cell crystals" was coined by Sten Samson, another pioneer in the field of intermetallic compounds [3]. Inside the giant unit cells a cluster substructure exists. For example, there is a large group of alloys based on the 55 atom Mackay icosahedron, another group is based on the 105 atom Bergman cluster [4]. Today hundreds of such compounds are known whose structures are based on giant unit cells [5]. However, taking into account that in each new ternary phase diagram

Complex Metallic Alloys: Fundamentals and Applications
Edited by Jean-Marie Dubois and Esther Belin-Ferré
Copyright © 2011 WILEY-VCH Verlag GmbH & Co. KGaA, Weinheim
ISBN: 978-3-527-32523-8

studied, more phases based on giant unit cells are found, their number should run into thousands.

Although crystallographers learned to solve these challenging giant unit cell structures the field was essentially abandoned during the nineteen seventies. The primary reason for this was that the tiny single crystals sufficient for structure determination were by far not enough for physical property studies which could have, on the long run, justified the effort. The growth of larger single crystals as a prerequisite for studies into the intrinsic properties of these compounds was far outside the scope of what metallurgy was able to do at that time. Furthermore, in the precomputer age, solid-state theory was not developed enough to justify hopes that such systems could ever be understood.

The momentum for the rediscovery of giant unit cell compounds was provided by Danny Shechtman's discovery of the quasicrystalline form of solid matter [6]. In fact quasicrystals and giant unit cell compounds share a number of common structural features, the most prominent being the internal cluster substructure. In a sense quasicrystals can be considered giant unit cell structures with an infinitely large unit cell. Indeed considerable progress in the development of models for the structure of quasicrystals was made starting from the known structure of related giant unit cell intermetallics. Nevertheless, giant unit cell intermetallics were for a long time not considered a field of materials science of its own. The term "approximants (to the quasicrystal structure)" indicated that the essentially crystallographic interest was limited to an auxiliary part to be played in the quest for a solution of the structure of quasicrystals. This changed with a programmatic lecture [7] on "Structurally Complex Alloy Phases" given on September 9th, 2002 on the "Eighth International Conference on Quasicrystals" in Bangalore, India, which was meant as an appeal to dedicate intensive research efforts to "one of the last white spots on the map of metal physics". Today, last but not least, after more than two decades of quasicrystal research, the tools are available to deal with giant unit cell materials. Large single crystals can be grown for many systems allowing intrinsic physical property studies, and the last 30 to 40 years have seen an extraordinary development of solid-state theory allowing today to tackle the difficult consequences of the particular atom arrangement for all kinds of properties.

For purely practical reasons it appeared useful to change, sometime later on, from the term "Structurally Complex Alloy Phases" to "Complex Metallic Alloys" and the acronym "CMA" during the application for a European Network of Excellence in the 6th framework Programme of the European Union. There is no clear division line between the family of *Complex Metallic Alloys* and more conventional small-unit cell alloys, on the one end, and quasicrystalline alloys, on the other end. The authors of the present volume find it useful to leave the "boarders" open and to benefit from ideas growing on a wide platform accommodating also Zintl-phases, skutterudides, clathrates and Heusler-alloys.

To what extent it is useful to treat quasicrystals as a part of the *Complex Metallic Alloy* family remains to be seen. From a phenomenological point of view it is useful to define two physical correlation lengths, one related to the lattice parameter, the other referring to correlation effects related to the cluster substructure inside the crystal unit cell. A division criterion between giant unit cell alloys and quasicrystalline phases can then be derived on the basis of the relative importance of the lattice-periodicity related correlation length for a particular feature or property of interest. Although this can help to better understand the properties of either type of atom arrangement, it has to be pointed out that things remain complicated. From an experimental point of view the attempts to construct a correlation between the size of the unit cell volume and the measured electronic density of states at the Fermi level provided valuable insight but they were not entirely successful. On the other hand, an ideal quasicrystal is fully long-range ordered. Its construction rules, although in principle simple in its six-dimensional reference lattice, are complicated in three-dimensional space. Therefore, to build up such a largely defect-free quasicrystal structure as observed experimentally, e.g. in the Al-Mn-Pd system, requires a long-range correlation of an even more stringent nature than that governing conventional crystal formation. In periodic crystals the symmetry of the individual building blocks is compatible with a periodic lattice and therefore both long-range and short-range atomic ordering are driving the system in the same direction. To build up a quasiperiodic lattice requires a long-range structural correlation length exceeding that occurring in conventional crystals.

The use of the term "complex" and the discussion of "complexity" in science is rapidly increasing in recent years. As already pointed out by Warren Weaver in his pioneering paper [8] this can be attributed, on the one hand, to an increasing awareness of the fact that sciences have in the past neglected complexity as a constitutive element of what is happening in nature and in society. On the other hand, sciences have developed to such an extent that phenomena too complex to be dealt with in the past can now be tackled employing the tools and techniques available today.

The field of *Complex Metallic Alloys* is typical for either group of arguments. *Complex Metallic Alloys* are characterised by their large crystal unit cells, by a pronounced cluster substructure with a large variety of coordination polyhedra and by inherent disorder both structurally and chemically as well as by partial site occupation, i.e. lattice sites are left vacant as a result of constraints of electronic origin and simple atom size effects. Furthermore, the recent work, in particular on plasticity, has shown that there is another very important feature of this class of materials and this is the existence of a high number of structurally similar phases within a very narrow region (a few atomic percent wide) of the thermodynamic phase diagram. These provide the system with an additional degree of freedom which is, for instance, used by nature in the formation of metadislocations and for plastic deformation. Often these phases differ so little from each other that it is difficult to isolate them and investigate them individually. And, typical for a complex system, the phenomenology of how these materials develop in time, temperature and composition space depends, in a way that is difficult to grasp, on kinetics as well as on the (difficult to define) starting and boundary conditions. A situation not much different from the situation described by Friedrich August von Hayek in his 1974 Nobel Prize Lecture for complexity in Economics [9] applies to such elementary physical phenomena as diffusion in giant unit cell materials. Although mass transport in *Complex Metallic Alloys* and quasicrystalline phases has been extensively studied experimentally, there is not even an idea how this transport occurs on an atomistic scale and of what nature the decisive variables are or how they can be properly defined maintaining the full complexity of the systems.

Seth Lloyd has compiled a list of about forty different measures of complexity that have been proposed in recent years. Melanie Mitchell, External Professor at the Santa Fe Institute dedicated to complexity research, comments: "...people are going to have to figure out how these diverse notions... are related to one another and how to most usefully refine the overly complex notion of *complexity*. This is work that largely remains to be done..." [10]. In the present volume, some of the authors directly or indirectly use as a measure for complexity the size of the crystal unit cell. Such type of measures have been discussed before and linked to Shannon Entropy, Algebraic Information Content, Thermodynamic Depth, Degree of Hierarchy and so forth [10]. As it is the case with all attempts to quantify complexity since Murray Gell-Mann's pioneering early paper [11] such definitions, although they may be useful at times, have still to stand the test of *practical applicability* and *usefulness*.

Apart from the general scientific insight gained by experimental and theoretical work on *Complex Metallic Alloys* this field has recently seen a number of surprising discoveries. Among these are particular transport properties, e.g. metallic, semiconducting electronic conductivity and isolating behaviour, the observation of superconductivity [12], the observation of a novel magnetic memory effect [13] and the observation of entirely new mechanisms of plastic deformation [14].

The current volume written by experts in the field of *Complex Metallic Alloys* gives in separate chapters an overview on the current state of research in this field as well as on first applications which provide excellent examples of the variety of properties. At the same time this book can serve as a comprehensive guide to the literature and as a starting point of further in-depth studies in the future.

References

1 Pauling, L. (1923) *J. Am. Chem. Soc.*, **45**, 2777.
2 Pauling, L. (1955) *American Scientist*, **43**, 285.
3 Samson, S. (1969) in *Developments in the Structural Chemistry of Alloys Phases* (ed B.C. Giessen), Plenum, New York, p. 65.
4 Bergman, G., Waugh, J.L.T., and Pauling, L. (1957) *Acta Crystallogr.*, **10**, 254.
5 Villars, P. and Calvert, L.D. (1986) *Pearson's Handbook of Crystallographic Data for Intermetallic Phases*, American Society of Metals, Metals Park, OH.
6 Shechtman, D., Blech, I., Gratias, D., and Cahn, J. (1984) *Phys. Rev. Lett.* **53**, 951.
7 Urban K. and Feuerbacher, M. (2004) *J. Non Cryst. Sol.*, **334–335**, 143.
8 Weaver, W. (1948) *American Scientist*, **36**, 536.
9 von Hayek, F.A. (1992) in *Nobel Lectures, Economics 1969–1980* (ed A. Lindbeck), World Scientific Publishing Co., Singapore, p. 258.
10 Mitchell, M. (2009) *Complexity – a Guided Tour*, Oxford University Press, New York.
11 Gell-Mann, M. (1995) *Complexity*, **1**, 16.

12 Bauer, E., Kaldarar, H., Lackner, R., Michor, H., Steiner, W., Scheidt, E.-W., Galatanu, A., Marabelli, F., Wazumi, T., Kumagai, K., and Feuerbacher, M. (2007) *Phys. Rev. B*, **76**, 014528.

13 Dolinsek, J., Feuerbacher, M., Jagodic, M., Jaglicic, Z., Heggen, M., and Urban, K. (2009) *J. Appl. Phys.*, **106**, 043917.

14 Heggen, M., Houben, L., and Feuerbacher, M. (2010) *Nature Materials*, **9**, 332.

Germany
September 2010

Knut W. Urban
JARA Senior Professor
RWTH Aachen
University & Research Centre Jülich

Contents

Foreword *V*
Preface *XIX*
List of Contributors *XXI*

1 **Introduction to the Science of Complex Metallic Alloys** *1*
Jean-Marie Dubois, Esther Belin-Ferré, and Michael Feuerbacher
1.1 Introduction *1*
1.2 Complex Metallic Alloy: What Is It? *2*
1.2.1 Composition and Varieties *3*
1.2.2 Complexity at a Glance *4*
1.2.3 Defects *6*
1.3 Complex Metallic Alloy: Why Is It Complex? *12*
1.3.1 Electronic Densities of States and Hume-Rothery Rules *12*
1.3.2 Self-Hybridization in Al-Mg Alloys *15*
1.4 A Brief Survey of Properties *18*
1.4.1 Transport Properties *18*
1.4.2 Surface Physics and Chemistry *20*
1.4.3 Surface Energy *22*
1.4.4 Plasticity *25*
1.5 Potential Applications *30*
1.5.1 Applications Related to Surface Energy *30*
1.5.2 Applications Related to Transport Properties *33*
1.5.3 Applications Related to Dispersion of Particles in a Matrix *35*
1.6 Conclusion and Introduction of the Following Chapters *36*
References *36*

2 **Properties of CMAs: Theory and Experiments** *41*
Enrique Maciá and Marc de Boissieu
2.1 Introduction *41*
2.2 Electronic-Structure-Related Properties *43*
2.2.1 Transport Properties of Quasicrystals and Approximants *43*

Complex Metallic Alloys: Fundamentals and Applications
Edited by Jean-Marie Dubois and Esther Belin-Ferré
Copyright © 2011 WILEY-VCH Verlag GmbH & Co. KGaA, Weinheim
ISBN: 978-3-527-32523-8

2.2.1.1	Inverse Matthiessen Rule	43
2.2.1.2	Current–Voltage Curves	44
2.2.1.3	Optical Conductivity	45
2.2.1.4	Seebeck Coefficient	49
2.2.1.5	-Wiedemann–Franz Law	49
2.2.2	Chemical Trends	53
2.2.3	Electronic Structure	56
2.2.3.1	Fermi-Level -Pseudogap	56
2.2.3.2	Fine Spectral Features	57
2.2.3.3	Spectral Conductivity Models	59
2.2.3.4	The Role of Critical States	60
2.2.4	Phenomenological Approaches	63
2.2.4.1	Kubo–Greenwood Formalism of Transport Coefficients	63
2.2.4.2	Application Examples	67
2.3	Phonons	71
2.3.1	Phonons: An Introduction	71
2.3.2	Measuring Phonons: Inelastic Neutron and X-Ray Scattering	75
2.3.2.1	Coherent Inelastic Neutron Scattering	75
2.3.2.2	Incoherent Inelastic Neutron Scattering	78
2.3.3	Beyond the Harmonic Approximation	78
2.3.4	Phonons in Quasicrystals and their Approximants	81
2.3.4.1	The Zn_2Mg Laves Phase	81
2.3.4.2	The i-Al-Pd-Mn Icosahedral Quasicrystal	86
2.3.4.3	The i-Zn-Mg-Sc Quasicrystal and its 1/1 Zn-Sc Approximant	88
2.3.5	Phonons in Cage Compounds and Thermoelectricity	95
2.3.5.1	Clathrates	96
2.3.5.2	Skutterudites	101
2.3.5.3	Zinc-Antimony Alloy Zn_4Sb_3	103
2.3.6	Phonon and Transport Properties: The Example of Thermoelectricity	105
2.3.6.1	Thermal Conductivity	105
2.3.6.2	Thermoelectric Figure of Merit	106
2.4	Conclusion	109
	References	109
3	**Anisotropic Physical Properties of Complex Metallic Alloys**	117
	Janez Dolinšek and Ana Smontara	
3.1	Introduction	117
3.2	Structural Considerations and Sample Preparation	118
3.2.1	Y-Al-Ni-Co	118
3.2.2	o-$Al_{13}Co_4$	119
3.2.3	Al_4(Cr,Fe)	119
3.3	Anisotropic Magnetic Properties	120
3.3.1	Y-Al-Ni-Co	120
3.3.2	o-$Al_{13}Co_4$	120

3.3.3	$Al_4(Cr,Fe)$	122
3.4	Anisotropic Electrical Resistivity	124
3.4.1	Y-Al-Ni-Co	124
3.4.2	o-$Al_{13}Co_4$	125
3.4.3	$Al_4(Cr,Fe)$	126
3.5	Anisotropic Thermoelectric Power	130
3.5.1	Y-Al-Ni-Co	130
3.5.2	o-$Al_{13}Co_4$	131
3.5.3	Al4(Cr,Fe)	132
3.6	Anisotropic Hall Coefficient	132
3.6.1	Y-Al-Ni-Co	132
3.6.2	o-$Al_{13}Co_4$	134
3.6.3	$Al_4(Cr,Fe)$	135
3.7	Anisotropic Thermal Conductivity	136
3.7.1	Y-Al-Ni-Co	136
3.7.2	o-$Al_{13}Co_4$	136
3.7.3	$Al_4(Cr,Fe)$	138
3.8	Fermi Surface and the Electronic Density of States	140
3.8.1	Y-Al-Ni-Co	140
3.8.2	o-$Al_{13}Co_4$	142
3.8.3	$Al_4(Cr,Fe)$	142
3.9	Theoretical *Ab Initio* Calculation of the Electronic Transport Coefficients	144
3.9.1	Anisotropic Hall Coefficient of Y-Al-Ni-Co	146
3.9.2	Anisotropic Transport Coefficients of o-$Al_{13}Co_4$	148
3.9.2.1	Electrical Resistivity	148
3.9.2.2	Electronic Thermal Conductivity	149
3.9.2.3	Hall Coefficient	150
3.10	Conclusion	151
	References	152

4	**Surface Science of Complex Metallic Alloys**	**155**
	Vincent Fournée, Julian Ledieu, and Jeong Y. Park	
4.1	Introduction	155
4.2	Surface-Structure Determination	156
4.2.1	Surface Preparation	156
4.2.2	Structure from Real-Space Methods	159
4.2.3	Structure from Reciprocal-Space Methods	165
4.2.4	Structure from *Ab Initio* Methods	167
4.2.5	Stability of Alloy Surfaces	169
4.3	Electronic Structure	170
4.3.1	The Pseudo-Gap Feature	170
4.3.2	Nature of the Electronic States	175
4.4	Thin-Film Growth on CMA Surfaces	177
4.4.1	Low-Coverage Regime	178

4.4.1.1	Nucleation Mechanism	*178*
4.4.1.2	Identification of the Trap Sites	*179*
4.4.1.3	Pseudomorphic Layers	*181*
4.4.2	Multilayer Regime	*184*
4.4.2.1	Twinning of Nanocrystals	*184*
4.4.2.2	Intermixing and Alloying	*185*
4.4.2.3	Electron Confinement	*186*
4.5	Adhesion, Friction and Wetting Properties of CMA Surfaces	*188*
4.5.1	Wetting Properties	*188*
4.5.2	Atomic-Scale Adhesion Properties of Complex Metallic Alloys	*190*
4.5.2.1	Continuum-Mechanics Models	*190*
4.5.2.2	Adhesion on Clean and *In-Situ* Oxidized Quasicrystal Surfaces	*191*
4.5.2.3	Adhesion Measured in the Elastic and Inelastic Regime	*192*
4.5.2.4	Adhesion on Air-Oxidized Quasicrystal Surfaces	*193*
4.5.3	Atomic-Scale Friction Properties	*194*
4.5.3.1	Friction-Measurement Apparatus – FFM and Tribometer	*194*
4.5.3.2	Friction on Atomically Clean and *In-Situ* Oxidized Quasicrystal Surfaces	*196*
4.5.4	Friction Anisotropy	*197*
4.5.4.1	Friction Anisotropy of Clean 2-Fold Al-Ni-Co Surface	*198*
4.5.4.2	Friction Anisotropy After Surface Modification	*200*
4.5.4.3	Low Friction of Quasicrystals and Its Relation with Wetting and Adhesion	*200*
4.6	Conclusion	*201*
	References	*202*
5	**Metallurgy of Complex Metallic Alloys**	*207*
	Saskia Gottlieb-Schoenmeyer, Wolf Assmus, Nathalie Prud'homme, and Constantin Vahlas	
5.1	Introduction	*207*
5.2	Basic Concepts of Crystal Growth	*208*
5.2.1	Bridgman Method	*210*
5.2.2	Zone Melting	*212*
5.2.3	Czochralski Technique	*212*
5.2.4	Flux Growth Technique	*214*
5.3	Examples of Single-Crystal Growth of CMAs	*215*
5.3.1	$Al_{13}Co_4$ and $Al_{13}Fe_4$ Using the Czochralski Technique	*215*
5.3.2	Single-Crystal Growth of β-Al-Mg	*217*
5.3.2.1	Bridgman Growth	*218*
5.3.2.2	Czochralski Growth	*218*
5.3.2.3	Self-Flux Growth	*218*
5.3.3	Single-Crystal Growth of $Mg_{32}(Al,Zn)_{49}$	*219*
5.3.3.1	Bridgman Growth	*219*
5.3.3.2	Czochralski Growth	*220*
5.3.4	Single-Crystal Growth of Al-Pd-Mn Approximants	*221*

5.3.5	Crystal Growth of Yb-Cu Superstructural Phases	222
5.3.6	Single-Crystal Growth of MgZn$_2$	224
5.3.6.1	Bridgman Technique	224
5.3.6.2	Liquid-Encapsulated Kyropoulus Technique	224
5.4	Introduction to Chemical Vapor Deposition of Coatings Containing CMAs	225
5.5	MOCVD Processing of Al-Cu-Fe Thin Films	227
5.5.1	Precursors Selection	229
5.5.2	Deposition of Aluminum	232
5.5.3	Deposition of Copper	233
5.5.4	Deposition of the Al$_4$Cu$_9$ Approximant Phase	235
5.5.5	Deposition of Iron	237
5.6	Concluding Remarks	239
	References	241

6 Surface Chemistry of CMAs 243
Marie-Geneviève Barthés-Labrousse, Alessandra Beni, and Patrik Schmutz

6.1	Introduction	243
6.2	Surface Chemistry of CMAs Under UHV Environment	244
6.2.1	Interaction with Oxygen	244
6.2.2	Interaction with Other Molecules	248
6.3	Atmospheric Aging	251
6.3.1	Atmospheric Oxidation	251
6.3.2	Surface Properties in Atmospheric Conditions	254
6.4	Surface Chemistry and Reactions in Aqueous Solutions	255
6.4.1	Thermodynamic Stability	255
6.4.2	Oxide Electronic Properties	259
6.4.3	Localized Degradation Reactions	262
6.4.4	Summary: Localized Corrosion Model for the Al$_4$(Fe,Cr) Compound	265
6.5	High-Temperature Corrosion	266
6.5.1	Bulk Samples	266
6.5.2	Powders, Thin Films and Oxidation-Induced Phase Transformations	268
6.6	Conclusion	270
	References	270

7 Mechanical Engineering Properties of CMAs 273
Jürgen Eckert, Sergio Scudino, Mihai Stoica, Samuel Kenzari, and Muriel Sales

7.1	Introduction	273
7.2	Structure and Mechanical Properties of CMAs	274
7.2.1	Single-Phase Intermetallics	274
7.2.2	Multiphase Intermetallics	280

7.3	Metal Matrix Composites Reinforced with CMAs 290
7.3.1	Processing of Aluminum Matrix Composites Reinforced with CMAs 291
7.3.1.1	Thermal Stability of CMAs in Al-Based Matrix Composites 291
7.3.1.2	Preserving Complex Phases in Al-Based Matrix Composites 292
7.3.2	Mechanical Properties of Al-Based Composites Reinforced with CMAs 294
7.4	Surface Mechanical Testing and Potential Applications 299
7.4.1	Fretting Tests (Cold Welding) of CMAs 299
7.4.2	Friction Properties of Composites 306
7.4.2.1	CMA Matrix Composites 306
7.4.2.2	Al-Based Composites Reinforced with CMAs 309
7.5	Conclusions 311
	References 312
8	**CMA's as Magnetocaloric Materials** *317*
	Spomenka Kobe, Benjamin Podmiljšak, Paul John McGuiness, and Matej Komelj
8.1	Introduction 317
8.2	Materials 320
8.2.1	Theoretical Investigation of the Magnetocaloric Effect 320
8.2.1.1	$Gd_5Si_2Ge_2$ 321
8.2.1.2	$LaFe_{13-x}Si_x$ 321
8.2.2	Elemental Magnetocalorics 322
8.2.3	Intermetallic Compounds 323
8.2.3.1	Laves Phases 323
8.2.3.2	CMAs [$Gd_5(Si_{1-x}Ge_x)_4$ Alloys and Related 5:4 Materials] 324
8.2.4	Mn-Based Compounds 326
8.2.4.1	$Mn(As_{1-x}Sb_x)$ Alloys 326
8.2.4.2	$MnFe(P_{1-x}As_x)$ Alloys 327
8.2.4.3	Ni_2MnX (X = Ga, In, Sn, Sb) Heusler Alloys 328
8.2.4.4	Miscellaneous Compounds 329
8.2.5	$La(Fe_{13-x}M_x)$-Based Compounds 330
8.2.6	Manganites 332
8.2.7	Miscellaneous Intermetallic Compounds 334
8.2.8	Nanocomposites 336
8.2.9	Comparison of MCE Materials 336
8.2.10	Conclusions 337
8.3	Magnetocaloric Effect and Hysteresis Losses of CMAs 337
8.3.1	Substituting Ge and Si with Various Elements to Reduce the Hysteresis Losses 339
8.3.2	Phase Formation and Magnetic Properties of $Gd_5Si_2Ge_2$ with Fe Substitutions 339
8.3.3	X-Ray Diffraction Measurements 344
8.3.4	Magnetic Measurements 344

8.3.5	Hysteresis Losses	346
8.4	TEM Investigation of CMAs	346
8.4.1	TEM Research on - $Gd_5Si_2Ge_2$ Alloys	346
8.4.2	Imaging in the TEM	349
8.4.2.1	Mass-Thickness Contrast	350
8.4.2.2	TEM Images	350
8.4.2.3	STEM Images	351
8.4.2.4	TEM Diffraction Contrast	351
8.4.2.5	STEM Diffraction Contrast	352
8.4.3	Experimental Results of TEM Analyses	353
8.4.4	TEM of Low Loss Samples	354
8.5	Conclusions	358
	References	358

9 Recent Progress in the Development of Thermoelectric Materials with Complex Crystal Structures *365*

Silke Paschen, Claude Godart, and Yuri Grin

9.1	Introduction	365
9.2	Thermoelectric Figure of Merit	365
9.3	Design Principles	368
9.3.1	Phonon Engineering	368
9.3.2	Electron Engineering	371
9.4	Thermoelectric Materials	372
9.4.1	Zintl Phases	373
9.4.2	Skutterudites	373
9.4.3	Clathrates	376
9.4.4	Zn_4Sb_3	377
9.4.5	MgAgAs-type (half-Heusler) Phases	379
9.5	Concluding Remarks	380
	References	380

10 Complex Metallic Phases in Catalysis *385*

Marc Armbrüster, Kirill Kovnir, Yuri Grin, and Robert Schlögl

10.1	Introduction	385
10.2	Why Use Intermetallic Compounds – The General Concept	387
10.2.1	Chemical Bonding	388
10.2.2	Investigating the Stability *In Situ*	390
10.2.2.1	Bulk Stability	390
10.2.2.2	Surface Stability	393
10.3	The Semihydrogenation of Acetylene	395
10.4	Complex Metallic Phases as Platform Materials for Heterogeneous Catalysis	397
	References	398

Index *401*

Preface

The European Network of Excellence CMA, for Complex Metallic Alloys, was active during the five years from July 2005 until the end of June 2010. It has assembled in a joint effort twenty research institutions based in twelve different European countries, with more than 500 individuals on board. The areas of focus were metallurgy and crystal growth, crystallography and defects, electronic, phononic and mechanical properties, surface physics and chemistry, surface treatment and coating technologies, as well as a number of applied topics such as composites, thermoelectricity, magnetocaloric materials, and catalysts.

The present book is an attempt to summarize the knowledge gained by the network over this short period of time. Addressing specifically beginners in the field, it complements the more detailed Series of Books on Complex Metallic Alloys,[1] which was edited by one of us (EBF) in combination with the annual sessions of the so-called EuroSchool of the network. It is organised in ten self-contained chapters, with the view to begin with the more general notions, explain how complex metallic alloys may be grown using standard metallurgical routes, see how specific their properties are, either in bulk or at the surface, and finish with application-driven issues: coatings, magnetocalorics, thermoelectrics and catalysts.

The editors are deeply indebted to the authors of the chapters, who have accepted – within a tight schedule – to describe in a pedagogical way the many facets of the science and engineering of complex metallic alloys. They are grateful to the many colleagues, throughout Europe, who have contributed, in one way or another, to the life of the network, especially as task leader or as responsible of the Virtual Integrated Laboratories of CMA-NoE, and who took care to achieve a very high degree of integration of our research teams in order to counterbalance the fragmentation of research within the European Research Area.

Special thanks go to Annemarie Gemperli, the secretary general of CMA-NoE, to Mr Karl Hoehener and Prof. Louis Schlapbach, who acted with one of us (JMD) as executive officers to lead the CMA-NoE. In this respect, the inspiration and dedication gained from Prof. Knut Urban during the early period of the network was instrumental in the success of the whole process. As a pioneer in the field of complex

---

1) Book Series on Complex Metallic Alloys, Vols. I, II, III & IV, World Scientific, Singapore, 2008–2010.

metallic alloys, Prof. Urban has kindly accepted to write the foreword of the present book. Last, but not least, we acknowledge the financial support offered by the European Commission under contract N° NMP3 – CT – 2005 – 500140.

Nancy, July 14, 2010

Jean-Marie Dubois
Esther Belin-Ferré
Directors of Research at CNRS

List of Contributors

Marc Armbrüster
Max-Planck-Institute for Chemical
Physics of Solids
Nöthnitzer Str. 40
01187 Dresden
Germany

Wolf Assmus
University of Frankfurt
Physikalisches Institut
Max-von-Laue-Str. 1
60438 Frankfurt
Germany

Marie-Geneviève Barthés-Labrousse
Institut de Chimie Moléculaire et des
Matériaux d'Orsay
Laboratoire d'Etude des Matériaux Hors
Equilibre (UMR CNRS-UPS 8182)
Bâtiment 410, Université Paris Sud
91405 Orsay Cedex
France

Esther Belin-Ferré
Université Pierre et Marie Curie
Laboratoire de Chimie Physique-
Matière et Rayonnement (UMR 7614
CNRS-UPMC)
11 rue Pierre et Marie Curie
75231 Paris Cedex 05
France

Alessandra Beni
Swiss Federal Laboratories for Materials
Testing and Research (EMPA)
Laboratory for Corrosion and Materials
Integrity
Abt. 136, Ueberlandstrasse 129
8600 Dübendorf
Switzerland

Marc de Boissieu
Grenoble INP/CNRS
Sciences de l'Ingéniérie des Matériaux
et des Procédés (SIMaP)
UJF 1130 rue de la piscine, BP 75
38402 Saint Martin d'Hères Cedex
France

Janez Dolinšek
Jožef Stefan Institute
Solid State Physics Department
Jamova 39
1000 Ljubljana
Slovenia
University of Ljubljana
Faculty of Mathematics and Physics
Jadranska 19
1000 Ljubljana
Slovenia

Complex Metallic Alloys: Fundamentals and Applications
Edited by Jean-Marie Dubois and Esther Belin-Ferré
Copyright © 2011 WILEY-VCH Verlag GmbH & Co. KGaA, Weinheim
ISBN: 978-3-527-32523-8

Jean-Marie Dubois
Nancy-Université UPV-Metz
Institut Jean Lamour (UMR 7198 CNRS)
Ecole des Mines, Parc de Saurupt
CS 14234 54042 Nancy Cedex
France

Jürgen Eckert
IFW Dresden
Institut für Komplexe Materialien
Postfach 270116
01171 Dresden
Germany

Michael Feuerbacher
Forschungszentrum Juelich GmbH
Institut fuer Festkoerperforschung
52425 Juelich
Germany

Vincent Fournée
Nancy Université-UPV Metz
Institut Jean Lamour (UMR 7198 CNRS)
Ecole des Mines, Parc de Saurupt
54042 Nancy Cedex
France

Claude Godart
ICMPE – CMTR
CNRS UMR 7182
2–8 Rue H. Dunant
94320 Thiais
France

Saskia Gottlieb-Schoenmeyer
University of Frankfurt
Physikalisches Institut
Max-von-Laue-Str. 1
60438 Frankfurt
Germany

Yuri Grin
Max-Planck-Institute for Chemical
Physics of Solids
Nöthnitzer Str. 40
01187 Dresden
Germany

Samuel Kenzari
Nancy-Université UPV-Metz
Institut Jean Lamour (UMR 7198 CNRS)
Ecole des Mines de Nancy, Parc de
Saurupt
CS 14234 54042 Nancy Cedex
France

Spomenka Kobe
Jozef Stefan Institute
Department for Nanostructured
Materials
Jamova 39
1000 Ljubljana
Slovenia

Matej Komelj
Jozef Stefan Institute
Department for Nanostructured
Materials
Jamova 39
1000 Ljubljana
Slovenia

Kirill Kovnir
Fritz-Haber-Institute of the Max-Planck-
Society
Faradayweg 4-6
14195 Berlin
Germany

Julian Ledieu
Nancy Université-UPV Metz
Institut Jean Lamour (UMR 7198 CNRS)
Ecole des Mines, Parc de Saurupt
54042 Nancy Cedex
France

Enrique Maciá
Universidad Complutense de Madrid
Fac. CC. Físicas
Dpto. Física de Materiales
28040 Madrid
Spain

Paul John McGuiness
Jozef Stefan Institute
Department for Nanostructured
Materials
Jamova 39
1000 Ljubljana
Slovenia

Jeong Y. Park
Korean Advanced Institute of Science
and Technology (KAIST)
Graduate School of EEWS
Daejaon, 305-701
Republic of Korea

Silke Paschen
Vienna University of Technology
Institute of Solid State Physics
Wiedner Hauptstr. 8–10
1040 Vienna
Austria

Benjamin Podmiljšak
Jozef Stefan Institute
Department for Nanostructured
Materials
Jamova 39
1000 Ljubljana
Slovenia

Nathalie Prud'homme
CIRIMAT
ENSIACET
118 Route de Narbonne
31077 Toulouse Cedex 4
France

Muriel Sales
Austrian Institute of Technology
Aerospace and Advanced Composites
2444 Seibersdorf
Austria

Robert Schlögl
Fritz-Haber-Institute of the Max-Planck-
Society
Faradayweg 4-6
14195 Berlin
Germany

Patrik Schmutz
Swiss Federal Laboratories for Materials
Testing and Research (EMPA)
Laboratory for Corrosion and Materials
Integrity
Abt. 136, Ueberlandstrasse 129
8600 Dübendorf
Switzerland

Sergio Scudino
IFW Dresden
Institut für Komplexe Materialien
Postfach 270116
01171 Dresden
Germany

Ana Smontara
Institute of Physics
Laboratory for the Study of Transport
Problems
Bijenička 46, POB 304
10001 Zagreb
Croatia

Mihai Stoica
IFW Dresden
Institut für Komplexe Materialien
Postfach 270116
01171 Dresden
Germany

Constantin Vahlas
CIRIMAT
ENSIACET
118 Route de Narbonne
31077 Toulouse Cedex 4
France

1
Introduction to the Science of Complex Metallic Alloys

Jean-Marie Dubois, Esther Belin-Ferré, and Michael Feuerbacher

1.1
Introduction

Complex metallic alloys (or CMA for short), also called SCAPs (for structurally complex alloy phases) for some time [1] encompass a broad family of crystalline compounds made of metals, alloyed with metalloids or rare earths or chalcogenides. They exhibit large unit cells, containing up to thousands of atoms. The periodicity of the crystal unit cell is no longer a relevant property since it becomes much larger than the average first-neighbor distance, and more specifically exceeds the distance that characterizes basic interactions in the crystal. As a consequence, most physical properties, and especially transport of electrons and phonons, depart significantly from the ones observed in conventional metals and alloys that are characterized by smaller unit cells, containing just a few atoms. The conduction of heat for instance in an Al-Cu-Fe CMA of appropriate composition, although made of good heat conductors, is as low as in zirconia, a typical heat insulator used in aerospace industry [2].

Several types of CMAs exist, depending on the nature of the constitutive elements and their respective concentrations. The most widely studied so far are based on aluminum [3]. They comprise quasicrystals, which were first pointed out by Shechtman in the period 1982–1985 in metastable, melt-spun ribbons of an Al-Mn alloy [4, 5]. Later, the existence of stable quasicrystals was revealed by different groups, among which the most decisive contributions came from Tsai *et al.* who demonstrated that quasicrystals may be grown in various Al-based alloys at very specific compositions, but by slow cooling of the molten alloy [6–9]. A large research effort was dedicated to understanding better the crystalline structure of quasicrystals and its relationship to their properties, see reference 3 for a review of this work.

The challenge was indeed, and to a large extent still is, first to describe the atomic positions in quasicrystals (where are the atoms? as Bak used to write [10]) and secondly to discover the building principle that leads quasicrystals to abandon

Complex Metallic Alloys: Fundamentals and Applications
Edited by Jean-Marie Dubois and Esther Belin-Ferré
Copyright © 2011 WILEY-VCH Verlag GmbH & Co. KGaA, Weinheim
ISBN: 978-3-527-32523-8

translational periodicity, although it is seemingly[1] at work in the whole world of ordered solid matter known so far (why are the atoms where they are?). The first part of the challenge has been met very successfully and the new tools developed to match this goal have proven to apply to many complex materials, including the CMAs that we shall address in this book. This is not true as far as the second aspect of the challenge is considered. To the best of our knowledge, complexity is observed in metallic alloys, but no unifying principle has yet clearly emerged to show us why it is so in any given alloy, and help us design new alloys of controlled complexity based on a fully self-consistent theory. Empirical rules, and fragmentaiy understanding, nevertheless, do exit and will be explained in this book.

Most of this knowledge emerged from careful studies of a large variety of complex metallic alloys, which is the major focus of the present book. It started long ago, when the most famous chemist Linus Pauling was still a PhD student in the mid-1930s [12]. This research stopped, however, about 40 years later, when it appeared that the most complex crystalline structures known at that time could no longer be solved with the tools then available. The discovery of quasicrystals, which unfortunately arose in an undue controversy [13], forced crystallographers and physicists to reconsider very complex atomic architectures and revisit CMAs for their properties and potential applications. More than two decades after the discovery of quasicrystals, the crystallographic tools to describe complexity in metallic alloys were available. The moment had come to see whether the field of CMAs, to a very large extent unexploited until then, was offering a new frontier in condensed-matter physics or not.

1.2
Complex Metallic Alloy: What Is It?

A complex metallic alloy is characterized by (i) its chemical formula, (ii) the size of its unit cell, and (iii) the variety of atomic clusters that this unit cell contains. The composition often, but not always, is that of a ternary, or a quaternary, and so on, alloy, that is, a multicomponent alloy. Yet, quite a few binary complex crystalline compounds have been studied in depth in recent years, which led to the remarkable discovery of a stable binary quasicrystal [14]. The upper limit of the unit cell size is of course infinite in quasicrystals. It is more difficult to assess the lower limit. Conventionally, a crystal that contains a few tens of atoms in its unit cell is considered a CMA, which corresponds to lattice parameters of a few angströms. Very frequently, the atom clusters exhibit icosahedral symmetry. This leads to incorporation of Laves phases, with a small unit cell size and only 12 atoms per unit cell into the family of CMAs.

A simple definition of complexity in CMAs is difficult to express and different equivalent definitions may apply. For instance, the diffraction pattern of a CMA is characterized in reciprocal space by its many spots, some being close to the origin,

---

1) A natural Al-Cu-Fe quasicrystal discovered recently in a million-year old rock tends to prove that Nature had discovered quasicrystals long before human beings [11].

and the closer as the size of the unit cell increases. If one is looking for a single parameter, the distance from origin to the first diffraction spot could be used. Theoretically, it would be equal to 0 in a perfect quasicrystal. However, due to the necessarily limited experimental resolution on diffracted intensities, the first measurable spot with nonzero intensity is located at some distance from origin, which opens a possibility to confuse a true quasicrystal with a high-order approximant.[2] For this reason, we will rely on a more reliable definition in Section 1.2.2. We will see also in Section 1.2.3 how specific defects signal complexity in CMAs.

1.2.1
Composition and Varieties

The best-studied example of a CMA system that comprises both quasicrystalline and crystalline CMAs is the Al-Cu-Fe system (for references, see ref. 3). An isothermal cut through the phase diagram is shown in Figure 1.1. Many different compounds appear: binary alloys comprise simple Al-Cu crystals, like the θ-Al_2Cu

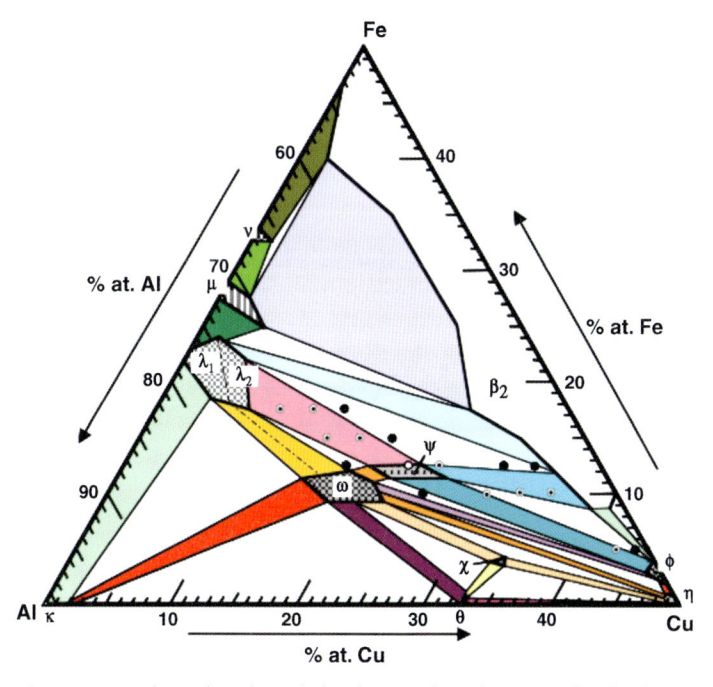

Figure 1.1 Isothermal cut through the Al-Cu-Fe phase diagram in the Al-rich corner. See text for the labels of the compounds.

2) An approximant is a crystalline CMA the structure of which may be derived from the same high-dimensional lattice as the quasicrystal that lies nearby, in the same composition field.

phase with 16 atoms per unit cell (at/uc in the following), or more complex ones like the γ-brass Al_4Cu_9 compound with 52 at/uc (not visible on this diagram because it is located outside the concentration region drawn here). The nearly binary $Al_{10}Cu_{10}Fe$ compound is labeled ϕ in this diagram. The letter η labels an hexagonal compound closely related to the B2-CsC1 type β-phase that shows a broad stability region in the ternary system and plays an important role since it is the primary solidification phase that appears first when growth from the liquid state is applied [15]. Similarly, Al-Fe binary compounds comprise the $λ$-$Al_{13}Fe_4$ and $μ$-Al_5Fe_2 compounds, which are intermetallics of quite different complexity with, respectively, 102 and 12 at/uc.

The stable, ternary quasicrystal, marked ψ, shows up approximately in the middle of the diagram, in a narrow composition region located around $Al_{62}Cu_{25.5}Fe_{12.5}$ (at. %). Within the region labeled ψ, but not visible at this enlargement scale, coexist at least 3 high-order approximants of very similar compositions [16]. They contain hundreds of atoms per unit cell. In contrast, the much simpler ω phase, with 48 at/uc, lies only slightly apart from the previous composition region, and shows definitely different electron-transport properties [17].

Such a variety of compositions and structures in a metallic alloy system is a characteristic of CMA-forming alloys. It is an indication for the formation of a stable quasicrystal that has guided Tsai *et al.* in the hunt for these compounds at the beginning of the history of the field. Unfortunately, it does not supply us with a number that may be used to quantify complexity in a very simple way.

1.2.2
Complexity at a Glance

Indeed, we need a single number, preferably, that could possibly be used to correlate some, if not all, properties of a given alloy system to the complexity of its CMAs. This number does exist at least in Al-based CMAs, like the ones encountered in the Al-Cu-Fe phase diagram of the previous subsection. This is:

$$\beta_C = \text{Ln}(N_{UC})$$

where N_{UC} is the number of atoms in the unit cell.

By definition, $N_{UC} \rightarrow \infty$ in a quasicrystal. However, since the size of the specimen is always finite, β_C is finite as well, and is close to $\ln(N)$, where N is Avogadro's number, that is, $23 \leq \beta_C \leq 55$ depending on the size of the sample. We will take arbitrarily $\beta_C = 23$ in the following, which does not change our conclusions.

To exemplify, we show in Figure 1.2, two very different properties plotted as functions of β_C for a large number of Al-TM (TM: transition metal) CMAs of varying complexity according to the choice of TM and Al concentrations. In quite a few examples, a second TM' element was added, like for instance in the Al-Cu-Fe ω-phase or in the quasicrystal. The first property, labeled $n(E_F)$ is the density of Al 3p states obtained by soft X-ray emission spectroscopy (SXES) measurements at room temperature [18]. We will come back to this data in Section 1.3.1. The second

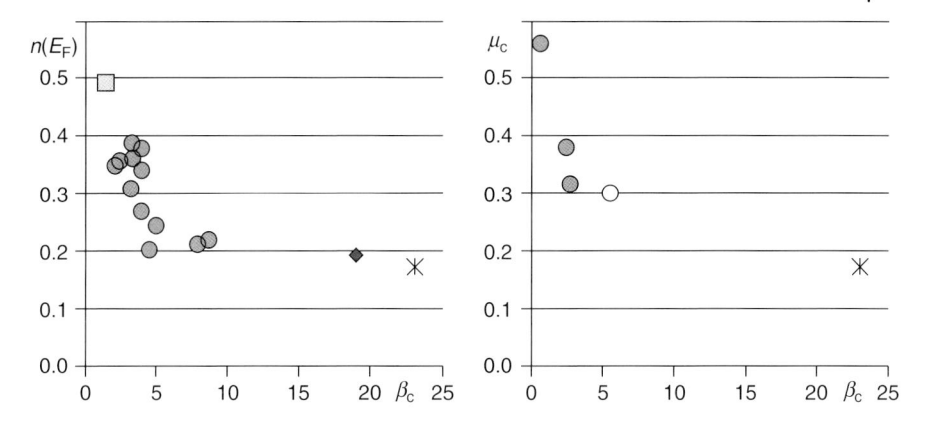

Figure 1.2 Variation with complexity index of the density of Al3p states at the Fermi energy E_F (left) and adhesion against steel in vacuum (right) for a variety of Al-TM CMAs. The data selected in the right hand side figure comes from CMA samples with nearly identical concentration of valence electrons, see Figure 1.20.

property is friction in vacuum against hard steel of the very same specimens, after correction for the plowing component of friction and noted μ_C. This correction is explained later in this chapter (Section 1.4.3).

The two properties, which cannot be related a priori, show the same decreasing dependence with increasing complexity of the crystal (in this system). Furthermore, a ln–ln plot of the data of Figure 1.2 demonstrates that they vary according to a same power law of β_C, that is, $n(E_F)$ and $\mu_C \approx \beta_C^{-1/3}$ (Figure 1.3).

To the best of our knowledge, this provides one of the scarce but clear indications available so far that friction in vacuum between two metals is dominated by adhesion

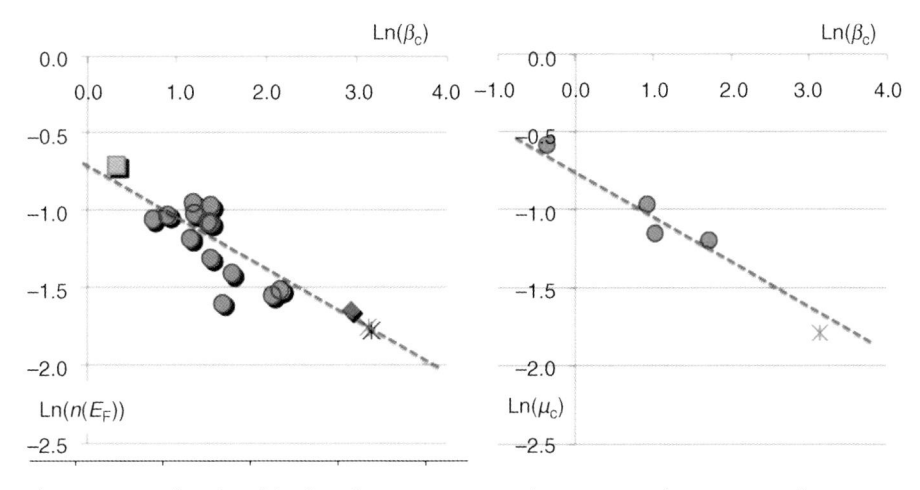

Figure 1.3 Log–log plot of the data shown in Figure 1.2, demonstrating the same power law dependence towards β_C. The dashed lines have a slope of $-1/3$.

of electronic origin. Most presumably, the underlying mechanism is the formation of a hybrid band between d states in the steel pin and sp states in the CMA sample (as is often observed in the CMA itself, see Section 1.3.1).

1.2.3
Defects

Generally, in CMAs all types of defects known from simple metallic structures, can occur. This includes zero-dimensional defects, for example, vacancies and interstitials, and planar defects, for example, stacking faults and antiphase boundaries. Line defects such as dislocations also exist in CMAs, but, due to the particular structural properties of CMAs, require more detailed discussion.

As a basic structural feature, CMAs possess large lattice parameters. A direct consequence is that conventional dislocation-based deformation mechanisms are prone to failure. The elastic line energy of a dislocation is proportional to b^2, where b is the length of the Burgers vector [19]. For most materials, Burgers vector lengths larger than about 5 Å are energetically unfavorable. Accordingly, perfect dislocations in CMAs, which frequently would require Burgers vectors exceeding 10 Å in length, are highly unfavorable and not likely to form. In common materials, dislocations with large Burgers vectors split into partials. However, for the case of very large lattice constants as in CMAs, splitting into a high number of partials, each possessing its individual energy cost would be required. Moreover, the corresponding planar faults, which necessarily have to be introduced as soon as partials are involved, cost additional energy.

On the other hand, it has been shown that CMAs, at least at high temperatures, are ductile and that in all cases studied dislocations mediate plastic deformation [20–23]. It is therefore a central question in CMA research, to explore the deformation mechanisms structure of the defects involved.

Furthermore, in some CMA structures, salient one-dimensional defects exist. These defects, referred to as "phason lines" do not exist in simple metals. Even though they are linear in character, they are not dislocations. Nevertheless, they are pivotally involved in the deformation process of some CMA structures (see below) but also in phase transitions between related CMAs.

As a result of the structural hindrances for the formation of regular dislocations, in many CMAs novel types of dislocations are formed. Taking advantage of structural features of the host CMA, they allow, via their particular construction, for the accommodation of Burgers vectors corresponding to energetically acceptable strain fields. Particular examples of such defects in CMAs are the metadislocations.

Metadislocations were discovered in 1999 [24] and are today known to exist in various forms in different CMA structures. They mediate the plastic deformation process in these CMAs, which additionally involves phason defects. Metadislocations were firstly observed in orthorhombic ε-phases. These are based on ε_6-Al-Pd-Mn [25], which has an orthorhombic unit cell (space group *Pnma*) with cell parameters $a = 23.541$ Å, $b = 16.566$ Å and $c = 12.339$ Å. The structure can be represented in terms of a tiling of flattened hexagons arranged in two different orientations. The

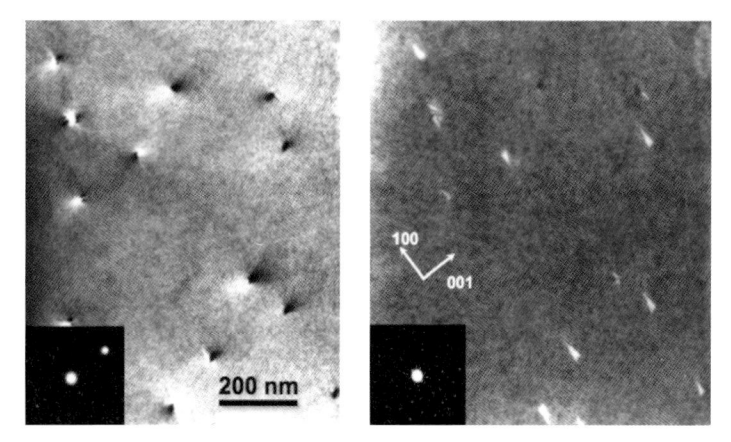

Figure 1.4 Transmission electron microscopy images of dislocations in a deformed ε_{28}-Al-Pd-Mn single crystal.

vertices of the hexagons are decorated with 52-atom clusters of local icosahedral order, the so-called pseudo-Mackay clusters.

The compound ε_6-Al-Pd-Mn is the basic phase of a family of superstructures with equal a and b but larger c lattice parameters referred to as the orthorhombic ε-phases. The most prominent of these possesses a c parameter of 57 Å and has been termed ε_{28}-Al-Pd-Mn. Other superstructures with $c = 32.4$ Å, 44.9 Å, and 70.1 Å have been reported [26].

Figure 1.4 displays transmission electron micrographs of dislocations in a deformed ε_{28}-Al-Pd-Mn single crystal. Figure 1.4a is taken under two-beam Bragg conditions using the (10 0 0) reflection close to the [0 1 0] axis (inset). A high density of dislocations is seen in end-on orientation. No stacking-fault contrast is seen in the image, that is, the dislocations appear to be perfect dislocations. Contrast-extinction analysis [27] shows that the Burgers vectors of the dislocations are parallel to the [0 0 1] direction, that is, they are pure edge dislocations. Figure 1.4b shows the same sample area imaged using a symmetric selection of reflections of the (0 1 0) zone axis (inset). Under these conditions, it can be seen that each dislocation position is decorated by a small area of bright contrast. Figure 1.5a shows one of the single dislocations at a higher magnification. It consists of a dislocation-like structure with six associated half-planes. Note, however, that the dislocation-like structure resides on a length scale that is larger than the atomic scale by about one order of magnitude.

Figure 1.5b is a schematic representation of the defect structure of Figure 1.5a in terms of a tiling description [28]. The dislocation core is represented by the dark-gray polygon in the image center. The upper and lower edges as well as the right-hand side of the figure are represented by a tiling composed of pentagons, banana-shaped nonagons, and flattened hexagons in two different orientations. This is the representation of the ideal ε_{28}-Al-Pd-Mn structure [29]. The pentagon–nonagon pairs represent phason lines, which periodically arrange along the [1 0 0] direction and form planes referred to as "phason planes."

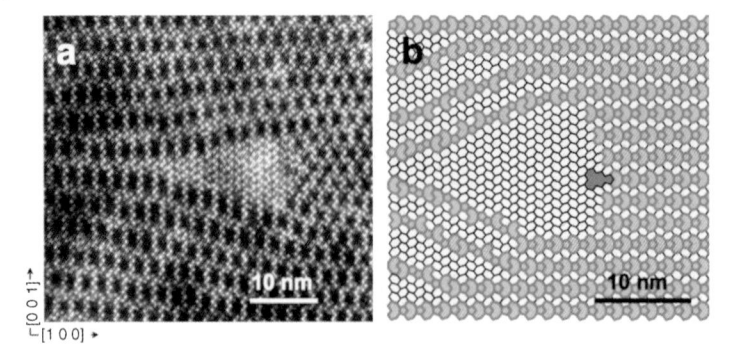

Figure 1.5 Core region of a single dislocation at high magnification (a) and schematic representation of the defect structure in terms of a tiling description, see text (b).

On the left-hand side of the metadislocation core is a triangle-shaped area, the tiling of which consists of flattened hexagons in alternating orientations. This tiling represents the ideal ε_6-Al-Pd-Mn structure [25]. In the upper and lower vicinity of the ε_6-Al-Pd-Mn triangle, the phason-plane arrangement, which in the undistorted ε_{28}-Al-Pd-Mn structure forms straight (0 0 1) planes, relaxes around the ε_6'-Al-Pd-Mn triangle.

The Burgers vector of the dislocation can be determined as $\vec{b} = c/\tau^4 (0\ 0\ 1)$ by forming a Burgers circuit around the dark-gray core. The Burgers-vector length amounts to 1.83 Å, that is, the dislocation is a small irrational partial. The complete defect structure is inseparably formed by the partial dislocation on the atomic scale and the dislocation-like structure on the larger length scale formed by the associated phason half-planes. The latter accommodate the partial dislocation to the lattice in such a way that the ideal ε_{28}-Al-Pd-Mn structure can be continued above and below the dislocation core. As a direct consequence, the defect structure as a whole can move through the lattice without introducing any additional planar defects. Later, other types of metadislocation with 4, 10, and 16 associated phason half-planes were discovered [30]. It was demonstrated that metadislocations mediate the plastic deformation process in ε_6- and ε_{28}-Al-Pd-Mn [20]. The mode of dislocation motion has not been directly identified, but strong evidence was found that the movement takes place by a pure climb process [31].

Closely related but structurally different metadislocations were observed in ε-type phases in the system Al-Pd-Fe [32]. Figure 1.6a is a micrograph of a metadislocation in ξ-Al-Pd-Fe. The metadislocation core is located in the lower-left part of the image. It is associated with three planar defects extending to the upper right (dark contrast), which can be identified as phason half-planes. A Burgers circuit around the metadislocation core reveals a closure failure of $1/2\tau^4$ [1 0 1] in terms of the ξ lattice, which corresponds to the metadislocations Burgers vector. The Burgers vector length is 1.79 Å. Figure 1.6b shows two metadislocations in the structure ε_{22}-Al-Pd-Fe. The left metadislocation is associated with five phason half-planes and that on the right with eight phason half-planes. In the areas directly below both metadislocation cores, regions of ξ structure can be identified. The metadislocations in Figure 1.6 are typical

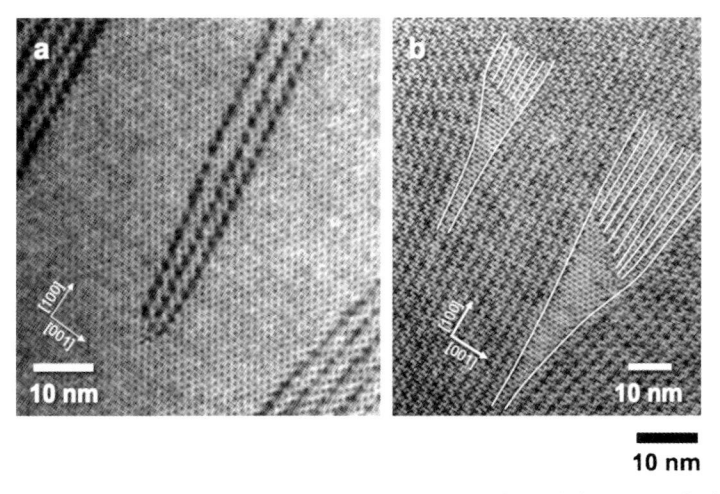

Figure 1.6 Transmission electron microscopy images of metadislocations in ξ-phases.

for monoclinic host structures. As for the case of the metadislocations in orthorhombic ε-phases described above, we find series of metadislocation with different numbers of associated phason half-planes, corresponding to different Burgers vectors. While for the orthorhombic ε-phases we find that the numbers of associated phason half-planes follow double Fibonacci numbers, for the monoclinic case we find single Fibonacci numbers. In a tiling representation, the metadislocation cores are for both types described by the same tiles, which of course, correspond to the same Burgers vector length with respect to the individual lattice constant.

In the orthorhombic $Al_{13}Co_4$ phase, metadislocations that are less closely related are found. $Al_{13}Co_4$ is an orthorhombic phase with space group *Pmn21* and lattice parameters $a = 8.2$ Å, $b = 12.3$ Å, and $c = 14.5$ Å [33]. The main structural features are pair-connected pentagonal-prismatic channels extending along the [1 0 0] direction. Within the (1 0 0) plane, the structure can be matched by a tiling consisting of regular pentagons and rhombs, where the rhombs are arranged in an antiparallel manner. Figure 1.7 displays a transmission electron micrograph of a deformed $Al_{13}Co_4$ single crystal [34]. A high density of dislocations (black arrow) and trailing planar defects (white arrow) can be seen. The dislocations have [0 1 0] Burgers vectors and [1 0 0] line direction, and their movement takes place in (0 0 1) planes. That is, the dislocations are of pure edge type and move by pure glide.

Figure 1.8a is a micrograph of the dislocation core at higher magnification. A tiling representing the unit-cell projection along the [1 0 0] direction is superposed. Rectangular tiles represent the orthorhombic $Al_{13}Co_4$ phase and rhomb-shaped tiles represent a closely related monoclinic modification [33, 35]. The dislocation core is localized in the open center, and the stacking fault stretches out to the right. It can clearly be seen that the planar fault consists of a slab of monoclinic structure within the otherwise orthorhombic lattice. Figure 1.8b is a schematic of the defect in terms of a pentagon tiling [36]. The superposed unit-cell projections correspond to those

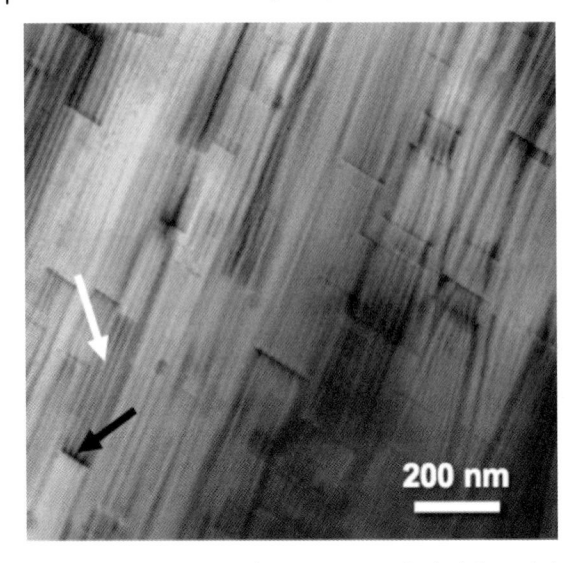

Figure 1.7 Transmission electron micrograph of a deformed $Al_{13}Co_4$ single crystal. The black arrow shows a dislocation, trailing planar defect (white arrow).

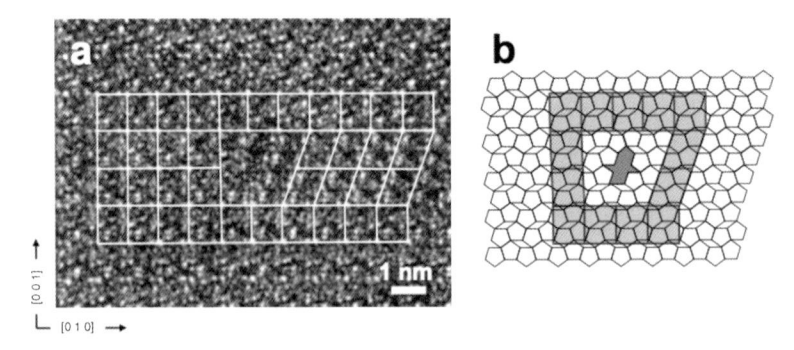

Figure 1.8 Dislocation core at high magnification onto which a tiling representing the unit cell projection along the [1 0 0] direction is superimposed (a) and corresponding tiling representation (b).

shown in the experimental image. The dislocation core is represented by the dark-gray tile. The planar defect corresponding to the slab of monoclinic phase stretches out to the right and is represented by a parallel arrangement of pentagon and rhomb tiles, while the surrounding orthorhombic phase is represented by an alternatively oriented arrangement of pentagon and rhomb tiles.

Other types of metadislocations were observed in the orthorhombic Taylor phases based on Al_3Mn. A review, describing all types of metadislocations in great depth, is given by Feuerbacher and Heggen [37].

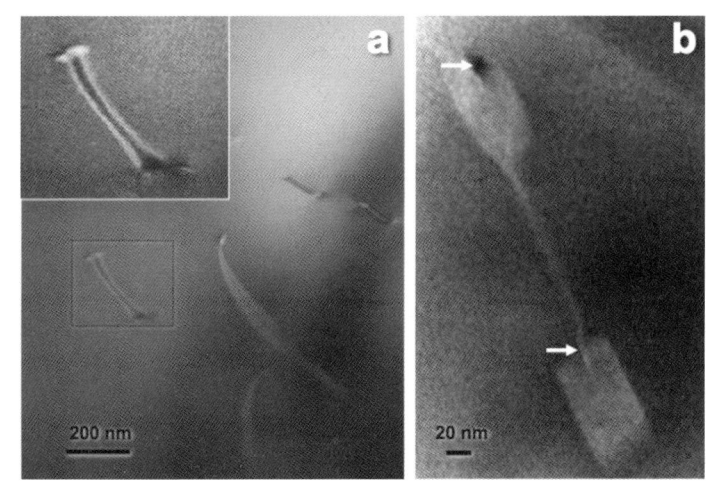

Figure 1.9 Bright-field Bragg-contrast image of a typical defect arrangement in plastically deformed C_2-Al-Pd-Fe (a) (the insert is an enlargement of part of the figure) and (b) high-resolution image of the arrangement shown in the insert of (a) along the [−1 1 0] direction.

An apparently different mechanism was observed in C_2-Al-Pd-Fe. The structure of this phase was determined by Edler *et al.* [38]. It is cubic with a lattice constant of 15.5 Å and a unit cell containing 248 atoms. The structure can be described in terms of icosahedral cages generated by Pd atoms. These cages are alternatively filled with two different cluster motifs. The resulting two types of clusters are distributed on a cubic lattice such that the different fillings lead to a face-centered ordering.

Figure 1.9a shows a bright-field Bragg-contrast image of a typical defect arrangement in plastically deformed C_2-Al-Pd-Fe. The presence of dislocations attached to planar faults showing fringe contrast is clearly seen. The inset presents a contrast-enhanced enlargement of the boxed area, showing two dislocations separated by about 300 nm terminating a stacking-fault fringe contrast. Figure 1.9b shows a high-resolution TEM image of such an arrangement along the [−1 1 0] direction. The defect appears as a dumbbell-shaped object with two almost rectangular-shaped extremities. The stacking fault is seen in edge-on orientation. The terminating dislocations, marked by white arrows, are located at the upper ends of the rectangular brighter-contrast areas.

The defect can be analyzed by back-transform Fourier filtering analysis. This yields a Burgers vector $a_0/2$ [0 0 1] for both terminating dislocations. The whole arrangement hence has Burgers vector a_0 [0 0 1] and consists of a perfect [0 0 1] dislocation with a Burgers vector length of 15.5 Å, which is split into two energetically more favorable partials with Burgers vector lengths of 7.8 Å at the cost of a stacking fault between the latter.

Fourier filtering analysis reveals that the rectangular bright-contrast areas are of body-centered structure, that is, a structure locally differing from the face centered host. Phase-diagram investigations [39] revealed that in the Al-Pd-Fe system indeed a body-centered phase, C_1-Al-Pd-Fe, exists in the compositional range around $Al_{63}Pd_{31}Fe_6$. This phase possesses a slightly smaller lattice constant of 15.4 Å.

These findings are consolidated in the following interpretation: In order to lower the elastic line energy of the perfect dislocation the latter splits into two partials. These, however, still possess rather large Burgers vector lengths of 7.8 Å, which have to be accommodated by the structure. The experimental findings suggest that it is energetically favorable to transform a small portion of the structure, which is lying in the compressive part of the dislocation strain field, to a body-centered structure possessing a slightly smaller lattice constant. By this means, the dislocation with the large Burgers vector can be more easily accommodated into the face-centered structure at the cost of a portion of "wrong" structure considering the composition of the crystal.

In CMAs, hence, dislocations of different nature from that in simple metals occur. Besides metadislocations in a number of CMA phases, we also find other unusual mechanisms, such as the dislocations in cubic Al-Pd-Fe described above. While these mechanisms at a first glance seem completely different, they have one major common property: in all cases we find that accommodation of the dislocation in the host structure involves a local area of different but related structure. This locally differing structure occurs in the form of a slab in the wake of the moving dislocation (e.g., in the case of $Al_{13}Co_4$) or in the form of a bar around the dislocation core. We should, however, note that in some CMA phases investigated, for example, the Samson phase β-Al_3Mg_2[22] or the Bergman phase in the system Mg-Zn-Al [40], rather conventional deformation mechanisms, involving ordinary partial dislocations trailing stacking faults, have been observed.

1.3
Complex Metallic Alloy: Why Is It Complex?

For many years, the key point in understanding CMAs, and especially quasicrystals, was to know where the atoms are [10]. This question is now solved to a very satisfactory degree [41], using various techniques derived from high-dimensional crystallography and, as often as possible, comparison to a known, high-order approximant when it exists [42]. Nowadays, the central question is to understand "why the atoms are where they are," which by the way is a question that is simply not solved in all crystals of simpler structure known so far. We address this issue in the following section from the electronic structure standpoint, in a way that is very much reminiscent of the early works by Hume-Rothery [43], Jones [44] and Blandin [45] and in more recent years Friedel and Dénoyer [46] and Mizutani [47].

1.3.1
Electronic Densities of States and Hume-Rothery Rules

Does the particular atomic structure of complex metallic systems engender characteristic features in their electronic structure? A theoretical approach is most often pretty uneasy because of the too large number of atomic sites in the unit cell that, in addition, may not be all fully occupied. Information is gained from experimental results derived from resistivity and specific heat measurements or densities of states (DOS) investigations using spectroscopic techniques, but results are available only for a restricted number of theses compounds. Hence, to figure out what the electronic structure may be in CMAs, one may rely on data obtained either for stable quasicrystalline compounds (QCs), which we shall consider as the ultimate state of complexity in CMAs, with a single unit cell containing an infinite number of sites, or their approximants, also with large unit cells, but of finite size.

For quasiperiodic lattices, theoretical studies of the energy spectrum pointed out that the wavefunctions should be critical, hence the corresponding electronic states differ from conventional systems since they are neither extended nor localized [48, 49]. Due to the lack of periodicity, exact DOS calculations cannot be carried out for QCs. However, they are available for structures whose atomic order mimics the local arrangement of the QCs [50–52] and in some cases, also for true QC systems owing to the Rietveld method applied to experimental diffraction patterns combined with first-principles methods [53]. Many calculations have been also carried out for series of Al-TM conventional intermetallics [54, 55]. In all cases, a *pseudogap* was found at or nearby the Fermi energy (E_F).

Hume-Rothery found empirically that in many systems such as Cu-Zn, Al-Cu, and so on, for specific values of the valence electron to atom ratio (e/a), there exist alloys displaying different crystal structures that do not behave as free-electron systems do, but are stabilized owing to the creation of a depletion in the DOS at E_F, namely a pseudogap [43]. These specific alloys are denoted Hume-Rothery (H-R) phases. The mechanism by which such a pseudogap occurs may be summarized as follows [56]. Electronic waves in the alloy are scattered by the Bragg planes of the Brillouin zone. This mechanism opens up gaps in reciprocal space. For specific electron concentrations, the Fermi sphere overlaps with the Brillouin-zone boundary, which in turn produces a depletion in the DOS at the Fermi energy after integration over all directions in reciprocal space of the scattered waves. As a consequence, electronic states from the top of the valence band (VB) are repelled towards lower binding energies as shown in Figure 1.10, thus stabilizing the crystal structure for that specific electron concentration.

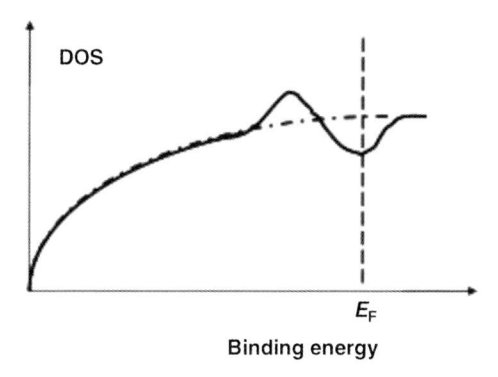

Figure 1.10 Total density of states (DOS) of a free-electron system (dotted line) and of a H-R alloy (full line).

Friedel and Dénoyer [46] noticed that (i) many quasicrystals and approximants are made from elements with small differences in atomic radii and electronegativities and (ii) there is a large number of spots, several with high intensity, in their X-ray diffraction patterns the position of which in reciprocal space demonstrates that the Hume-Rothery rule introduced above is obeyed. Actually, a pseudo-Brillouin zone (PBZ) can be constructed from the location of the most intense peaks in reciprocal space, although a true Brillouin zone can not be defined due to the loss of translational periodicity. This assessment was successfully taken as a thumb rule for the search of new stable quasicrystalline compounds and, indeed, many QCs and approximants were found this way with e/a values around 1.86 e^-/at (electrons per atom) and 2.04 e^-/at, mainly in Al-based systems as for example Al-Pd-Mn, Al-Cu-Ru, Al-Mg-Zn, ... QCs [7–9].

The pseudogap at the Fermi energy is perhaps one of the most emblematic features of QCs as well as of CMAs and is of great importance to interpret many of their physical properties [3, 57, 58]. Its existence was checked experimentally for a number of Al-based QCs and CMAs as well as for conventional H-R crystals using spectroscopic techniques among which soft X-ray emission spectroscopy (SXES). SXES scans separately occupied partial and local DOSs in a compound, whatever it is [18, 59–63]. It was ascertained that the intensity of the Al 3p states distribution at E_F (I(Al3p/E_F) reflects the metallic character of the specimens and is directly connected to the importance (depth and width) of the pseudogap [18].

Furthermore, it was established that the Hume-Rothery mechanism alone cannot explain the formation of the pseudogap since in genuine H-R alloys such as θ-Al$_2$Cu or ϕ-Al$_{10}$Cu$_{10}$Fe, it is rather faint as compared to Al-based approximants and QCs that contain the same transition metals. This is shown in Figure 1.11 as a plot of I(Al3p/E_F) in various simple and complex phases versus the e/a ratio [64]. The data is the same as the one in Figure 1.2, but is shown versus the e/a ratio that is directly related to the position of the CMA specimen in the composition field, assuming a contribution to the valence band of +3 electrons for Al, +1 for Cu and a negative valence of Fe of -2.6 e^-/at. Clearly, I(Al3p/E_F) is almost constant for the true Al-Cu H-R alloys and decreases progressively when going to approximants and QCs, a signature of increasing complexity of the lattice, as we pointed out earlier in this chapter. From the study of all electronic distributions in the VB of these samples, it was pointed out that sp states at the top of the VB are mixed to TM d states in agreement with the above-mentioned DOS calculations done on series of conventional Al-transition metal (TM) alloys [18].

Further studies of complex alloys highlighted the importance of hybridization between p and d states [63]. A beautiful assessment of the respective weights of H-R and hybridization mechanisms was given recently by Mizutani and coworkers [65–68] while studying a series of γ-brass phases to determine whether they are stabilized following or not the e/a rule (or H-R mechanism). The authors pointed out that the H-R stabilization mechanism produces a pseudogap across E_F as a result of a resonance between electron waves and particular sets of lattice planes that differ from each other, depending on the studied specimens. They also pointed out that e/a does not keep the canonical 21/13 value for all the γ-brass samples, but rather varies from 1.6 e^-/at for example in Cu$_9$Al$_4$ to 1.8 e^-/at in Fe$_2$Zn$_{11}$ and

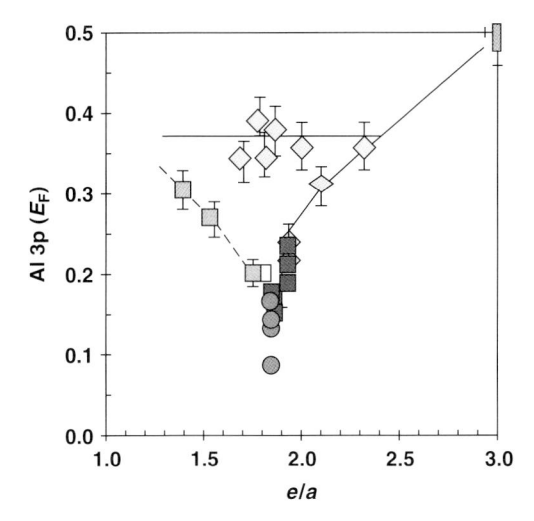

Figure 1.11 Al 3p intensity at E_F as a function of the valence electron to atom ratio e/a for a series of Al based alloys as obtained from SXES experiments. The symbols are as follows: narrow rectangle at $e/a = 3$ for pure fcc Al, diamonds for genuine Al-Cu Hume-Rothery phases, narrow diamond for ω-Al$_7$Cu$_2$Fe, gray squares for B2 cubic Al-Cu-Fe phases with structural complexity due to vacancies, increasing from left to right, dark squares for approximants of icosahedral Al-Cu-Fe QC, dark dots are for icosahedral QCs, namely from top to bottom Al-Cu-Fe, Al-Cu-Ru and Al-Pd-Mn, whereas the lowest dot corresponds to Al-Pd-Re system. The value of $n(E_F)$ for specimens represented by diamonds, squares, and the top circle were already shown in Figures 1.2 and 1.3, left, but as a function of β_C.

1.46 e$^-$/at in Al$_8$V$_5$. For this latter alloy, using LMTO-ASA DOS calculations, they clearly demonstrated the key role played by V 3d – Al 3p hybridization in the formation of the pseudogap across the Fermi level (Figure 1.12, left side). This theoretical result was confirmed experimentally, again using soft X-ray emission spectroscopy, as displayed in the same figure (Figure 1.12, right side) [63]. The interaction between the Al and TM states can be viewed as a Fano-like interaction, namely an interaction between extended and localized states, as described by Terakura [69].

To summarize, in CMAs made of Al (or Mg) alloyed with other elements among which TMs, the pseudogap present in the DOS at E_F results from the combination of two mechanisms, on the one hand, the so-called H-R mechanism and on the other hand, hybridization (interaction) between extended and localized states at E_F and nearby, which results in the presence of states with a localized character at the top of the VB.

1.3.2
Self-Hybridization in Al-Mg Alloys

What is the situation in CMAs made only of sp elements, for which there are normally no real localized states at the top of the VB? Let us consider the case of an emblematic

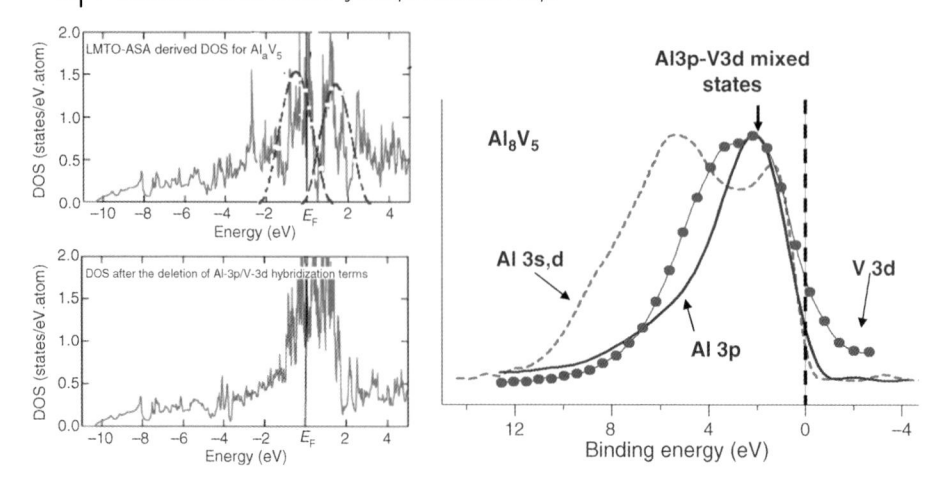

Figure 1.12 Left: DOS calculations by the LMTO-ASA method for Al_8V_5 with (top panel) and without (low panel) accounting for the V3d-Al 3p hybridization. In the low panel, clearly the pseudogap has disappeared as compared to the top panel. Right: Partial DOS as obtained from SXES experiments. The various distribution curves are normalized to their own maximum intensity. The mixed Al3p-V3d states are found at about 2 eV below E_F whereas in the calculation the V states are set at 1 eV below E_F.

CMA, namely the so-called Samson phase [70]. This is cubic β-Al_3Mg_2, with cell parameter $a = 28.24$ Å, containing 1168 at/uc distributed over icosahedra and Friauf polyhedra. Many atomic sites are not fully occupied, which induces an important degree of disorder. As far as transport properties are concerned, this compound behaves similarly to a simple mixture of the pure metals. This result is to be contrasted to the extremely complex structure of the Samson phase ($\beta_C = 7.07$), which raises the question to understand why the two simplest ways of staking metal atoms (Al is fcc and Mg is hcp) end at this specific composition in such a complex architecture. So, what is the electronic structure? Clearly, exact DOS calculations are extremely difficult to carry out or even impossible in this system, at least for the moment. On the contrary, experiments, again using X-ray emission spectroscopy that analyzes separately Al and Mg contributions to the VB, may give qualitatively valuable information about the bulk specimen.

The data presented in Figure 1.13 may be summarized as follows [71, 72]. First, Al 3p and Mg 3p states overlap over the extent of the VB, which points to the strong covalency in this energy range of the VB. However, chemical bonding is not so simple, since the maximum of the Mg 3p curve coincides with a depletion of the Al 3p curve, which indicates some degree of repulsion existing between Mg and Al 3p states in this energy range [69]. Secondly, the intensity of both Al and Mg 3p curves at E_F is about $25 \pm 2\%$ of the maximum intensity (set to 100%) and the edges are shifted towards lower binding energies since they are distant from E_F by 0.3 ± 0.05 eV at half-maximum intensity. Note that in both pure metals, the intensity at E_F is 50% of the maximum intensity and the edges very steeply crossing the Fermi level axis at half-

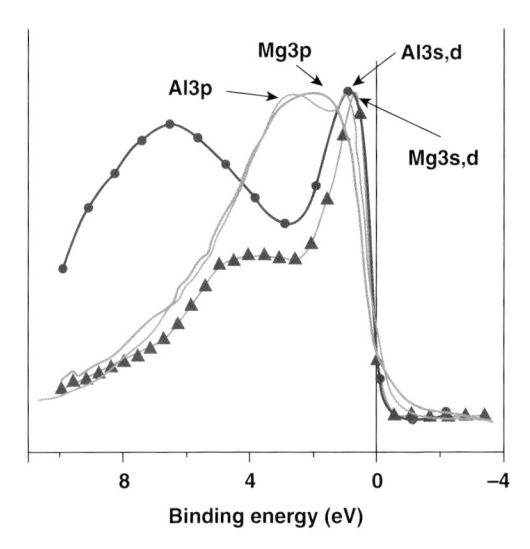

Figure 1.13 Partial DOS distributions in β-Al$_3$Mg$_2$. Al 3s,d: stared line, Mg 3s,d: line with triangles, Al 3p full line, Mg 3p: line with open dots.

maximum intensity of their respective distribution curves. All these observations suggest that in cubic β-Al$_3$Mg$_2$, a faint pseudogap is present at E_F. The maximum of the Al 3p curve is found at the same energy as a depletion of the Al 3s,d curve and of a plateau on the Mg 3s,d curve, indicating repulsive interaction between the Al 3p and the 3s,d states.

The most striking feature is the rather narrow peaks present at about 1.5 eV below E_F in the 3s,d distribution curve for both Al and Mg in the compound. These narrow peaks are distant by about 0.2 eV from each other, which suggests that some repulsion takes place between the spectral distributions at the top of the VB, whereas the peak in the Al 3s,d curve overlaps totally the Al 3p and Mg 3p curves. Here, we shall mention that SXES studies of compounds containing Al and two TM that are neighbors in the classification of the elements, have shown that the d states are located at the top of the VB and repel each other [62, 63]. Thus, the narrow shapes of the Al and Mg 3s,d curves close to E_F, their slight separation along the binding energy scale and the presence of the faint pseudogap suggest that these states are to some extent localized, namely have a d-like character. Let us mention that in fcc Al, as well as in hcp Mg, only a very small fraction of states with a d-like character is found near E_F [73]. In β-Al$_3$Mg$_2$ we see that the proportion of the d-like states is strongly enhanced, in line with the results described in the previous section where we reported that in Al-TM CMAs, localized states present at the top of the VB hybridize to sp states for a pseudogap is formed and stabilizes the system. Therefore, the mechanism in β-Al$_3$Mg$_2$ is not the same as in compounds containing a TM bringing d states to the VB.

Instead, achieving stabilization of the β-Al$_3$Mg$_2$ system is a little more complex. First, Al and Mg states self-hybridize in such a way that states with a localized character are found at the top of the VB. Secondly, the localized Al and Mg states thus formed slightly repel each other and interact with Al-Mg covalently bonded states. It is

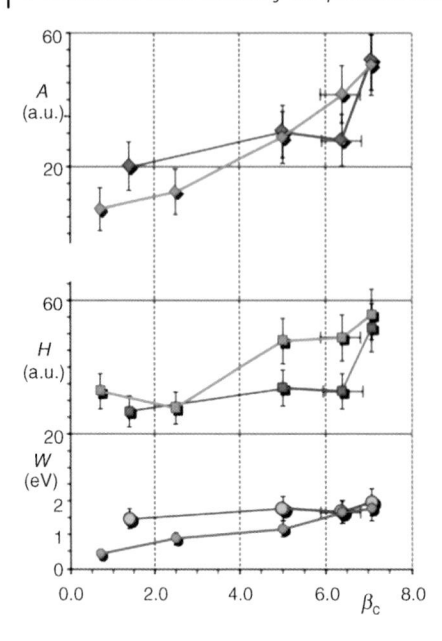

Figure 1.14 Change with complexity index β_C of the area A, amplitude H and width W of the d-like peak that is exhibited by (from left to right) Laves phase, Bergman phase, low-temperature variant and high-temperature variant of the Samson phase. The symbols below $\beta_C = 2$ are for hcp Mg and fcc Al (left to right).

interesting to mention that the same mechanism occurs in other Al-Mg compounds, but of lesser complexity such as the Zn_2Mg Laves phase ($N_{UC} = 12$, $\beta_C = 2.48$), the low-temperature variant of the Samson phase [74], or the Al-Mg-Zn Bergman phase ($N_{UC} = 148$, $\beta_C = 5$). In these latter compounds, the same peak of localized d states is pointed out by SEXS at the same energy position, but with intensity (amplitude) varying smoothly with the complexity index and almost unchanged width [72]. The result of this analysis is presented in Figure 1.14. It has an interesting issue, namely that Nature selects preferably a complex structure when the self-hybridization mechanism is at work, although far simpler atom packings exist nearby in the phase diagram (not taking into account the role of configuration entropy that is able to further reduce the free energy of the system).

1.4
A Brief Survey of Properties

1.4.1
Transport Properties

Resistivity measurements performed for specimens with increasing structural complexity up to quasicrystalline compounds have pointed out a different behavior

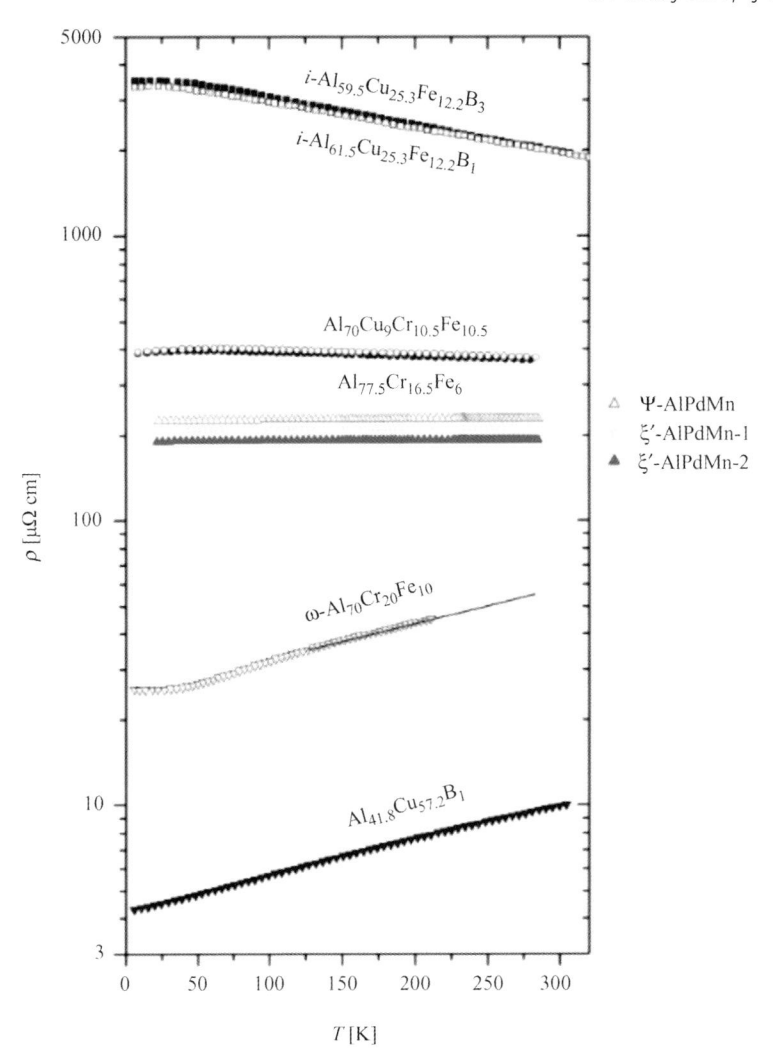

Figure 1.15 Variation of the resistivity coefficient versus temperature for Al-based alloys ranging from simple structures to CMAs and quasicrystals.

from that of conventional alloys. This is exemplified in Figure 1.15 as a plot of the variation of the electrical resistivity against temperature for a series of Al-based alloys. The curves at the bottom of the figure, with resistivity below 100 μΩ cm, correspond to structurally simple alloys, namely CsCl-cubic Al-Cu-B and tetragonal ω-Al$_7$Cu$_2$Fe. They show a normal increase of the resistivity with temperature, characteristic of conventional alloys and metals. The curves in the middle of the figure were obtained from approximants of quasicrystalline compounds. Their resistivity values are much larger than for conventional alloys and the variation with temperature is almost zero or very weak. In strong contrast to normal alloys, genuine quasicrystals exhibit resistivity values at low temperature that are high to very high and the variation of the resistivity with increasing temperature is of opposite sign to that of simple alloys and metals [58].

Hence, attention was paid to the electronic properties of CMAs in relation to their structures. Magnetic, electrical and thermal transport properties have been probed so far for series of CMAs [75]. Recent data refer to Al_4TM and $Al_{13}TM_4$ families whose atomic structure can be viewed as a stacking of flat and corrugated atom layers with structural complexity and unit cell size increasing from Al_4TM to $Al_{13}TM_4$ families. For completeness, the investigations were carried out perpendicular as well as along the stacking direction. These data are detailed in Chapter 3.

Basically, magnetic susceptibility, electrical resistivity, thermoelectric power, Hall coefficient and thermal conductivity all displayed anisotropic behavior that is more marked with increasing structural complexity from Al-Ni-Co to $Al_{13}Co_4$ and $Al_4(Cr,Fe)$. The temperature coefficient is of Boltzmann-type in Al-Ni-Co and $Al_{13}Co_4$ for all crystallographic directions and non-Boltzmann as far as the temperature-dependent in-plane resistivity is concerned for $Al_4(Cr,Fe)$. The thermal conductivity of the same specimens behaves similarly to the electric conductivity. It is about one order of magnitude lower when measured along the stacking axis than in-plane [76]. The Hall coefficient is also anisotropic. It is the lowest when the external field is applied along the stacking direction and higher when it is applied in-plane. However, no clear connection with structural complexity was evidenced in these series of samples. All these investigations pointed out strong differences in the respective contributions of the lattice and conduction electrons. They led to the conclusion that the anisotropy of the atomic structure of CMAs extends to their electronic properties, thus departing from those of conventional alloys and their elemental constituents. These results are detailed in Chapter 3.

1.4.2
Surface Physics and Chemistry

The physical and chemical properties of the surfaces of CMAs depend on their preparation, namely whether it is a clean surface worked out in ultrahigh vacuum (UHV) or a "dirty" surface kept in ambient atmosphere with an important contribution of the native oxides lying above the top layers of the CMA material.

The outermost layers of clean CMAs surfaces (including quasicrystals) as prepared in UHV by successive cycles of sputtering and *in situ* annealing are similar to simple bulk terminations with steps and islands. So far, no reconstruction of the surface and no chemical segregation effects were observed on highly complex CMAs. The atomic density is high and the topmost layers host the elements that have the lowest surface energy. This latter characteristic governs nucleation and growth on CMA surfaces used as templates for nanostructuration of foreign atoms.

The electronic structure of such sputtered-annealed CMAs surfaces has been investigated using several spectroscopic techniques. A main issue is that the pseudogap that exists at the Fermi level in the electronic density of states of the bulk specimen is also present in the surface top layers. However, this is no longer true for surfaces obtained from mechanical fracture in UHV with no further treatment.

Actually, such cleaved surfaces display a more metallic character, as shown by the reduction of the importance of the pseudogap with respect to the bulk sample. Band gaps have also been predicted by theoretical means in the phonon density of states, but this is still a matter of experimental investigations. These characteristics of the electronic and phononic structures at the surface of CMAs influence physical properties such as adhesion, friction and energy dissipation [77]. Chemical reactivity is also affected by the electronic structure and atomic structural complexity, but to a far lesser extent.

Oxidation in UHV of CMA surfaces shows selective oxidation of Al and the formation of a passive amorphous overlayer that is similar to the one that forms on pure Al. The thickness of this oxide layer depends on the conditions under which oxidation is carried out [78]. Note that the oxide layer formed on Al-Cr-Fe CMAs after water immersion is relatively thin, giving this CMA a significant corrosion resistance, especially with respect to the action of water [79].

Wetting by ultrapure water of Al-based CMA surfaces kept in ambient air and cleaned using a strict protocol avoiding contamination was studied by measurements of contact angles. Strong differences were pointed out from sample to sample, despite their outermost oxide layer being the same from the chemical point of view (Figure 1.16). It was concluded [59] that the reversible adhesion energy of water

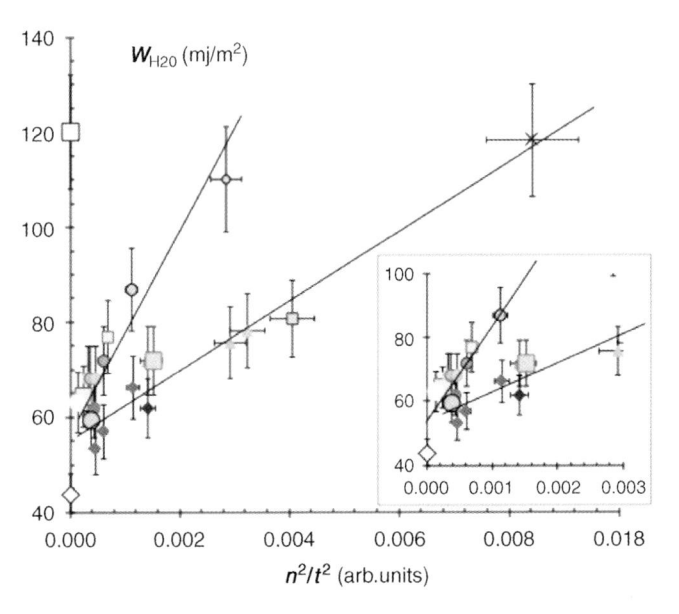

Figure 1.16 Variation of the reversible adhesion energy of water, W_{H2O}, against $n2/t2$ where $n = n(E_F)$ is the intensity of the Al 3p partial DOS at E_F and t is the oxide thickness. The two straight lines correspond to specimens with different contributions of 3d states at EF. The line with the largest slope corresponds to Al−Cr−Fe(−Cu) samples, whereas the other is for Al−Cu(−Fe) specimens. The open symbols located on the y-axis are for Teflon (diamond) and alumina (square). They define, respectively, the lowest and largest values that can be observed with the present set of Al-based samples. The inset presents an enlargement of the data in the region $n2/t2$ below 0.003.

(W_{H_2O}) deduced from contact angle depends on the Al 3p density of states at the Fermi level of the bulk CMA substrate and on the inverse of the squared thickness of the oxide layer. Using image force theory, the wetting properties of the oxidized surfaces of Al-based CMAs appeared to be dependent on long-range forces between the dipole sitting on water molecules above the oxide top layer and their image dipoles developed into the conduction sea, far beneath the surface oxide. For the *i*-Al-Pd-Mn quasicrystalline phase, the oxidation, low adhesive properties and reduced surface reactivity as compared to pure Al were found to be consistent with the more ionic character of alumina that grows on this sample in ambient air [80]. It was also found that the oxidation kinetics of CMAs significantly departs from that of classical intermetallics [81]. In addition, the adsorption behavior of molecules other than oxygen or water suggests that CMAs surfaces can be more reactive than their pure metal constituents or related conventional intermetallics.

The high chemical reactivity of CMAs makes them good candidates for catalysis purposes. Actually, the performance of powdered quasicrystals was investigated for several specific catalytic reactions after various chemical treatments. It has been shown that, for example, Al-Cu-Fe or Al-Pd-Mn CMAs are good candidates for replacing at low cost, expensive catalysts that are necessary to the chemical industry [82].

1.4.3
Surface Energy

Surface energy is an essential property of a material. It determines the equilibrium shape of a single crystal, it is directly related to its cleavage energy, it is involved in the height of the crystal nucleation barrier when the crystal grows from the liquid state, it is related to its adhesion energy to another solid, and so on. It is, however, very difficult to measure experimentally and as far as CMAs are concerned, it is nearly impossible to compute owing to the limited power of present computers. A variety of experimental facts, based on friction measurements or on contact-angle measurements of small droplets deposited at the surface of CMAs, indicates that the surface energy of CMAs with large unit cell may be characterized by a reduced surface energy. Experiments performed in ultrahigh vacuum show the same trend, namely that the surface energy of icosahedral compounds might be much smaller than that of the constituent species.

A series of pin-on-disk friction experiments performed in high vacuum allows us to quantitatively estimate the surface energy of a large number of CMAs [3, 83]. A pin-on-disk experiment is housed in a vacuum chamber in which the residual pressure of oxygen is small enough to forbid the growth of a complete oxide layer between two successive passages of the indenter. Under such conditions, and if the applied load is small enough to produce no third body, or equivalently, if friction is performed at equilibrium, friction takes place between the two naked bodies (pin and surface of interest) after just a few rotations of the disk during which the native oxide layer is destroyed by the contact and escapes from the trace due to the rotation of the sample. The friction coefficient is given by $\mu = F_T/F_N$, where F_T stands for the force that works against the movement of the sample relative to the pin, in the plane of the contact, and F_N is the applied load. To first order, for hard enough samples, we may write:

$$\mu = \alpha/H_V + \beta\, W_{SP}$$

with H_V, the (Vickers) hardness of the disk material, W_{SP} the work of adhesion between surface of interest S and pin P and α and β, two fit parameters that can be determined experimentally for that specific experimental setup, using materials of known, or measurable, hardness and surface energy (see why below). Assuming friction does indeed take place at equilibrium, which can easily be ascertained after the end of the experiment by inspection of the trace, W_{SP} becomes the reversible adhesion energy between S and P:

$$W_{SP} = \gamma_S + \gamma_P - \gamma_{SP}$$

with γ_S and γ_P the surface energy, respectively, of the studied CMA surface and pin, and γ_{SP} the interfacial energy developed between (naked) S and P bodies.

In a very crude assumption, we will take the term $\gamma_P - \gamma_{SP}$ equal to 0, which leads to an overestimate of μ by combining the equation above to the previous one:

$$\mu \geq \alpha/H_V + \beta\, \gamma_S \tag{1.1}$$

and therefore estimate the upper limit of the surface energy of the CMA specimen:

$$\gamma_S \leq (\mu - \alpha/H_V)/\beta$$

Despite this model being very crude, compared to more sophisticated ones published in the literature [84], it turns out to fit very satisfactorily the hardness and surface energy of many reference samples like transition metals, window glass, aluminum, aluminum oxide, and so on, which were used to calibrate the pin-on-disk apparatus used in the present study, thus delivering reliable values for α and β (Figure 1.17). It must be stressed, however, that a few metals, like Co and W, do not obey the same simple model, essentially because the sticking coefficient of oxygen on those metals is so high that friction is always lubricated in the conditions of the experiment, and therefore forbids naked surfaces to come into contact.

Application of the previous equation to CMAs of unknown γ_S, but easily measurable hardness, delivers an upper value for their surface energy [83]. The results are summarized versus the experimental m data in Figure 1.18. They are consistent with other experimental data supplied by ultrahigh vacuum growth experiments [85] and show that the surface energy of a quasicrytal of high crystal perfection is smaller than that of its pure constituents by a factor comprised between 2 (Al) and 4 (Cu, Fe).

Furthermore, the term $\mu_C = \mu - \alpha/H_V$ represents the adhesive part of friction, when μ is corrected for the mechanical deformation of the specimen under the stress developed by the contact to the pin. Relevant values of μ_C were presented earlier in this chapter in the right side of Figure 1.2 for a series of Al-Cu-Fe CMAs of varying crystal structure and complexity. Another view at the adhesive part of friction against hard steel in vacuum is given in Figure 1.19 as a function of the partial densities of Al 3p, Al 3s,d and TM 3d states probed separately at the Fermi energy by XSES. The

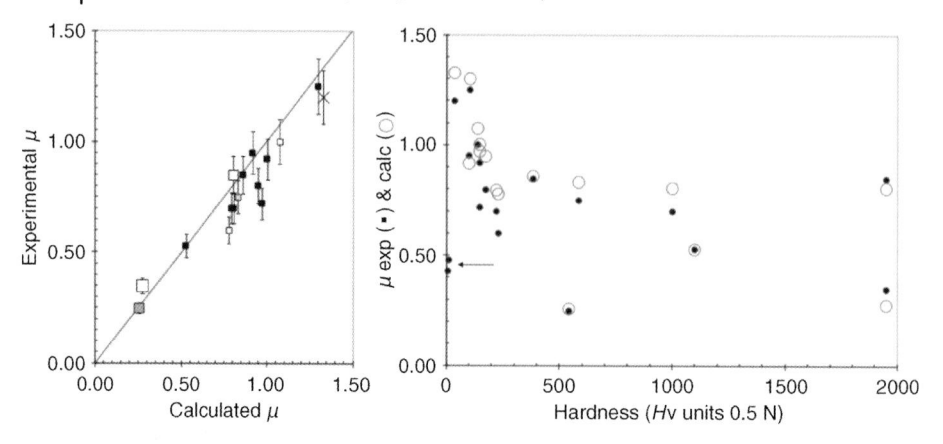

Figure 1.17 A set of pin-on-disk experiments in vacuum allows us to measure the friction coefficient μ against hard steel for various reference samples of known hardness and surface energy. Right: the plot shows that μ values calculated according to Equation 1.1 fit very well the experimental data, except when the hardness is small (arrow) since Equation 1.1 diverges when $H_v \to 0$. Left: same data, from which calibration parameters α and β applicable to our specific experiment are deduced with satisfactory accuracy.

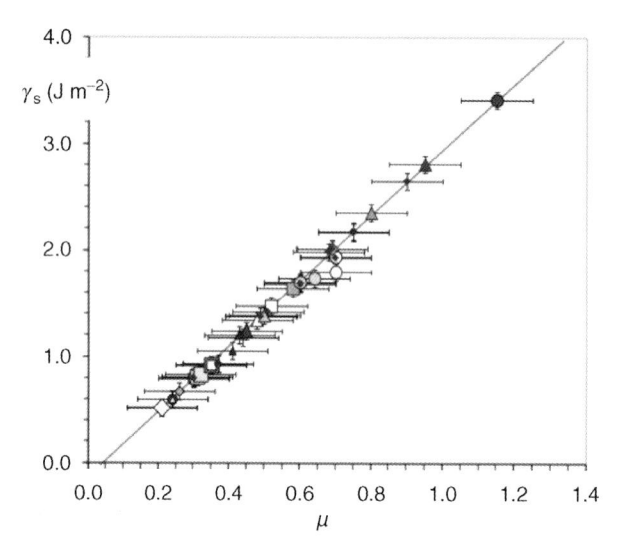

Figure 1.18 Summary in a single chart of all upper limits of the surface energy data deduced from friction experiments in vacuum against hard steel. The lower data is for an Al-Cu-Fe quasicrystal of high lattice perfection and falls in the range 0.5–0.6 J m^{-2}, well below the surface energy of the pure constituents (Al: 1.2 J m^{-2}; Cu: 1.8 J m^{-2}; Fe: 2.4 J m^{-2}). The other data are for CMAs of decreasing complexity and increasing density of d states at the Fermi level (see below).

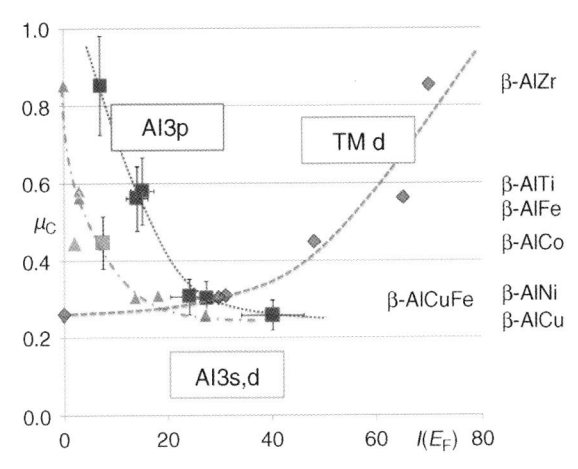

Figure 1.19 Adhesive part of friction in vacuum against hard steel versus partial density of states at E_F for a series of CMAs of (nearly) identical complexity ($\beta_C = \ln 2$), but varying nature of the TM constituent.

CMA crystals are all of about the same complexity and belong to the B2 CsCl-type of cubic phases. The overall trend observed on the figure is that the adhesive part of friction increases with increasing TM 3d DOS in the CMA sample, whereas it decreases with increasing Al 3p and Al 3s,d intensities. Since the DOS in the pin is determined by the nature of the hard steel used for the pin ball, this behavior is equivalent to that expected from the formation of a band between CMA sample and pin when the two bodies come into contact.

The tendency to decrease adhesion in vacuum (once more, against hard steel) with the filling of the band is further exemplified in Figure 1.20 where we show an overview on all surface energy data gained from the same specimens as in Figure 1.18, but plotted as a function of VCE, the total number of valence electrons of the sample. The surface energy decreases with increasing VCE, until a minimum is reached at VCE = 8 e^-/at. Beyond this value, a copper-rich sample and fcc Cu itself show the opposite trend. It is worth noting that the quasicrystal in this family of CMAs (star in Figure 1.20) is found significantly below the other specimens. This result, again, stresses the electronic origin of adhesion in vacuum between the solids considered herein. Empirical rules may be derived from the present study to optimize the choice of materials that are placed in contact under severe load, whereas vacuum hinders the formation of a diffusion barrier at the contact interface. We will come back to this aspect later in this chapter.

1.4.4
Plasticity

The plasticity of CMAs is a novel field of materials science. To date, only a very limited number of different CMA phases have been experimentally investigated. It is a

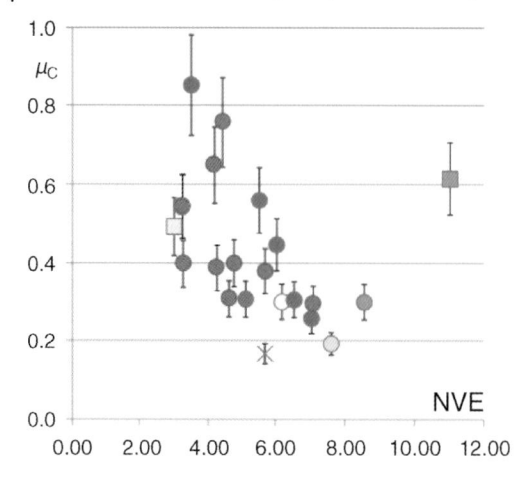

Figure 1.20 Adhesive part of friction in vacuum against hard steel as a function of the total number of valence electrons per atom. The squares are for fcc Al (left) and fcc Cu (right), respectively. The Al-Cu-Fe icosahedral CMA is shown by a star and is found below the other data at identical NVE. The values of μ_C presented as a function of complexity index in the right side of Figures 1.2 and 1.3 were taken in the range $5.5 \leq NVE \leq 6.2$.

common property of all these materials that they are brittle at room temperature. Ductility sets in at temperatures of the order of 70% and higher of the melting point of the individual materials. This is a feature discriminating CMAs from essentially all other simple crystalline metals – the latter typically show ductile behavior at room temperature and even below. In this respect, the plastic behavior of CMAs rather resembles that of covalently bond crystals such as for example, silicon.

In the following, we will discuss experimental results on three CMA phases, ε-Al-Pd-Mn, β-Al$_3$Mg$_2$, and Al$_{13}$Co$_4$. The phases ε_6-Al-Pd-Mn and Al$_{13}$Co$_4$ are orthorhombic and have 320 and 102 atoms per unit cell, respectively. Their structures were described in Section 1.2.3. The compound ß-Al-Mg is cubic, space group $Fd\bar{3}m$. We already mentioned that the lattice parameter is $a = 2.82$ nm and the unit cell contains about 1168 atoms [70]. The coordination polyhedra in the structure comprise 672 icosahedra (ligancy 12), 252 Friauf polyhedra (ligancy 16), 24 polyhedra of ligancy 15, 48 polyhedra of ligancy 14 and 172 more or less irregular coordination shells of ligancy 10–16. Because of incompatibilities in the packing of the Friauf polyhedra, this structure features a high amount of inherent disorder, which is apparent as displacement disorder, substitutional disorder and fractional site occupation.

Figure 1.21 shows the stress–strain curve of an ε_6-Al-Pd-Mn sample, deformed at 700 °C with a strain rate of 10^{-5} s^{-1}. The compression axis was chosen parallel to the [0 1 0] lattice direction. For a discussion of the main features of the stress–strain curve, ignore the three sharp dips, which are the result of stress-relaxation tests to be discussed below. The course of the curve is at the positions of the relaxation dips interpolated by dotted lines. At very small strains ε, the curve shows an almost linear behavior. This is the elastic regime, where the deformation is reversible and,

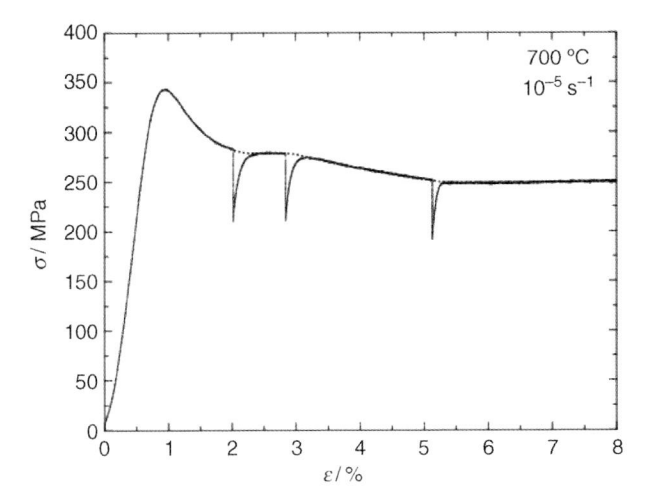

Figure 1.21 True stress–true strain curve recorded in compression at 700 °C for an ε_6-Al-Pd-Mn single grain sample. The three dips visible on the curve come from relaxation experiments performed at different stages of deformation.

according to Hooke's law, the stress σ is proportional to the strain. Plastic deformation sets in at about 0.70% strain, where first deviations from a linear course occur. At 1% strain the curve reaches an upper yield point at about 350 MPa. Subsequently, the stress decreases down to a value of about 280 MPa, where it reaches a lower yield point at about 2.5%. After the lower yield point, the stress–strain curve goes through two further stages. First, from about 3 to 5% strain, the stress decreases with strain, that is, the material shows a work-softening stage. Second, from about 5% to the termination of the experiment at 8%, the stress–strain curve is essentially horizontal. In this stage, the material is in a dynamic equilibrium, corresponding to a steady state where hardening and softening processes in the microstructure balance.

Figure 1.22 displays a set of stress–strain curves of $Al_{13}Co_4$ samples, deformed along the $[6\ \bar{4}\ 5]$ direction at a strain of $10^{-5}\,s^{-1}$ and temperatures between 650 and 800 °C [23]. Each curve shows signatures of additional temperature cycling tests and a stress-relaxation test, marked "TC" and "R," respectively, in the uppermost curve. The corresponding results will be considered below.

At all temperatures the curves have common qualitative features. After the elastic regime, a strong yield-point effect is observed in the strain range between 0.25 and 0.55%. At 700 °C, for instance, a stress difference as large as 45% between the lower and upper yield stress was measured. Additional yield-point effects are seen after the temperature changes and after stress relaxation. At high strains, above about 2%, the curves show an almost constant flow stress or, at some temperatures, a very weak work-hardening stage. The deformation behavior is strongly temperature dependent: the stress strongly decreases with increasing temperature, leading to high-strain flow stresses between about 320 MPa at 650 °C to 120 MPa at 800 °C.

Figure 1.23 displays stress–strain curves of β-Al_3Mg_2 samples at $10^{-4}\,s^{-1}$. The black curves represent deformations of single crystalline samples along the [1 0 0]

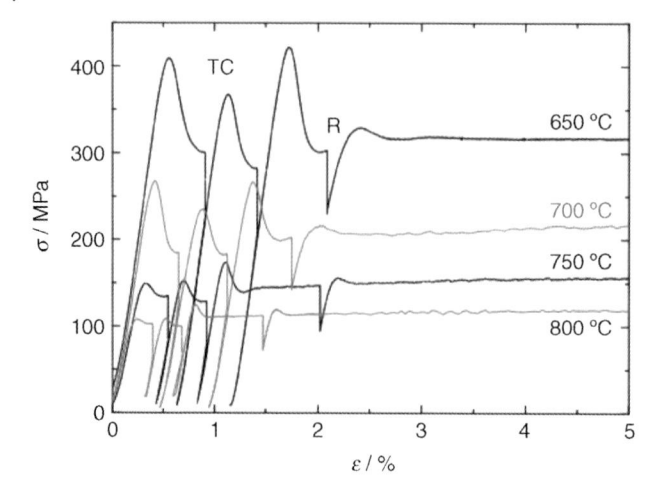

Figure 1.22 Stress–strain curves for single grains of $Al_{13}Co_4$ deformed at various temperatures as indicated.

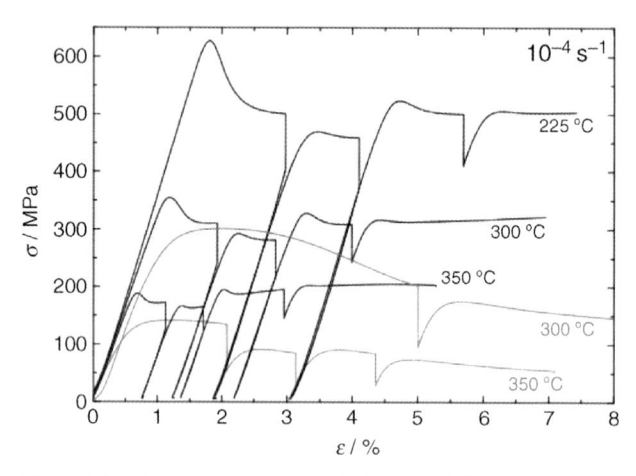

Figure 1.23 Stress–strain curves of β-Al_3Mg_2 at different temperatures as indicated.

direction, the gray curves represent deformations of polycrystalline samples (grain size about 20 μm). Temperature changes and relaxations were carried out during most of the experiments. The single-crystal deformations show similar features to those of $Al_{13}Co_4$, with a generally smaller yield-point effect. The polycrystals deformations, on the other hand, show considerably different behavior. The yield points are much broader, the curves show work softening, and the high-strain flow stresses are considerably smaller than for the single-crystalline case. Also, the single-crystalline samples can be deformed at temperatures down to about 225 °C, while the polycrystals are ductile only above 300 °C.

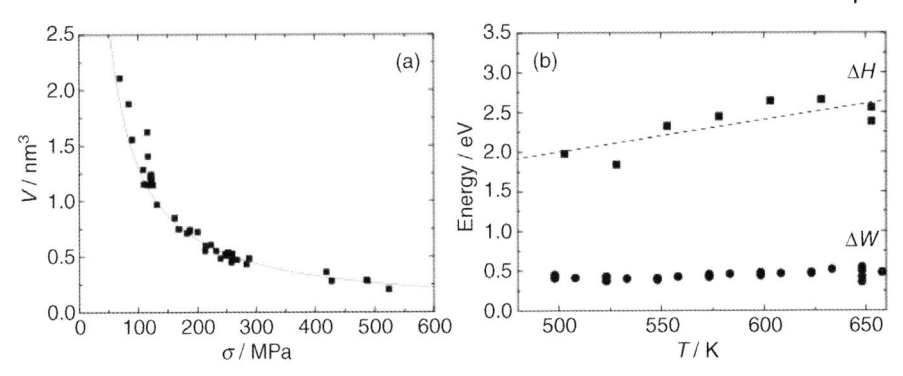

Figure 1.24 Activation volume (a), activation enthalpy and work term (b) deduced from relaxation experiments performed on ε-Al-Pd-Mn (a) and β-Al$_3$Mg$_2$ (b) single crystals, respectively.

The activation parameters are determined by dedicated incremental tests, that is, stress-relaxation tests (R in Figure 1.22) and temperature changes (T in Figure 1.22). Figure 1.24a displays the activation volume of ε-Al-Pd-Mn, determined by stress relaxation as a function of stress. The activation volume is strongly stress dependent. It decreases with increasing stress, following a hyperbolic curve. The absolute values vary within the range of about 0.5–2 nm^3. This stress dependence and the absolute values of V are typical for CMAs. Let us compare different CMA phases for a given stress value of 300 MPa: Values of $V = 0.45$ nm^3 (ε-Al-Pd-Mn), $V = 0.8$ nm^3 (Al$_{13}$Co$_4$), and $V = 0.6$ nm^3 (β-Al$_3$Mg$_2$) are found. Scaled by the respective atomic volumes, we find $V/V_a = 30$ for ε-Al-Pd-Mn. For Al$_{13}$Co$_4$ and β-Al$_3$Mg$_2$ we find $V/V_a = 53$ and $V/V_a = 32$, respectively.

The values found for the activation volume of different CMAs obviously exceed the atomic volumes by more than an order of magnitude, which indicates that large obstacles containing some tens of atoms control dislocation motion. Recall that we have accounted for the presence of a cluster substructure as a distinct structural feature of CMAs. Accordingly, it was concluded for several CMA phases, that the cluster substructure provides the rate-controlling obstacles for dislocation motion [20, 22].

Figure 1.24b shows the activation enthalpy and the work term of β-Al$_3$Mg$_2$ single crystals as a function of deformation temperature, obtained from combined temperature-change and stress-relaxation tests. The values for the activation enthalpy ΔH are shown as solid squares. Values increasing with temperature from about 1.8 to 2.6 eV are found. A linear fit under the boundary condition $\Delta H(T = 0\,\text{K}) = 0\,\text{eV}$ is shown as a dashed line. The work term, corresponding to the part of the energy, which is supplied by the applied stress, is shown by circles. It is roughly constant in the observed temperature range and amounts to about 0.4 eV.

The activation enthalpy ΔH is larger by about a factor of six than the work term. It can hence be concluded that the deformation is a thermally activated process. Similar behavior of the energetic activation parameters is also found for other CMAs. For Al$_{13}$Co$_4$ $\Delta H = 2.2$ eV[23] and for ε-Al-Pd-Mn $\Delta H = 5$ eV is found [20]. The activation

1 Introduction to the Science of Complex Metallic Alloys

enthalpy is always much larger than the work term, and the enthalpies are considerably larger than the corresponding self-diffusion energies. As the latter finding indicates that the deformation-rate-controlling mechanism is not given by a single-atom diffusion mechanism, this is consistent with the conclusions drawn from the results for the activation volumes.

1.5
Potential Applications

1.5.1
Applications Related to Surface Energy

It is a well-known fact that technological developments often anticipate a full understanding of the property they are based upon. This has been the case for Al-Cu-Fe-Cr quasicrystals and approximants, which were shown to yield appreciate antistick properties and interesting corrosion resistance, making them suitable for a new generation of cookware [86]. This type of utensil was combining low adhesion and excellent mechanical resistance to scratch, in contrast to many modern devices that do not offer both performances together. Figure 1.25 summarizes the basics behind the performance of such utensils. It is based on the reduction of the apparent surface energy of a CMA coating, equipped with its native Al_2O_3 oxide that forms in ambient conditions. The reversible adhesion energy of water (taken as a model material representative of food, although the chemical reactions that take place during cooking are far more complicated) is then related to the contact angle, as discussed earlier in this chapter. It may be divided into two parts, coming respectively from fluctuations of the electric charges on both sides of the water/oxide interface

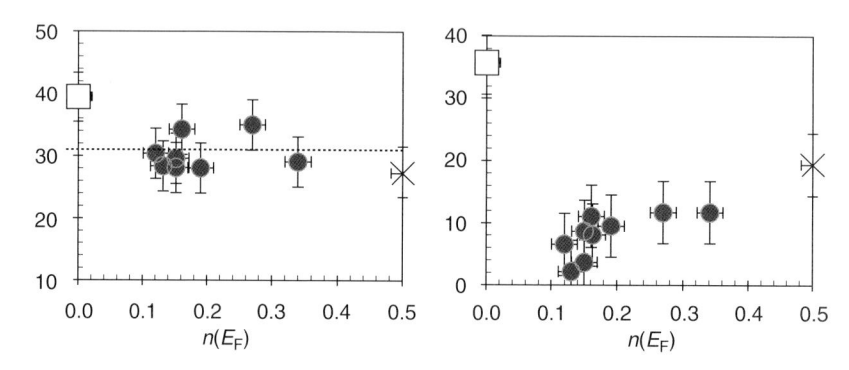

Figure 1.25 Lifshitz–Van der Waals (left) and I^{AB} (right) components measured using various liquids for pure alumina (square), fcc aluminum (star) and a series of Al-Cu-Fe-Cr CMA materials of changing complexity and therefore, different Al3p DOS at E_F ($n(E_F)$). Observe that I^{AB} cancels for selected samples when $n(E_F) \approx 0.12$, which corresponds to quasicrystals and high-rank approximants in Figure 1.11.

and from a component accounting for the presence of permanent electric charges at the interface and its vicinity. The first component is called the Lifshitz–Van der Waals term, and the second is often termed simply I^{AB}.

Then, the reversible adhesion energy of water reads:

$$W_{H2O} = \gamma_L(1 + \cos\theta) = 2(\gamma_S^{LW} + \gamma_L^{LW})^{1/2} + I^{AB}$$

where the subscripts S and L stand for solid and liquid, respectively, θ is the contact angle, and LW is for Lifshitz–Van der Waals. The part of the surface energy of water γ_L that accounts for LW interactions is well known. The equation above is valid only if the film pressure of water on the surfaces of interest is negligible, which is the case in the present work for all CMA samples of large complexity that were studied. Now, using various liquids, which allows us to vary the ratio γ_L/γ_L^{LW}, it is possible to assess the respective weights of γ_S^{LW} and I^{AB} for a given solid (Figure 1.25). It turns out that I^{AB} vanishes for CMA materials of high complexity, like the Al-Cu-Fe quasicrystal and the Al-Cu-Fe-Cr orthorhombic approximant of the decagonal phase that was designed for this purpose [87], in strong contrast to fcc aluminum or conventional Al-based intermetallics. Since γ_S^{LW} is merely constant in all alloys, the low stick property pointed out for those CMAs is an intrinsic property related to complexity, via the I^{AB} component and its reduction with the decreasing Al3p DOS at E_F (or equivalently, increasing β_C, Figure 1.2).

Definite attempts were made by one of the present authors to transfer this discovery to industry. A process able to produce large amounts of atomized powder (up to 1000 kg/day) was designed in association with the preparation at large scale of coatings by plasma spray (Figure 1.26). Demonstrators were supplied to restaurants and to a number of participants in the study. Unfortunately, when marketing started, the pans that were sold had not undergo the thermal treatment designed to stabilize the mixture of icosahedral and β-CsCl type phase that are quenched in a metastable state during the fast cooling that follows the projection from liquid state [88]. As a consequence, the high corrosion resistance characteristic of the alloy selected for this application was lost, especially in washing machines. It resulted in a massive return of the products, and customers claimed for reimbursement, which sadly concluded the story. The inventor, who had never been informed of the change in the process, gave up with this type application of quasicrystals and approximant CMAs.

Other possible applications related to the surface energy of Al-based CMAs were looked at over the years. They revolve around friction and adhesion in vacuum, which is a concern for the aerospace industry and microelectronics. Satellites, for instance, make a broad use of mechanical devices, which must be kept closed under high mechanical stress during launch and travel, to avoid uncontrolled movement, and must open when they arrive at their final destination. A lot of vibration during the travel phase, and event later, produces what is called cold welding in the vacuum of space. Metallic alloys usually bond under such conditions, and the parts do not separate when demanded from earth, which causes the loss of the mission, that is, billions of US Dollars or Euros. To avoid such problems, surface coatings are often

Figure 1.26 One of the many 1-kg ingots of $Al_{62}Cu_{18}Fe_{10}Cr_{10}$ (at.%) CMA alloy that was prepared by a conventional metallurgical route for feeding a powder atomization tower and then a plasma torch for the sake of producing low-stick coatings for frying pans. A utensil then had to be thermally treated and polished before sale.

used, with the purpose to forbid direct contact between the naked metal surfaces. Fretting tests performed at Austrian Institute of Technology, in Seibersdorf, Austria [89], have shown that Al-based CMAs are excellent candidates for this purpose (Figure 1.27). They exhibit both high hardness, which is mandatory to sustain the high applied stress during launch of the rocket, and do not bond against hard steel or aluminum alloys. For the moment, the limitation to the use of these coatings is basically the process of covering complex shapes like the ones encountered in this field with a coating made of a CMA of excellent lattice perfection, and no macroscopic defects like cracks or pores. A later chapter deals with this side of CMA metallurgy.

In order to insist a little more on the care that must be taken to produce coatings of excellent microstructural quality to achieve the level of performance expected for CMA materials, we show in Figure 1.28 a successful attempt made at the Josef Stefan Institute by Cekada *et al.* [90]. The starting point (Figure 1.28a) is a multilayer stacking of Al, Cu and Fe films the respective thicknesses of which are selected in order to reproduce the adequate stoichiometry of the material. After thermal mixing (Figure 1.28b), a homogeneous thick coating is formed. It can be used for example, to coat cutting inserts made of a WC-Co sinter, covered by the Al-Cu-Fe film. Alternatively, physical vapor deposition (PVD) can process the same quality of

1.5 Potential Applications

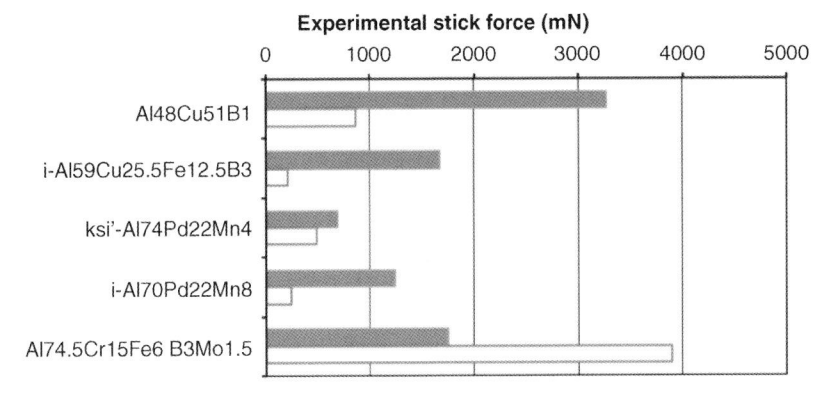

Figure 1.27 Stick-force measured in high vacuum against an aluminum alloy (dark gray) and hard steel (light gray) for a number of CMAs of varying complexity and composition. Whereas the performance against the Al-based alloy is not outstanding, the one against hard steel, at least for the two icosahedral phases, is quite attractive.

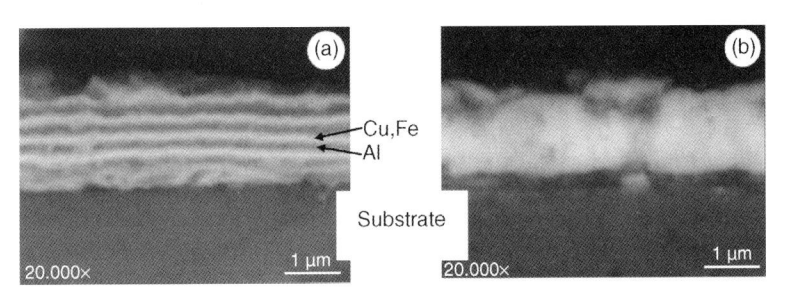

Figure 1.28 Al, Cu and Fe layers stacked on top of each other on a WC-Co substrate (a) transform upon thermal treatment in a homogeneous coating (b).

coatings, starting from a target the composition of which was designed to take into account the shift from stoichiometry that results from preferential sputtering rates encountered for different elements like Al, Cu and Fe. Standard tests defined according to the state-of-the-art prove that the lifetime of the tools under representative machining conditions is increased by 25% (Figure 1.29), which represents a very significant saving for the profession, provided the production costs are kept low. This step for the time being is not yet achieved.

1.5.2
Applications Related to Transport Properties

Applications related to transport properties are of four types: heat insulation, light absorption in view of solar heating of houses, applications connected with the

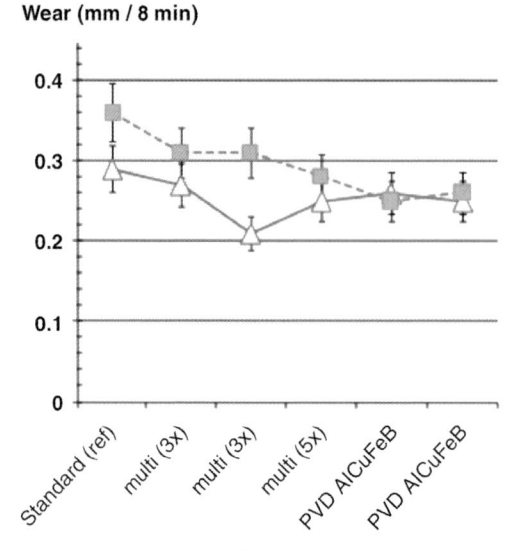

Figure 1.29 Standardized wear rate of cutting insert, according to the state-of-the-art (left), and for various multilayers after thermal mixing (center of the figure) or two PVD coatings (right).

magnetocaloric effect (MCE), namely the heating or cooling of a magnetic material in response to the variation of an external magnetic field and finally, applications referring to the thermoelectric effect, namely the transformation of caloric energy into electric energy or vice versa. These two latter aspects are treated in Chapters 8 and 9, hence we will not detail them any further. Let us just mention that as far as the magnetocaloric effect is concerned, much effort is done nowadays with the purpose of finding materials with a high MCE around room temperature [91], easy to prepare and suitable for designing the first magnetic refrigerator usable in everyday life. Also, an investigation is being carried out with the goal of discovering new environmentally friendly energy sources, as for example, for achieving electricity generation from waste of heat via thermoelectric modules. These two potential applications of CMAs are dealt with in Chapters 8 and 9, respectively.

The potential of CMAs for applications in the domain of heat insulation, namely the production of thermal barriers for the automotive and aeronautic industries, and solar-light absorption for the purpose of low-cost house heating was recognized and secured by one of the present authors quite some time ago (see references in Ref. 3). Thermal barriers were produced by plasma spray technology and demonstrators were submitted to tests [92]. A limitation comes from the too low melting point of the CMA coatings known so far. However, for certain niche applications like turbine blades of helicopter engines or military aircrafts, the potential of CMA coatings has been recognized and studied to some extent (Figure 1.30).

Successful attempts were also carried out by Eisenhammer [93] to replace the so-called $TiNO_x$ technology by Al-Cu-Fe films of equivalent light-absorption performance, but much higher working temperature, and therefore better thermodynamic

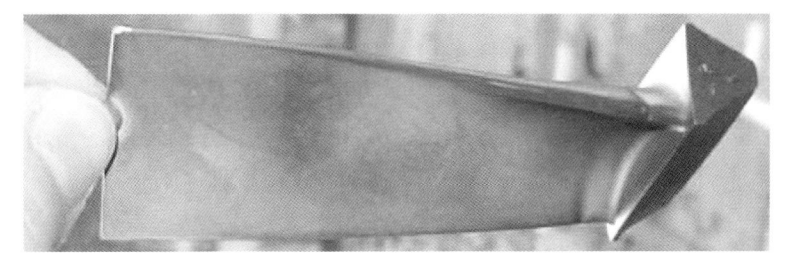

Figure 1.30 A small helicopter turbine blade covered by magnetron plasma sputtering with a 0.3-mm thick Al-Co-Fe-Cr thermal barrier. The thumbnail on the left gives an approximate scale for the figure (Courtesy S. Drawin, Onera, France).

efficiency. In fact, all trials have failed so far due to the nonmature film deposition industrial processes of CMAs layers and coatings, which increase the cost of the layers above the thresholds that industry may accept.

1.5.3
Applications Related to Dispersion of Particles in a Matrix

In Section 1.4.4, we have seen that CMAs are ductile only at elevated temperatures. In temperature ranges of 70% of the melting temperature and above, the flow stresses are of the order of some hundreds of MPa. The flow stress is strongly temperature dependent and increases with decreasing temperature. Accordingly, at lower temperatures, for example, at room temperature, the yield strength of CMA materials is very high, albeit at very low ductility. Consistently with the low surface energy, the toughness constants are small, if not negligible.

These properties can be taken as an advantage for using CMA particles to harden ductile metals and alloys of lower yield strength. For instance, *in situ* precipitation of nanoparticles of icosahedral symmetry was used long ago to produce maraging steels [94] of amazingly large yield strength that are used in a commercial application by Philips (razor blades). Other metal–matrix composites can be produced under such conditions that CMA particles precipitate in a soft matrix, for example, in Al-based alloys, or by mechanical alloying and sintering. It has, for example, been demonstrated that volume fractions of β-Al_3Mg_2 as low as 20% spread in an Al-matrix lead to an increase of strength by 400% at a still high ductility of about 40% (Eckert *et al.*, Chapter 7). Particle strengthening using CMA composites is treated in detail in Chapter 6.

A real breakthrough was recently achieved using laser selective sintering, a rapid prototyping method able to produce a variety of composites based either on polymer or aluminum matrices [95, 96]. Figure 1.31 shows an example of a toy part made according to this process, with very high mechanical properties that exceed the current state-of-the-art and may find application in many areas of the transport industry.

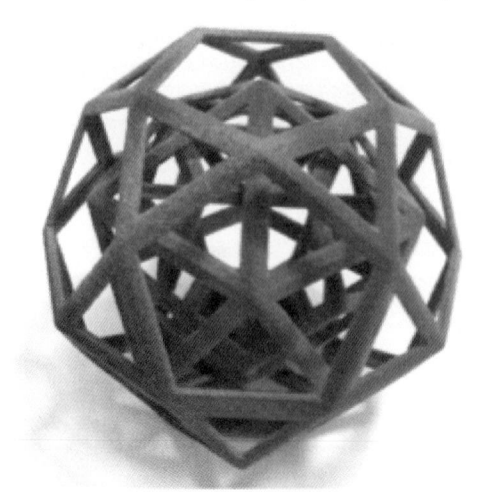

Figure 1.31 Example of a complex shape produced by laser selective rapid prototyping and sintering of a very complex shape of embedded polyhedra made of CMA powder in a polymer matrix. The object is a few cm in diameter.

1.6
Conclusion and Introduction of the Following Chapters

This book is intended to introducing the reader to a state-of-the-art comprehension of the most salient features of the science of CMAs, which the editors selected in view of their relevance to potential technological applications. The book is organized in 10 self-contained chapters. In addition to the present introduction, the following chapters are dedicated to the study of the properties of CMAs from theoretical and experimental standpoints (Chapters 2 and 3), to the surface science and surface chemistry of CMAs (Chapters 4 and 6), to metallurgy, preparation, processing and engineering properties of CMAs (Chapters 5 and 7), to magnetocaloric properties and thermoelectricity (Chapters 8 and 9) and finally to CMAs as catalysts (Chapter 10).

Acknowledgments

The authors would like to acknowledge the financial support of the European Network of Excellence "Complex Metallic Alloys" (CMA) in FP6 (Contract no NMP3-CT-2005-500140 of the European Commission).

References

1 Urban, K. and Feuerbacher, M. (2004) *J. Non Cryst. Solids*, **334–335**, 143.
2 Dubois, J.M. (1993) *Scr. Phys.*, **T49**, 17.
3 Dubois, J.M. (2005) *Useful Quasicrystals*, World Scientific, Singapore.
4 Shechtman, D., Blech, I., Gratias, D., and Cahn, J. (1984) *Phys. Rev. Lett.*, **53**, 951.

5. Shechtman, D. and Blech, I. (1985) *Metall. Trans. A*, **16**, 1005.
6. Dubost, B., Lang, J.M., Tanaka, M., Sainfort, P., and Audier, M. (1986) *Nature*, **324**, 48.
7. Tsai, A.P., Inoue, A., and Masumoto, T., (1987) *Jpn. Appl. Phys.*, **27**, L1524.
8. Tsai, A.P., Inoue, A., and Masumoto, T. (1990a) *Mater. Trans. Jpn. Inst. Metals*, **30**, 463.
9. Tsai, A.P., Inoue, A., Yokohama, Y., and Masumoto, T. (1990b) *Mater. Trans. Jpn. Inst. Met.*, **31**, 98.
10. Bak, P. (1986) *Phys. Rev. Lett.*, **56**, 861.
11. Bindi, L., Steinhardt, P.J., Yao, N., and Lu, P.J. (2009) *Science*, **324**, 1306.
12. Pauling, L. (1923) *J. Am. Chem. Soc.*, **45**, 2777.
13. Pauling, L. (1985) *Nature*, **317**, 512.
14. Tsai, A.P., Guo, J.Q., Abe, E., Takakjura, H., and Sato, T.J. (2000) *Nature*, **408**, 537.
15. Dong, C., Dubois, J.M., de Boissieu, M., and Janot, C. (1989) *J. Phys.: Condens. Matter*, **2**, 6339.
16. Quiquandon, M., Quivy, A., Devaud, J., Lefebvre, S., Bessière, M., and Calvayrac, Y. (1996) *J. Phys.: Condens. Matter*, **8**, 2487.
17. Berger, C. (1994) *Lectures on Quasicrystals* (eds F. Hippert and D. Gratias) Les Editions de Physique, Les Ulis, p. 463.
18. Belin-Ferré, E. (2002) *J. Phys.: Condens. Matter*, **14**, R789, and references therein.
19. Hirth, J.P. and Lothe, J. (1982) *Theory of Dislocations*, Wiley, New York.
20. Feuerbacher, M., Klein, H., and Urban, K. (2001) *Philos. Mag. Let.*, **81**, 639.
21. Feuerbacher, M., Heggen, M., and Urban, K. (2004) *Mater. Sci. Eng. A*, **375**, 84.
22. Roitsch, S., Heggen, M., Lipinska, M., and Feuerbacher, M. (2007) *Intermetallics*, **15**, 833.
23. Heggen, M., Deng, D., and Feuerbacher, M. (2007) *Intermetallics*, **15**, 1425.
24. Klein, H., Feuerbacher, M., Schall, P., and Urban, K. (1999) *Phys. Rev. Lett.*, **82**, 3468.
25. Boudard, M., Klein, H., de Boissieu, M., Audier, M., and Vincent, H. (1996) *Philos. Mag. A*, **74**, 939.
26. Yurechko, M., Grushko, B., Velikanova, T., and Urban, K. (2002) *Phase Diagrams in Materials Science*, MSI GmbH, p. 92.
27. Williams, D.B. and Carter, C.B. (1996) *Transmission Electron Microscopy*, Plenum Press, New York.
28. Beraha, L., Duneau, M., Klein, H., and Audier, M. (1997) *Philos. Mag. A*, **76**, 587.
29. Klein, H. (1997) PhD Thesis Inst. Nat. Polytech. de Grenoble.
30. Klein, H. and Feuerbacher, M. (2003) *Philos. Mag.*, **83**, 4103.
31. Feuerbacher, M. and Heggen, M. (2006) *Philos. Mag.*, **86**, 985.
32. Feuerbacher, M., Balanetskyy, S., and Heggen, M. (2008) *Acta Mater.*, **56**, 1849.
33. Grin, J., Burkhardt, U., Ellner, M., and Peters, K. (1994) *J. Alloys Compd.*, **2006**, 243.
34. Heggen, M., Deng, D., and Feuerbacher, M. (2007) *Intermetallics*, **15**, 1425.
35. Hudd, R.C. and Tailor, W.H. (1962) *Acta Crystallogr.*, **15**, 441.
36. Saito, K., Sugiyama, K., and Hiraga, K. (2000) *Mater. Sci. Eng. A*, **294–296**, 279.
37. Feuerbacher, M. and Heggen, M. (2010) *Dislocations in Solids*, vol. 16 (ed. J.P. Hirth and L. Kubin) Elsevier, Den Haag, p. 111.
38. Edler, F., Gramlich, V., and Steurer, W. (1998) *J. Alloys Compd.*, **269**, 7.
39. Edler, F. (1997) PhD Thesis ETH Zürich.
40. Roitsch, S. (2008) PhD Thesis RWTH Aachen.
41. Takakura, H., Gomez, C.P., Yamamoto, A., de Boissieu, M., and Tsai, A.P. (2007) *Nature Mater.*, **6**, 58.
42. Janot, C. (1994) *Quasicrystals: A Primer*, 2nd edn, Oxford Science Pub., Oxford.
43. Hume-Rothery, W. (1926) *J. Inst. Metals*, **35**, 295.
44. Jones, H. (1937) *Proc. Roy. Soc. A*, **49**, 250.
45. Blandin, A.P. (1967) *Phase Stability in Metals and Alloys* (eds P.S. Rudman, J. Stringer, and R.I. Jaffee) McGraw Hill, New York, p. 115.
46. Friedel, J. and Dénoyer, F. (1987) *C. Rend. Ac. Sci. Série 2*, **305**, 171.
47. Mizutani, U. (2001) *Introduction to the Electron Theory of Metals*, Cambridge University Press, Cambridge.
48. Sire, C. (1994) *Lectures on Quasicrystals* (eds F. Hippert and D. Gratias) Les Editions de Physique, Les Ulis, p. 505, and references therein.
49. Macia, E. and Dominguez-Adame, F. (2000) *Electrons, Phonons and Excitons in Low Dimensional Aperiodic Systems*, Editorial Complutense, Madrid.
50. Fujiwara, T. (1989) *Phys. Rev. B*, **40**, 942.

51. Fujiwara, T. and Yokokawa, T. (1991) *Phys. Rev. Lett.*, **66**, 333.
52. Hafner, J. and Krajci, M. (1992) *Phys. Rev. Lett.*, **68**, 2321.
53. Mizutani, U., Takeuchi, T., Banno, E., Fournée, V., Takata, M., and Sato, H. (2001) *Mater. Res. Soc. Symp. Proc*, Materials Research Society, Warrendale, p. 643, pK13.1.
54. Trambly de Laissardière, G., Nguyen Manh, D., Magaud, L., Julien, J.P., Cyrot-Lackmann, F., and Mayou, D. (1995) *Phys. Rev. B*, **52**, 7920.
55. Trambly de Laissardière, G. and Mayou, D. (1997) *Phys. Rev. B*, **55**, 2890.
56. Mott, N.F. and Jones, H. (1932) *Theory of the Properties of Metals and Alloys*, Clarendon Press, Oxford.
57. Belin-Ferré, E., Klanjšek, M., Jaglic, Z., Dolinšek, J., and Dubois, J.M. (2005) *J. Phys.: Condens. Matter*, **17**, 6911.
58. Dubois, J.M., Fournée, V., Thiel, P.A., and Belin-Ferré, E. (2008) *J. Phys.: Condens. Matter*, **20**, 314011.
59. Traverse, A., Dumoulin, L., Belin, E., and Sénémaud, C. (1988) *Quasicrystalline Materials* (eds C. Janot and J.M. Dubois), World Scientific, Singapore, p. 399.
60. Belin, E. and Traverse, A. (1991) *J. Phys.: Condens. Matter*, **3**, 2157.
61. Belin-Ferré, E., Dankhazi, Z., Fournée, V., Sadoc, A., Berger, C., Mueller, H., and Kirchmayr, H. (1996) *J. Phys.: Condens. Matter*, **8**, 6213.
62. Belin-Ferré, E., Dankhazi, Z., Fontaine, M.-F., Thirion, J., de Weerd, M.-C., and Dubois, J.M. (2004) Quasicrystals 2003-preparation, properties and applications, in *MRS Proceedings*, vol. 805 (eds E. Belin-Ferré, M. Feuerbacher, Y. Ishii, and D.J. Sordelet) Warrendale, p. 143.
63. Belin-Ferré, E., Dankhazi, Z., Fontaine, M.-F., de Weerd, M.-C., and Dubois, J.M. (2010) *Croat. Chem. Acta*, **83**, 55.
64. Belin-Ferré, E., Fournée, V., and Dubois, J.M. (2001) *Mater. Trans. JIM*, **42–6**, 911.
65. Sato, H., Takeuchi, T., and Mizutani, U. (2004) *Phys. Rev. B*, **70**, 024210.
66. Asahi, R., Sato, H., Takeuchi, T., and Mizutani, U. (2005) *Phys. Rev.*, **B72**, 125102.
67. Mizutani, U., Asahi, R., Sato, H., and Takeuchi, T. (2006) *Phys. Rev.*, **B74**, 235119.
68. Mizutani, U. (2010) *Series on Complex Metallic Alloys*, vol. 3 (ed. E. Belin-Ferré) World Scientific, Singapore, p. 323.
69. Terakura, K. (1977) *J. Phys. F: Metal Phys.*, **7**, 1773.
70. Samson, S. (1965) *Acta Crystallogr.*, **19**, 401.
71. Fournée, V., Belin-Ferré, E., Sadoc, A., Donnadieu, P., Flank, A.-M., and Mueller, H. (1999) *J. Phys.: Condens. Matter*, **11**, 191.
72. Belin-Ferré, E. and Dubois, J.M. (2008) *Philos. Mag.*, **88**, 2163.
73. Papaconstantopoulos, D.A. (1986) *Handbook of the Band Structure of Elemental Solids*, Plenum Press, New York.
74. Feuerbacher, M. *et al.* (2007) *Z. Kristallogr.*, **222**, 259.
75. Bihar, Z., Bilusik, A., Lukatela, J., Smontara, A., Jeglic, P., McGuiness, P., Dolinsek, J., Jaglicic, Z., Janovec, J., Demange, V., and Dubois, J.M. (2006). *J. Alloys Compd.*, **407**, 65.
76. Smiljanic, I., Smontara, A., Bilusic, A., Barisic, N., Stanic, D., Lukatela, J., Dolinsek, J., Feuerbacher, M., and Grushko, B. (2008) *Philos. Mag.*, **88**, 2155.
77. Park, J.Y., Ogletree, D.F., Salmeron, M., Ribeiro, R.A., Canfield, P.C., Jenks, C.J., and Thiel, P.A. (2005) *Science*, **309**, 1354.
78. Demange, V., Anderegg, J.W., Ghanbaja, J., Machizaud, F., Sordelet, D.J., Besser, M., Thiel, P.A., and Dubois, J.M. (2001) *Appl. Surf. Sci.*, **173**, 327.
79. Veys, D., Weisbecker, P., Domenichini, B., Weber, S., Rapin, C., Fournée, V., and Dubois, J.M. (2007) *J. Phys.: Condens. Matter*, **19**, 376207.
80. Dubot, P., Cenedese, P., and Gratias, D. (2003) *Phys. Rev. B*, **68**, 033403.
81. Rouxel, D. and Pigeat, P. (2006) *Prog. Surf. Sci.*, **81**, 488.
82. Tsai, A.P. and Yoshimura, M. (2001) *Appl. Catal. A, General*, **214**, 237.
83. Dubois, J.M., de Weerd, M.-C., and Brenner, J. (2004) *Ferroelectrics*, **305**, 159.
84. Darque-Ceretti, E. and Felder, E. (2003) *Adhésion et Adhérence*, CNRS Editions, Paris, and references therein.
85. Fournée, V., Ross, A.R., Lograsso, T.A., Evans, J.W., and Thiel, P.A. (2003) *Surf. Sci.*, **537**, 5.
86. Dubois, J.M. and Weinland, P. (1988) French Patent no. 2635117.

87 Kang, S.S., Malaman, B., Venturini, G., and Dubois, J.M. (1992) *Acta Crystallogr. B*, **48**, 770.

88 Sordelet, D.J., Widener, S.D., Tang, Y., and Besser, M.F. (2000) *Mater. Sci. Eng.*, **294–296**, 834.

89 Sales, M., Merstallinger, A., Brunet, P., de Weerd, M.-C., Khare, V., Traxler, G., and Dubois, J.M. (2006) *Philos. Mag.*, **86-6-8**, 849.

90 Cekada, M. *et al.* (2008) Private communication.

91 Pecharsky, K. and Gschneidner, K.A. Jr. (1997) *Phys. Rev. Lett.*, **78**, 4494.

92 Dubois, J.M., Archambault, P., and Colleret, B. (1991) French Patent no. 2685349.

93 Eisenhammer, T. (1995) *Thin Solid Films*, **270**, 1.

94 Liu, P. and Nilsson, J.O. (1996) *New Horizons in Quasicrystals, Research and Applications* (eds. A.I. Goldman, D.J. Sordelet, P.A. Thiel, and J.M. Dubois), World Scientific, Singapore, p. 263.

95 Kenzari, S. and Fournée, V. (2008) French Patent FR2929541.

96 Kenzari, S. and Fournée, V. (2009) PCT WO2009/144405.

2
Properties of CMAs: Theory and Experiments
Enrique Maciá and Marc de Boissieu

2.1
Introduction

CMA encompasses different kinds of structurally complex materials sharing a basic property: a full description of their atomic arrangement requires the consideration of more than one spatial scale. Thus, on the scale of several nanometers, these alloys exhibit a well-defined atomic long-range order, whereas on a shorter scale, they locally resemble cluster aggregates.

In order to gain some understanding about the role played by this multiscale feature on the physical properties of CMAs it is convenient to broadly classify them by two main criteria: the nature of the long-range order present in the sample and the size and local atomic distribution of its unit cell. The extreme case corresponds to quasicrystals (QCs) which can be regarded as a natural extension of the notion of a crystal to structures with quasiperiodic, rather than periodic, long-range order. As a consequence, ideal three-dimensional QCs exhibit a self-similar distribution of icosahedral clusters at all scales and are characterized by an effective unit cell of infinite size. In the case of decagonal QCs, two kinds of long-range order simultaneously coexist in the same sample, namely periodic order along one direction and quasiperiodic order in the planes perpendicular to the previous one. Due to this fact, these alloys exhibit highly anisotropic effects in most of their physical properties, and they will be discussed in a separate chapter. The next step corresponds to the so-called approximant alloy phases exhibiting a well-defined, huge unit cell that periodically repeats through the three directions of space, though the local distribution of atoms inside this unit cell is completely isomorphous to that corresponding to closely related QCs in the phase diagram. Finally, we have those CMAs with giant unit cells that are not related to any QC structure. According to this approximate classification scheme the role played by the local symmetry of the structural clusters progressively increases from the non-QC related CMAs to the QC-related ones.

Complex Metallic Alloys: Fundamentals and Applications
Edited by Jean-Marie Dubois and Esther Belin-Ferré
Copyright © 2011 WILEY-VCH Verlag GmbH & Co. KGaA, Weinheim
ISBN: 978-3-527-32523-8

From a fundamental point of view one reasonably expects that the presence of two physically relevant length scales – one defined by the unit-cell parameters and the other by the cluster substructure – will have a significant impact on the physical properties of these materials. In this chapter we will focus on those properties determined by their electronic structure and lattice dynamics. Following the structural approach previously described we will start by considering the transport properties of QCs and their related approximants. As we will describe in Section 2.2.1, it is now well established that transport properties of thermodynamically stable QCs of high structural quality are quite unusual by the standard of common metallic alloys, as most of their transport properties resemble a more semiconductor-like than metallic character. Thus, high-quality QCs provide an intriguing example of solids made of typical metallic atoms that do not exhibit most of the physical properties usually signaling the presence of metallic bonding, a topic that will be addressed in Section 2.2.2. Subsequently, the main features of the electronic structure close to the Fermi level will be discussed along with the nature of the so-called critical electronic wavefunctions and their role in the resulting charge transport efficiency. Finally, in Section 2.2.4 we will briefly describe a phenomenological approach that allows for a unified description of different transport coefficients, providing some illustrative application examples.

In Section 2.3 we turn our attention to the lattice dynamics related properties. The section opens with a brief review introducing the basic notions and experimental procedures usually considered in the study of phonons in solid-sate physics. A Particular attention to the role of anharmonic effects is devoted in Section 2.3.3, since these effects have a relevant impact in the thermal conductivity of the considered systems. In the following sections these basic tools are systematically applied to the derivation and analysis of the phonon dispersion relations of several CMAs representatives. For the sake of comparison the considered samples are arranged according to the structural classification scheme previously introduced. Thus, the properties of phonons in QCs and approximant phases are described in detail in Section 2.3.4. The samples considered include the high-quality icosahedral Al-Pd-Mn and Zn-Mg-Sc phases, for which accurate structural models have been recently reported. Afterwards, in Section 2.3.5 we consider the lattice dynamics of the so-called cage compounds, typically including clathrates and skutterudites. These compounds are characterized by the presence of a framework of cages or large structural vacancies that can be filled with heavy atoms which interact with the propagating phonons by activating "rattling" modes, hence reducing the resulting thermal conductivity. The potential of these compounds in the field of thermoelectric materials research is also discussed, complementing the thermoelectric figure of merit results presented in Section 2.2.2 for QCs and approximant phases.

The emerging view is that attending to their physical properties CMAs appear as very promising alloys that can be efficiently used in order to obtain materials with novel capabilities, like a combination of metallic electrical conductivity with low thermal conductivity, tuneable electrical and thermal resistances by varying the composition, or an improved thermoelectric efficiency.

2.2
Electronic-Structure-Related Properties

2.2.1
Transport Properties of Quasicrystals and Approximants

The first QCs obtained were metastable, preventing a significant study of several physical properties, in particular the temperature dependence of their transport properties. Even the first thermodynamically stable QCs, obtained in the systems Al-Cu-Li and Ga-Mg-Zn, were unsuitable to this end, since they were usually contaminated with small crystalline inclusions and exhibited a relatively large number of structural imperfections. Nevertheless, shortly after the discovery of thermodynamically stable quasicrystalline alloys of high structural quality in the Al-Cu(Fe,Ru,Os), Al-Pd(Mn,Re), Zn-Mg(RE), and Cd(Yb,Ca) icosahedral systems, as well as the Al-Co (Cu,Ni) decagonal system, it was progressively realized that these materials occupy an odd position among the well-ordered condensed-matter phases. In fact, since QCs consist of metallic elements one would expect they should behave as metals. Nonetheless, as we will describe below, it is now well established that transport properties of stable QCs are quite unusual by the standard of common metallic alloys, as most of their transport properties resemble a more semiconductor-like than metallic character [1].

2.2.1.1 Inverse Matthiessen Rule
For typical metals resistivity decreases as the temperature is decreased and it can even completely vanish at low enough temperatures for those materials reaching the superconducting state. Conversely, the electrical resistivity of QCs progressively increases as the temperature is decreased, suggesting the possibility of reaching a metal–insulator transition in high-quality icosahedral quasicrystals at low temperatures [2–5]. On the other hand, the electrical conductivity steadily increases as the temperature increases up to the melting point, and its value very sensitively depends on minor variations of the sample stoichiometry (Figure 2.1).

Quite remarkably, the conductivity curves of different quasicrystalline samples are nearly parallel up to about 1000 K, so that one can write [6]

$$\sigma(T) = \sigma_0 + \Delta\sigma(T) \tag{2.1}$$

where σ_0 measures the sample-dependent residual conductivity, and $\Delta(T)$ is proposed to be a general function. According to this expression the contribution to the sample conductivity due to different sources of scattering seems to be additive. This is just the opposite to what happens to normal metals, where the resistivities due to different sources of disorder are additive. This unexpected behavior, referred to as the inverse Matthiessen rule [6], has been also observed in quasicrystalline approximants [7], and even in amorphous phases prior to their thermally driven transition to the QC phase (see Figure 2.7) [8].

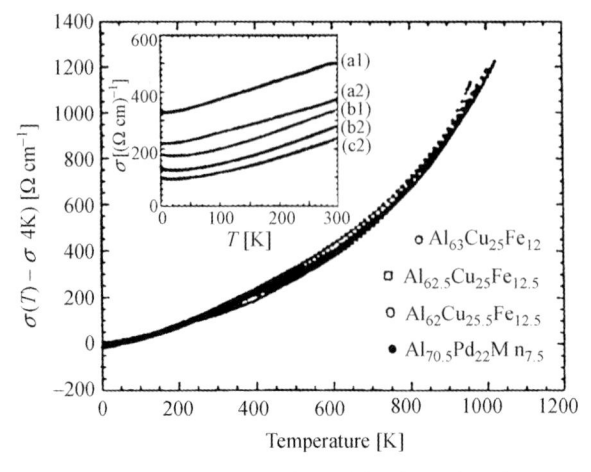

Figure 2.1 Temperature dependence of the electrical conductivity for four different quasicrystalline samples up to 1000 K. The inset illustrates the sensitivity of the residual conductivity value to minor variations in the sample composition. (Adapted from reference [6]. Courtesy of C. Berger).

Accordingly, the inverse Matthiessen rule may be a quite general property of CMAs and the question concerning the possible existence of a suitable physical mechanism supporting the presumed universality of the $\Delta\sigma(T)$ function naturally arises. In fact, the parallelism of the $\sigma(T)$ curves is difficult to understand in terms of a classical thermally activated mechanism, since the temperature dependence of $\sigma(T)$ does not follow an exponential law of the form $\exp(\sim E_g = k_B T)$; where k_B is the Boltzmann constant, which implies the absence of a conventional semiconducting-like gap in QCs [9]. In addition, the $\sigma(T)$ curves do not decrease at high enough temperatures, as one should expect if QCs were comparable to heavily doped semiconductors, which show up a conductivity saturation when all the impurity levels have become ionized.

2.2.1.2 Current–Voltage Curves

Another clear indication that QCs cannot be regarded as standard semiconducting materials came from the fact that their characteristic current–voltage (I–V) curves exhibit a perfect Ohmic behavior at low temperatures ($T \simeq 4$ K) for bias voltages that vary by several orders of magnitude [10]. Such a linear behavior holds as the sample temperature is progressively increased (Figure 2.2), clearly indicating that a linear I–V behavior is not restricted to low-temperature regimes. This behavior lends support to the possible presence of relatively extended states close to the Fermi level and should be interpreted in the light of the electronic structure of icosahedral QCs (see Sections 14.3 and 1.2.3) which is characterized by the presence of three relevant energy scales close to the Fermi level, namely: (i) a broad pseudogap on the energy scale of about 1 eV (related to the Hume-Rothery stabilization mechanism); (ii) a narrow dip of about 0.1 eV (due to hybridization effects among d-states and sp-states); and (iii) some narrow features in the density of states (DOS), on the scale of about

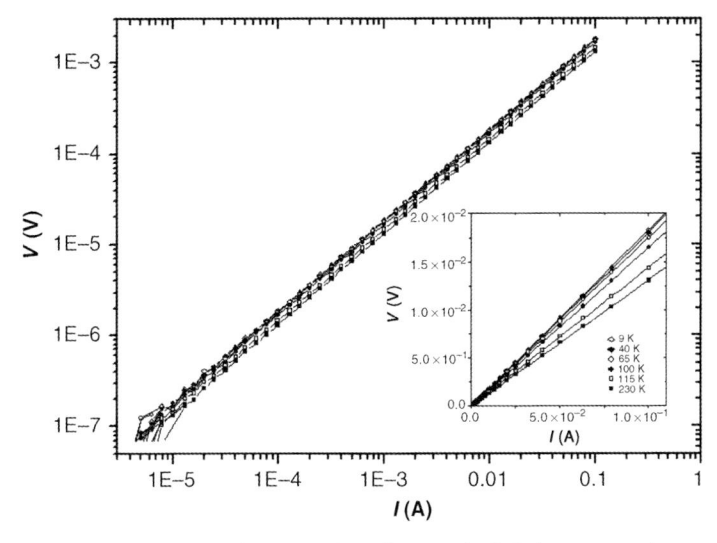

Figure 2.2 Double logarithmic I–V plots of an icosahedral $Al_{63}Cu_{25}Fe_{12}$ phase sample (kindly provided by Jean Marie Dubois) at T = 9; 45; 65; 100; 175 and 230 K. The inset shows the linear representation of the same data. (Courtesy of Javier García-Barriocanal).

0.01 eV (stemming from resonant effects among quasiperiodically distributed transition-metal atoms) [11]. Thus, one would expect to observe some nonlinearity related to the presence of these spectral features as soon as the energy change of the charge carriers involved in the measurement process is in the range 0.01–1 eV.

Now, the highest electric fields applied in these experiments are in the range $E = 50$–100 V/cm, so that we get the electron energy $\varepsilon \simeq eEl_0 \simeq 10^{-5}$ eV, where $l_0 \simeq 20$ Å, is a rough estimate of the electronic mean free path in these materials [10]. Certainly, this figure is small enough to play a subsidiary role in the considered I–V measurements. Consequently, stronger electric fields should be applied in order to observe any possible effect related to finer electronic structure features in these materials.

2.2.1.3 Optical Conductivity

Suitable information about relatively fine details of the electronic structure and the spectrum of excitations of a solid can be gained from the study of its optical properties. To this end, one experimentally obtains the reflectivity curve as a function of the incoming electromagnetic radiation frequency, $R(\omega)$, and derives from it the optical conductivity curve $\sigma(\omega)$ by means of the so-called Kramers–Krönig transformation of the reflectivity spectrum. The $\sigma(\omega)$ curve of metallic alloys is determined by several contributions. First, we have intraband transitions involving conduction electrons that can be analyzed using the Drude model for free electrons

$$\sigma(\omega) = \frac{\sigma(0)}{1 + (\omega\tau)^2} \tag{2.2}$$

where $\sigma(0)$ is the dc conductivity and is the relaxation time. This contribution dominates the optical response at low frequencies and results in a characteristic Lorentzian function centered at the zero frequency, known as the Drude peak, followed by a rapid decay of the optical conductivity at low frequencies. A second contribution (mainly affecting the far-infrared region of the spectrum) is related to the presence of optical phonon modes, which are activated when the incoming radiation frequency is equal to or exceeds the necessary excitation energy. Additional contributions come from transitions involving both the valence and conduction bands (interband transitions) in the visible spectral range. Accordingly, good conductors show a reflectance close to 100% at frequencies below the onset of absorption due to interband transitions and a characteristic sudden decay (known as the plasma edge) as the frequency increases approaching the so-called plasma frequency value determining the free electrons coupling to the oscillating electromagnetic field of incoming photons. Thus, a metal is basically transparent to light for wavelengths smaller than the plasmon cut-off, and absorbing and reflecting above.

On the other hand, in semiconducting materials the absorption of a photon of energy $\hbar\omega_g$ is possible as soon as it equals the gap width E_g (direct transitions) or if the top of the valence band and the minimum of the conduction band in reciprocal space are separated by a wavevector belonging to the lattice (indirect transitions).

As a matter of fact, the optical conductivity of icosahedral QCs studied so far is quite different from that of either a metal or a semiconductor. Thus, reflectance of high-quality icosahedral samples was found to be significantly small in a wide wavelength region from about 300 nm (UV region) to 15 μm (IR region), and the following unusual features were observed in the optical conductivity:

1) The far-infrared $\sigma(\omega)$ response is very weak and no Drude peak appears at low frequencies (Figure 2.3), though extrapolation to the zero frequency yields conductivity values in good agreement with the measured dc conductivity [12–16]. Two different explanations have been proposed to account for the unusual absence of a Drude peak: (a) the low $\sigma(\omega)$ would be related to an extremely low density of states at the Fermi level due to the presence of a pseudogap in the band structure of QCs, hence leading to a substantially small value of $\sigma(0)$ in Equation 2.2 [17] (b) the localization of charge carriers due to the quasiperiodicity of the structure would lead to an anomalous diffusion mechanism. In that case, Drude's formula for the optical conductivity would adopt the more general form

$$\sigma(\omega) = Ae^2 N(E_F)\Gamma(2\beta+1)\left(\frac{\tau}{1-i\omega\tau}\right)^{2\beta-1} \qquad (2.3)$$

where A is a constant, Γ is the Gamma function, $N(E_F)$ is the density of states at the Fermi level, and β is a diffusion exponent that depends on the energy [18]. The real part of this expression reduces to Equation 2.2 in the case $\beta = 1$. Quite remarkably, values as low as $\beta = 0:07$ and $\beta = 0:03$ were found from fitting analysis of the $\sigma(\omega)$ curves of Al-Cu-Fe-B and Al-Pd-Mn QCs, respectively, whereas the value $\beta = 0:4$ was obtained for the O_1/O_2- Al-Cr-Fe CMA (Figure 2.3) [16].

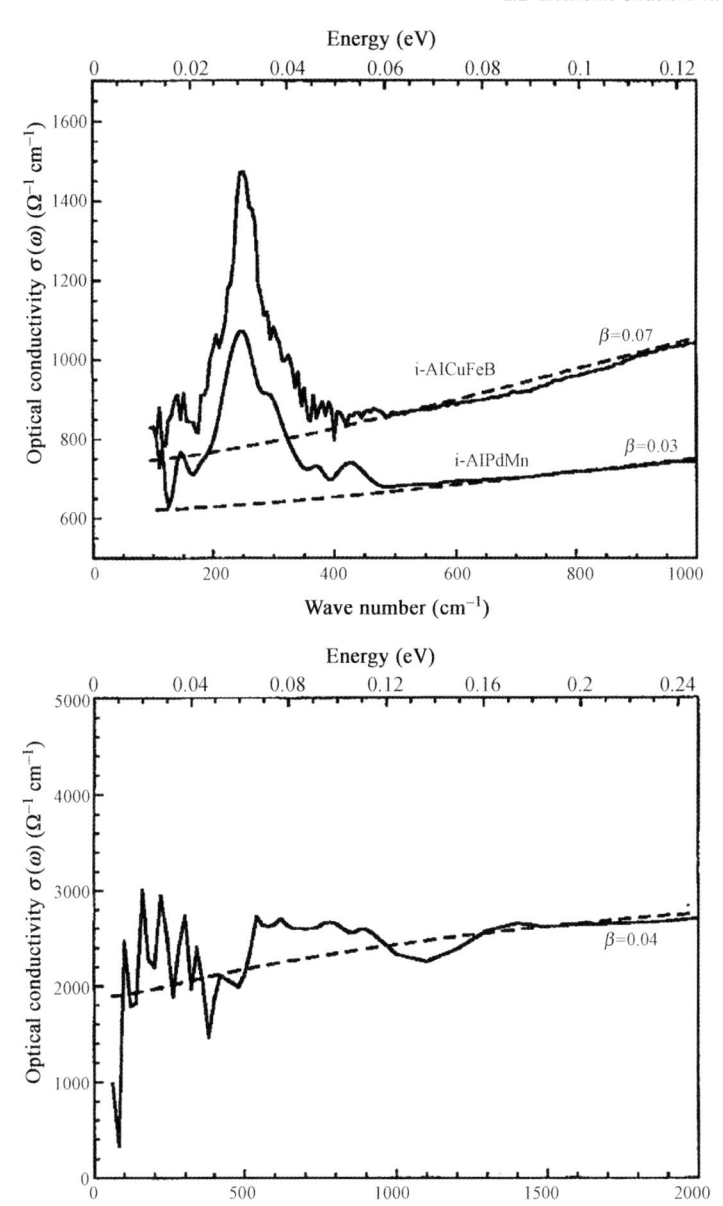

Figure 2.3 (a) Optical conductivity in i-Al-Cu-Fe-B and i-Al-Pd-Mn (solid lines). The fit curves (dashed lines) are obtained after Equation 2.3. The peaks around 0.03 eV are associated to phonons. (b) Same as in (a) for a O_1/O_2- Al-Cr-Fe sample. (Adapted from reference [16]. By courtesy of J. M. Dubois).

2) All the studied QCs exhibit a typical absorption feature overlapping the low-frequency tail of the far-infrared region (Figure 2.3). This relatively broad feature (which splits into two separate contributions at about 25 and 35 meV in several cases) is ascribed to phonon effects. At higher energies (~0.4 eV) the optical conductivity progressively rises reaching a peak at about 0.7 eV (i-Zn-Mg-Y, i-Zn-Mg-Tb), 1.2–1.5 eV (i-A-Cu-Fe, i-Al-Pd-Mn), or 2.6–2.9 eV (i-Al-Pd-Re), after which the conductivity decreases. This absorption feature is commonly ascribed to excitations across a characteristic pseudogap related to the Hume-Rothery stabilization mechanism

In summary, unlike disordered metals (where a Drude model is applicable) or semiconductors (with a well-developed conductivity gap), the reflectivity spectra of icosahedral phases display low optical conductivity on the far-infrared energy range and a marked absorption in the visible.

These characteristic features are also observed in typical approximant phases, such as 1/1 Al-Mn-Si, but for decagonal phases different behaviors of the $\sigma(\omega)$ curve can be clearly established between the quasicrystalline and the periodic directions (Figure 2.4) [19]. In fact, a Drude peak is present when light is irradiated within a narrow area parallel to the periodic axis, whereas no peak is detected in a plane

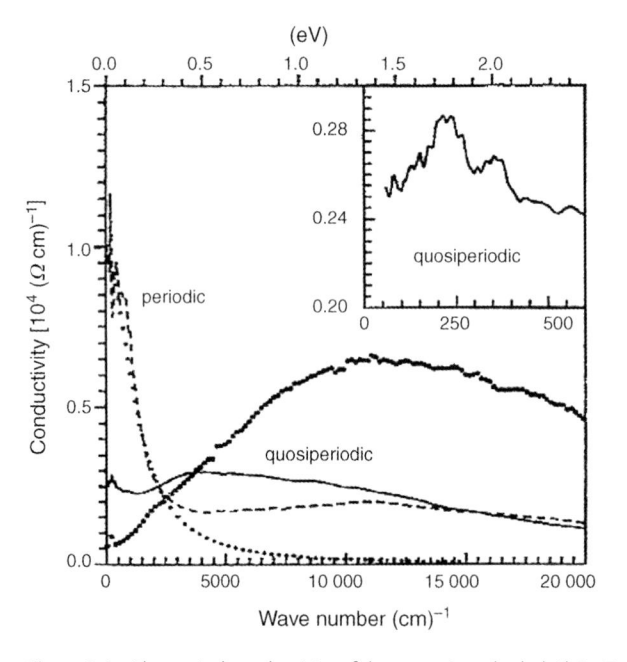

Figure 2.4 The optical conductivity of the decagonal Al-Co-Cu-Si quasicrystal for the periodic (short-dashed line) and the quasiperiodic (solid line) directions is compared with the conductivity of the icosahedral Al-Cu-Fe studied in reference [12] (solid dots). In the inset the quasiperiodic conductivity in the far-infrared part of the spectrum is shown. (From reference [19]. With permission).

perpendicular to it. The analysis of the optical data shows that contrary to the case of icosahedral QCs, there is no clear evidence for the presence of a marked pseudogap at the Fermi level.

2.2.1.4 Seebeck Coefficient

Thermoelectric power describes the electric response of a sample due to the application of an external temperature gradient through the relationship $\Delta V = S(T)\Delta T$, where $S(T)$ is the so-called Seebeck coefficient. During the last decade the thermoelectric power of samples belonging to different icosahedral families has been measured. Reported data refer to a broad range of stoichiometric compositions and cover different temperature ranges in the interval from 1 K to 900 K. From the collected data the following general conclusions can be drawn for high-quality QCs containing transition metals [20–22].

- Room-temperature thermoelectric power usually exhibits large values (50–120 µV K^{-1}) when compared to those of both crystalline and disordered metallic systems (1–10 µV K^{-1}).
- The temperature dependence of the Seebeck coefficient usually deviates from the linear behavior, exhibiting pronounced curvatures (either positive or negative) at temperatures above ∼50–100 K. This behavior is at variance with that exhibited by ordinary metallic alloys where the $S(T)$ curve is dominated by electron diffusion yielding a linear temperature dependence.
- Small variations in the chemical composition (of just a few atomic per cent) can give rise to sign reversals in the thermopower curve.
- The $S(T)$ curves exhibit well-defined extrema in several cases. Both the magnitude and position of the extrema observed in the thermoelectric power curves are extremely sensitive to minor variations in the chemical stoichiometry of the sample.

On the other hand, thermopower measurements of rare-earth-bearing QCs in the system i-Zn-Mg(Y,Tb,Ho,Er) exhibit markedly linear temperature dependences above ∼50 K [23]. An analogous behavior has been reported for the thermodynamically stable Cd-Yb QC, which also contains rare-earth atoms [24]. Such different behaviors among the i-Al-Cu-(Fe,Ru,Os) and i-Al-Pd-(Mn,Re) families (bearing transition metals) and the i-Zn-Mg(RE) and i-Cd-Yb families (bearing rare-earth atoms), strongly suggest that chemical effects may be playing a significant role.

2.2.1.5 -Wiedemann–Franz Law

In the study of the thermal transport properties of CMAs the Wiedemann–Franz law (WFL) is routinely applied in order to estimate the phonon contribution to the thermal conductivity, $\kappa_{ph}(T)$. As is well known, this law links the electrical conductivity, $\sigma(T)$, and the charge carriers contribution to the thermal conductivity, $\kappa_{\varepsilon}(T)$, of a substance by means of the relationship $\kappa_{\varepsilon}(T) = L_0 T \sigma(T)$, where T is the temperature and $L_0 = (k_B/e)^2 \eta_0$ is the Lorenz number, where k_B is the Boltzmann constant, e is the electron charge, and η_0 depends on the sample's nature (for metallic systems

Table 2.1 Values of the ratio κ_e/κ_{ph} at T = 300 K for different CMAs derived from the experimental transport curves reported in the literature making use of Equation (2.4).

κ_e/κ_{ph}	Quasicrystals	Approximants	Clathrates
5–1	i-Zn-Mg-Y [25]	Bergman [29]	$Y_3Ir_4Ge_{14}$ [32]
~1	i-Ag-In-Yb [26]	Cd_6Yb [30]	$Eu_8Ga_{16}Ge_{30}$ [33]
0.5–0.01	i-Al-Pd(Mn,Re) [27, 28]	1/1-Al-Re-Si [31]	—

$\eta_0 = \pi^{2/3}$, and we get the Sommerfeld's value $L_0 = 2{:}44 \times 10^{-8}$ V K^{-2}, whereas for semiconductors we have $\eta_0 \simeq 2$). Accordingly, by subtracting to the experimental data, $\kappa_m(T)$, the expected electronic contribution, one gets

$$\kappa_{ph}(T) = \kappa_m(T) - L_0 T\sigma \tag{2.4}$$

In so doing, the ratio κ_e/κ_{ph} at room temperature has been determined for several CMA representatives (Table 2.1). Keeping in mind that this ratio takes on values within the range 100–10 for conventional alloys, one realizes that the thermal transport of CMAs is largely dominated by phonons at room temperature. By inspecting Table 2.1 we also see that the presence of transition-metal atoms is related to smaller κ_e/κ_{ph} ratios in the studied samples, suggesting the possible existence of some chemical bonding effect.

Physically, the WFL expresses a transport symmetry arising from the fact that the motion of the carriers determines both the electrical and thermal currents at low temperatures. As the temperature of the sample is progressively increased, the validity of WFL will depend on the nature of the interaction between the charge carriers and the different scattering sources present in the solid. In general, the WFL applies as far as elastic processes dominate the transport coefficients, and usually holds for arbitrary band structures provided that the change in energy due to collisions is small compared with $k_B T$ [34]. Accordingly, one expects some appreciable deviation from WFL when electron–phonon interactions, affecting in a dissimilar way to electrical and heat currents, start to play a significant role. On the other hand, at high enough temperatures the heat transfer is dominated by the charge carriers again, due to Umklapp phonon scattering processes, and the WFL is expected to hold as well. Nonetheless, since transport properties of most CMAs are quite unusual by the standard of common metallic alloys, it seems convenient to check up on the validity of this law for these materials, since our understanding of thermal properties in these materials should be substantially revised if it does not hold [35–37].

A suitable experimental measure of the WFL validity over a given temperature range can be gained from the study of the magnitude $\kappa_m(T)/\sigma(T) = TL(T) + \varphi(T)$, where the so-called Lorenz function is defined by the relationship

$$L(T) \equiv \frac{\kappa_e(T)}{T\sigma(T)} \tag{2.5}$$

Table 2.2 Values of the enhancement parameter ε for different CMAs reported in the literature. The ε values were obtained from a fitting analysis of the thermal conductivity experimental curves making use of Equation (2.6).

ε	Sample	Ref.
~0:00	γ-AlCrFe	[40]
0:03	Ψ-AlPdMn	[41]
0:14	$1/O_2$-AlCrFe	[40]
0:16	ξ'-AlPdMn	[41]
0:43	$i\text{-}Zn_{57}Mg_{34}Y_9$	[25]
1:1	$i\text{-}Al_{64}Cu_{23}Fe_{13}$	[42]

and φ(T) accounts for the phonon contribution to the heat transport. A study of the temperature variation of the κ_m/σ ratio in several intermetallic compounds showed that the experimental data may be fitted by a linear temperature dependence of the form $\kappa_m/\sigma = LT + B$ over the temperature range 350–800 K [38, 39]. By comparing the slopes obtained for pure aluminum and icosahedral Al-Cu-Fe samples the ratio $L_{QC}/L_{Al} \simeq 1.21$ was obtained, hence indicating an enhanced Lorenz number for quasicrystalline alloys at high temperatures. In a similar way, room-temperature $L(T)$ values larger than the Sommerfeld's value L_0 were experimentally reported for other CMAs (ranging from $L_{300}/L_0 = 1.15$ [40], to from $L_{300}/L_0 = 1.43$ [25]), hence suggesting the convenience of introducing a slightly modified WFL of the form

$$\kappa_e(T) = (1+\varepsilon)L_0 T\sigma(T) \tag{2.6}$$

By inspecting Table 2.2 we see that the enhancement parameter ε is roughly related to the structural complexity of underlying lattice, progressively increasing as a fully three-dimensional quasiperiodic order is attained in the considered sample. Following this trend, a generalized WFL of the form $\kappa_e(T) = L(T)T\sigma(T)$, which is characterized by a nonlinearly temperature dependent Lorenz number (Figure 2.5), has been recently proposed on a theoretical basis [37].

The impact of Lorenz's function temperature dependence in a proper analysis of the phonon contribution to the thermal conductivity is illustrated in Figure 2.6. In this figure we compare the measured thermal conductivity (including contributions from both charge carriers and phonons) with the phonon contribution derived from the application of the WFL by either assuming a constant value for the Lorenz number (Equation 2.4, circles) or explicitly taking into account its temperature dependence through the expression

$$\kappa_{ph}(T) = k_m(T) - L(T)\sigma(T)T \tag{2.7}$$

One can clearly appreciate that the temperature dependences of the resulting $\kappa_{ph}(T)$ curves substantially differ in both cases. In fact, when one considers $L(T) \simeq L_0$ one obtains an anomalous behavior, characterized by a smooth increase of the phonon contribution as the temperature increases. Conversely, when experimental

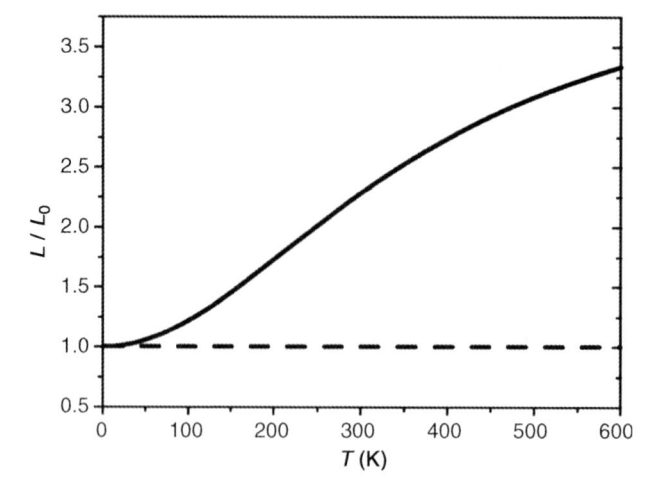

Figure 2.5 Temperature variation of the normalized Lorenz function. The WFL is obeyed at very low temperatures. At high temperatures $L(T)$ approaches the asymptotic limit value $2^{1} = 5$. A significant enhancement of the Lorenz number with respect to the Sommerfeld value takes place over a wide temperature range.

data are properly corrected from the $L(T)$ enhancement effect one gets a physically sound phonon contribution to the thermal conductivity that steadily decreases with the temperature starting at $T \gtrsim 100$ K, as expected on general physical principles involving phonon–phonon interactions.

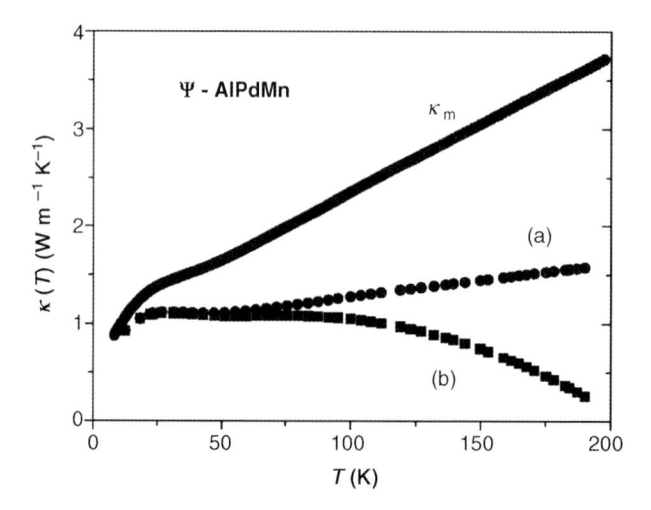

Figure 2.6 The phonon contribution to the thermal conductivity is derived by subtracting to the experimentally measured thermal conductivity (κ_m, experimental data by courtesy of J. Dolinšek) the charge carrier contribution ($\sigma(T)$ was reported in reference [43] by assuming (a) the validity of the WFL according to Equation 2.4 (circles), or (b) by considering a temperature-dependent Lorenz function as given by Equation 2.7 (squares).

Table 2.3 Comparison between the physical properties of QCs and typical metallic systems.

Metallic systems property	Metals	Quasicrystals
Mechanical	ductility malleability	brittle (I)
Tribological	relatively soft easy corrosion	very hard (I) low friction coefficient corrosion resistant
Electrical	high conductivity resistivity increases with T small thermopower	moderate–low conductivity (S) resistivity decreases with T (S) large thermopower (S)
Magnetic	paramagnetic	diamagnetic
Thermal	high conductivity high specific heat values high melting points	very low conductivity (I) low specific heat values
Optical	metallic luster, Drude peak	metallic luster, IR absorption (S)

2.2.2
Chemical Trends

As is well known, metallic substances exhibit a number of characteristic physical properties that are directly related to the presence of a specific kind of chemical bond among their atomic constituents: the so-called metallic bond [44]. For the sake of comparison in Table 2.3 we list a number of representative physical properties of both metals and QCs. By inspecting this table one realizes that quasicrystalline alloys significantly depart from metallic behavior, resembling either ionic or semiconducting materials (respectively labeled I or S in Table 2.3). Thus, QCs are an intriguing example of solids made of typical metallic atoms that do not exhibit most of the physical properties usually signaling the presence of metallic bonding.

Therefore, the fundamental question arises concerning whether these anomalous properties should be mainly attributed (or not) to the characteristic quasiperiodic order of QCs structure. In this regard, several experimental evidences strongly suggest that the nature of the chemical bonding determining the local atomic arrangements would play a significant role in most physical properties of these materials [45–47], namely:

1) Transport measurements show that the structural evolution from the amorphous to the quasicrystalline state (Figure 2.7) is accompanied by a parallel evolution of the electronic transport anomalies, clearly indicating the importance of short-range effects on the emergence of some transport anomalies.
2) Many unusual physical properties of QCs are also found in approximant phases, though transport measurements also indicate that these anomalies are more pronounced for QCs than for their related approximant phases, hence suggesting that the relative intensity of the anomalous behavior is significantly emphasized due to the presence of long-range quasiperiodic order.

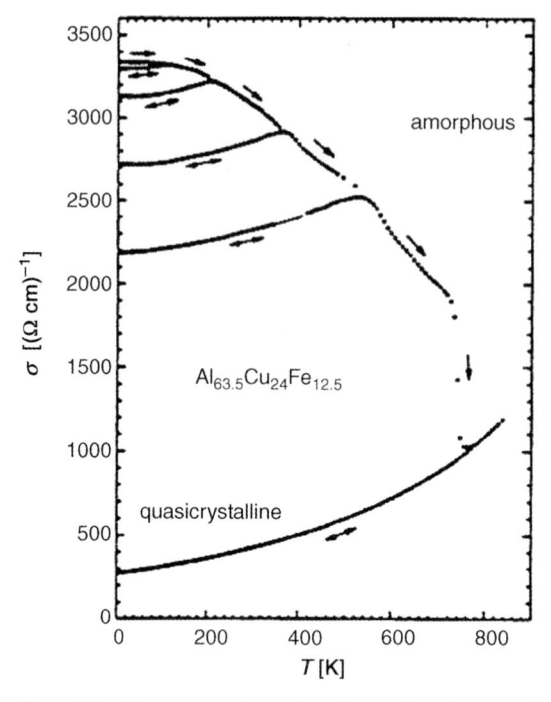

Figure 2.7 Temperature-dependent electrical conductivity of an Al-Cu-Fe Film for different annealing states leading from the amorphous to the icosahedral quasicrystalline phase. (From reference [48]. Courtesy of Peter Häussler).

3) Certain anomalous transport properties, like a high resistivity value or a negative temperature coefficient, are also observed in some crystalline alloys consisting of normal metallic elements whose structure is unrelated to the structure of QCs (as, for instance, the Hëusler-type Fe_2VAl alloy) which share with them certain characteristic feature in the electronic structure (i.e. a narrow pseudogap) [49].
4) Transport properties of metallic alloys with complex unit cells, having a similar number of atomic species to those of approximant phases, but not exhibiting the local isomorphism property, are typically metallic [29].
5) Other kinds of aperiodic crystals, like incommensurately modulated phases and composites do not show the physical anomalies observed in QCs.

According to (1)–(3) the emergence of physical anomalies in QCs should be traced back to chemical bonding effects (short-range), giving rise to some characteristic features in the electronic structure close to the Fermi level (such as the presence of a narrow pseudogap), which are generic but not specific of QCs [50]. Thus, chemical effects may ultimately become more important than quasiperiodic order effects in explaining the unusual behavior of these materials. Accordingly, crystalline approximants, which exhibit a local atomic environment very similar to their related QC alloys, appear as natural candidates to investigate the relative importance of short-range versus long-range order effects on the transport properties. This conclusion is

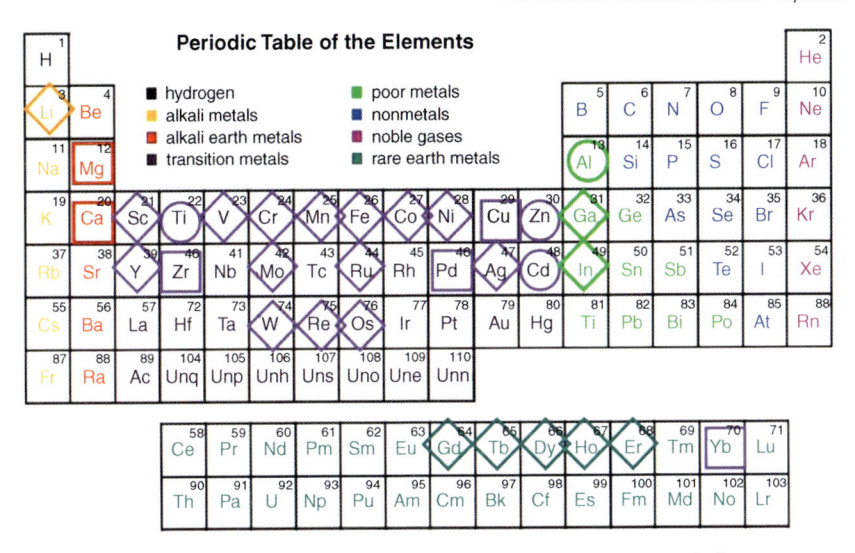

Figure 2.8 Chemical elements found in thermodynamically stable quasicrystal alloys. Main forming elements (Al, Ti, Zn, and Cd) are circled. The second major constituents are squared. Minor constituents are marked with a diamond.

further supported by (4) and (5), which indicate that mere structural complexity is not a sufficient condition to give rise to the emergence of anomalous transport properties in complex metallic alloys.

Most atomic elements composing thermodynamically stable quasicrystalline alloys observed to date belong to the chemical family of metals, located at either alkaline, earth-alkaline, transition metals, or rare-earth groups (Figure 2.8). From this chart we see that most metallic atoms are able to participate in the formation of quasicrystalline phases under the proper stoichiometric conditions. On the other hand, certain chemical trends can also be appreciated in different QC families. For instance, the minor atom constituent in the systems $Al_{63}Cu_{25}(Fe,Ru,Os)_{12}$ and $Al_{70}Pd_{20}(Mn,Re)_{10}$ belongs to the same group of the Periodic Table, hence indicating the importance of their chemical valence for the stability of the compound. This fact has been successfully exploited in order to obtain the family of stable quaternary QCs given by the formula [51] $Al_{70}Pd_{20}(V, Cr, Mn, W)_5(Co, Fe, Ru, Os)_5$.

Several chemical trends are also observed in the transport properties of QCs belonging to the Al-Cu(Fe,RuOs) and Al-Pd(Mn,Re) families. Thus, it is seen that increasing the atomic number of the third (incomplete d band) transition metal significantly increases the low-temperature electrical resistivity of the sample as well as its temperature dependence as measured in terms of the ratio $R = \varrho(4\text{ K})/\varrho(300\text{ K})$ [52]. This trend may be due to the relativistic contraction of the s and p states relative to the d and f states. As a consequence of this contraction the orbital energies of s and p states are lowered which, in turn, screens the nucleus, causing the outer d electrons to experience less binding and therefore a larger spatial extent. Thus, the relativistic lowering of the energy of the s and p bands, and the associated raising of the energy of the d bands brings these bands closer to each other, hence favoring sp-d hybridization effects leading to an increase of cohesive energy.

2.2.3
Electronic Structure

2.2.3.1 Fermi-Level -Pseudogap

It was pointed out by William Hume-Rothery (1899–1968) that certain metallic compounds with closely related structures but apparently unrelated stoichiometries exhibit the same ratio of number of valence electrons to number of atoms (the so-called e/a ratio) [53]. This fact is explained as resulting from a perturbation of the energy of the valence electrons by their diffraction by the crystal lattice. The perturbation is of such a nature as to stabilize electrons with energy just equal to or less than that corresponding to Bragg reflection and to destabilize electrons with a larger energy. Hence, special stability would be expected for metals with just the right number of electrons. This number is proportional to the volume of a polyhedron in reciprocal space (the so-called Brillouin–Jones zone), corresponding to the crystallographic planes giving rise to the perturbation.

Although QCs have a dense reciprocal space, only a few diffraction peaks have very strong intensities. The Hume-Rothery criterion can then be applied to QCs by introducing a pseudo-Brillouin zone determined by the most intense diffraction spots [54, 55]. Due to their great symmetry, in the case of icosahedral QCs this zone is quite close to spherical shape, so that the diffraction condition can be expressed in the form

$$K_{hkl} = 2 k_F \qquad (2.8)$$

where K_{hkl} is the reciprocal vector of the considered diffraction plane, $k_F = \sqrt[3]{3\pi^2 n}$ is the radius of the Fermi sphere, and n is the electron number per unit volume. Equation 2.8 has been successfully used to explain the stability of i-QCs containing elements with a full d-band, like $Al_{56}Li_{33}Cu_{11}$ ($e/a = 2.12$), $Zn_{43}Mg_{37}Ga_{20}$ ($e/a = 2.:2$); $Zn_{60}Mg_{30}(RE)_{10}$ ($e/a = 2.1$) or $Zn_{80}Sc_{15}Mg_5$ ($e/a = 2.15$), by adopting the valence values Li $= 1$, Mg $= 2$, Sc $= 3$; Ga $= 3$, and RE $= 3$. In all these samples the redistribution of electronic states due to the Fermi-sphere–pseudo-Brillouin zone interaction gives rise to a significant reduction of the density of states (pseudogap) close to the Fermi energy [46, 50, 56].

For alloys containing a small concentration of a transition element one can properly extend the Hume-Rothery mechanism by assuming a negative effective valence arising from a combined effect of strong hybridization between the sp states and the transition metal d orbitals along with the diffraction of sp states by Bragg planes. As a consequence, there is an increase of the sp component of the DOS below the Fermi energy as compared to the free electron DOS [11]. Thus, in QCs bearing transition-metal atoms, such as Al-Cu(Fe,Ru,Os) or Al-Pd(Mn,Re), the presence of hybridization effects between sp aluminum states and $3d$ transition metal states enhances the (structure related) Fermi-surface–Brillouin-zone diffraction effect, further deepening the pseudogap close to the Fermi level [57–60].

Making use of the values Fe $=-2.66$, Mn $=-3.66$, and Pd $=0$ one obtains $e/a = 1.75$ and $e/a = 1.73$ for $Al_{65}Cu_{20}Fe_{15}$ and $Al_{70}Pd_{20}Mn_{10}$, respectively. The binary i-Cd(Yb,Ca) family, which is composed of divalent atoms, has $e/a = 2.0$, a value that lies close to that of the full d band representatives. Notwithstanding this, the role played by hybridization effects in the stability of the i-Cd(Yb,Ca) phase is significantly larger than that coming from the Fermi-surface–Brillouin-zone mechanism in this binary QC [61, 62]. In this case, the orbitals involved in the hybridization process come from occupied Cd-5p and unoccupied Yb-5d (or Ca-3d) orbitals, which highlights the importance of chemical bonding aspects in these quasicrystalline compounds.

In summary, two main features can be observed in the DOS close to the Fermi energy in high-quality, thermodynamically stable QCs containing transition-metal atoms: a structurally induced broad minimum (1 eV width) due to the Hume-Rothery mechanism and a narrow and sharply confined dip (0.1 eV width) due to hybridization effects involving the transition-metal bands. The physical existence of a relatively broad pseudogap has received strong experimental support during the last decade, as indicated by measurements of the specific heat capacity [63], photoemission [64], soft X-ray spectroscopies [65, 66], magnetic susceptibility and nuclear magnetic resonance probes [67]. In addition, experimental investigation of Al-Cu-Fe quasicrystalline films by scanning tunneling spectroscopy at low temperatures gave evidence for a narrow, symmetric gap of about 60 meV wide located around the Fermi level [68]. The existence of a sharp DOS valley of about 20 meV at the Fermi level in both quasicrystalline and approximant phases has also been confirmed by nuclear magnetic resonance studies, which probe the bulk properties of the considered samples [69]. All these observations indicate that the dip centered at the pseudogap is not a surface feature and that both its width and depth are sample dependent. The dependence of the pseudogap structure with the temperature was also investigated by means of tunneling and point-contact spectroscopy, and it was reported that the width of the broad pseudogap remains essentially unmodified as the temperature is increased from 4 K to 77 K. On the contrary, the dip feature centered at the Fermi level exhibits a significant modification, deepening and narrowing progressively as the temperature is decreased [70].

For the sake of illustration, in Figure 2.9 we show low-temperature tunneling spectroscopy measurements corresponding to the quasicrystalline sample i-$Al_{63}Cu_{25}Fe_{12}$. These measurements reveal a broad pseudogap extending over an energy scale of about 0.6 eV (shown in the inset) along with some fine structure close to the Fermi level (labeled 1 and 2 in the main frame). The broad pseudogap stems from the Fermi-surface–pseudo-Brillouin-zone interaction, while the dips may be respectively related to hybridization effects between d-Fe states and sp-states (feature labeled 1 in Figure 2.9) and d-orbital resonance effects (feature labeled 2 in Figure 2.9).

2.2.3.2 Fine Spectral Features

The possible existence of very narrow features in the electronic DOS over an energy scale of about 10 meV, obtained in self-consistent *ab initio* calculations dealing with

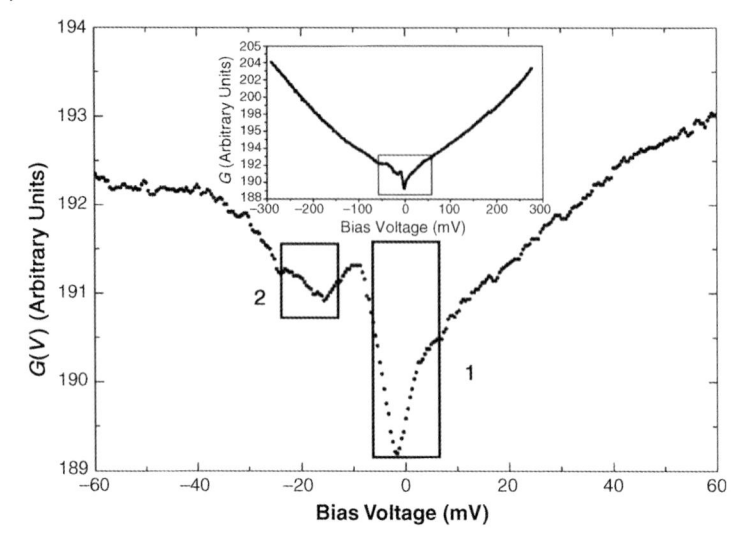

Figure 2.9 The differential conductance for the $Al_{63}Cu_{25}Fe_{12}$-Al tunnel junction at a temperature of $T = 2\,K$ at two different energy scales: 60 meV (main frame) and 300 meV (inset). (Adapted from reference [70]. Data file courtesy of R. Escudero).

several suitable quasicrystalline approximants [71], was considered as a possible characteristic feature of quasiperiodic crystals DOS. In this sense, it was argued that these peaks may stem from the structural quasiperiodicity of the substrate due to cluster aggregation [72], or d-orbital resonance effects [73]. However, several STM investigations of Al-Cu-Fe and Al-Pd-Re quasicrystalline ribbons confirming the presence of a dip of about 50 meV wide around the Fermi level, did not show evidence for finer structures in the DOS over the energy region extending about 0.5 eV from the Fermi level [74].

Accordingly, the possible existence of the spiky component of the DOS is still awaiting for a definitive experimental confirmation [76, 77]. In fact, difficulties in the experimental investigation of fine structure in the DOS arise form the requirement of a high energy resolution, as the peaks and gaps to be observed are only a few meV wide. Thus, both high-resolution photoemission and tunneling spectroscopies have failed to detect the theoretically predicted dense distribution of spiky features around the Fermi level. Several reasons have been invoked in order to explain these unsuccessful results. Among them the existence of some residual disorder present even in samples of high structural quality has been invoked as a plausible agent to smear out the finer details of the DOS [78]. It has also been argued that photoemission and STM techniques probe the near-surface layers, so that sharp features close to the pseudogap could be removed by subtle structural deviations near the surface from that of the bulk, as those reported for annealed QC surfaces [79].

On the other hand, detailed analysis of higher-resolution, extensive *ab-initio* calculations of several QC approximants suggests that a significant contribution to the spiky DOS component may probably stem from numerical artifacts [80]. Notwithstanding this, recent tunneling spectroscopy measurements performed in icosahedral QCs at low temperature (5.3 K) have provided some experimental support for the existence of a large number of energetically localized features close to the Fermi level in the electronic structure of the fivefold surface of an *i*-Al-Pd-Mn sample at certain local regions [81].

2.2.3.3 Spectral Conductivity Models

Anyway, in order to make a meaningful comparison between band-structure calculations and experimental measurements one should take into account possible phason, finite lifetime and temperature broadening effects. In so doing, it is observed that most finer details in the DOS are significantly smeared out and only the most conspicuous peaks remain in the vicinity of the Fermi level at room temperature [49]. These considerations suggest to reduce the number of main spectral features necessary to capture the most relevant physics of the transport processes. To this end, it is useful to consider the spectral conductivity function, $\sigma(E)$; defined as the $T \to 0$ conductivity with the Fermi level at energy E. Generally speaking, the conductivity spectrum should take into account both the DOS structure and the diffusivity, $D(E)$ of the electronic states, according to the relationship $\sigma(E) \propto N(E)D(E)$: Thus, although it may be tempting to assume that the $\sigma(E)$ function should closely resemble the overall structure of the DOS, it has been shown that dips in the $\sigma(E)$ curve can correspond to peaks in the DOS at certain energies [82, 83]. This behavior is likely to be related to the peculiar nature of critical electronic states close to the Fermi level [83–86].

Two fruitful results have been reported regarding the main features of the spectral conductivity function in QCs. On the one hand, it has been shown that the main qualitative features of the $\sigma(T)$, $S(T)$; and $R_H(T)$ curves, can be accounted for by considering an *asymmetric* spectral conductivity function characterized by a broad minimum exhibiting a pronounced dip within it, hence encompassing the transport properties of both amorphous phases and QCs within a unified scheme [8]. On the other hand, a series of *ab-initio* studies have shown that the electronic structure of both QCs and approximant phases belonging to the Al-Cu(Fe,Ru) and Al-Pd(Mn,Re) icosahedral families can be satisfactorily described in terms of a spectral resistivity, $\varrho(E) = \sigma^{-1}(E)$, exhibiting two basic spectral features close to the Fermi level, namely, a wide and a narrow Lorentzian peaks, according to the expression [82],

$$\sigma(E) = \bar{\sigma} \left\{ \frac{\gamma_1}{(E-\delta_1)^2 + \gamma_1^2} + \frac{\alpha \gamma_2}{(E-\delta_2)^2 + \gamma_2^2} \right\}^{-1} \qquad (2.9)$$

where the wide Lorentzian peak is related to the Hume-Rothery mechanism and the narrow Lorentzian peak is related to *sp-d* hybridization effects. This model includes six parameters, determining the Lorentzian's heights ($\bar{\sigma}/\gamma_i$) and widths ($\sim \gamma_i$), their positions with respect to the Fermi level, δ_i, and their relative weight in the overall structure, $\alpha > 0$. The parameter $\bar{\sigma}$ is a scale factor measured in (Ω cm eV)$^{-1}$ units. Suitable values for these electronic model parameters can be

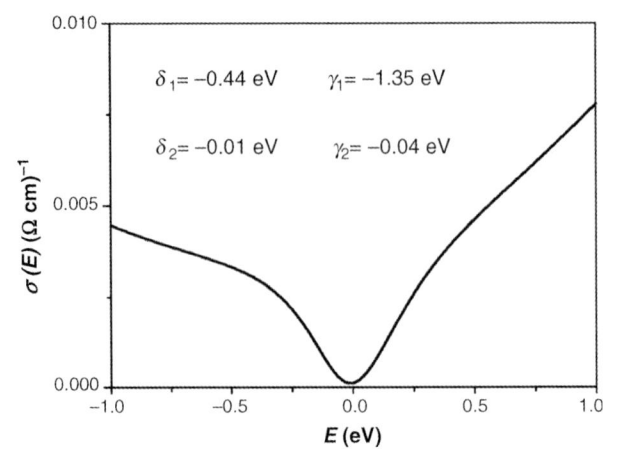

Figure 2.10 Spectral conductivity curve in the energy interval 1 eV around the Fermi level as obtained from Equation (2.9) for the electronic model parameter values γ_i and δ_i indicated in the frame.

Values shown in figure: $\delta_1 = -0.44$ eV, $\gamma_1 = -1.35$ eV, $\delta_2 = -0.01$ eV, $\gamma_2 = -0.04$ eV.

obtained by properly combining *ab-initio* calculations of approximant phases with experimental transport data of icosahedral samples within a phenomenological approach [75, 87]. In Figure 2.10 the overall behavior of the $\sigma(E)$ curve is shown for a suitable choice of the model parameters. By comparing this figure with Figure 2.9 we see that Equation 2.9 properly captures the main spectral features of realistic samples.

2.2.3.4 The Role of Critical States

An important open question in the field regards whether the purported anomalies in the transport properties observed in high-quality quasicrystals can be satisfactorily accounted for by merely invoking band-structure effects or, conversely, they must be traced back to the critical nature of the electronic states. At this stage, it seems quite reasonable that the proper answer should likely require a proper combination of both kinds of effects.

Generally speaking, critical states exhibit a rather involved oscillatory behavior, displaying strong spatial fluctuations that show distinctive self-similar features in some instances (Figure 2.11). As we can see, the wavefunction is peaked on short chain sequences but reappear far away on chain sequences showing the same lattice ordering. This is a direct consequence of the underlying lattice self-similarity and, as a consequence, the notion of an envelope function, which has been most fruitful in the study of both extended and localized states, is mathematically ill-defined in the case of critical states, and other approaches are required to properly describe them and to understand their structure.

From a rigorous mathematical point of view the nature of a state is uniquely determined by the measure of the spectrum to which it belongs. In this way, since it has been proven that Fibonacci lattices have purely singular continuous energy spectra [92], we must conclude that the associated electronic states cannot, strictly

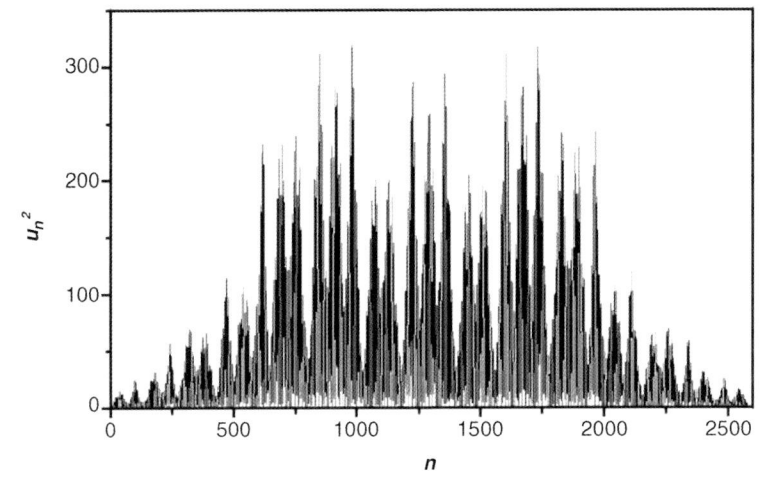

Figure 2.11 Squared amplitude distribution of a critical phonon normal mode in a Fibonacci lattice composed of $N = 2584$ atoms with a mass ratio $m_A/m_B = 34/21$.

speaking, be extended in Bloch's sense. This result holds for other aperiodic lattices (Thue–Morse, period doubling) as well, and it may be a general property of the spectra of self-similar aperiodic systems [93]. However, this fact does not necessarily imply that all these critical states behave in exactly the same way from a physical viewpoint. In fact, physically states can be classified according to their transport properties. Thus, conducting states in crystalline systems are described by periodic Bloch functions, whereas insulating systems exhibit exponentially decaying functions corresponding to localized states.

A first step towards a better understanding of critical states was provided by the demonstration that the amplitudes of critical states in a Fibonacci lattice do not tend to zero at infinity, but are bounded below through the system [94]. This result suggests that the physical behavior of critical states might be more similar to that corresponding to extended states than to localized ones, supporting the convenience of widening the very notion of extended state in aperiodic systems to include critical states that are not Bloch functions [86]. Accordingly, the possible existence of extended critical states in several kinds of aperiodic systems, including both quasiperiodic [86, 95, 96, 98–100] and nonquasiperiodic ones [97, 101], has been discussed in the last years spurring the interest on the precise nature of critical wavefunctions and their role in the physics of aperiodic systems.

In more precise terms one can describe a critical state in a quasiperiodic system in the following qualitative way [102]. Let us assume that a given state L spreads over a region of characteristic length L. Then, Conway's theorem implies that a similar region must exist at a distance $\leq 2L$. If L is sufficiently long, then both regions will be good candidates for a tunneling process between them, so that we might express $\psi^{2L} = z \psi^{L}$, where z is a damping factor roughly measuring the probability amplitude of the tunneling event. Within such a description the case $z = 0$ corresponds to strictly localized states, whereas $|z|=1$ is the signature of extended states. For intermediate localization cases, one can write

$$\psi^L \simeq L^{-\ln|z|/\ln 2} \simeq L^{-\alpha} \qquad (2.10)$$

where the precise value of $|z|$ will be dependent on the parameters of the considered model. In this way, the spatial structure of the wavefunction amplitudes is directly related to the topological properties of the quasiperiodic substrate. In particular, the self-similar properties of most critical wavefunctions can be traced back to the self-similarity of the lattice itself, through a series of hierarchical tunneling events involving the overlap of different subsystems at different length scales. Accordingly, one of the main results concerning electronic localization in quasiperiodic chains is the power-law behavior of the envelope of the wavefunction $\psi^N \simeq N^{-a}$ that characterizes most critical states.

In a quasiperiodic system the algebraic localization of typical wavefunctions, as described by Equation 2.10, gives rise to a scaling behavior of the bandwidths of the form $W \simeq tL^{-\beta}$, where the exponent $\beta > 1$ is related to the distribution of α's [102]. An overall estimation of the influence of critical states in the transport properties of quasiperiodic systems can be inferred from this expression by taking into account that the mean group velocity for a critical state can be approximated as $v \simeq LW \simeq tL^{1-\beta}$. This expression indicates that the mobility of the charge carriers goes to zero as the system size grows, but this asymptotic limit is reached more slowly than it is achieved in the case of exponentially localized states, whose mobility vanishes at a rate determined by the relationship $v \simeq t L^D e^{-L/\xi}$. This qualitative result provides strong support to the view of critical states as occupying an intermediate position between localized and extended states, although one may be tempted to consider them closer to the last from a physical point of view.

In fact, among the broad diversity of critical states belonging to general aperiodic and fractal systems one can find a class of critical wavefunctions that are extended from a physical point of view. These states arise from the very existence of resonant effects and correspond to specific energy values related to certain model parameters in the considered system. For instance, in the case of Fibonacci chains the energy values energy satisfying the relation

$$E_* = \epsilon \frac{1+\gamma^2}{1-\gamma^2} \qquad (2.11)$$

where $\gamma \equiv t_{AA}/t_{AB}$ measures the ration between the transfer integrals and the origin of on-site energies is defined in such a way that $\varepsilon_A = \epsilon = -\varepsilon_B$ correspond to extended states whose transmission coefficient equals unity irrespectively of the chain length (i.e. $T_N(E) = 1, \forall N$) [86]. The presence of theses states widens the notion of extended wavefunction to include electronic states that *are not* Bloch functions, and it is a relevant first step to clarify the precise manner in which the aperiodic order of these systems influences their transport properties. Subsequent numerical studies of the energy spectrum of mixed Fibonacci lattices have shown that a significant number of electronic states exhibiting very large transmission coefficients ($T_N(E) = 0.99999$) are located around the transparent states given by Equation 2.11 [103]. This result suggests that these critical states behave in a way quite similar to conventional extended states from a physical viewpoint, albeit they can not be rigorously described in terms of Bloch functions. To further analyze this important issue the study of the ac conductivity at zero temperature is very convenient, since it is very sensitive to the distribution nature of eigenvalues and the localization properties of the wavefunction close to the Fermi energy. In this way, by comparing the ac conductivities corresponding to periodic and mixed Fibonacci lattices it was concluded that both systems exhibit a similar behavior, though the value of the ac conductivity takes on systematically smaller values in the Fibonacci case, due to the fact that the ac conductivity involves the contribution of nontransparent states within an interval of $\hbar\omega$ around the Fermi level in this case [103]. In this way, although *most* critical functions exhibit rather low transmission coefficients, it is possible to find *transparent* states exhibiting a physical behavior completely analogous to that corresponding to usual Bloch states in periodic systems for a given choice of the model parameters, prescribed by Equation 2.11.

The rich variety of critical states in general Fibonacci systems suggests the appealing possibility of *modulating* the transport properties of normal modes propagating through a Fibonacci lattice by properly selecting the values of the masses composing the chain (isotopic effect). In fact, when studying band-structure effects in the thermal conductivity of Fibonacci quasicrystals a great variety of *critical normal modes* are found [104]. These modes exhibit quite different physical behaviors, which range from highly conducting extended states to critical states whose transmission coefficient oscillates periodically between two extreme values, depending on the system's length [98, 104]. Similar results concerning the existence of extended states in other kinds of self-similar structures, like Thue–Morse chains and hierarchical lattices, have been reported in the literature [95, 105], and its role in the transport properties has been analyzed in detail in terms of multifractal formalism on the basis that fractal dimension is directly associated to the localization degree of the eigenstates [106, 107].

2.2.4
Phenomenological Approaches

2.2.4.1 Kubo–Greenwood Formalism of Transport Coefficients

From the knowledge of the spectral conductivity function introduced in Section 2.3.3 the temperature-dependent transport coefficients can be obtained by means of the Kubo–Greenwood version of the linear response theory [108–110]. Within this approach the electrical, **j**, and thermal, **h**, current densities are, respectively, related to the voltage and temperature gradients according to the expression:

$$\begin{pmatrix} j \\ h \end{pmatrix} = \begin{pmatrix} \mathcal{L}_{11} & \mathcal{L}_{12} \\ \mathcal{L}_{21} & \mathcal{L}_{22} \end{pmatrix} \begin{pmatrix} -\nabla V \\ -\nabla T \end{pmatrix} \qquad (2.12)$$

The central information quantities are the kinetic coefficients

$$L_{ij}(T) = (-1)^{i+j} \int \sigma(E)(E-\mu)^{i+j-2}\left(-\frac{\partial f}{\partial E}\right) dE \tag{2.13}$$

where $f(E, \mu, T)$ is the i-Fermi–Dirac distribution function, E is the electron energy, and μ is the chemical potential. In this formulation all the microscopic details of the system are included in the $\sigma(E)$ function. From the knowledge of the kinetic coefficients one obtains the electrical conductivity

$$\sigma(T) = \mathcal{L}_{11}(T) \tag{2.14}$$

The thermoelectric power,

$$S(T) = \frac{\mathcal{L}_{12}(T)}{|e|T\sigma(T)} \tag{2.15}$$

the electronic thermal conductivity,

$$\kappa_e(T) = \frac{1}{e^2 T}\mathcal{L}_{22}(T) - T\sigma(T)S(T)^2 \tag{2.16}$$

and the Lorenz function

$$L(T) \equiv \frac{\kappa_e(T)}{T\sigma(T)} \tag{2.17}$$

in a unified way. As a first approximation one generally assumes (T) E_F: Then, by expressing Equations 2.14–2.17 in terms of the scaled variable $x \equiv (E-\mu)\beta$, where $\beta \equiv (k_B T)^{-1}$, the transport coefficients can be rewritten as [36, 37]

$$\sigma(T) = \frac{J_0}{4} \tag{2.18}$$

$$S(T) = -\frac{k_B}{|e|}\frac{J_1}{J_0} \tag{2.19}$$

$$k_e(T) = \frac{k_B^2 T}{4e^2}\left(J_2 - \frac{J_1^2}{J_0}\right) \tag{2.20}$$

$$L(T) = \left(\frac{k_B}{eJ_0}\right)^2 \begin{vmatrix} J_0 & J_1 \\ J_1 & J_2 \end{vmatrix} \tag{2.21}$$

in terms of the reduced kinetic coefficients

$$J_n(T) = \int x^n \sigma(x) \operatorname{sech}^2(x/2) dx \tag{2.22}$$

2.2 Electronic-Structure-Related Properties

Making use of Equation 2.9 these kinetic coefficients can be expressed in the form

$$J_0 c_0^{-1} = \frac{4\pi^2}{3}\beta^{-2} + a_3\beta^{-1}H_1 + a_4 H_0 + 4a_0,$$

$$J_1 c_0^{-1} = \frac{4\pi^2}{3}a_1\beta^{-1} + a_5 H_1 + a_3\beta(4-q_0 H_0), \tag{2.23}$$

$$J_2 c_0^{-1} = \frac{28\pi^4}{15}\beta^{-2} + a_6\beta H_1 + a_5(4-q_0 H_0)\beta^2 + \frac{4\pi^2}{3}a_0$$

where $c_0 \equiv \bar{\sigma}(\gamma_1 + \alpha\gamma_2)^{-1}$, and the coefficients a_i were defined in reference [36]. We have introduced the auxiliary integrals

$$H_k(\beta) \equiv \int_{-\infty}^{\infty} \frac{x^k}{\beta^{-2}x^2 - 2\beta^{-1}q_1 x + q_0}\operatorname{sech}^2(x/2)dx \tag{2.24}$$

where $q_0 \equiv \varepsilon\varepsilon_1^2\varepsilon_2^2(\gamma_1 + \alpha\gamma_2)^{-1}$, $q_1 = (\gamma_1\delta_2 + \alpha\delta_1\gamma_2)(\gamma_1 + \alpha\gamma_2)$, $\varepsilon_i^2 \equiv \gamma_i^2 + \delta_i^2$ and $\varepsilon \equiv \gamma_1\varepsilon_1^{-2} + \alpha\gamma_2\varepsilon_2^{-2}$.

By inspecting Equation 2.24 we realize that the auxiliary integral H_1 identically vanishes in the case $q_1 = 0$, due to the odd parity of the integrand. In that case, taking into account the Fourier transform relationship

$$\frac{1}{x^2 + a^2} = \frac{1}{2a}\int_{-\infty}^{\infty} e^{-a|\omega|}e^{i\omega x}d\omega. \tag{2.25}$$

the auxiliary integral H_0 can be properly rearranged in the form

$$H_0(\beta) \equiv \frac{\beta^2}{2a}\int_{-\infty}^{\infty} e^{-a|\omega|}d\omega \int_{-\infty}^{\infty} e^{i\omega x}\operatorname{sech}^2\left(\frac{x}{2}\right)dx \tag{2.26}$$

where $a^2 \equiv q_0\beta^2$. Now, the second integral in Equation 2.26 is just the Fourier transform of the function $4\pi\omega\operatorname{cosech}(\pi\omega)$, so that one finally obtains [37]

$$H_0 \equiv \frac{2\pi\beta^2}{a}\int_{-\infty}^{\infty} e^{-a|\omega|}\omega\operatorname{cosech}(\pi\omega)d\omega = 4q_0^{-1}\tilde{\beta}\varsigma_H(2, 1/2 + \tilde{\beta}) \tag{2.27}$$

where $\tilde{\beta} \equiv \sqrt{q_0}\beta/2\pi$ is a scaled variable and $\varsigma_H(s, a) \equiv \sum_{k=0}^{\infty}(k+a)^{-s}$ is the Hurwitz Zeta function, which reduces to the Riemann Zeta function in the case $a = 1$ [111]. Making use of these analytical expressions Equation 2.23 can be rearranged in the matrix form

$$\begin{pmatrix} J_0 \\ J_1 \\ J_2 \end{pmatrix} = \frac{4\pi^2 c_0}{3}\begin{pmatrix} \frac{3}{\pi^2}\tilde{J}_{00} & 0 & 1 \\ 0 & \tilde{J}_{11} & 0 \\ \tilde{J}_{20} & 0 & \frac{7\pi^2}{5} \end{pmatrix}\begin{pmatrix} 1 \\ \beta^{-1} \\ \beta^{-2} \end{pmatrix} \tag{2.28}$$

where $\tilde{J}_{00} \equiv a_0 + a_4 q_0^{-1}\tilde{\beta}\zeta_H, \tilde{J}_{11} \equiv a_1 + 12 a_3 q_0^{-1} f(\tilde{\beta}), \tilde{J}_{20} \equiv a_0 + 12 a_4 q_0^{-1} f(\tilde{\beta})$ with $f(\tilde{\beta}) \equiv \tilde{\beta}^2 (1-\tilde{\beta}\zeta_H)|$. In this way, under the assumption that q_1 is negligible in Equation 2.24, one obtains closed analytical expressions for the different transport coefficients. It turns out that this assumption is a reasonable one for several QCs of interest. In fact, the values $q_1 = -0.025$ eV, $q_1 = -0.015$ eV, and $q_1 = -8.8 \times 10^{-5}$ eV, are, respectively, obtained for Al-Mn-Si approximant phases [88], i-Al-Cu-Fe QCs [75] and i-AlPdRe QCs [112]. We notice that the smaller q_1 value corresponds to higher structural quality QCs, whereas the largest one is obtained for an approximant crystal. Accordingly, we can confidently assume the limiting behavior $q_1 \to 0$ properly applies to *ideal* QCs.

In the more realistic case $q_1 \neq 0$ we can obtain useful information by expanding Equation 2.24 in Taylor series around the Fermi level to get

$$H_0 \simeq \frac{4}{q_0}\left(1 + \frac{\pi^2}{3}\frac{4q_1^2 - q_0}{q_0^2}\beta^{-2}\right),$$

$$H_1 \simeq \frac{8\pi^2 q_1 \beta^{-1}}{3 q_0^2}\left(1 + \frac{14\pi^2}{5}\frac{2q_1^2 - q_0}{q_0^2}\beta^{-2}\right)$$

(2.29)

In this way, one obtains approximate analytical expressions for the electrical conductivity and Seebeck coefficient curves [113],

$$\sigma(T) = \sigma(0)[1 + bT^2 \Lambda(T)] \tag{2.30}$$

with

$$\Lambda(T) = \xi_2 + \xi_4 bT^2 + \xi_6 b^2 T^4 \tag{2.31}$$

and

$$S(T) = -2|e|\mathcal{L}_0 T \frac{\xi_1 + \xi_3 bT^2}{1 + \xi_2 bT^2 + \xi_4 b^2 T^4} \tag{2.32}$$

where $b \equiv e^2 \mathcal{L}_0$, $\mathcal{L}_0 = \pi^2 k_B^2 / 3e^2 = 2.44 \times 18^{-8}$ V^2 K^{-2} is the Lorenz number. These expressions are valid in the low temperature regime, up to about 50–100 K [36]. The coefficients ξ_n can be explicitly expressed in terms of the electronic model parameters and contain detailed information about the electronic structure of the sample. For instance, the first-order phenomenological coefficients are defined in terms of the electronic model parameters as [113]

$$\xi_1 \equiv -\frac{\gamma_1 \delta_1 \varepsilon_2^4 + \alpha \gamma_2 \delta_2 \varepsilon_1^4}{\varepsilon \varepsilon_1^4 \varepsilon_2^4} \tag{2.33}$$

$$\xi_2 \equiv \frac{\gamma_1 \varepsilon_2^6 (\varepsilon_1^2 - 4\delta_1^2) + \alpha \gamma_2 \varepsilon_1^6 (\varepsilon_2^2 - 4\delta_2^2)}{\varepsilon \varepsilon_1^6 \varepsilon_2^6} + 4\xi_1^2 \tag{2.34}$$

and can be related to the topology of the spectral conductivity function $\sigma(E)$ by means of the following expressions,

$$\xi_1 = \frac{1}{2}\left(\frac{d\ln\sigma(E)}{dE}\right)_{E_F} \tag{2.35}$$

and

$$\xi_2 = 2\xi_1^2 + \frac{1}{2}\left(\frac{d^2\ln\sigma(E)}{dE^2}\right)_{E_F} \tag{2.36}$$

Thus, from the knowledge of the phenomenological coefficients ξ_1 and ξ_2 we can obtain suitable information concerning the slope and curvature of the DOS close to E_F.

For instance, in the low-temperature limit Equation 2.32 reduces to the linear form

$$S(T \to 0) = -2|e|\mathcal{L}_0\xi_1 T \equiv m_0 T \tag{2.37}$$

The sign of the slope m_0 is determined by the sign of the parameter ξ_1 that, in turn, depends on the electronic structure of the sample according to Equation 2.35. Therefore, Equation 2.37 reduces to the well-known Mott's formula

$$S = -|e|\mathcal{L}_0\left(\frac{d\ln\sigma(E)}{dE}\right)_{E=\mu}$$

in the low-temperature limit: It then follows that Mott's formula will properly describe the thermoelectric power of QCs as far as the remaining coefficients ξ_2, ξ_3 and ξ_4 in Equation 2.32 are negligible as compared to ξ_1. Since these coefficients are multiplied by the temperature-dependent factors bT^2 and b^2T^4, respectively, it is clear that the range of validity of Mott's formula will be strongly dependent on the electronic structure of the sample.

2.2.4.2 Application Examples

In this section we will illustrate the phenomenological framework introduced in the previous one by relating the main topological features of the experimental $\sigma(T)$ and $S(T)$ curves to certain characteristic features of the electronic structure of the samples. The key point of this approach relies on the analytical coefficients ξ_n, which can be regarded as phenomenological parameters containing information about the electronic structure of the sample. Since the values of the ξ_n coefficients can be *also* determined from the analysis of the experimental transport curves, one can obtain useful information about the spectral conductivity function (E) from the topological features present in these curves. The first step consists in determining the values of the ξ_n coefficients from suitable fits to the experimentally obtained transport curves. The next step will be then to determine the electronic model parameters γ_i, δ_i, and α, from the obtained n values making use of previously derived analytical formulae. Due to the involved nature of the analytical expressions relating the phenomenological coefficients to the model parameters, this is a rather cumbersome task. Fortunately,

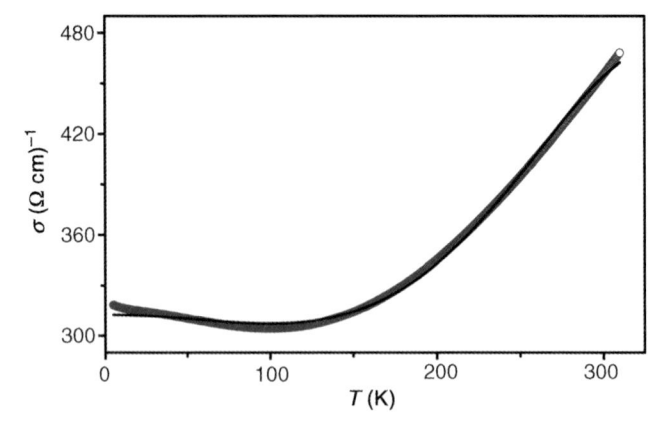

Figure 2.12 Electrical conductivity as a function of temperature for the $Al_{73.6}Mn_{17.4}Si_9$ cubic approximant (open circles). The solid line corresponds to the best fit curve $\sigma(T) = \sigma_0(1 + BT^2 + CT^4 + DT^6)$ with $\sigma_0 = 312.6 \pm 0.2$ $(cm)^{-1}$; $B = (-3.50 \pm 0.08) \times 10^{-6}$ K^{-2}; $C = (1.91 \pm 0.02) \times 10^{-10}$ K^{-4}; $D = (-1.07 \pm 0.02) \times 10^{-15}$ K^{-6}, with a correlation coefficient $r = 0.9824$ (From reference [88]. With permission).

even the partial knowledge of some phenomenological coefficients suffices to gain some physical insight onto certain relevant features of the electronic spectrum of the sample.

As a suitable sample let us first consider the $Al_{82.6-x}Mn_{17.4}Si_x$ ($x = 9$) -phase [114], which is a well-documented representative of the 1/1-cubic approximants class. In Figure 2.12 we show the temperature dependence of the electrical conductivity for the $Al_{73.6}Mn_{17.4}Si_9$ cubic approximant. The curve exhibits a typical metallic behavior up to ~100 K, where the conductivity attains a minimum and then it progressively increases as the temperature is further increased. In Figure 2.13 we show the temperature dependence of the thermoelectric power for the same approximant phase. The thermopower shows a remarkable nonlinear behavior, exhibiting a broad minimum at about $T_1 = 160$ K, and changes its sign twice at about $T_0 = 50$ K and 260 K, respectively. This anomalous behavior resembles that observed for several icosahedral QCs [21–23].

From the knowledge of the complete set of phenomenological parameters listed in the corresponding figure captions one can derive the corresponding electronic model parameters following the algebraic procedure described in reference (92). Finally, making use of Equation 2.9 one determines the spectral conductivity function, which is shown in Figure 2.16 along with the (E) curves derived for other CMA representatives. By inspecting this figure we see that the spectral conductivity of the quasicrystalline phase is both deeper and broader than that corresponding to the approximant phase, thus indicating a less effective Hume-Rothery mechanism for the approximant crystal. On the other hand, the presence of a well-defined spectral feature at about ~0.03 eV may be indicative of hybridization effects likely related to bond formation in the approximant sample.

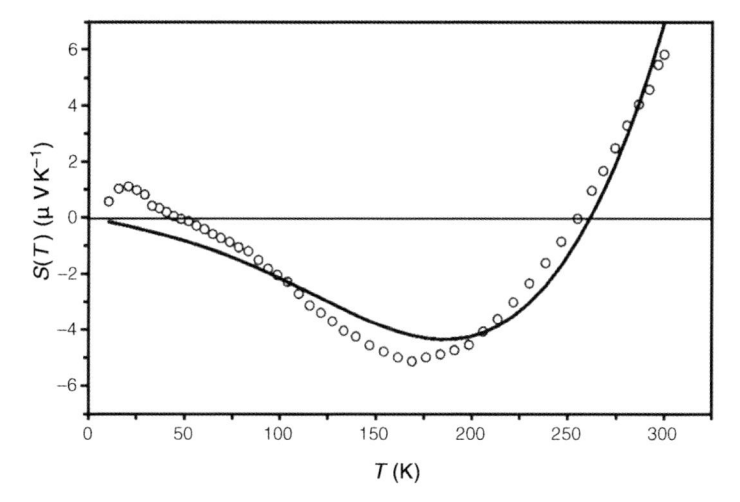

Figure 2.13 Thermoelectric power as a function of temperature for the $Al_{73.6}Mn_{17.4}Si_9$ cubic approximant (open circles). The solid line corresponds to the best fit curve given by $S(T) = -0.0488T(a + fT^2 + gT^4)/(1 + BT^2 + CT^4 + DT^6)$ with $a = 0.29 \pm 0.05$ (eV)$^{-1}$; $f = (6 \pm 2) \times 10^{-5}$ K^{-2}; and $g = (-1.1 \pm 0.3) \times 10^{-9}$ K^{-4}; with Pearson $\chi^2 = 0.562$. (From reference [88]. With permission).

Another illustrative example is provided by the orthorhombic 0 phase of the Al-Pd-Mn alloys system exhibiting a complex unit cells, composed of 258 atoms. The electrical resistivity of this phase shows an almost negligible temperature dependence between 4 and 300 K (Figure 2.14) [115]. Whereas weakly temperature-dependent resistivities are not uncommon for both amorphous alloys and bulk

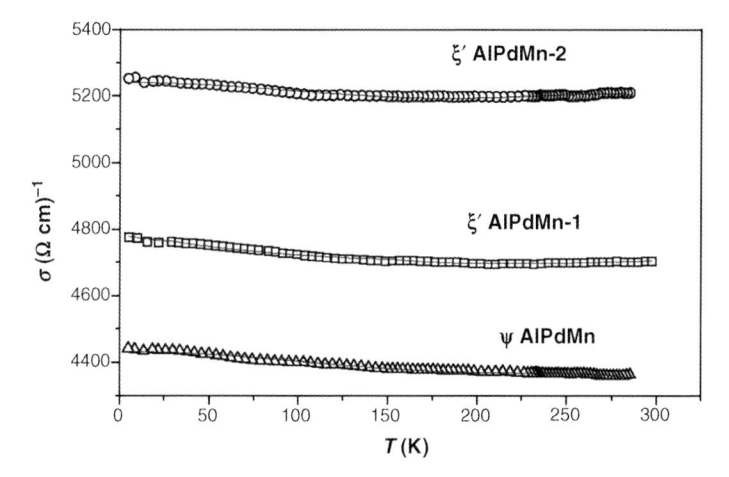

Figure 2.14 Electrical conductivity of Al-Pd-Mn complex alloys as a function of temperature. Solid curves are best fits obtained by a simultaneous analysis of the conductivity and thermopower data. (From reference [89]. With permission).

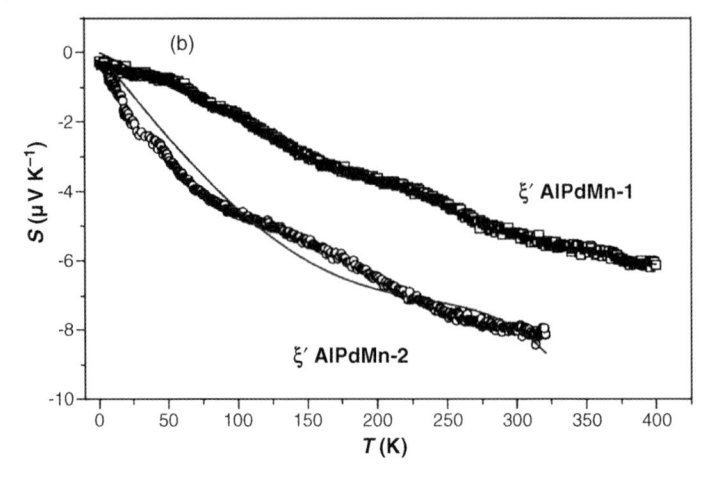

Figure 2.15 Thermoelectric power of Al-Pd-Mn complex alloys as a function of temperature. Solid curves are best fits obtained by a simultaneous analysis of the conductivity and thermopower data. (From reference [89]. With permission).

metallic glasses lacking long-range ordered crystalline lattices [50], the temperature independent resistivity of ξ'-Al-Pd-Mn was observed on monocrystalline samples of good lattice perfection and structural homogeneity. The corresponding thermopower curves are displayed in Figure 2.15. Their values are small and show a rather smooth behavior with several changes of the slope within the investigated temperature range. Following the procedure previously described one obtains the spectral conductivity functions shown in Figure 2.16.

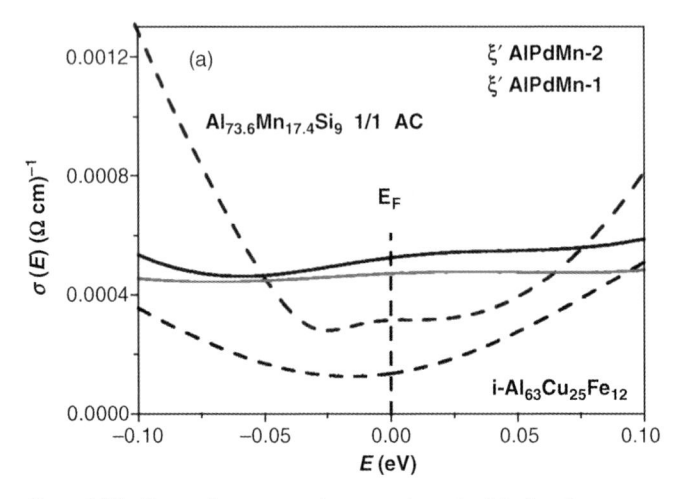

Figure 2.16 Comparison among the spectral conductivity functions corresponding to quasicrystals, approximant phases, and complex metallic alloys. (From reference [89]. With permission).

The absence of a pseudogap in the case of ξ'-Al-Pd-Mn samples is clearly appreciated, indicating that the Hume-Rothery mechanism there is less effective and the electrical conductivity is consequently higher. The (E) curves of the ξ'-Al-Pd-Mn samples are relatively flat as compared to those corresponding to $Al_{63}Cu_{25}Fe_{12}$ and $Al_{73.6}Mn_{17.4}Si_9$ compounds. Thus, the origin of the almost temperature-independent electrical conductivity of the ξ'-Al-Pd-Mn complex alloys can then be traced back to the specific form of the spectral conductivity, which exhibits very weak variation over the energy scale of several meV around the Fermi level. Yet, they show some fine structure that yields observable effects in the temperature-dependent thermoelectric power curves. These electronic structure related effects highlight the difference between ξ'-Al-Pd-Mn phase and conventional free-electron alloys.

2.3 Phonons

2.3.1 Phonons: An Introduction

The study of phonons in CMA bears some similarities with what has been presented for electrons. The large unit cell, the eventual aperiodic character will influence the vibrational properties of the material. The very notion of Bloch waves even is questionable and the nature of the eigenmodes is still an open question.

There are, however, important differences. On one hand, although eigenmodes are no longer scalar but vectors for phonons, the situation is somewhat simpler since we do not deal with the difficulties of electrons bands and interaction. On the other hand, it is possible to measure experimentally the dispersion relation using inelastic neutron or X-ray scattering, which provides an extremely powerful experimental tool for phonon studies.

As a simple toy model let us first consider a one-dimensional system of identical atoms with mass m located on a periodic lattice with a lattice constant a and connected by identical springs characterized by their stiffness K. We consider only first-neighbor interactions [116, 117].

Because of the long-range periodic order, the solution to the dynamical problem is a superposition of plane waves, characterized by their wavevector **q** (module $q = 2\pi/\lambda$) an energy $E = \hbar\omega$ and a polarization **e**. In the one-dimensional case there is only one "longitudinal" polarization, whereas in a three-dimensional case one has to consider three polarizations, one longitudinal and two transverse.

Because of the periodicity we only need to consider plane waves whose wavevectors lie in the first Brillouin zone, that is, $q < \pi/a$. The dynamics of the lattice is thus entirely characterized by the knowledge of the plane waves or vibrational modes. One key parameter is the dispersion relation, which relates q and E for each vibrational mode. The other important parameter is the eigenmodes that characterize the pattern of atomic vibration. These are eigenmodes that may carry the eventual signature of structural complexity. Finally, it is important to recall that phonons are quantum excitations that obey the Bose–Einstein statistics.

If we now come back to the monoatomic 1D system, the solution of the dynamical problem is easily solved and the dispersion relation writes:

$$E = 2\sqrt{\frac{K}{m}}|\sin(qa/2)| \qquad (2.38)$$

In the limit of long-wavelength excitations (or when q goes to zero) the dispersion relationship is linear with a slope related to the sound velocity: this is the acoustic regime. In this regime all atoms vibrates almost in phase. As q increases, the dispersion relation departs from the linear regime to get a zero slope at the Brillouin zone boundary: at this particular point the solution is a stationary wave.

Let us now consider a system with two atoms with masses m_1 and m_2 regularly placed on a periodic lattice and connected with identical springs. The crystal lattice parameter is now $2a$, that is, twice that of the single atom, and the corresponding Brillouin zone is half. There are now two branches in the dispersion relation, which means that for each wavevector q there are two modes with different energies. When q is close to zero, the two modes are (i) an acoustic mode where all atoms are moving in phase and (ii) an optic mode where two neighboring atoms are moving in phase opposition. There is also a gap in the dispersion relation, which defines an energy region for which there are no phonons. This gap is related to the mass difference in this simple example (Figure 2.17).

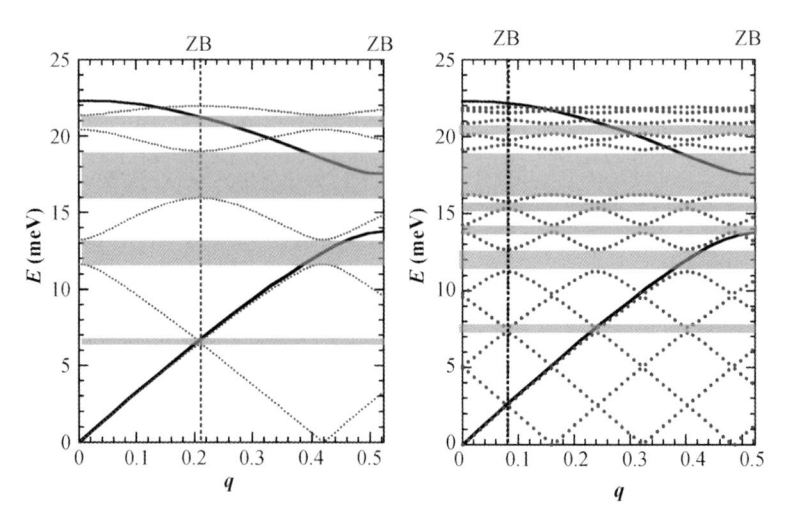

Figure 2.17 Extended zone scheme dispersion relation for 5 atoms (left) and 13 atoms (right) Fibonacci approximant. The gray areas indicate the largest gap in the dispersion. The Brillouin zone boundaries are shown by vertical dashed lines. The dispersion relation of the two-atom model is shown by a black solid line up to the Brillouin zone boundary.

The next step toward a complex system is to increase the number of atoms in the unit cell. This can be achieved by using an approximant to the Fibonacci chain. If we consider the two masses m_1 and m_2 with $m_2/m_1 = \tau$ where τ is the golden mean, the distribution of masses in the unit cell can be described by successive approximant *LS*, *LSL*, *LSLLS*.... If we consider the *LSLLS* approximant with 5 atoms in the unit cell the dispersion relation, shown in Figure 2.17, contains now 5 branches. The Brillouin zone is now smaller, and to exemplify similarities with the 2-atom model the dispersion relation is displayed in an extended zone scheme. There are now 4 gaps, shown by gray areas, but some of them are very small: this is the case for the gap around 7 meV. One can also notice some similarities with the simple 2-atom model: in particular the largest gap is located in the same energy range around 17 meV. The phonon eigenvectors and the corresponding pattern of displacement is, however, much more complex than in the simple 2 atoms per cell. As the number of atoms increases, with for instance 13 atoms in the unit cell, shown on the right panel, the number of gaps will of course increase. There are now 12 gaps, but some of them are extremely small and cannot be seen in the figure and the similarities between the two models are quite clear.

When the system goes to infinity for the one-dimensional quasicrystal with the Fibonacci chain structure, the number of gaps goes to infinity. Some of them are of course very small, but this large number of gaps will play a role in the nature of the vibrational modes. In fact, an acoustic regime still exists in the 1D quasicrystal, although with a smaller range, which of course will affect thermal conductivity properties. The most characteristic signature of the long-range aperiodic order, as for electrons, is that in some energy range the modes are critical: phonons do not behave as in a simple periodic structure, with propagating extended plane waves. Modes are neither like in disordered solid, where the modes decay exponentially with the distance around a defect. Modes are "critical:" the phonon wave propagates but with a decay that follows a power law. This is related to the quasiperiodic distribution of similar local environments in the structure: the phonon wave somehow "propagates" from one such environment to another similar one with which it will "resonate" (for an introduction to aperiodic crystals and their dynamics see reference [118]).

Within the harmonic approximation, the above simple results can be generalized to three-dimensional systems. The solution of the dynamical problems is written in the form of plane waves or Bloch waves. If both the structure and the interactions are known the dynamical problem may be written in a matrix form (dynamical matrix), the solution being the eigenvalues (energies) and eigenvectors (pattern of vibration). If there are n atoms in the unit cell, there are $3n$ modes, 3 of which are acoustic, namely one longitudinal and two transverse acoustic modes.

For a 3D quasicrystal, there are no exact solutions, and the nature of the phonon modes is still an open question. In the low-energy range it has been shown that there is an acoustic regime, although in a limited energy range. The definition of pseudozone boundaries is also important: this defines the most important points in reciprocal space, for which one may expect a gap opening in the dispersion relation. At higher energy, simulations show that only a limited number of atoms participate in a given mode, and it has been postulated that modes should also be critical, as in the one-dimensional case, but this has not been proved yet even on large simulations.

We will show in the following that in several complex systems it is now possible to carry out realistic simulations that compare well with the experimental results. This is a rather demanding simulation, which requires knowledge of both the structure and of the interaction Hamiltonian. Even for a "simple" structure, the atomic structure might not be as simple, for in most cases disorder is present (partially occupied sites, split positions, mixed occupancy....) which in most cases is not taken into account properly. The Hamiltonian is derived from *ab initio* methods using the DFT approach, but this is restricted to periodic cells with only a few hundred atoms. For larger numbers of atoms one generally uses adapted Hamiltonian, such as the embedded atom method (EAM potentials) or pair potentials, etc.

Besides the dispersion relation, the vibrational density of states (VDOS), $n(E)$, is an important quantity. It represents the number of vibrational modes whose energy lies between E and $E + dE$. The VDOS is, for instance, of importance to compute macroscopic thermodynamical quantities such as the specific heat. As this will be used in the following we give here two important properties of the VDOS: (i) in the acoustic regime where the dispersion relation is linear the VDOS goes like E^2; when the dispersion relation slope is equal to zero, there is a so-called van Hove singularity in the dispersion. This will be the case at the Brillouin zone boundary but also at each tile there is a gap opening.

The Debye approximation, frequently used in the field of complex systems, is a very crude approximation that replaces the complex VDOS shape, by a simple quadratic behavior. In other words, the Debye approximation considers that the vibrational modes are only in the acoustic regime.

Figure 2.18 illustrates the above concepts in the case of a relatively simple system, CaF_2 with three atoms in the unit cell (9 modes altogether). The dispersion relation

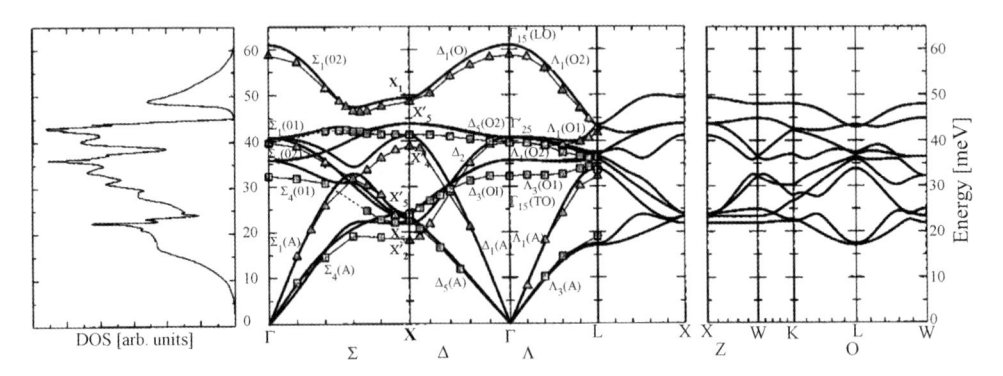

Figure 2.18 Middle and right: Phonon-dispersion curves from inelastic neutron scattering (data points with thin connecting lines) at RT and from *ab initio* theory (thick lines). Triangles refer to longitudinal, and squares to transverse polarization. Left: Phonon density of states from *ab initio* theory (From reference [132]).

has been measured along high symmetry axes by inelastic neutron scattering on a single grain, and is compared to DFT calculations. The left panel presents the corresponding VDOS. Note the ω^2 dependence at low energy, and van Hove singularities corresponding to dispersionless branches generally located close to Brillouin zone boundaries.

2.3.2
Measuring Phonons: Inelastic Neutron and X-Ray Scattering

Measuring macroscopical quantities such as the specific heat allows one to get indirect information on the vibrational density of states but the inelastic neutron scattering method, invented in the 1950s by the Nobel Prize winner B. Brockhouse, remains the experimental method of choice for the study of lattice dynamics. We will show in the following that this method not only allows determination of the dispersion relation, but also provides some insight into the nature of the modes and eigenvectors, which makes them a unique tool.

Neutron sources (research nuclear reactors or spallation sources), can deliver thermal neutrons whose energy is of the order a few tens of meV, that is, of the same order of magnitude as phonon excitations. The incoming neutron can thus interact inelastically with the sample, resulting in a neutron energy change that can be analyzed.

Two main techniques are used to study phonons: the coherent inelastic neutron scattering on a single crystal using the triple axis instrument, allowing the direct measurement of the dispersion relation, and the vibrational DOS measurement using incoherent neutron scattering on polycrystalline samples.

The neutron–matter interaction is a complex phenomenon, where the neutron interacts with the nucleus of the atoms via strong nuclear interaction. The interaction is characterized by the scattering cross-section, which contains two terms, the coherent scattering cross-section and the incoherent (or self-) scattering cross-section. The coherent scattering is the important one for processes such as the diffraction, and is characterized by the scattering length b, which is the equivalent of the atomic scattering factor used for X-rays.

2.3.2.1 Coherent Inelastic Neutron Scattering
The principle of the coherent inelastic neutron scattering is relatively simple. A monochromatic neutron beam with a wavevector $\mathbf{k}_i$, is sent onto a single-crystal sample. During the interaction the neutron can exchange energy with the crystal [120, 121]. It can be shown that the energy transfer corresponds to the "creation" when the neutron loses energy, or the annihilation, when the neutron gain energy, of a phonon. This is illustrated in Figure 2.19, where the incoming neutron has "created" a phonon in the system, the scattered neutron now having a wavevector $\mathbf{k}_f$. In order to detect only those scattered neutron a crystal analyzer is installed between the sample and the detector.

There are two important relations given by the momentum and energy conservation law:

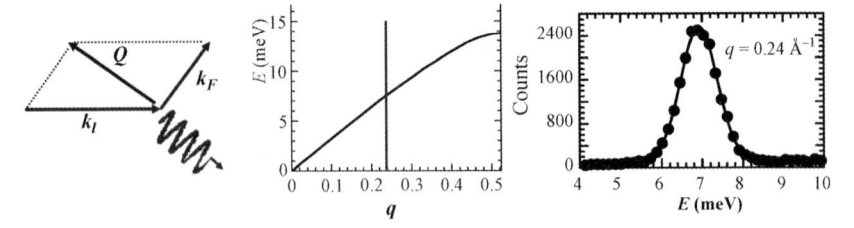

Figure 2.19 Illustration of the inelastic scattering process. Left panel: the incoming neutron exchanges one phonon with the sample. Middle panel: a constant-Q scan and the resulting scan in the right panel.

$$\mathbf{k}_F - \mathbf{k}_I = \mathbf{Q} = \mathbf{Q}_{Bragg} \pm \mathbf{q} \tag{2.39}$$

where $\mathbf{Q}_{Bragg}$ is the closest Bragg peak. During the scattering process there has been an energy exchange between the neutron and a phonon characterized by its wavevector $\mathbf{q}$ and mode s, with an energy $E = \hbar\omega$. This is expressed by the energy conservation law, where E_F and E_I are the final and incoming energy of the neutron:

$$E_F - E_I = \pm\hbar\omega_s(\mathbf{q}) \tag{2.40}$$

These two relations thus express the fact that the inelastic scattering signal is directly related to a particular phonon characterized by its wavevector q, its mode label s and energy $\hbar\omega_s$.

The measured inelastically scattered intensity is given by the differential scattering cross-section related to the coherent scattering law $S(Q, E)$ by:

$$\left(\frac{\partial^2 \sigma}{\partial \Omega \times \partial E_F}\right)_{coh,n} = \frac{k_F}{k_I} \times S_{coh,n}(\mathbf{Q}, E) \tag{2.41}$$

For the case of the creation or annihilation of a single phonon characterized by the quantum number q, s the scattering law is related to the inelastic structure factor by the following relation:

$$S_s(\mathbf{Q}, E)|_{\pm 1} \alpha \left|F_{inel,s}(\mathbf{Q}, \mathbf{q})\right|^2 \delta(\mathbf{Q} - \mathbf{Q}_{Bragg} \pm \mathbf{q}) \frac{1}{E} n(E)|_{\pm 1} \delta(E \pm E_s(\mathbf{q})) \tag{2.42}$$

This is a very important relation expressing that a delta peak appears in the measured signal for both a wavevector $\mathbf{q}$ and an energy E_s. The inelastically measured signal is thus directly related to the phonon dispersion relation. This expression also contains the inelastic structure factor defined by:

$$F_{inel,s}(\mathbf{Q}, \mathbf{q}) = \sum_j^{Maille} b_j \left(\mathbf{Q} \cdot \frac{\mathbf{e}_{q,s,j}}{\sqrt{M_j}}\right) \exp(i\mathbf{Q} \cdot \mathbf{R}_j) \exp(-w_j(\mathbf{Q})) \tag{2.43}$$

where $\mathbf{e}$ is the polarization of the mode (q,s) for the atom j, $\mathbf{R}_j$ the coordinate of the atom inside the unit cell, b_j its scattering length and w_j the corresponding Debye–Waller factor.

This last expression is particularly important since it relates the observed inelastically scattered intensity to a Fourier transform where the polarization or eigenmodes of the phonon are included. This means that the measured intensity contains information on the eigenmodes, which we emphasize again, is a unique experimental tool. The above expressions stand for a single mode. For a crystal with $3n$ modes the resulting measured signal is obtained by summing up on the $3n$ modes that is,

$$S(\mathbf{Q}, E) = \sum_{s=1,\ldots,3N\varphi} S_s(\mathbf{Q}, E) \tag{2.44}$$

The scalar products $\mathbf{Q}.\mathbf{e}$ in the expression (2.43) of the inelastic structure factors, is a selection rule and allows, in favorable cases, to single out a single mode. This is particularly true for the three acoustic modes: the relative position of $\mathbf{Q}$ and $\mathbf{q}$ and thus $\mathbf{e}$, allows one to select a position in reciprocal space such that two of the three acoustic signals are equal to zero. This is a particularly important tool for measuring accurately acoustic dispersions. Moreover, because all atoms are vibrating in phase, the integrated intensity for the acoustic inelastic signal (in the limit of high temperature and for a linear dispersion) is given by the expression:

$$I_{\text{Int}} \approx \frac{k_B T}{\hbar} (\mathbf{Q} \cdot \mathbf{e}_{q,s})^2 \frac{I_{\text{Bragg}}(\mathbf{Q})}{(E_s(\mathbf{q}))^2} \tag{2.45}$$

where I_{Bragg} is the Bragg peak intensity and T the temperature of the measurement. This relation implies that acoustic phonons are best measured close to strong Bragg peaks and at high value of Q since the signal grows quadratically with Q. On the other hand, the intensity decays rapidly with the energy of the mode, so that high-energy modes are generally difficult to measure. We will show in the following that the relation (2.45) is particularly useful for checking the acoustic character of a mode. In particular, we define a normalized intensity as

$$I_{\text{norm}} = I_{\text{int}} \cdot E^2 \tag{2.46}$$

which is a constant as long as the phonon has an acoustic character.

Experimentally, the measurement is carried out by performing energy scans while keeping the $\mathbf{Q}$ vector constant. This is exemplified in Figure 2.20 in the case of a single

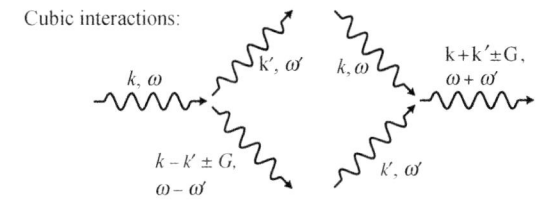

Figure 2.20 Illustration of the phonon–phonon interaction for a cubic term in the anharmonic expansion (From reference (130)).

phonon mode. In the energy scan, a peak appears for the energy of the phonon. In a triple-axis experiment, the determination of the experimental resolution is complex and depends on the incoming neutron energy, on the monochromator and crystal analyzers, on the sample mosaic and on the dispersion relation. As an order of magnitude, the energy resolution is equal to about 1 meV on a thermal neutron source and 0.1 meV on a cold neutron source.

Recently, with the advent of third-generation synchrotron radiation X-ray sources, the inelastic X-ray scattering technique has been greatly developed [122]. It requires a tremendous resolution of the crystal analyzer: indeed for an incoming beam of the order 20 keV, energy resolution of the order 1.5 meV is achieved with a Si (11 11 11) crystal monochromator and analyzer in backscattering geometry, that is, an exceptional value of dE/E is equal to 10^{-7}. The principle of measurement is similar to the one explained for the neutron case, although the main advantages is that small sample size of the order 0.1 mm can be used, whereas inelastic neutron scattering requires sizes of the order of 1 cm.

2.3.2.2 Incoherent Inelastic Neutron Scattering

For polycrystalline samples, time-of-flight spectrometers are generally used. One measures the response function $S(Q,E)$ but averaged over all orientations, which makes it more difficult to interpret. On the other hand, this technique is very efficient for measuring the vibrational density of states. In effect, for a monoatomic system, within the incoherent approximation (i.e. for an incoherent scattering, where only the incoherent cross-section plays a role) it can be shown that the integral over Q of the measured signal is proportional to the vibrational density of state. In the case of a polyatomic system the measured signal is the generalized vibrational density of states (GVDOS), which is a sum of each single atomic partial vibrational density of state weighted by the incoherent scattering length divided by the mass of each constituent. This approach also applies, within some approximation, to the case of a purely coherent signal.

2.3.3
Beyond the Harmonic Approximation

So far, all the presented results have been obtained under the harmonic approximation. The energy of the system can be expressed as a Taylor expansion as a function of the atomic displacements, as shown in expression (2.47). In the harmonic approximation only the first quadratic term is retained. This is of course most of the time a valid approximation, for potential interactions are generally close to this quadratic dependence near their minimum. It is important to point out that the harmonic approximation is the only one allowing an exact calculation of the dynamical matrix and thus of phonon dispersion relations and eigenmodes determination.

$$E = E_0 + \frac{1}{2} \sum_{u,u',\alpha,\alpha'} \frac{\partial^2 E}{\partial u_{\alpha,j} \partial u'_{\alpha',j'}} u_{\alpha,j} u'_{\alpha',j'} + \ldots \tag{2.47}$$

When higher-order terms are taken into account, an exact solution to the dynamical problem is no longer possible and approximate solutions have to be derived. This is what is called anharmonic processes. An elegant way of dealing with anharmonic processes is using the Feynman diagrams. Indeed, it can been shown that anharmonic perturbations are equivalent to phonon–phonon "collisions" together with creation or annihilation of phonons. In the case of the cubic term in the Taylor expansion 3 phonons are involved, whereas in the case of the term of order 4, 4 phonons are involved. In this process, which is summarized Figure 2.20 for the cubic term, there is of course energy and momentum conservation: created phonons thus have energy and a wavevector different from the initial ones as illustrated in Figure 2.20. For periodic crystals changes of the wavevector can be accomplished modulo a vector of the reciprocal vector **G** so that the resulting wavevector writes:

$$\mathbf{k_f} = \mathbf{k} \pm \mathbf{k'} \pm \mathbf{G} \tag{2.48}$$

When the final wavevector is close to a Brillouin zone boundary, this can lead to a wavevector having a direction opposite to the two initial one: the phonon will propagate "backward," and thus limit the phonon thermal conductivity. This is called the Umklapp process, which plays a major role in the understanding of phonon thermal conductivity. This Umklapp process has been generalized to quasicrystals and used to interpret the temperature dependence of the thermal conductivity.

Using perturbation theory it can be shown that anharmonic interactions have two consequences on the observed phonon spectrum: (i) phonons have a finite lifetime. (ii) The energy of the phonon is displaced towards the low energy. Both effects can be measured experimentally. If the phonon has a finite lifetime τ, the amplitude of the mode will decay exponentially has $\exp(-t/\tau)$. This will produce a broadening of the observed excitation in the energy domain, which is the Fourier transform of the time domain. The Fourier transform of an exponential decay is a Lorentzian function, so that the observed phonon peak now has the shape of a Lorentzian with a half-width at half-maximum Γ equal to $1/\tau$, instead of being a delta peak in the harmonic case. More accurately, the observed signal is that of a damped harmonic oscillator, which can be approximated by a Lorentzian for small broadening.

It is interesting to give hand-waving arguments concerning the expected evolution of anharmonic effects as the temperature is increased. When the temperature is increased, the population of the different energy levels increases and thus the probability of phonon–phonon interactions increases also. One thus expects the observed signal to broaden as the temperature increases. In the meantime, as the temperature increases the interaction potential generally becomes slightly softer so that a displacement of the phonon mode towards lower energy is expected. In general, the anharmonic broadening is rather small and a high-resolution setup is required for a detailed measurement of the anharmonic effect.

We illustrate the above results with the case of the CaF_2 crystal. A particularly large anharmonic effect has been observed in this case, as illustrated in Figure 2.21. As the temperature is increased a clear broadening is observed, together with a change of the

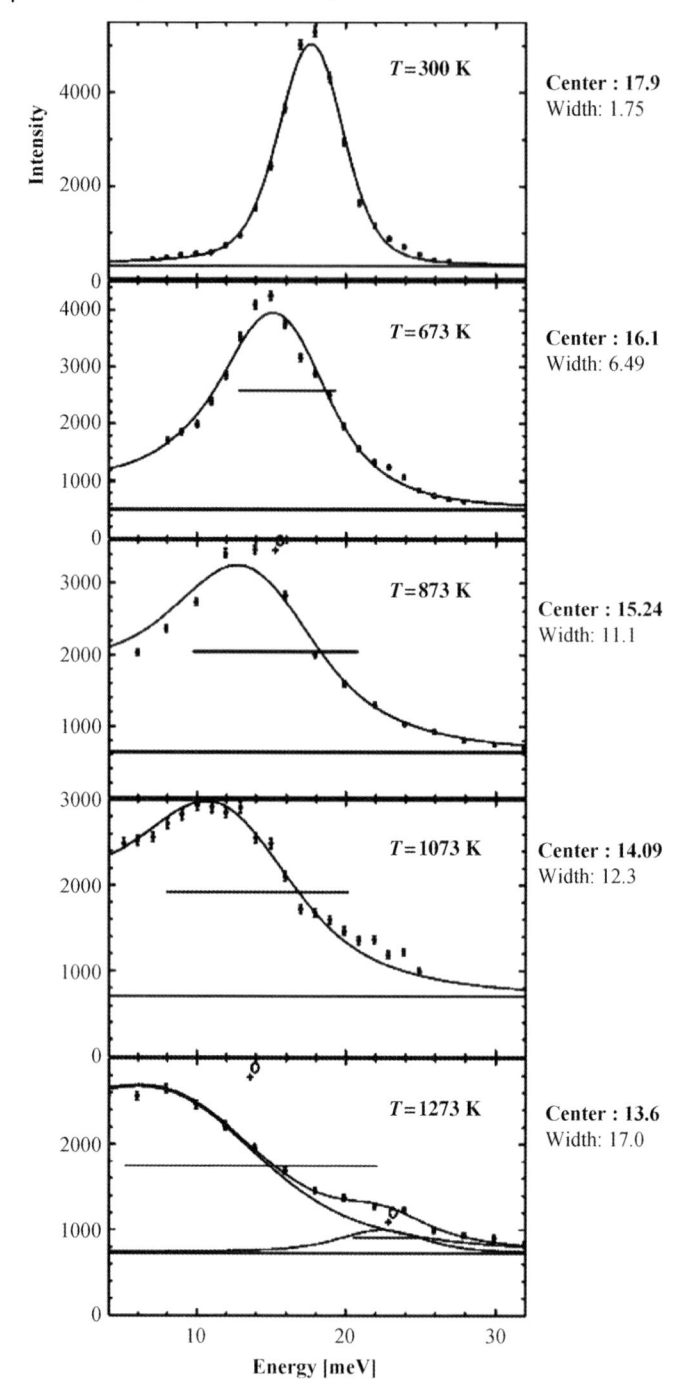

Figure 2.21 Evolution of the inelastic scattering signal measured for an optical phonon as a function of the temperature in CaF_2. The position and width of each fitted excitation is given on the right part of the panel (From reference [132]).

shape of the measured signal that is characteristic of a damped harmonic oscillator response [119]. The position, together with the width (in meV) is given on the right side of each panel: as expected when the temperature increases, the width is increasing and the position of the excitation is lower in energy.

Although it is not an anharmonic effect, it is interesting to briefly discuss the effect of disorder on the phonon modes. The "simplest" disorder in that respect is the so-called mass defect, which has been intensively studied. If we consider a monoatomic crystal, one atom can be randomly substituted by an isotope: the interatomic potential interaction is in principle not changed, but the mass is changed randomly on the periodic lattice. If the concentration of the substituted atom is small, a perturbative calculation can be carried out. In that case the mass defect is an isolated defect, and because it has a different mass it will vibrate as an isolated oscillator with its own frequency. This localized vibrational mode is seen as a dispersionless horizontal line in the dispersion curve. If the substituting atom is heavier than the atom it is replacing, the localized mode occurs at rather low energy (remember that roughly the energy scales as $1/\sqrt{m}$) and will interact with the acoustic branch. When the dispersionless branch crosses the acoustic branch, the two modes interact and resonate, giving rise to an anticrossing scheme (see below). This mass defect is thus a new channel for a finite lifetime of the phonon. Indeed, simulations and measurements show that the acoustic phonon lifetime is reduced.

If the substituting atom is lighter, it will vibrate at high energy, outside the energy range of the dispersion curve.

Similar arguments can also be applied for chemical disorder, a frequently encountered case in CMA. The chemical disorder certainly will reduce the phonon lifetime, or will "broaden" the dispersion curve with new branches.

In summarizing the previous paragraphs we have shown that phonons can be measured by inelastic neutron or X-ray scattering. This allows the determination of the dispersion curve, but also to get some insight into the nature of the modes. In particular, the broadening of the observed signal is directly related to the finite lifetime of the phonon. Finally, it is important to recall that a lot of information is also contained in the intensity distribution of the measured $S(Q,E)$ function, since it is indirectly related to the eigenvectors of the phonon modes. This is a particularly important point in order to determine the influence of long-range aperiodic order on vibrational properties. It also constitutes a very severe test of any modeling.

In the following we present some experimental results. We will first present quasicrystal and their approximants and then some results obtained in thermoelectric compounds in particular for cage compounds.

2.3.4
Phonons in Quasicrystals and their Approximants

2.3.4.1 The Zn$_2$Mg Laves Phase
Although the Laves phase is not exactly a quasicrystal approximant, it is a phase of moderate complexity whose atomic structure can be described by a periodic packing of the Friauf polyhedron, a building block common to the Franck–Kasper-type

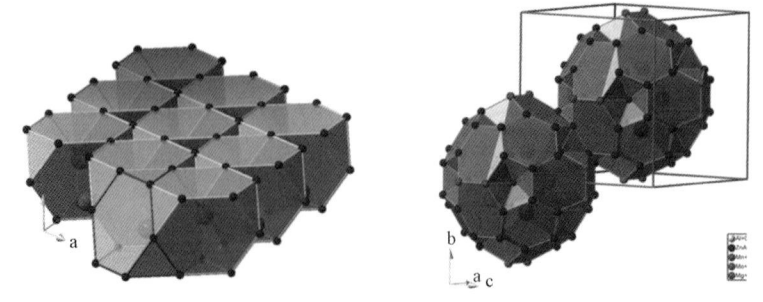

Figure 2.22 Structure of the Zn$_2$Mg Laves phase (left) and the Bergman Zn-Al-Mg phase (right). The "soccer ball" cluster in the Bergman phase is obtained by the union of 20 Friauf polyhedra.

quasicrystal or approximant such as Al-Li-Cu or Zn-Al-Mg. The structure is characterized by the packing of a large (Mg) and a small (Zn) atom. The Friauf polyhedron consists of a central Mg atom, surrounded by 12 Zn atoms located at the vertices of a truncated tetrahedron and 4 Mg atoms that cap their hexagonal face. A layer of the Zn$_2$Mg phase is displayed on the Figure 2.22, the next layer being obtained by mirror symmetry. The unit cell is hexagonal and contains 12 atoms. This leads to 36 branches, which can be dealt with by DFT calculations.

The phonon dispersion relations have been measured by inelastic neutron scattering from single grains [101] along high-symmetry axes [123]. Excitations have been measured along the main high-symmetry directions Δ, T, T′ and Σ of the hexagonal Brillouin zone, parallel to the (001), (110) and (100) directions of the hexagonal reciprocal lattice, respectively. To enhance the acoustic part of the signal, measurements have been carried out for excitations originating from the strong (006), (220) and (300) Bragg reflections.

When considering general features characterizing the behavior for all the measured acoustic excitations, only 2 different behaviors are observed among the four acoustic modes studied.

The first one is represented by the dispersion curve of transverse acoustic (TA) modes propagating along the (T) direction and polarized along the (001) one. In the dispersion curve shown in Figure 2.23, full symbols correspond to the acoustic mode and open symbols to the optical excitations. The acoustic dispersion curve rapidly bends over and departs significantly from linearity, becoming almost flat at the Brillouin zone boundary. At the same time, the intensity of the optical excitation located around 2.7 THz increases, while the intensity of the TA mode vanishes progressively. This is exemplified in Figure 2.23 (right panel), which shows the evolution of the normalized integrated intensity (expression (2.45) and (2.46)) for both excitations. As previously explained, the intensity should remain constant as long as the signal is purely acoustic. The strong intensity variation is reminiscent of an "anticrossing" of two branches. When considering the width of the acoustic mode as a function of the wavevector, we observe a slight increase, going as q^2 (Figure 2.19a). The width of the TA excitation remains, however, small when compared to what is observed in quasicrystals.

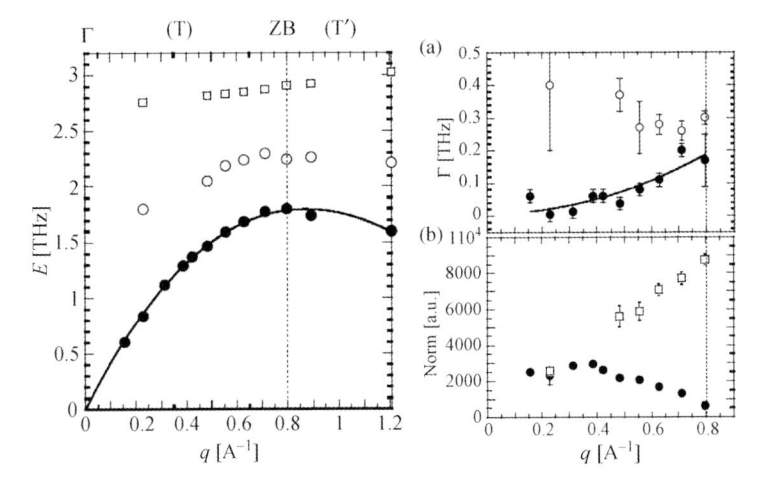

Figure 2.23 Left panel: Dispersion relation for modes propagating along the TT′ direction and polarized along c^*. The acoustic excitation is shown with a full circle; open symbols are for the two observed optical modes. Error bars are smaller than the symbols size. The solid line is a guide for the eye. Right panel: (a) Evolution of the width of the acoustic excitation (full circle) and of the optical one (open circle) of Figure 2.18. The acoustic excitation width increases as q^2 (solid line). (b) Same for the evolution of the normalized integrated intensity. There is a clear intensity exchange between the acoustic and optical mode.

The second behavior is represented by the dispersion of TA modes propagating along the direction Δ and polarized along the (110) direction. Unlike the previous case, the dispersion of the acoustic mode only slightly departs from linearity up to relatively high energy values, of the order of 2.6 THz, and for wavevectors values up to 0.8 Å^{-1}. For q values beyond the first Brillouin zone, the measured transverse excitation is no longer an acoustic mode but an optical mode. However, the excitation remains of acoustic character, as shown by the evolution of its normalized integrated intensity: the norm is almost constant up to 0.6 Å^{-1}. Two optical excitations have also been measured. The evolution of the width of the acoustic mode as a function of the wavevector shows a broadening rate going as q^4.

These experimental results have been compared to simulations using two different methods to compute both the dispersion relation and the intensity distribution, which is a much tougher test than the dispersion relation alone. In the first method, the dynamical structure factor can be obtained from the eigenvalues of the dynamical matrix determined in harmonic approximation. In principle, this method can be used also with classical interaction potentials, but in our case we determine the dynamical matrix from *ab initio* forces, which should result in the best possible accuracy. In a second method, the dynamical structure factor is interpreted as a certain correlation function, which can be measured in a molecular dynamics (MD) simulation. This requires much larger samples, prohibiting the direct use of *ab initio* methods. Classical interaction potentials are therefore necessary. For the best possible reliability, these potentials are fitted to reproduce *ab initio* data, however. The main advantage of this approach is that it does not rely on the harmonic approximation. Moreover, the possibility of using larger samples also has the advantage that structural or occupational disorder can be taken into account, which may prove useful for the more complex structures, as will be shown later.

The Laves phase of $MgZn_2$ has 12 atoms in the primitive unit cell, so that *ab initio* calculations can be performed relatively easily, using the VASP code with the PAW method. The relaxed unit cell is promoted to an orthorhombic supercell with 48 atoms and relaxed again. The dynamical matrix in harmonic approximation is then determined by displacing one atom at a time (by 0.05 Å), and computing the resulting forces on all other atoms. This has to be repeated for 12 independent displacements and is most conveniently done by using the PHONON package, which generates the configurations for VASP with the required displacements, and computes and diagonalizes the dynamical matrix from the VASP forces. The response function can then be calculated using expressions (2.42), (2.43) and (2.44).

In order to account for the instrumental resolution, $S(Q, E)$ has been convoluted with a Gaussian, whose width was chosen so as to obtain the best fit with experiment. The final comparison is shown for the direction $(\xi, \xi, 6)$ in reciprocal space (Figure 2.24). This means that the scans in Q-space are started at the Bragg peaks (0 0 6) and proceed along the directions $q = (\xi \xi 0)$. The graphs show the results for selected values of ξ; the corresponding magnitudes of q are indicated in the upper right corner of each subgraph. To obtain this comparison, a constant background was added to the calculated intensities, which were also uniformly rescaled for each direction. Furthermore, the energies had to be rescaled by a constant factor of 1.14. This seems to indicate that the sound velocities are not accurately reproduced by the *ab initio* calculations. Apart from this, the agreement is extremely good, not only for the overall dispersion but also for the intensity distribution. In particular, the bending of the dispersion and the intensity transfer from the acoustic to the optic mode is well reproduced for the TT' direction, as shown in Figure 2.24.

Experimental data have also been compared with EAM potentials calculations using a molecular dynamic simulation, which, as already pointed out, does not rely on the harmonic approximation. Potentials of EAM type have been fitted to reproduce forces, energies and stresses computed *ab initio*. This so-called force-matching method, which is implemented in the potential-fitting code potfit, ensures that even classical potentials make best possible use of quantum-mechanical information. In Figure 2.25, the dynamical structure factor is shown in an intensity diagram for the TT' and Δ directions. Again, there is a good qualitative agreement both for the dispersion relation and for the intensity distribution. However, since the EAM potentials are a bit too "soft" the energies have been rescaled proportionally. These results thus validate the use of EAM potentials to compute the lattice dynamics of larger samples either to include disorder or to study structurally complex phases such as quasicrystals.

From the *ab initio* simulation it is also possible to compute the eigenmodes and look for the pattern of vibrations in the unit cell. Although there are only 12 atoms per unit cell, this is already quite complex. As a general trend, the light Mg atoms are participating in high-energy modes, whereas in this high-energy range Zn atoms do

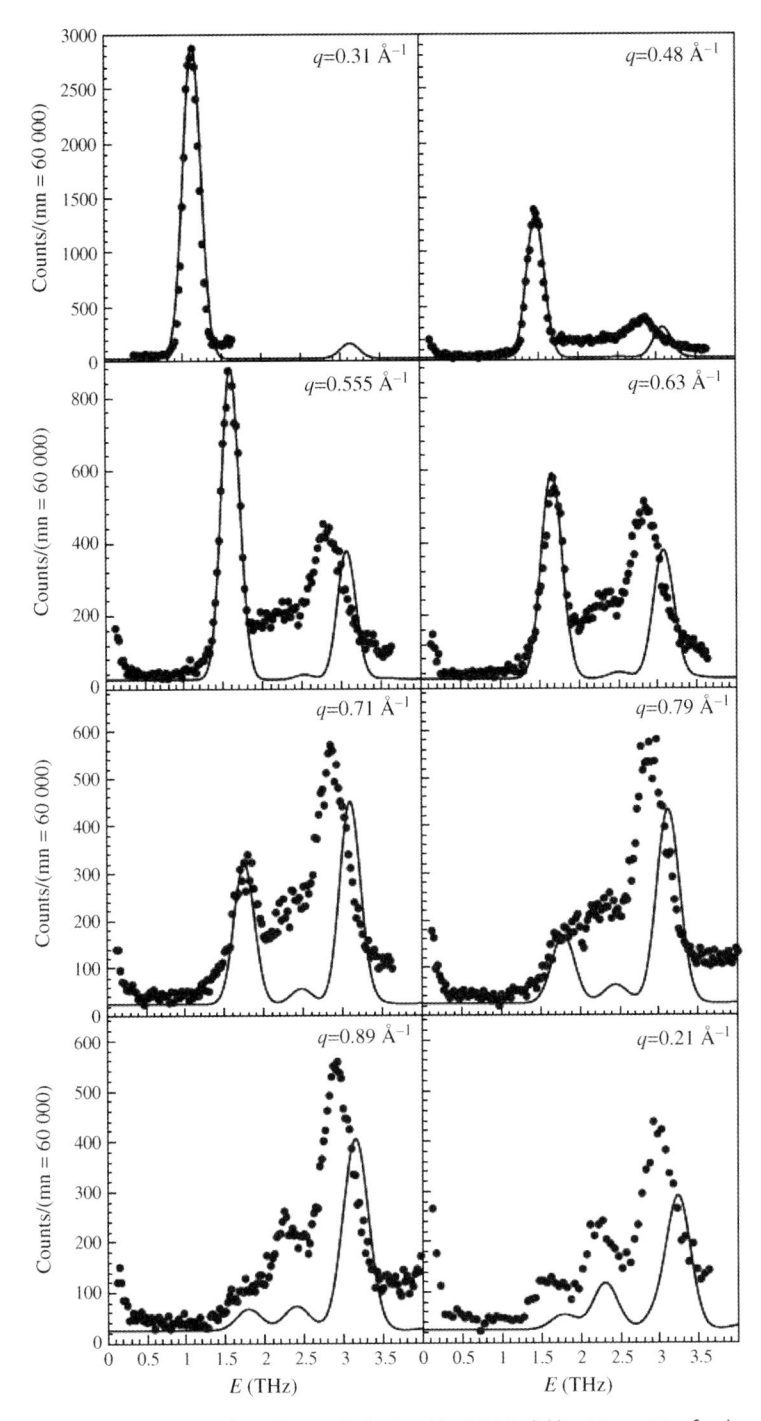

Figure 2.24 Measured (circles) and calculated by DFT (solid line) intensities for direction ($\xi\,\xi\,6$), (TT') for eight different q values. The error bars in the measurement are smaller than the symbol size.

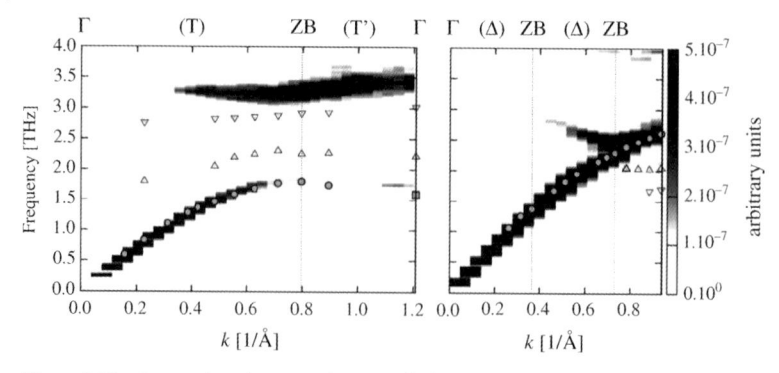

Figure 2.25 Comparison between data (symbols) and the simulation using EAM potentials and molecular dynamic along the TT' and Δ directions. The intensity transfer between the acoustic and the optic excitation along the TT' direction is well accounted for.

not move much. A detailed analysis has to be conducted though, so as to extract a clear picture of vibrations.

2.3.4.2 The i-Al-Pd-Mn Icosahedral Quasicrystal

The icosahedral Al-Pd-Mn phase was the first quasicrystal that had a high structural quality [123] and could be grown as large single crystals of several cm^3 [124, 125]. It was thus in this system that the first detailed study of the lattice dynamic could be achieved, although previous measurements were carried out in the i-Al-Li-Cu [126, 127] and i-Al-Cu-Fe phases [128, 129].

Figure 2.26, left panel, displays a typical measurement of the scattering law $S(\mathbf{Q}, E) = S(\mathbf{Q}_{Bragg} + \mathbf{q}, E)$ carried out in the transverse geometry, and going away from a strong twofold Bragg reflection [130, 131]. Close to the Bragg peak there is a well-defined acoustic signal, whose width is limited by the instrumental resolution. Up to $q = 0.3$ Å^{-1}, this signal remains resolution limited and its normalized intensity, as defined in 2.45 remains constant, as a signature of a purely acoustic mode. This is what is expected in the long-wavelength limit also for quasicrystals. Above $q = 0.3$ Å^{-1}, the signal broadens very rapidly. In the meantime, the normalized intensity is no longer exactly constant, and slightly increases, as a signature of mode mixing. Nevertheless, the signal was interpreted as a single damped harmonic oscillator (see Section 3.3) whose width is increasing. This is already clearly visible at $q = 0.45$ Å^{-1} for instance, where the signal is significantly broader than at $q = 0.15$ Å^{-1}. The width is reported as open symbols on the right panel, which displays the extracted dispersion relation, with the strong twofold reflections chosen as zone center. Up to $q = 0.7$ Å^{-1} the signal can be considered to be mainly acoustic in character, although their width is quite large at this point. At higher q, the signal can be analyzed as a set of broad dispersionless excitations centered at 7, 12, 16 and 24 meV. The width of these excitations is of the order 4 meV and certainly corresponds to the mixing of several excitations. This can be viewed in a similar way to what is represented for the Fibonacci chain in Figure 2.17, where the signal around 15 and 20 meV will be broad as a result of the superposition of several modes.

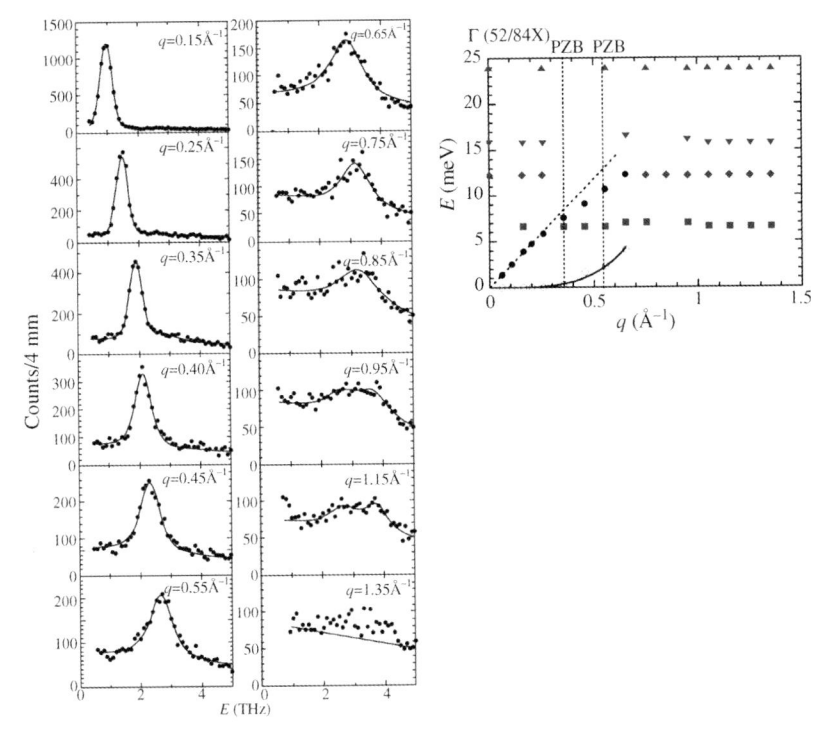

Figure 2.26 Experimental study of the lattice dynamics of the *i*-Al-Pd-Mn phase. Left panel: successive constant q-scans measured around the strong twofold reflection. The distance from the Bragg peak is indicated inside each panel. Right panel: Dispersion relation extracted from different measurements. Filled black symbols are for the TA acoustic modes, gray symbols for optic like excitations. The open circle stands for the FWHM of the acoustic excitations, the solid line is a q^4 fit. The vertical dashed lines indicate the first pseudo-Brillouin zone boundaries.

A useful concept, already introduced, is that of pseudo-Brillouin zone boundaries in quasicrystals, as proposed by Niizeki [132, 133]. Although the quasicrystal is nonperiodic, the long-range quasiperiodic order results in a distribution of strong Bragg peaks or strong Fourier component. Phonons are no longer Bloch waves, but there is some quasiperiodic distribution of the eigenvectors. This is true for acoustic modes, whose signature will be identical around strong Bragg peaks. The concept of pseudo-Brillouin zone boundary (PBZB) can be derived within a weak-coupling-type theory [133–137]. Within this framework one can show that the acoustic dispersion curve should display a gap opening at pseudo-Brillouin zone boundaries (PBZB), whose position in reciprocal space is defined by $\mathbf{q}_{PBZB} = \mathbf{Q}_{Bragg}/2$, where $\mathbf{Q}_{Bragg}$ is a reciprocal lattice vector of the quasicrystal. Although the reciprocal space is densely filled in a QC, only the strongest Fourier components are relevant, the width of the gap being proportional to the amplitude of the structure factor $F(\mathbf{Q}_{Bragg})$. At the PZB, the phonon acoustic wave is "Bragg reflected" as in a standard Brillouin-zone boundary. To select the main zone boundaries we have to consider the intensity of the Bragg peaks structure factors. Unlike the case of electrons, where the important Bragg planes are those with a length close to the Fermi sphere radius, in the case of phonons one has to consider rather small Q_{Bragg} wavevectors in the range 0 to 2 Å^{-1}, or q_{PBZB} in the range 0 to 1 Å^{-1}, since the acoustic regime only holds for q smaller than 0.6 Å^{-1}. These PBZB are placed around the strong Bragg peaks acting as zone centers and are displayed as vertical dashed line in Figure 2.26. This approximation is only valid in the vicinity of the zone center, where the acoustic mode is well defined.

As seen in Figure 2.26, the dispersionless modes at 7 and 12 meV do correspond to the crossing of the transverse acoustic mode with the PZB: they can thus be viewed as the trace of the mode interaction occurring at the PZB. The higher energy dispersionless excitations correspond to the crossing of longitudinal acoustic excitation with the PZB.

From the broadening rate and the slope of the acoustic excitation one can extract a phonon lifetime and more interestingly a phonon mean free path. At the limit of the acoustic regime, both the mean free path and the wavevector points towards a characteristic length of the order of 10 Å [138].

Similar measurements have been carried out for the i-Zn-Mg-Y quasicrystal [139]. In this case a similar situation has been observed with acoustic modes showing an abrupt broadening for wavevectors larger than 0.3 Å^{-1}. However, the broadening rate is smaller and goes as q^2 instead of q^4 in the i-Al-Pd-Mn case. Moreover, for the same wavevector the broadening is significantly smaller in i-Zn-Mg-Y than in i-Al-Pd-Mn, leading to a mean free path of 20 Å. The atomic structure is different, but also the structural quality is higher in Zn-Mg-Y, which may explain the observed differences.

2.3.4.3 The i-Zn-Mg-Sc Quasicrystal and its 1/1 Zn-Sc Approximant

The most detailed study so far has been conducted in the Zn(Mg)Sc system [140]. Indeed the discovery of a binary Cd-Yb quasicrystal by the group of An-Pang Tsai in 2000 [141] was a real breakthrough in the field that led to the first accurate structural determination of a quasicrystal. The stable binary i-Cd-Yb [141] icosahedral (i-) phase and the isostructural i-Zn-Mg-Sc [142] phase are particularly interesting in that respect since both a quasicrystal and a periodic approximant can be synthesized, with almost the same chemical composition [143, 144]. As shown recently [145], their structure can be described by a packing of a large rhombic triacontahedral (RTH) unit with a diameter of about 1.52 nm and containing about 158 atoms arranged on successive shells (Figure 2.27). In both the quasicrystal and the 1/1 approximant the RTH units are connected along their twofold axes by shared faces, and along their threefold axes where they overlap. In the 1/1 approximant the RTH are located on the vertices of a BCC lattice (lattice parameter $a = 1.57$ and 1.38 nm for Cd–Yb and Zn–Sc, respectively). In the QC 94% of the atoms are parts of the RTH units that are located on vertices of a quasiperiodic network, with two small structural units filling the gaps between the clusters. A section of the quasicrystalline structure perpendicular to a fivefold axis is shown in Figure 2.27: only the cluster centers are represented. Clusters are arranged in a hierarchical way forming a cluster of clusters (orange decagonal disk), which in turn forms a cluster of "cluster of clusters" at an

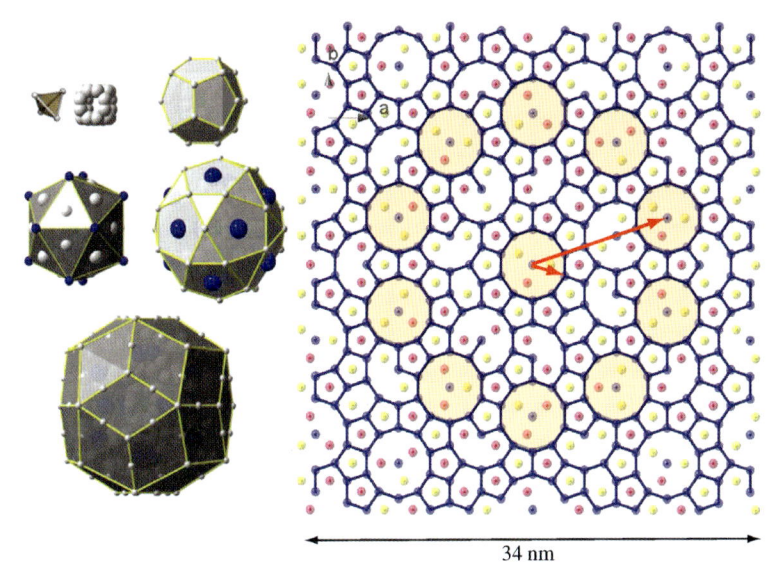

Figure 2.27 Structure of the CdYb icosahedral quasicrystal. Left panel: successive shells of the cluster found both in the QC and approximants. White and blue color stand for Cd and Yb atoms respectively. Right panel: 5 fold section showing the cluster distribution. Only the clusters centers have been represented. Yellow and pink colors are for cluster centers slightly above (or below) the plane. The hierarchical packing of the clusters is shown with the orange disks.

inflated scale. This hierarchical organization, typical of the quasiperiodic long-range order, has been invoked as an important characteristic for physical properties [146–148].

The QC and its approximant thus offer a unique possibility of comparing the respective effect of the short-range order (RTH units) and long-range order (periodic versus quasiperiodic) on physical properties and on their lattice dynamics. The similarity and differences of the QC and its approximant are also visible in reciprocal space, as shown in Figure 2.28. Whereas the intensity distribution and the position in reciprocal space of the strongest Bragg peaks are similar, there are significant differences for weaker reflections, and of course a much larger number of Bragg peaks in the case of the quasicrystal.

The inelastic scattering function $S(\mathbf{Q}, E) = S(\mathbf{Q}_{Bragg} + \mathbf{q}, E)$ has been measured by inelastic neutron scattering and X-ray scattering for both the QC and its 1/1 approximant. Some results are reported in Figures 2.29 and 2.30 for transverse excitations. The extracted dispersion relation has an overall similar shape in both cases. In particular, there is an acoustic branch, separated by a pseudogap from an optical excitation located around 14 meV. However, significant differences are also observed. In particular, the width of the pseudogap is larger and better defined in the 1/1 approximant than in the quasicrystal: this is visible in the dispersion relation but also in the intensity distribution displayed in Figure 2.30: the scan recorded at

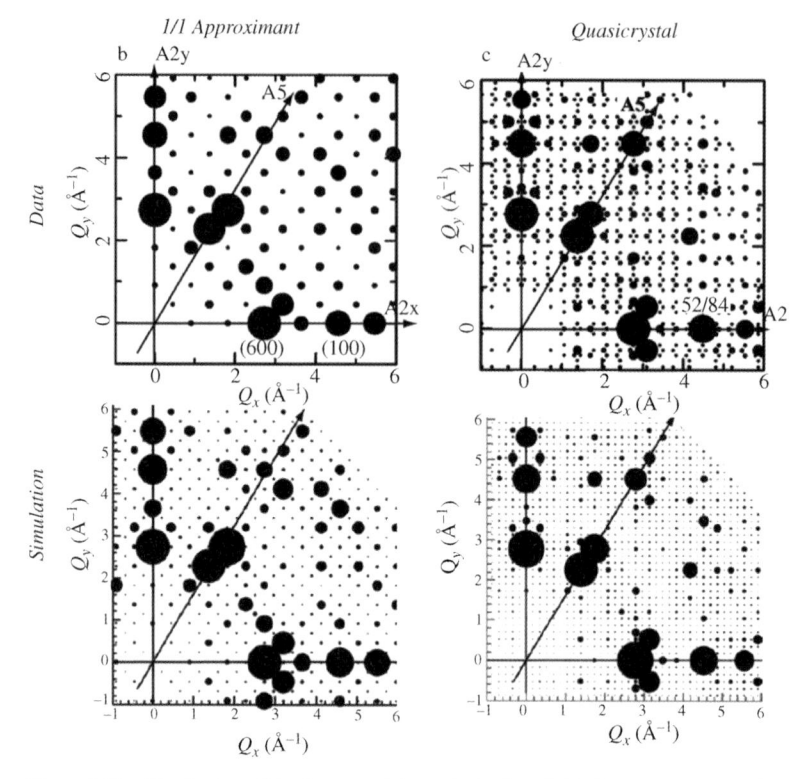

Figure 2.28 Distribution of the Bragg peak intensity in the 1/1 Zn-Sc approximant and Zn-Mg-Sc quasicrystal. The top panels correspond to the experiment, the bottom panel to the simulation.

$q = 0.54$ Å^{-1} is a good example of the differences, where the two excitations are much better separated in the 1/1 approximant. The second difference is a slightly lower energy for the 1/1 approximant optical modes than for the quasicrystal. This is the first clear evidence of a difference in physical properties and in particular lattice dynamics between an approximant and a quasicrystal.

These differences can be qualitatively interpreted using the concept of PBZB previously introduced in the case of the quasicrystal. Indeed, as already explained the important PZB have rather small wavevector and corresponds to Bragg peaks having a high Q_{per} component. As a result, the single ZB in the 1/1 approximant is replaced by two PZB shown as vertical dashed lines in the figure. This means that the "efficiency" of the phonon wave Bragg reflection is smaller in the QC than in the 1/1 approximant. This is also confirmed by the intensity of the corresponding Fourier components, which is smaller in the QC than in the 1/1 approximant. As a result, the pseudogap between the acoustic and optical excitation is smaller in the QC.

Finally similarly to other quasicrystals one observes a rapid broadening of the acoustic excitations for q larger than 0.3 Å^{-1} in both cases, although excitations in the QC are slightly broader than the one in the 1/1 approximant.

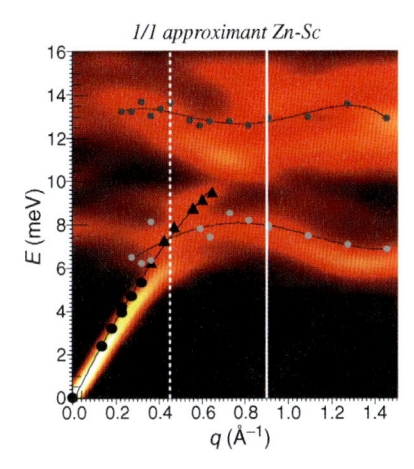

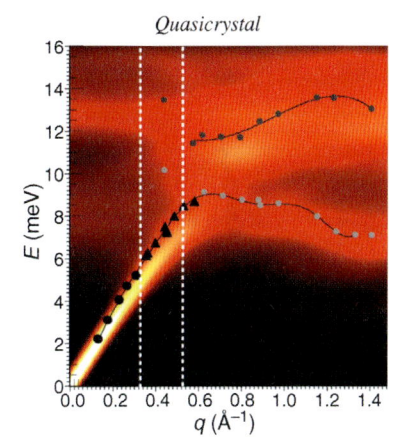

Figure 2.29 Transverse excitations: comparison between the measured dispersion relation (symbols) and the simulated response function $S(Q,E)$ (temperature color-coded) in the 1/1 approximant (left panel) and the quasicrystal (right panel). The figure shows the intensity distribution of the simulated response function on a temperature color-coded scale. The experimental positions of the excitations, as measured by neutron inelastic scattering, are shown by symbols: the black closed circles and triangles stand for the acoustic signal, whereas other symbols correspond to optical excitations. Black triangles indicate the q positions for which the normalized acoustic intensity is no longer constant. Vertical white dashed lines indicate the position of the (pseudo) Brillouin zone boundaries. The simulation reproduces both the general trend and the differences observed between the QC and its approximant.

The above experimental results have been compared with simulations on realistic atomic models. This is possible thanks to the detailed knowledge of the atomic structure. To model the interactions in the Zn–Sc system, oscillating pair potentials that have been fitted against *ab initio* data [149] have been used and are shown Figure 2.31. Note that the minimum of Sc–Sc distances is larger than the Zn–Zn, as expected for those two atoms with different sizes. The right panel of Figure 2.31 illustrates the quality of the force-matching fitting using oscillating pair potentials.

A difficulty encountered in the modeling of the 1/1 approximant and the quasicrystal is the orientational correlations of the central tetrahedra of the atomic cluster (see Figure 2.27). Indeed in the 1/1 approximant whereas tetrahedra have a disordered orientations at room temperature, an ordering takes place below 150 K leading to a superstructure [150, 151]. Structural analysis of the 1/1 ordered low-T phase showed that the tetrahedron orientation induces a strong distortion of the successive icosahedral shells, breaking their icosahedral symmetry [151]. To model the room-temperature random tetrahedron orientations, a larger supercell has been used, containing 8 clusters and with lattice parameters equal to $a_1 = 2a$, $b_1 = c_1 = \sqrt{2}a$, where a is the lattice parameter of the 1/1 approximant and equals to 13.76 Å. Tetrahedra were randomly placed at each of its 8 cluster centers, followed by a molecular dynamics annealing of the structure at room temperature, followed by quench to $T = 0$ K. Introducing this disorder has been found to be crucial in order to have a reasonable comparison with experimental data.

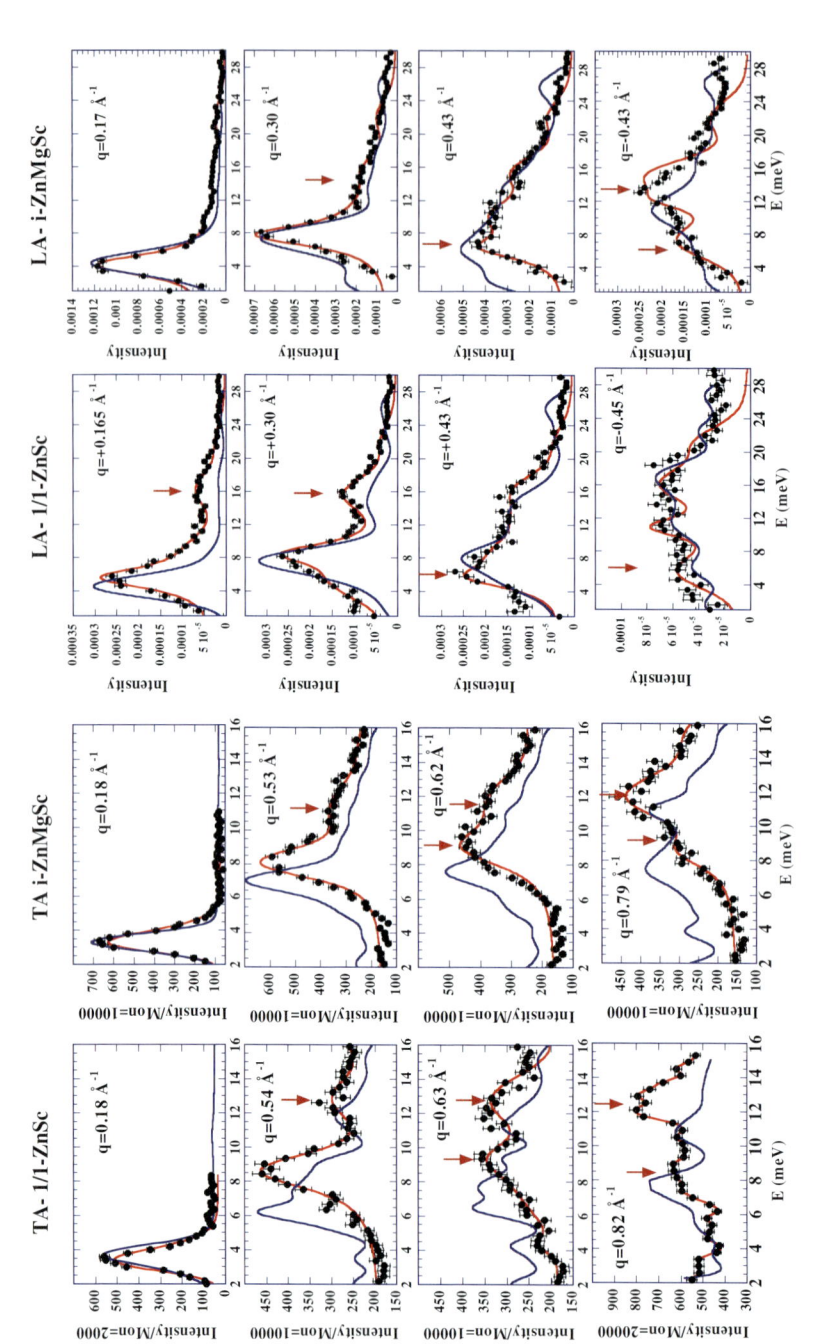

Figure 2.30 Comparison between the experiment and the simulation for a few representative constant-Q energy scans. Circles correspond to the measurement, red lines are fit to the data (from which dispersion relations are extracted), blue lines show the simulated response function $S(Q, E)$. The q value is indicated in insert. Left panels: transverse acoustic modes measured in the 1/1 approximant and the quasicrystal by inelastic neutron scattering. Right panels: longitudinal excitations measured in the 1/1 approximant and quasicrytsal by inelastic X-ray scattering (For clarity the elastic contribution has been subtracted from the data). The simulated data have been convoluted by a Gaussian with a FWHM of 1 meV and 3 meV for neutron and X-ray data, respectively. The energy position of the two transverse calculated spectra at 0.18 Å^{-1} have been artificially translated at higher energy in order to illustrate the good reproduction of excitation whose width is limited by the instrumental resolution. Notice the visible broadening, as q increases in transverse geometry. The intensity distribution is well reproduced by the simulation. Note in particular the good reproduction of the longitudinal scans at + or − 0.43 Å^{-1}. Red arrows point to remarkable similarities and differences.

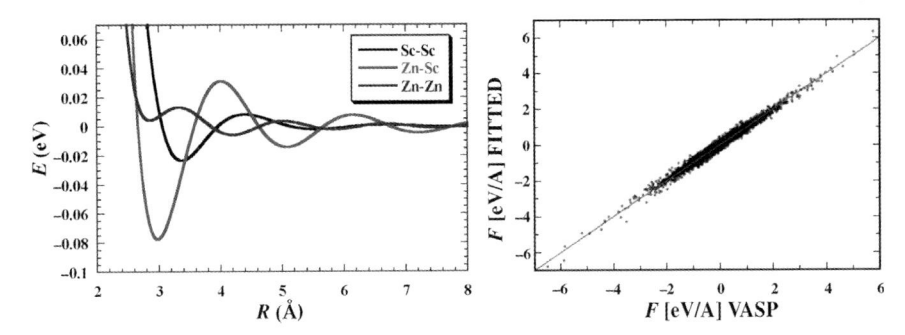

Figure 2.31 Left panel: Radial dependence of the energy of the oscillating pair potentials used in the simulation of the Zn-Sc approximant and QC dynamics. Right panel: Force matching fit, showing the forces calculated with *ab initio* (VASP) as a function of the one calculated with the oscillating pair potentials. There is very good agreement.

To compute the dynamical response of the QC, we needed a unique realization of the structure with periodic boundary condition. A 3/2 cubic approximant, which contains 32 clusters in the unit cell, and has lattice parameter equal to 36.13 Å has been used. A binary decoration Zn–Sc, based on the related i-CdYb atomic structure [145] and on the ternary 2/1 Mg-Sc-Zn [152] has been used. The modeling was achieved by positioning the RTH units on the vertices of the so called canonical cell tiling [153] of the 3/2 approximant with a decoration procedure as described i [154]. Ambiguous atomic sites positions were determined by total-energy minimization. As for the 1/1 approximant, the tetrahedral orientations were obtained by a molecular dynamic annealing followed by a quench. Induced deviations from icosahedral symmetry of the successive shells around the tetrahedron are, as for the 1/1 approximant, a crucial parameter. The resulting model has a composition $Zn_{2528}Sc_{456}$, contains 2984 atoms per unit cell and presents a diffraction pattern which compares well to the QC one (see Figure 2.28).

Using these models and the fitted pair potentials the inelastic response function $S(\mathbf{Q}, E)$ has been calculated either in the harmonic approximation and by direct diagonalization of the dynamical matrix, or from atom trajectories generated by room-temperature molecular dynamics, using the method described in ref [155]. The latter approach does not rely on the harmonic approximation, which might have been an issue due to suspected shallow minima in the energy landscape related to tetrahedron librations. However, the two approaches did not show any significant differences.

Figures 2.29 and 2.30 display the calculated $S(\mathbf{Q}, E)$ in the transverse geometry for both the approximant and the quasicrystal. The calculation is temperature color coded, and does not include the term $n(E)/E$ (where $n(E)$ is the Bose occupation factor), so that the acoustic mode presents a constant intensity. In both figures, the simulation has been convoluted with a Gaussian distribution with full width at half-maximum intensity (FWHM) of 1 meV. Although the calculated transverse acoustic excitations are slightly too soft, the main features are nicely reproduced by the calculation: acoustic modes and pseudogap, high-energy optical excitations, low lying optical excitation in longitudinal geometry. Even the detailed differences between the QC and the approximant are reproduced: a larger pseudogap in the approximant than in the QC, the interaction with an optical excitation around 7.5 meV at the BZB in transverse geometry for the approximant, lower positions of the optical bands in the QC than in the approximant.

Not only are the dispersion relations matched by the calculation, but also the intensity distribution is qualitatively well reproduced. To illustrate this point, simulated spectra have been superimposed on the experimental data for a few characteristic scans in Figure 2.30. Neutron and X-ray simulated spectra have been convoluted with a Gaussian distribution (1 and 3 meV FWHM, respectively), to account for the effect of instrumental resolution. Although transverse modes are too soft, the intensity distribution is well reproduced by the calculation. In particular, the larger gap observed in the approximant is well accounted for. For longitudinal excitations, the intensity distribution is also well reproduced. Note in particular the good reproduction of the different intensity distribution at $+$ and -0.45 Å^{-1} for the mode located at 6 meV. The broadening of TA mode is also well reproduced, demonstrating that it corresponds to mixing with low-lying excitations, since the simulation is carried out in the harmonic approximation.

We insist on the fact that the very good quantitative agreement for the intensity distribution is exceptional. Indeed, remembering expressions (2.43) and (2.44), this means that the simulation reproduces correctly the eigenvectors or the pattern of vibrations of both the QC and its 1/1 approximant. In fact, there are extremely few examples, even for simpler systems, for which such a quantitative comparison has been achieved.

This has been only possible because a detailed structure analysis is available on the one hand and because the oscillating pair potentials are a good approximation of the Hamiltonian on the other hand. This is certainly because (i) there is a strong size effect with a large and small atom, (ii) the oscillations of the pair potential are a good approximation for the electronic stabilization and in particular the so-called Friedel oscillations.

Now that the simulations are validated on a very firm basis, it is possible to analyze in more details the results. In particular, it is interesting to check the eventual role of clusters; are they clusters modes? Is the quasiperiodic order bringing in a signature? What about critical modes? For this last question a 3/2 approximant might be too small and larger-scale simulations are probably necessary. But one of the main challenges is finding new tools for the simulation analysis. Indeed, the simulations are done on a periodic lattice, although with a rather large unit cell. This means that all wavefunctions are described as plane waves, a tool that might not be adequate to grab the physics of quasicrystals. We must admit that theoretical tools are still missing in that respect.

Some preliminary analysis is, nevertheless, available. A first simple piece of information can be gained in computing the partial density of states [156] projected on the different shells of the atomic cluster. An example is given in Figure 2.32. The partial density of states for the tetrahedral central cluster displays low-energy modes, certainly related to the possibility of having different orientations. This might be related to the phason modes that are also present in these QC phases [157, 158]. It is

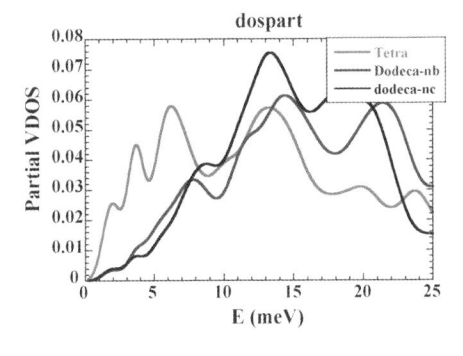

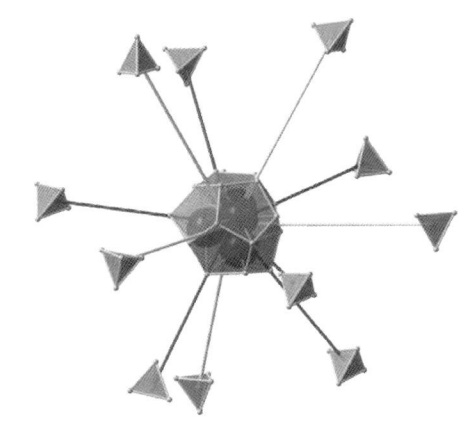

Figure 2.32 Left panel: partial vibrational density of states for the tetrahedral and the dodecahedra. Sites on the dodecahedra are split in two families corresponding to the bonding scheme. Right panel: first dodecahedra shell around the tetrahedron as obtained in the simulation. Notice the strong distortion induced on the dodecahedron. The neighboring cluster-center tetrahedra are shown together with the twofold and threefold bonds.

also interesting to look for the detailed configuration of the clusters as they appear in the simulation. The right panel of Figure 2.32 shows the first shell around the tetrahedron. As expected, the tetrahedron induces a strong distortion of the first dodecahedral shell that is visible on the figure and which is then also "transmitted" to the next shell. The orientation of the neighboring tetrahedra, connected along 3- **and twofold** axis is also displayed. That, locally, the clusters no longer have a quasicrystalline structure is one of the surprising results already seen in the structure analysis and in the low-temperature phase of the 1/1 approximant [151]. Of course, the overall icosahedral symmetry is restored by the different orientations of the clusters.

To conclude this section we would like to make the link with the following section on thermoelectrics. In thermoelectrics, a poor phonon heat transport or phonon glass-like state is a crucial parameter [159]. Most quasicrystals have been found to be poor thermal conductors [160–162], which has been attributed to a generalized Umklapp process [163] or to a hierarchically variable-range hopping on clusters [147]. The present simulations and results show that there is a strong acoustic-optic mode mixing for wavevectors larger than $0.3 \, \text{Å}^{-1}$. Indeed, the observed broadening of the acoustic excitation, which is certainly a signature for the low thermal conductivity, has been attributed to a mode mixing in the simulation. We suggest that this mode mixing is another route for producing low thermal conductivity, as already pointed out [138, 164].

2.3.5
Phonons in Cage Compounds and Thermoelectricity

The "tailoring" of physical properties has been the subject of intense research in the field of thermoelectrics. In particular, it has been shown that a very powerful way of

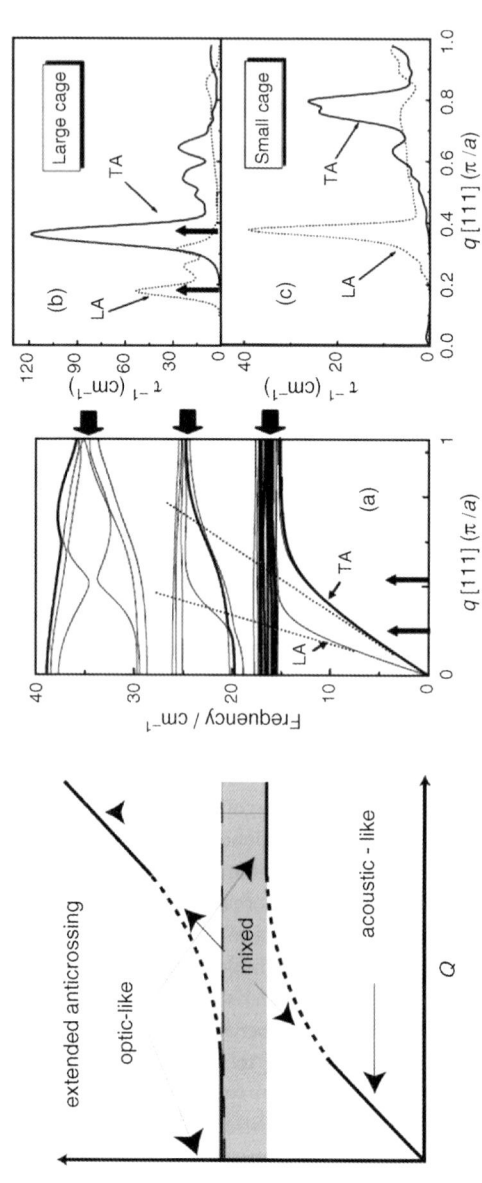

Figure 2.34 Left panel: Principle of the anticrossing between two branches with the same symmetry (From reference [155]). Middle panel: simulation of the dispersion relation of the Xe clathrate hydrate for modes propagating along the [110] directions. The horizontal arrows on the energy scale point to the three localized modes observed in the GVDOS (Figure 2.33). Vertical arrows point to the q values for which the LA and TA modes cross the low-lying optical mode around 17 cm^{-1}. Right panel: Inverse of the phonon lifetime calculated in a perturbative approach (see text) as a function of q for We atoms in the large cage (top) and small cage (bottom) (From reference [156]).

case of the interaction between an acoustic branch and a localized, dispersionless, optic excitation. In the vicinity of the crossing, the two branches "repel" themselves, which produces an important bending of the acoustic branch, while the optic branch departs from a horizontal dispersion. In the meantime the characters of the modes are exchanged. In the crossing area there is a strong hybridization or mixing between the modes that reduces the phonon lifetime.

The phonon dispersion relation along the [110] direction of the Xe clathrate hydrate has been calculated and is shown Figure 2.34. The three arrows on the side give the position of the low-energy bands observed experimentally in the GVDOS and located at 17 cm^{-1} (2.12 meV), 23 cm^{-1} (2.9 meV) and 32 cm^{-1} (3.97 meV). The longitudinal and transverse acoustic branches (labeled LA and TA) are flat and do not show any dispersion for wavevectors larger than the crossing point with the lower localized energy mode at 17 cm^{-1}. This is thus a characteristic of the anticrossing that will reduce the phonon lifetime. Using a perturbation approach, treating the Xe atoms as an impurity, it is possible to compute the phonon lifetime for the acoustic phonons that are carrying the heat. The framework of such a calculation was set up at the end of the 1960s by Maradudin [170] calculating the effect of a point defect or of disorder on lattice vibrations. Using Green's function, and the vibrational mode of the lattice without defects, the phonon lifetime is calculated. Figure 2.34 presents the results of the calculation for atoms located in the large cage or in the small one. The inverse lifetime τ^{-1} is represented as a function of the wavevector q. For atoms in the large cages, the inverse lifetime is zero for small wavevectors (i.e. large or infinite phonon lifetime) corresponding to propagative phonons, whereas it become extremely large at two particular wavevectors (black arrows), which correspond to the position where the LA and TA branches cross the low-lying excitation and where there is a strong resonant effect. The same calculation has been done for Xe atoms in the smaller cage, and as expected, the peaks in the inverse phonon lifetime occur for larger wavevectors since the corresponding localized mode has a larger energy. From these channels of phonon decay it is in principle possible to estimate the thermal conductivity, although it is quite complicated. Nevertheless, the overall picture of the phenomenon inducing a low thermal conductivity is now quite clear.

Recently, a similar anticrossing effect has been evidenced experimentally in $Ba_8Ga_{16}Ge_{30}$, which shows potential thermoelectric applications. Large single crystals could be grown that allowed the measurement of dispersion relation and of the $S(Q, E)$ function by inelastic neutron scattering en [171]. This was one of the first experiments that evidenced the anticrossing without ambiguity. Figure 2.35 gathers the main experimental results, which have been obtained. The left panel displays results obtained for longitudinal modes propagating along the [109] direction and measured from the strong Bragg peak (330). The figure displays the intensity distribution as a function of q. Close to the Bragg peak ($q=0$), the acoustic signal is very strong and then decays as $1/E^2$. There is also an optical excitation around 5 meV, whose intensity increases as q increases, especially near the crossing point. The anticrossing of the two branches is clearly visible, and in particular the acoustic mode has a flat dispersion at high q values. In the meantime there is an intensity "transfer" from the acoustic to the optical mode, as expected in this scheme. The

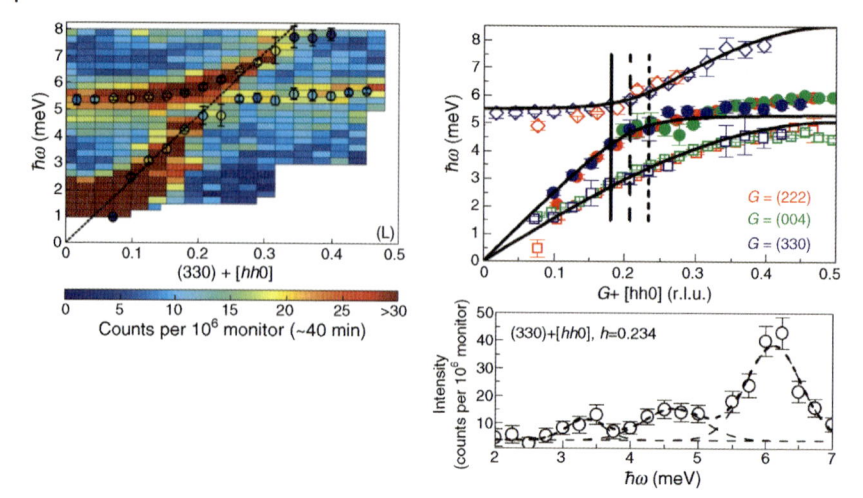

Figure 2.35 Results of the inelastic neutron scattering measurement in the $Ba_8Ga_{16}Ge_{30}$ clathrate. Left panel: Intensity distribution of longitudinal modes propagating along the [109] direction and measured around the (330) Bragg peak. Right top panel: Comparison between the measured dispersion relation (symbols) and a simulation using a coupling between two modes (solid line). Right bottom panel: constant-Q energy scan measured close to the crossing point (vertical small dashed line). Three modes are visible whose width gives the phonon lifetime (From reference [158]).

second panel gathers on the same curve the position of the excitations measured around different zone center and for different orientations. Both the longitudinal and transverse acoustic modes have a flat dispersion around 5 meV, while the localized optical excitation has a weak dispersion after the crossing. The data could be quite nicely reproduced by a simple model of mode coupling, shown as the solid black line on the figure. The good agreement between the model and the experiment thus validates this hypothesis of coupling between the guest atom and the framework structure.

Measuring the width of the different signals it is also possible to gain an estimate of the corresponding phonon lifetime. A characteristic energy scan extracted from the data is shown in Figure 2.35. A careful study of the phonon lifetime shows that it goes from 2.6 ps before the crossing to 1.3 ps at the crossing point. This thus evidenced the lifetime reduction due to the mode coupling. This phonon lifetime has been used to estimate the lattice thermal conductivity, using a rather crude approximation. Within this scheme, this lifetime reduction is not enough to explain the observed low thermal conductivity. The authors invoke the Umklapp process, which is another channel, as explained above, for a thermal conductivity reduction, which will occur at rather low T because of the low energy of the optical excitations. One should note that some other processes should also be investigated: in particular the lifetime reduction of transverse modes, which might be different from the longitudinal one, the effect of chemical disorder, another important channel for lifetime reduction and finally the structural complexity of the framework itself. In any case, the understanding of the law thermal conductivity certainly calls for a more detailed analysis in the spirit of what has been carried out by Tse *et al.* [169].

2.3.5.2 Skutterudites

Skutterudites have also been studied for their potential application as thermoelectrics. Their atomic structure contains large structural vacancies that can be "filled" by heavy atoms in order, as for clathrates, to reduce their thermal conductivity. Skutterudites are thus cage compounds, for which the concept of PGEC has been first applied with some success. The first inelastic neutron-scattering measurements have evidenced low-energy modes related to the filler or guest atoms [166]. More recently, a complete study has been carried by inelastic X-ray scattering by Tsuitsui *et al.*, on a single crystal of the $SmRu_4P_{12}$ [172]. With the inelastic X-ray scattering only small samples are required to measure accurately the dispersion relation, which makes it a very powerful technique. Moreover, using nuclear resonance they could extract directly the partial Sm density of state, that is, the density of state of the guest atom.

The Sm partial density of state displays a single peak at about 9 meV, which corresponds to the localized mode. The dispersion relation extracted from the inelastic X-ray measurement, shown in Figure 2.36, displays a clear anticrossing effect: the acoustic branch bends over rapidly, while the optical mode presents a small dispersion after the crossing. The study of the intensity distribution of the two modes close to the crossing shows that the there is a mode hybridization. Again we point out the importance of a careful intensity distribution study, which gives decisive clues on the nature of the modes. *Ab initio* calculations, shown on the right panel, have confirmed this hypothesis. The anticrossing is clearly visible, and the main character of the mode (transverse or longitudinal) in good agreement with this general scheme.

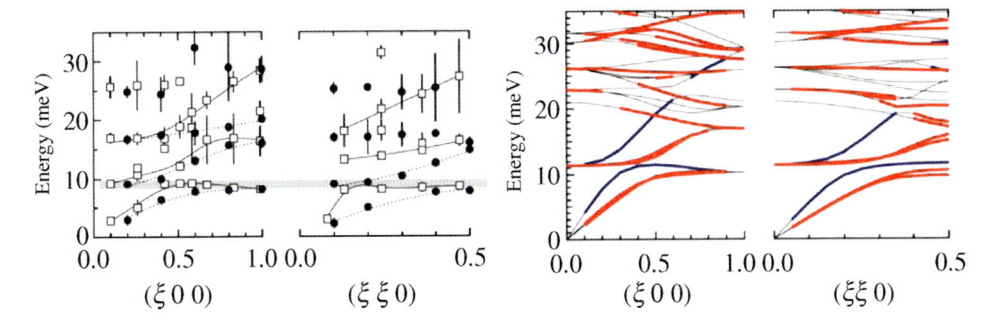

Figure 2.36 Left panel: Dispersion relation measured by inelastic X-ray scattering on a $SmRu_4P_{12}$ single grain. The gray horizontal line shows the position of the Sm localized mode. Right panel: *Ab initio* calculations. The transverse (longitudinal) character of the mode is indicated by a red (blue) color. The anticrossing is clearly visible (From reference [159]).

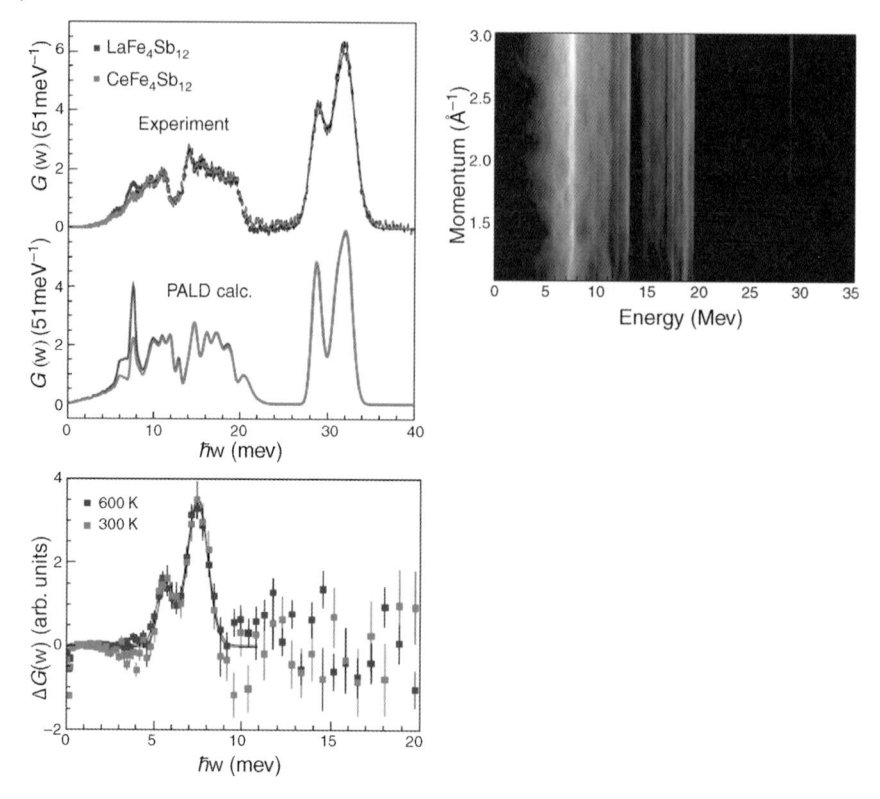

Figure 2.37 Left panel: comparison between the measured GVDOS and the calculated one for the La (black) and Ce (grey) Fe_4Sb_{12} skutterudite. The bottom panel shows the La partial density of states. Right panel: simulated La partial scattering function. The gray scale is on a logarithmic scale to enhance the weakest contributions (From reference [160]).

More recently, Koza et al. [173] claimed to have evidenced a breakdown of the PGEC scheme and in particular of the "rattling" scenario. They have carried out a detailed study of the lattice dynamics in Fe_4Sb_{12} skutterudites filled with La or Ce. The experiment has been carried out on powder samples, and the $S(Q,E)$ function has been compared to ab initio simulations. Using the different contrast of the La and Ce atoms, they have extracted the partial density of states of La atoms (Figure 2.37). La atoms have a density of states peaked at two energies 5.7 et 7.5 meV, most likely corresponding to localized vibration. These results have been compared to the ab initio simulation, using a powder-averaged calculation that reproduces conditions similar to the experimental one. The results of the simulation are in quite good agreement with the experiment, as shown in Figure 2.37. One can note, however, that around 6–10 meV the simulation presents a more pronounced structure in the signal, than what is observed in the experiment. This is most likely due to chemical disorder that has not been incorporated in the model (see reference [140] for supplementary information). They have also carried out a detailed comparison of the measured and simulated powder averaged $S(Q, E)$ function and in particular of the intensity distribution. As already pointed out for quasicrystal [140], this is a very important and extremely sensitive test since it is related to the eigenmodes of the system. The agreement between simulation and experiment is good, thus validating the approach so that the simulations can be used to analyze vibrational modes in more detail.

Of particular interest is the nature of the mode of the guest atoms. Indeed in the PGEC scenario, those atoms should behave as isolated Einstein oscillators. We have seen above that this result in an anticrossing scheme. The right panel of Figure 2.37 shows the partial $S(Q, E)$ function for the guest atom, powdered averaged. There is a clear dispersionless mode at 7 meV, but there are also weak contributions corresponding to acoustic excitation, as a signature of more collective-type modes resulting from the coupling with the rest of the structure. The guest atoms thus do not behave as "rattler" atoms. They conclude that the low lattice thermal conductivity is rather due to a Umklapp process, with Brillouin zone boundary modes at rather low energies and having an almost flat dispersion on a large portion of the Brillouin zone.

These results and conclusion seem thus contradictory with that of Tsuitsui et al. [172]. This is most likely because the rattling model has been considered in its extreme simplicity by Koza et al. [173]. Introducing a reasonable coupling between the host and guest structures allows both results to be reconciled. Indeed, the simulated dispersion relation by Koza et al., clearly shows an anticrossing scheme that strongly flattens the acoustic dispersion relation. This is a very peculiar situation, which is not encountered usually in crystals: this is because there is a low-energy mode so that the flattening can occur.

As already said above, going further would require a detailed simulation or best calculation (in a perturbative approach) of the thermal conductivity. An important step in that direction would be a detailed analysis of the nature of the lower-energy modes.

2.3.5.3 Zinc-Antimony Alloy Zn_4Sb_3

The Zn_4Sb_3 alloy is known to have very good thermoelectric properties, with a ZT value equal to 1.3 at 670 K. Although it is not a cage compound, it is interesting to look at the origin of the low thermal conductivity, since recent inelastic neutron-scattering results have been obtained. The atomic structure of this alloy is characterized by vacancies, interstitial disorder but also by the formation of Sb dimers, as shown in Figure 2.38 [174].

The specific heat measurements display a clear peak at low temperature, which can be interpreted as resulting from an Einstein mode with energy centered around 5.3 meV. One simple hypothesis is of course to associate this low-energy mode to the vibration of the Sb dimer, which is equivalent to a heavy atom. In order to analyze this mode Schweika et al. have carried out a measurement of the $S(Q,E)$ function on polycrystalline samples [175]. In order to get reasonable information on the nature of the vibrational modes it is important to carry out the measurement on a large Q area. The inelastic spectrum, measured at 300 K, is shown in Figure 2.37. The white central

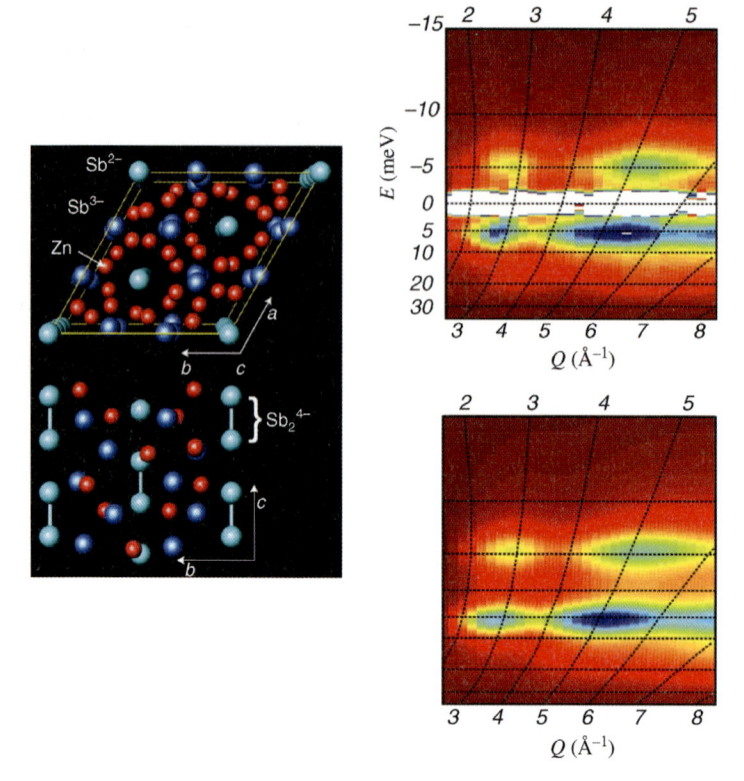

Figure 2.38 Left panel: illustration of the Zn_4Sb_3 atomic structure, without considering chemical disorder. Sb dimers are aligned along the c-axis [60]. Right panel: top $S(Q, E)$ function measured by inelastic neutron scattering for Zn_4Sb_3. Bottom: simulated $S(Q,E)$, for dimers vibrating along the c-axis and with a weak coupling between them (From reference [162]).

part corresponds to the elastic region, with on each side the neutron energy gain and loss spectrum.

The spectrum has two main characteristics: first there is a well-defined excitation with an energy of 5.2 meV; secondly the intensity distribution of the signal strongly depends on the Q wavevector. This Q variation is a signature of a collective motion of atoms. In effect, a single atom motion would give rise to a monotonic increase of the signal as Q^2. This again points to the importance of analyzing properly the intensity distribution.

Schweika et al. [175] used a simple model to interpret the data. They considered that the Sb dimer moves as a ball, both atoms moving exactly in phase as for a dumbbell. They considered different cases: isotropic motion of Sb or motion confined along the crystal c-axis. This last solution gave a much better agreement, as expected from the anisotropy in the structure. The best agreement has been obtained when introducing a weak coupling between two neighboring dimers. There are only two parameters in this simple model: the distance between the atoms in the dimer, and the vibration amplitudes, which give rise to a Debye–Waller factor. The result of the simulation is shown in Figure 2.37: both the position and the Q dependence of the intensity are well reproduced, with refined parameters that are in perfect agreement with the crystallographic data.

Moreover, the 5 meV localized mode displays a width larger than the instrumental resolution from which it is possible to extract a phonon lifetime equal to 0.39 ps. When associating this lifetime with the sound velocity for acoustic modes one gets a phonon mean free path equal to 9.6 Å. By considering a simple phonon gas, they could extract a thermal conductivity $\kappa = c_v v_s l_p/3$ a 13.2 mW/K/cm^{-1} at 300 K, in good agreement with the experimental κ, equal to 13 mW/K/cm^{-1}.

The results thus give quite a complete picture of the phenomena leading to a low thermal conductivity.

2.3.6
Phonon and Transport Properties: The Example of Thermoelectricity

2.3.6.1 Thermal Conductivity
The thermal conductivity of QCs belonging to different families has been measured, covering different temperature ranges, and the following general conclusions can be drawn from the collected data:

- Although most metallic alloys are good heat conductors, the thermal conductivity of QCs is unusually low, even lower than that observed for thermal insulators used extensively in the aeronautical industry, such as titanium carbides or nitrides, doped zirconia, or alumina. For example, in Al-Pd-Mn icosahedral phases the thermal conductivity at room temperature is comparable to that of zirconia (1 W m^{-1} K^{-1}), and this value decreases to about 10^{-4} W m^{-1} K^{-1} below 0.1 K [39, 40].
- Assuming that QCs obey the Wiedemann–Franz law (see Section 2.1.5) one estimates that the contribution of electrons to the thermal transport is, at least, one order of magnitude lower than that due to phonons over a wide temperature range (0.1 K $\leq T \leq$ 200 K) [90].
- The thermal diffusivity of these alloys is extraordinarily low, even lower than that of zirconium oxide [91].

The low thermal conductivity of QCs can be understood in terms of two main facts. In the first place, due to the presence of the pseudogap (see Section 2.3.1), the charge carrier concentration close to the Fermi level is low, so that heat must propagate by means of atomic vibrations (phonons). In the second place, in the energy window where lattice thermal transport is expected to be most efficient the frequency spectra of quasiperiodic systems is highly fragmented. As a consequence (see Section 2.3.4), the corresponding eigenstates become more localized and thermal transport is further reduced. Physically, this effect can be attributed to the fact that quasicrystal lattices have a fractal reciprocal space, lacking a well-defined lower bond as that provided by the lattice parameter in the case of periodic crystals. Consequently, the transfer of momentum to the lattice is not bounded below, which gives rise to a significant degradation of thermal current through the sample.

During the last decade some specific treatments, aimed to exploit the physical implications of the quasiperiodic order notion, have been introduced in the theory of thermal transport in these materials. Thus, the lack of a well-defined reciprocal lattice in quasicrystals leads to a power-law (instead of the usual exponential) dependence of the umklapp processes that can be expressed in terms of a scattering rate of the form $\tau^{-1} \propto \omega^2 T^4$, where ω is the phonon frequency [176]. These processes are expected to be dominant in the temperature range $20 \lesssim |T| \lesssim 100$. Making use of this scattering rate value in the expression

$$\kappa_{ph}(T) \sim T^3 \int_0^{\Theta/T} \frac{x^4 e^x}{(e^x-1)^2} \tau(x) dx \qquad (2.49)$$

where $x \equiv \hbar\omega/k_B T$, one gets $\kappa_{ph} \sim T^{-3}$. According to the data listed in Table 2.4, however, fitting analyzes to experimental κ_{ph} curves do not agree with theoretical expectations, leading to a temperature dependence of the form $\kappa_{ph} \sim T^{-1}$ in all considered cases. On the other hand, starting at $T \gtrsim 100$ K one expects the variable-range hopping mechanism introduced by Janot will play an increasingly significant role [177]. According to this model the thermal conductivity curve should rise following a power law of the form $\kappa_{ph} \sim T^{-3/2}$ within the temperature interval $100 \gtrsim T \gtrsim 400$. Nevertheless, the fitting analysis to experimental ph curves shown in Table 2.5 indicates that most considered samples significantly deviate from the expected behavior in this temperature range. Thus, a fundamental understanding of the thermal conductivity processes in these materials is still awaiting a definitive clarification.

2.3.6.2 Thermoelectric Figure of Merit

During the last few years we have witnessed a growing interest in searching for novel, high-performance thermoelectric materials for energy conversion. The efficiency of

Table 2.4 Values of the umklapp scattering rate and the phononic thermal conductivity temperature dependence $\kappa_{ph}(T)$, for different CMAs reported in the literature. The values were obtained from a fitting analysis of the thermal conductivity experimental curves, and the κ_{ph} temperature dependence was derived from Equation (2.38).

SAMPLE	τ^{-1}	$\kappa_{ph}(T)$	Reference
i-Zn$_{57}$Mg$_{34}$Y$_9$	$\omega^3 T$	T^{-1}	[25]
i-Al$_{64}$Cu$_{23}$Fe$_{13}$	$\omega^3 T$	T^{-1}	[42]
i-Al$_{72}$Pd$_{19.5}$Mn$_{18.5}$	$\omega^2 T$	T^{-1}	[180]
γ-Al-Cr-Fe	$\omega^2 T$	T^{-1}	[40]
ε-Al-Pd(Fe,Co,Rh)	$\omega^3 T$	T^{-1}	[181]
ψ- Al-Pd-Mn	ω^4	T^{-1}	[41]
ξ'-Al-Pd-Mn	ω^4	T^{-1}	[41]

Table 2.5 Reported values of the power-law exponent describing the phononic thermal conductivity temperature dependence ph(T) Tn, for different CMAs. The n values were obtained from a fitting analysis of the experimental curves in the indicated temperature ranges.

Sample	T range (K)	n	Reference
i-Cd-Yb	110–300	1.2	[29]
i-Al-Pd-Re	150–300	1.4	[25]
i-Ag-In-Yb	150–300	1.5	[27]
i-Al-Pd-Mn	200–300	1.7	[26]
i-Zn-Mg-Y	140–300	1.7	[24]

thermoelectric devices depends on the transport coefficients of the constituent materials and it can be properly expressed in terms of the figure of merit given by the dimensionless expression

$$ZT = \frac{T\sigma S^2}{\kappa_e + \kappa_{ph}} \tag{2.50}$$

where T is the temperature, $\sigma(T)$ is the electrical conductivity, $S(T)$ is the Seebeck coefficient and $\kappa_e(T)$ and $\kappa_{ph}(T)$ are the thermal conductivities due to the electrons and lattice phonons, respectively. The appealing question regarding what electronic structure provides the largest possible figure of merit was addressed some time ago, concluding that (i) the best thermoelectric material is likely to be found among materials exhibiting a sharp singularity (Dirac delta function) in the density of states close to the Fermi level, and (ii), in that case, the effect of the DOS background contribution onto the figure of merit value may be quite dramatic, the figure of merit value being inversely proportional (in a marked nonlinear way) to the DOS value near the singularity [178].

Quite interestingly, the electronic structure of quasicrystalline alloys satisfies these requirements in a natural way, since its exhibits a pronounced pseudogap at the Fermi level as well as some narrow features on the DOS close to the Fermi level (see Section 2.3.2). At first sight it may seem surprising to propose a metallic alloy as a suitable thermoelectric material, since it is well known that metallic compound usually exhibit very low $ZT \sim 10^{-3}$. However, such a proposal makes sense due to the peculiar transport properties of QCs [179]. In fact, their electrical conductivity (i) is remarkably low (ranging from 100 to 5000 Ω^{-1} cm^{-1} at room temperature), (ii) it steadily increases as the temperature increases up to the highest temperatures of measurement ($T \simeq 900$ K) and (iii) it is extremely sensitive to minor variations in the sample composition. This sensitivity to the sample stoichiometry is also observed in other transport parameters, like the Hall or Seebeck coefficients, and resembles doping effects in semiconductors. In addition, the temperature dependence of the Seebeck coefficient: (i) is clearly nonlinear, exhibiting well-defined extrema in most instances, (ii) small variations in the chemical composition give rise to sign reversals in the $S(T)$ value, (iii) and for a given sample stoichiometry it shows a strong dependence on the heat treatments applied to the sample [180]. Therefore, the

Table 2.6 Room-temperature transport coefficients values for different quasicrystalline families, after.

Sample	$\sigma\ (\Omega^{-1}\,cm^{-1})$	$S\ (\mu V\,K^{-1})$	$\kappa\ (W\,m^{-1}\,K^{-1})$	ZT
Al-Cu-Ru	250[b]	27b	1:8[d]	0:003
Al-Cu-Fe	310[b]	44b	1:8[c]	0:01
Cd-Yb	7000[f]	16f	4:7[f]	0:01
Al-Cu-Ru-Si	390[b]	50b	1:8[d]	0:02
Al-Pd-Re	175[e]	95e	0:7[e]	0:07
Al-Pd-Mn	640[a]	85a	1:6[a]	0:08

(From reference [112]. With permission).
a) Reference [180].
b) Reference [182].
c) Reference [40].
d) estimated.
e) Reference [184].
f) Reference [183].

electronic transport properties of quasicrystalline alloys exhibit unusual composition and temperature dependences, resembling more semiconductor-like than metallic character. Furthermore, the thermal conductivity of QCs is unusually low for a metallic alloy and it is mainly determined by the lattice phonons (rather than the charge carriers) over a wide temperature range. This low thermal conductivity of QCs is particularly remarkable in the light of Slack's phonon-glass/electron-crystal proposal for promising thermoelectric materials [181], and it has considerably spurred the interest on the potential application of QCs as thermoelectric materials from an experimental viewpoint.

Consequently, according to their transport properties, quasicrystalline alloys are marginally metallic and should be properly located at the *border line* between metals and semiconductors. Thus, QCs bridge the gap between metallic materials and semiconducting ones, occupying a very promising position in the quest for novel thermoelectric materials. Thus, one of the main advantages of QCs over other competing thermoelectric materials is that one can try to modify both the electrical conductivity and the thermoelectric power, without losing the low thermal conductivity, by properly varying the sample stoichiometry [180].

According to their chemical composition QCs can be grouped into several families. In Table 2.6 we list the transport coefficients values for those representatives yielding the best figure of merit values at room temperature. From the listed data we appreciate a progressive trend towards larger values of ZT resulting from the synthesis of suitable QCs. Furthermore, significantly enhanced figure of merit values are obtained at higher temperatures. Thus, we have $ZT=0.25$ for i-Al-Pd-Mn samples [185] at $T=550$ K; $ZT=0.11$ for i-Al$_{71}$Pd$_{20}$Re$_9$ samples at $T=660$ K, and $ZT=0.15$ for i-Al$_{71}$Pd$_{20}$(Re$_{0.45}$Ru$_{0.55}$)$_9$ samples at $T=700$ K [184]. Consequently, it seems reasonable to expect that relatively high values of the figure of merit may be obtained by a judicious choice of sample composition, working temperature, and Peltier cell structural design [186].

2.4
Conclusion

We have seen that tremendous progress was achieved recently in the understanding of the lattice dynamics of complex metallic alloys for very different systems: Zn-based quasicrystal and approximant, clathrates, skutterudites, etc. This has been possible thanks to a combined approach where experiments and simulations can be achieved jointly. In that respect neutron (or X-ray) inelastic scattering is an extremely powerful technique, which of course allows the determination of the dispersion relation. We have also seen that the intensity distribution is a very sensitive tool related to the nature of the vibrational modes: it is only recently that this intensity analysis has been proposed in complex systems.

In quite different complex systems simulations using either adapted potentials (EAM or oscillating pair potentials) or *ab initio* calculation have been proved to be in excellent agreement with the experiment. This paves the way for a deeper understanding of the nature of the vibrational modes. This is certainly a challenging task, which will require the development (or rediscovery) of theoretical tools. Indeed, in most of the complex systems the plane-wave expansion is probably no longer the appropriate tool to analyze the physics. This is of course the case for aperiodic crystals such as quasicrystals, but might also be the case for cage compounds, especially when disorder is introduced in the simulations. Whether there are cluster modes in quasicrystals, although most likely, remains an open question. Similarly, the existence of critical modes in quasicrystal is still open. The advances achieved, open a new and fascinating area in this field, which allow those fascinating questions to be tackled.

The understanding of the detailed mechanisms leading to a low thermal conductivity of CMA is still at a very early stage. Simple models have been put forward, such as rattling, generalized umklapp, anticrossing, etc. They certainly already grasp some important physical parameters. For instance, although there are some contradictory results, the existence of a low-lying localized energy mode and its interaction with the acoustic branch has been proved to be one channel in reducing the lattice thermal conductivity in cage compounds. Structural complexity, with the existence of a large number of low-lying optical excitations is also a channel for such a reduction. The obtained results certainly call for more detailed calculation and simulation of the thermal conductivity.

Finally, progress made both for electrons and phonons in complex systems opens the way to the understanding of the mechanism stabilizing the complex long-range order and in particular the aperiodic long-range order.

References

1 Kirihara, K. and Kimura, K. (2000) *Sci. Technol. Adv. Mater.*, **1**, 227.

2 Rosenbaum, R., Lin, S.T., and Su, T.I. (2003) *J. Phys.: Condens. Matter*, **15**, 4169.

3. Delahaye, J., Berger, C., and Fourcaudot, G. (2003) *J. Phys.: Condens. Matter*, **15**, 8753.
4. Lay, Y.Y., Jan, J.C., Chion, J.W., Tsai, H.M., Pong, W.F., Tsai, M.H., Pi, T.W., Lee, J.F., Ma, C.I., Tseng, K.L., Wang, C.R., and Lin, S.T. (2003). *Appl. Phys. Lett.*, **82**, 2035.
5. Pierce, F.S., Guo, Q., and Poon, S.J. (1994) *Phys. Rev. Lett.*, **73**, 2220.
6. Mayou, D., Berger, C., Cyrot-Lackmann, F., Klein, T., and Lanco, P. (1993) *Phys. Rev. Lett.*, **70**, 3915.
7. Quivy, A., Quiquandon, M., Calvayrac, Y., Faudot, F., Gratias, D., Berger, C., Brand, R.A., Simonet, V., and Hippert, F. (1996) *J. Phys.: Condens. Matter*, **8**, 4223.
8. Häussler, P., Nowak, H., and Haberken, R. (2000) *Mater. Sci. Eng.*, **294–296**, 283; Roth, C., Schwalbe, G., Knöáer, R., Zavaliche, F., Madel, O., Haberkern, R., and Häussler, P. (1999) *J. Non-Cryst. Solids*, **252**, 869.
9. Pierce, F.S., Poon, S.J., and Guo, Q. (1993) *Science*, **261**, 737.
10. Klein, T. and Symko, O.G. (1994) *Phys. Rev. Lett.*, **73**, 2248.
11. Trambly de Laissardière, G., Nguyen-Manh, D., and Mayou, D. (2005) *Prog. Materials Sci.*, **50**, 679.
12. Homes, C.C., Timusk, T., Wu, X., Altounian, Z., Sahnoune, A., and Ström-Olsen, J.O. (1991) *Phys. Rev. Lett.*, **67**, 2694.
13. Degiorgi, L., Chernikov, M.A., Beeli, C., and Ott, H.R. (1993) *Solid State Commun.*, **87**, 721, 3738.
14. Wu, X., Homes, C.C., Burkov, S.E., Timusk, T., Pierce, F.S., Poon, S.J., Cooper, S.L., and Karlow, M.A. (1993) *J. Phys.: Condens. Matter*, **5**, 5975.
15. Bianchi, A.D., Bommeli, F., Chernikov, M.A., Gubler, U., Degiorgi, L., and Ott, H.R. (1997) *Phys. Rev. B*, **55**, 5730.
16. Demange, V., Milandri, A., de Weerd, M.C., Machizaud, F., Jeandel, G., and Dubois, J.M. (2002) *Phys. Rev. B*, **65**, 144205.
17. Burkov, S.E., Timusk, T., and Ashcroft, N.W. (1992) *J. Phys.: Condens. Matter*, **4**, 9447.
18. Mayou, D. (2000) *Phys. Rev. Lett.*, **85**, 1290.
19. Basov, D.N., Timusk, T., Barakat, F., Greedan, J., and Grushko, B. (1994) *Phys. Rev. Lett.*, **72**, 1937.
20. Pierce, F.S., Poon, S.J., and Biggs, B.D. (1993) *Phys. Rev. Lett.*, **70**, 3919; Biggs, B.D., Li, Y., and Poon, S.J. (1991) *Phys. Rev. B*, **43**, 8747; Pierce, F.S., Bancel, P.A., Biggs, B.D., Guo, Q., and Poon, S.J. (1993) *Phys. Rev. B*, **47**, 5670.
21. Haberken, R., Fritsch, G., and Härting, M. (1993) *Appl. Phys. A*, **57**, 431.
22. Bilušić, A., Pavuna, D., and Smontara, A. (2001) *Vacuum*, **61**, 345; Biluöic, A., Smontara, A., Lasjaunias, J.C., Ivkov, J., and Calvayrac, Y. (2000) *Mater. Sci. Eng. A*, **294–296**, 711.
23. Giannò, K., Sologubenko, A.V., Chernikov, M.A., Ott, H.R., Fisher, I.R., and Canfield, P.C. (2000) *Mater. Sci. Eng. A*, **294–296**, 715.
24. Pope, A.L., Tritt, T.M., Gagnon, R., and Strom-Olsen, J. (2001) *Appl. Phys. Lett.*, **79**, 2345.
25. Giannò, K., Sologubenko, A.V., Chernikov, M.A., Ott, H.R., Fisher, I.R., and Canfield, P.C. (2000) *Phys. Rev. B*, **62**, 292.
26. Kuo, Y.K., Sivakumar, K.M., Lai, H.H., Ku, C.N., Lin, S.T., and Kaiser, A.B. (2005) *Phys. Rev. B*, **72**, 054202.
27. Bilušić, A., Budrović, M.Z., Smontara, A., Dolinšek, J., Canfield, P.C., and Fisher, I.R. (2002) *J. Alloys Compd.*, **342**, 413.
28. Kuo, Y.K., Lai, J.R., Huang, C.H., Lue, C.S., and Lin, S.T. (2003) *J. Phys.: Condens. Matter*, **15**, 7555.
29. Smontara, A., Smiljanić, I., Bilušić, A., Jagličić, Z., Klanjšek, M., Roitsch, S., Dolinšek, J., and Feuerbacher, M. (2007) *J. Alloys Compd.*, **430**, 29.
30. Muro, Y., Sasakawa, T., Suemitsu, T., Takabatake, T., Tamura, R., and Takeuchi, S. (2002) *Jpn. J. Appl. Phys.*, **41**, 3787.
31. Takeuchi, T., Otagiri, T., Sakagami, H., Kondo, T., Mizutani, U., and Sato, H. (2004) *Phys. Rev. B*, **70**, 144202.
32. Strydom, A.M. (2007) *J. Phys.: Condens. Matter*, **19**, 386205 1–15.
33. Paschen, S., Carrillo-Carrera, W., Bentien, A., Tran, V.H., Baenitz, M., Yu, Grin, and Steglich, F. (2001) *Phys. Rev. B*, **64**, 214404–1-11.

34 Ashcroft, N.W. and Mermin, N.D. (1976) *Solid State Physics*, Saunders College Publ., Cornell, p. 255.

35 Mayou, D. (2000) *Quasicrystals Current Topics* (eds E. Belin-Ferré, C. Berger, M. Quiquandon, and A. Sadoc) World Scientific, London, p. 445.

36 Maciá F E. (2002) *Appl. Phys. Lett.*, **81**, 88; Landauro, C.V., Maciá, E., and Solbrig, H. (2003) *Phys. Rev. B*, **67**, 184206.

37 Maciá, E., and Rodríguez-Oliveros, R. (2007) *Phys. Rev. B*, **75**, 104210.

38 Dubois, J.M. (2005) *Useful Quasicrystals*, World Scientific, Singapore, p. 137.

39 Dubois, J.M., Kang, S.S., Archambault, P., and Colleret, B. (1993) *J. Mater. Res.*, **8**, 38; Perrot, A., and Dubois, J.M. (1993) *Ann. Chim. Fr.*, **18**, 501.

40 Bihar, Z., Bilušić, A., Lukatela, J., Smontara, A., Jeglič, P., McGuiness, P.J., Dolinšek, J., Jagličić, Z., Jamovec, J., Demange, V., and Dubois, J.M. (2006) *J. Alloys Compd.*, **407**, 65–73.

41 Dolinšek, J., Jagličić, Z., McGuiness, P.J., Jagluccíc, Z., Bilušić, A., Smontara, A., Landauro, C.V., Feuerbacher, M., Grushko, B., and Urban, K. (2005) *Phys. Rev. B*, **72**, 064208 1–11.

42 Dolinšek, J., Vrtnik, S., Klanjšek, M., Jagličić, Z., Smontara, A., Smiljanić, I., Bilušić, A., Yokoyama, Y., Inoue, A., and Landauro, C.V. (2007) *Phys. Rev. B*, **76**, 054201 1–9.

43 Dolinšek, J., Jagličić, Z., and Smontara, A. (2006) *Philos. Mag.*, **86**, 671.

44 Pauling, L. (1960) *The Nature of the Chemical Bond*, Cornell University Press, Ithaca.

45 Tamura, R., Asao, T., Tamura, M., and Takeuchi, S. (2001) *Mater. Res. Symp. Proc.*, **643**, K13.3.1; Tamura, R., Asao, T., and Takeuchi, S. (2001) *Mater. Trans.*, **42**, 928.

46 Sato, H., Takeuchi, T., and Mizutani, U. (2001) *Phys. Rev. B*, **64**, 094207; Takeuchi, T., Onogi, T., Banno, E., and Mizutani, U. (2001) *Mater. Trans.*, **42**, 933.

47 Kirihara, K., Nagata, T., Kimura, K., Kato, K., Takata, M., Nishibori, E., and Sakata, M. (2003) *Phys. Rev. B*, **68**, 014205.

48 Haberken, R., Khedhri, K., Madel, C., and Häussler, P. (2000) *Mater. Sci. Eng.*, **294–296**, 475.

49 Stadnik, Z.M. (1998) Physical Properties of Quasicrystals, in: *Springer Series in Solid-State Physics 126* (ed. Z.M. Stadnik) Springer-Verlag, Berlin.

50 Mizutani F U. (2001) *Introduction to the Electron Theory of Metals*, Cambridge University Press, Cambridge.

51 Tsai, A.P. (1998) Physical Properties of Quasicrystals, in *Springer Series in Solid-State Physics 126* (ed. Z.M. Stadnik) Springer-Verlag, Berlin, p. 5.

52 Grenet, T. (2000) *Quasicrystals Current Topics* (eds E. Belin-Ferré, C. Berger, M. Quiquandon, and A. Sadoc) World Scientific, Singapore.

53 Hume-Rothery, W. (1926) *J. Inst. Met.*, **35**, 295.

54 Friedel, J. (1988) *Helv. Phys. Acta*, **61**, 538; Friedel, J. and Dénoyer, F. (1987) *C. R. Acad. Sci. Paris*, **305**, 171.

55 Massalski, T.B. and Mizutani, U. (1978) *Prog. Mater. Sci.*, **22**, 151.

56 Asahi, R., Sato, H., Takeuchi, T., and Mizutani, U. (2005) *Phys. Rev. B*, **71**, 165103.

57 Trambly de Laissardière, G., Mayou, D., and NguyenManh, D. (1993) *Europhys. Lett.*, **21**, 25; Trambly de Laissardière, G., Nguyen Manh, D., Magaud, L., Julien, J.P., Cyrot-Lackmann, F., and Mayou, D. (1995) *Phys. Rev. B*, **52**, 7920.

58 Mizutani, U. (2001) *Mater. Trans.*, **42**, 901; Mizutani, U., Takeuchi, T., Banno, E., Fourneé, V., Takata, M., and Sato, H. (2001) *Mater. Res. Soc. Symp. Proc.*, **643**, K13.1.1.

59 Krajčí, M. and Hafner, J. (2002) *J. Phys.: Condens. Matter*, **14**, 1865.

60 Fournée, V., Belin-Ferré, E., Pécheur, P., Tobola, J., Dankhazi, Z., Sadoc, A., and Müller, H. (2002) *J. Phys.: Condens. Matter*, **14**, 87.

61 Ishii, Y. and Fujiwara, T. (2001) *Phys. Rev. Lett.*, **87**, 206408.

62 Tamura, R., Takeuchi, T., Aoki, C., Takeuchi, S., Kiss, T., Yokoya, T., and Shin, S. (2004) *Phys. Rev. Lett.*, **92**, 146402.

63 Klein, T., Berger, C., Mayou, D., and Cyrot-Lackmann, F. (1991) *Phys. Rev. Lett.*, **66**, 2907; Pierce, F.S., Bancel, P.A., Biggs, B.D., Guo, Q., and Poon, S.J. (1993) *Phys. Rev. B*, **47**, 5670; Chernikov,

M.A., Bianchi, A., Felder, E., Gubler, U., and Ott, H.R. (1996) *Europhys. Lett.*, **35**, 431.
64 Mori, M., Matsuo, S., Ishimasa, T., Matsuura, T., Kamiya, K., Inokuchi, H., and Matsukawa, T. (1991) *J. Phys.: Condens. Matter*, **3**, 767.
65 Belin, E., Dankhazi, Z., Sadoc, A., Calvayrac, A., Klein, T., and Dubois, J.M. (1992) *J. Phys.: Condens. Matter*, **4**, 4459.
66 Belin, E. (2004) *J. Non-Cryst. Solids*, **334&335**, 323.
67 Shastri, A., Borsa, F., Goldman, A.I., Shield, J.E., and Torgeson, D.R. (1993) *J. Non-Cryst. Solids*, **153&154**, 347; (1994) *Phys. Rev. B*, **50**, 15, 651.
68 Klein, T., Symko, O.G., Davydov, D.N., and Jansen, A.G.M. (1995) *Phys. Rev. Lett.*, **74**, 3656.
69 Tang, X.P., Hill, E.A., Wonnell, S.K., Poon, S.J., and Wu, Y. (1997) *Phys. Rev. Lett.*, **79**, 1070.
70 Escudero, R., Lasjaunias, J.C., Calvayrac, Y., and Boudard, M. (1999) *J. Phys.: Condens. Matter*, **11**, 383.
71 Fujiwara, T., Yamamoto, S., and Trambly de Laissardière, G. (1993) *Phys. Rev. Lett.*, **71**, 4166; Trambly de Laissardiére, G. and Fujiwara, T. (1994) *Phys. Rev. B*, **50**, 5999; (1994). *ibid.* **50**, 9843.
72 Janot, C. and de Boissieu, M. (1994) *Phys. Rev. Lett.*, **72**, 1674.
73 Trambly de Laissardière, G. and Mayou, D. (1997) *Phys. Rev. B*, **55**, 2890; Trambly de Laissardière, G., Roche, S., and Mayou, D. (1997) *Mater. Sci. Eng. A*, **226–228**, 986.
74 Davydov, D.N., Mayou, D., Berger, C., Gignoux, C., Neumann, A., Jansen, A.G.M., and Wyder, P. (1996) *Phys. Rev. Lett.*, **77**, 3173.
75 Maciá, E. (2004) *Phys. Rev. B*, **69**, 132201.
76 Zhang, G.W., Stadnik, Z.M., Tsai, A.P., and Inoue, A. (1994) *Phys. Rev. B*, **50**, 6696; Shastri, A., Baker, D.B., Conradi, M.S., Borsa, F., and Torgeson, D.R. (1995) *ibid.* **52**, 12681.
77 Lindqvist, P., Lanco, P., Berger, C., Jansen, A.G.M., and Cyrot-Lackmann, F. (1995) *Phys. Rev. B*, **51**, 4796.
78 Stadnik, Z.M., Purdie, D., Garnier, M., Baer, Y., Tsai, A.P., Inoue, A., Edagawa, K., and Takeuchi, S. (1996) *Phys. Rev. Lett.*, **77**, 1777; Stadnik, Z.M., Purdie, D., Garnier, M., Baer, Y., Tsai, A.P., Inoue, A., Edagawa, K., Takeuchi, S., and Buschow, K.H.J. (1997) *Phys. Rev. B*, **55**, 10, 938.
79 Ebert, Ph., Feuerbacher, M., Tamura, N., Wollgarten, M., and Urban, K. (1996) *Phys. Rev. Lett.*, **77**, 3827.
80 Zijlstra, E.S. and Janssen, T. (2000) *Mater. Sci. Eng. A*, **294–296**, 886.
81 Widmer, R., Gröning, O., Ruffieux, P., and Gröning, P. (2006) *Philos. Mag.*, **86**, 781.
82 Solbrig, H. and Landauro, C.V. (2000) *Physica B*, **292**, 47; Landauro, C.V. and Solbrig, H. (2000) *Mater. Sci. Eng. A*, **294–296**, 600; Landauro, C.V. and Solbrig, H. (2001) *Physica B*, **301**, 267.
83 Roche, S. and Mayou, D. (1997) *Phys. Rev. Lett.*, **79**, 2518.
84 Maciá, E. and Domínguez-Adame, F. (2000) *Electrons, Phonons, and Excitons in Low Dimensional Aperiodic Systems*, Editorial Complutense, Madrid.
85 Thiel, P.A. and Dubois, J.M. (2000) *Nature*, **406**, 570.
86 Maciá, E. and Domínguez-Adame, F. (1996) *Phys. Rev. Lett.*, **76**, 2957.
87 Maciá, E. (2003) *J. Appl. Phys.*, **93**, 1014.
88 Maciá, E., Takeuchi, T., and Otagiri, T. (2005) *Phys. Rev. B*, **72**, 174208.
89 Maciá, E. and Dolinöek, J. (2007) *J. Phys.: Condens. Matter*, **19**, 176212.
90 Chernikov, M.A., Bianchi, A., and Ott, H.R. (1995) *Phys. Rev. B*, **51**, 153; Kalugin, P.A., Chernikov, M.A., Bianchi, A., and Ott, H.R. (1996) *ibid.* **53**, 14 145.
91 Dubois, J.M. (2001) *J. Phys.: Condens. Matter*, **13**, 7753.
92 Bellissard, J., Iochum, B., Scoppola, E., and Testard, D. (1989) *Commun. Math. Phys.*, **125**, 527; Süto, A. (1989) *J. Stat. Phys.*, **56**, 525.
93 Bovier, A. and Ghez, J.M. (1995) *J. Phys. A: Math. Gen.*, **28**, 2313.
94 Jochum, B., Testard, D. (1991) *J. Stat. Phys.* **65**, 70.
95 Iochum, B. and Testard, D. (1991) *J. Stat. Phys.*, **65**, 715; Chakrabarti, A., Karmakar, S.N., and Moitra, R.K. (1992) *Phys. Lett. A*, **168**, 301; (1994) *Phys. Rev. B*, **50**, 13276.
96 Kumar, V. (1990) *J. Phys.: Condens. Matter*, **2**, 1349.

97 Severin, M., Dulea, M., and Riklund, R. (1989) *J. Phys.: Condens. Matter*, **1**, 8851.

98 Maciá, E. (1998) *Phys. Rev. B*, **57**, 7661.

99 Kumar, V. and Ananthakrishna, G. (1987) *Phys. Rev. Lett.*, **59**, 1476.

100 Xie, X.C. and Das Sarma, S. (1988) *Phys. Rev. Lett.*, **60**, 1585; Ananthakrishna, G. and Kumar, V. (1988) *Phys. Rev. Lett.*, **60**, 1586.

101 Sil, S., Karmakar, S.N., Moitra, R.K., and Chakrabarti, A. (1993) *Phys. Rev. B*, **48**, 4192.

102 Sire, C. (1994) *Lectures on Quasicrystals* (eds F. Hippert and D. Gratias) Les Editions de Physique, Les Ulis.

103 Oviedo-Roa, R., Pérez, L.A., and Wang, Ch. (2000) *Phys. Rev. B*, **62**, 13805; Sánchez, V., Pérez, L.A., Oviedo-Roa, R., and Wang, Ch. (2001) *ibid.* **64**, 174205; Sánchez, V. and Wang, Ch. (2004) *ibid.* **70**, 144207.

104 Maciá, E. (2000) *Phys. Rev. B*, **61**, 6645.

105 Liu, Y. and Riklund, R. (1987) *Phys. Rev. B*, **35**, 6034.

106 Anselmo, D.H.A.L., Dantas, A.L., and Albuquerque, E.L. (2005) *Physica A*, **349**, 259.

107 Naumis, G.G. (1999) *Phys. Rev. B*, **59**, 11315.

108 Chester, G.V. and Thellung, A. (1961) *Proc. Phys. Soc. London*, **77**, 1005.

109 Greenwood, D.A. (1958) *Proc. Phys. Soc. London*, **71**, 585.

110 Kubo, R. (1957) *J. Phys. Soc. Jpn.*, **12**, 570; Kubo, R., Yokota, M., and Nakajima, S. (1957) *ibid.* **12**, 1203.

111 Apostol, T.M. (1995) *Introduction to Analytic Number Theory*, Springer-Verlag, New York.

112 Maciá, E. (2004) *Phys. Rev. B*, **69**, 184202.

113 Maciá, E. (2002) *Phys. Rev. B*, **66**, 174203.

114 Takeuchi, T., Otagiri, T., Sakagami, H., Kondo, T., and Mizutani, U. (2004) *Mater. Res. Soc. Symp. Proc. 2003*, **805**, 105.

115 Dolinšek, J., Jaglić, P., McGuiness, P.J., Jagličić, Z., Bilušić, A., Bihar, Z., Smontara, A., Landauro, C.V., Feuerbacher, M., Grushko, B., and Urban, K. (2005) *Phys. Rev. B*, **72**, 064208.

116 Kittel, C. (1976) *Introduction to Solid State Physics*, Wiley, New York.

117 Dove, M.T. (2002) *Structure and Dynamics. An Atomic View of Materials*, Oxford University Press, Oxford.

118 Janssen, T., Chapuis, G., and de Boissieu, M. (2007) *Aperiodic Crystals. From Modulated Phases to Quasicrystals*, Oxford University Press, Oxford, Oxford.

119 Schmalzl, K., Strauch, D., and Schober, H. (2003) *Phys. Rev. B*, **68**, 144301.

120 Squires, G.L. (1978) *Introduction to the Theory of Thermal Neutron Scattering*, Cambridge University Press, Oxford.

121 Lovesey, W.L. (1984) *Theory of Neutron Scattering from Condensed Matter*, Clarendon Press, Oxford University Press, Oxford.

122 Baron, A.Q.R., Tanaka, Y., Goto, S., Takeshita, K., Matsushita, T., and Ishikawa, T. (2000) *J. Phys. Chem. Solids*, **61**, 461.

123 Tsai, A.P., Inoue, A., Yokoyama, Y., and Masumoto, T. (1990) *Mat. Trans., JIM*, **31**, 98.

124 de Boissieu, M., Durandcharre, M., Bastie, P., Carabelli, A., Boudard, M., Bessiere, M., Lefebvre, S., Janot, C., and Audier, M. (1992) *Phil. Mag. Lett.*, **65**, 147.

125 Yokoyama, Y., Miura, T., Tsai, A., Inoue, A., and Masumoto, T. (1992) *Mater. Trans., JIM*, **33**, 97.

126 Goldman, A.I., Stassis, C., Bellissent, R., Mouden, H., Pyka, N., and Gayle, F.W. (1991) *Phys. Rev. B. – Short Communication*, **43**, 8763.

127 Goldman, A.I., Stassis, C., de Boissieu, M., Currat, R., Janot, C., Bellissent, R., Mouden, H., and Gayle, F.W. (1992) *Phys. Rev. B*, **45**, 10280.

128 Quilichini, M., Heger, G., Hennion, B., Lefebvre, S., and Quivy, A. (1990) *J. Phys. France*, **51**, 1785.

129 Quilichini, M., Hennion, B., Heger, G., Lefebvre, S., and Quivy, A. (1992) *J. Phys. II France*, **2**, 125–130.

130 de Boissieu, M., Boudard, M., Bellissent, R., Quilichini, M., Hennion, B., Currat, R., Goldman, A.I., and Janot, C. (1993) *J. Phys.: Condens. Matter*, **5**, 4945.

131 Boudard, M., de Boissieu, M., Kycia, S., Goldman, A.I., Hennion, B., Bellissent, R., Quilichini, M., Currat, R., and Janot, C. (1995) *J. Phys.: Condens. Matter*, **7**, 7299.

132 Niizeki, K. (1989) *J. Phys A: Math Gen.*, **22**, 4295.
133 Niizeki, K. and Akamatsu, T. (1990) *J. Phys: Condens. Matter*, **2** (12), 2759.
134 Smith, A.P. and Ashcroft, N.W. (1987) *Phys. Rev. Lett.*, **59** (12), 1365.
135 Lu, J.P. and Birman, J. (1988) *Phys. Rev. B.*, **38** (12), 8067.
136 Quilichini, M. and Janssen, T. (1997) *Rev. Mod. Phys.*, **69**, 277.
137 Niizeki, K. and Akamuatsu, T. (1990) *J. Phys.: Condens. Matter*, **2** (33), 7043.
138 de Boissieu, M., Currat, R., Francoual, S., and Kats, E. (2004) *Phys. Rev. B*, **69** (5), 54205.
139 Shibata, K., Currat, R., de-Boissieu, M., Sato, T.J., Takakura, H., and Tsai, A.P. (2002) *J. Phys.: Condens. Matter*, **14**, 1847.
140 de Boissieu, M., Francoual, S., Mihalkovic, M., Shibata, K., Baron, A.Q.R., Sidis, Y., Ishimasa, T., Wu, D., Lograsso, T., Regnault, L.P., Gähler, F., Tsutsui, S., Hennion, B., Bastie, P., Sato, T.J., Takakura, H., Currat, R., and Tsai, A.P. (2007) *Nat. Mater.*, **6**, 977.
141 Tsai, A.P., Guo, J.Q., Abe, E., Takakura, H., and Sato, T.J. (2000) *Nature*, **408**, 537.
142 Kaneko, Y., Arichika, Y., and Ishimasa, T. (2001) *Philos. Mag. Lett.*, **81**, 777.
143 Gomez, C.P. and Lidin, S. (2001) *Angew. Chem. Int.*, **40**, 4037.
144 Gomez, C.P. and Lidin, S. (2003) *Phys. Rev. B*, **68**, 024203\1.
145 Takakura, H., Gomez, C.P., Yamamoto, A., de Boissieu, M., and Tsai, A.P. (2007) *Nat. Mater.*, **6**, 58.
146 Janot, C. and de Boissieu, M. (1994) *Phys. Rev. Lett.*, **72**, 1674.
147 Janot, C. (1996) *Phys. Rev. B*, **53**, 181.
148 Janot, C. and de Boissieu, M. (1996) *Physica B*, **219–220**, 328.
149 Mihalkovic, M., Henley, C.L., Widom, M., and Ganesh, P. (2008) http://arxiv.org/pdf/0802.2926.
150 Tamura, R., Nishimoto, K., Takeuchi, S., Edagawa, K., Isobe, M., and Ueda, Y. (2005) *Phys. Rev. B*, **71**, 092203.
151 Ishimasa, T., Kasano, Y., Tachibana, A., Kashimoto, S., and Osaka, K. (2007) *Phil. Mag.*, **87**, 2887.
152 Lin, Q.S. and Corbett, J.D. (2006) *J. Am. Chem. Soc.*, **128**, 13268.
153 Henley, C.L. (1991) *Phys. Rev. B*, **43**, 993.
154 Mihalkovic, M. and Widom, M. (2006) *Philos. Mag.*, **86**, 519.
155 Rog, T., Murzyn, K., Hinsen, K., and Kneller, G.R. (2003) *J. Comput. Chem.*, **24**, 657.
156 Mihalkovic, M., Francoual, S., Shibata, K., de Boissieu, M., Baron, A.Q.R., Sidis, Y., Ishimasa, T., Wu, D., Lograsso, T., Regnault, L.P., Gahler, F., Tsutsui, S., Hennion, B., Bastie, P., Sato, T.J., Takakura, H., Currat, R., and Tsai, A.P. (2008) *Phil. Mag.*, **88**, 2311.
157 de Boissieu, M., Francoual, S., Kaneko, Y., and Ishimasa, T. (2005) *Phys. Rev. Lett.*, **95**, 105503.
158 de Boissieu, M., Currat, R., and Francoual, S. (2008) *Handbook of Metal Physics: Quasicrystals* (eds T. Fujiwara and Y. Ishii) Elsevier Science, Amsterdam, p. 107.
159 Sales, B.C., Mandrus, D., Chakoumakos, B.C., Keppens, V., and Thompson, J.R. (1997) *Phys. Rev. B*, **56**, 15081.
160 Kuo, Y.K., Lai, J.R., Huang, C.H., Ku, W.C., Lue, C.S., and Lin, S.T. (2004) *J. Appl. Phys.*, **95**, 1900.
161 Chernikov, M.A., Bianchi, A., and Ott, H.R. (1995) *Phys. Rev B*, **51**, 153.
162 Smontara, A., Smiljanic, I., Bilusic, A., Grushko, B., Balanetskyy, S., Jaglici, Z., Vrtnik, S., and Dolinsek, J. (2008) *J. Alloys Compd.*, **450**, 29.
163 Kalugin, P.A., Chernikov, M.A., Bianchi, A., and Ott, H.R. (1996) *Phys. Rev. B*, **53**, 14145.
164 Takeuchi, T., Nagasako, N., Asahi, R., and Mizutani, U. (2006) *Phys. Rev. B*, **74**, 054206.
165 Slack, G.A. (1997) *Mater. Res. Soc. Symp. Proc.*, **478**, 47.
166 Keppens, V., Mandrus, D., Sales, B.C., Chakoumakos, B.C., Dai, P., Coldea, R., Maple, M.B., Gajewski, D.A., Freeman, E.J., and Bennington, S. (1998) *Nature*, **395**, 876.
167 Gutt, C., Baumert, J., Press, W., Tse, J.S., and Janssen, S. (2002) *J. Chem. Phys.*, **116**, 3795.
168 Schober, H., Itoh, H., Klapproth, A., Chihaia, V., and Kuhs, W.F. (2003) *Eur. Phys. J. E*, **12**, 41.

169 Tse, J.S., Shpakov, V.P., Belosludov, V.R., Trouw, F., Handa, Y.P., and Press, W. (2001) *Europhys. Letters*, **54**, 354.

170 Maradudin, A.A. (1966) in *Solid State Physics* (eds. F. Seitz and D. Turnbull) Academic Press Inc., vol. 18, p. 273.

171 Christensen, M., Abrahamsen, A.B., Christensen, N.B., Juranyi, F., Andersen, N.H., Lefmann, K., Andreasson, J., Bahl, C.R.H., and Iversen, B.B. (2008) *Nat. Mater.*, **7**, 811.

172 Tsutsui, S., Kobayashi, H., Ishikawa, D., Sutter, J.P., Baron, A.Q.R., Hasegawa, T., Ogita, N., Udagawa, M., Yoda, Y., Onodera, H., Kikuchi, D., Sugawara, H., Sekine, C., Shirotani, I., and Sato, H. (2008) *J. Phys. Soc. Jpn.*, **77**, 033601.

173 Koza, M.M., Johnson, M.R., Viennois, R., Mutka, H., Girard, L., and Ravot, D. (2008) *Nat. Mater.*, **7**, 805.

174 Snyder, G.J., Christensen, M., Nishibori, E., and Caillat, T. (2004) *Nat. Mater.*, **3**, 458.

175 Schweika, W., Hermann, R.P., Prager, M., Persson, J., and Keppens, V. (2007) *Phys. Rev. Lett.*, **99**, 125501.

176 Kalugin, P.A., Chernikov, M.A., Bianchi, A., and Ott, H.R. (1996) *Phys. Rev. B*, **53**, 14145.

177 Janot, C. (1996) *Phys. Rev. B*, **53**, 181.

178 Mahan, G.D. and Sofo, J.O. (1996) *Proc. Natl. Acad. Sci. USA*, **93**, 7436.

179 Maciá, E. (2001) *Phys. Rev. B*, **64**, 094206.

180 Pope, A.L., Tritt, T.M., Chernikov, M.A., and Feuerbacher, M. (1999) *Appl. Phys. Lett.*, **75**, 1854.

181 Slack, G.A. (1995) *CRC Handbook of Thermoelectrics*, (ed D.M. Rowe) CRC Press, Boca Raton, FL.

182 Biggs, B.D., Poon, S.J., and Munirathnam, N.R. (1990) *Phys. Rev. Lett.*, **65**, 2700; Pierce, F.S., Poon, S.J., and Biggs, B.D. (1993) *Phys. Rev. Lett.*, **70**, 3919; Biggs, B.D., Li, Y., and Poon, S.J. (1991) *Phys. Rev. B*, **43**, 8747; Pierce, F.S., Bancel, P.A., Biggs, B.D., Guo, Q., and Poon, S.J. (1993) *Phys. Rev. B*, **47**, 5670.

183 Pope, A.L., Tritt, T.M., Gagnon, R., and Strom-Olsen, J. (2001) *Appl. Phys. Lett.*, **79**, 2345.

184 Nagata, T., Kirihara, K., and Kimura, K. (2003) *J. Appl. Phys.*, **94**, 6560; Kirihara, K., and Kimura, K. (2002) *J. Appl. Phys.*, **92**, 979.

185 Pope, A.L., Zawilski, B.M., Gagnon, R., Tritt, T.M., Ström-Olsen, J., Schneidmiller, R., and Kolis, J.W. (2001) *Mater. Res. Soc. Symp. Proc.*, **643**, K14.4.1.

186 Maciá, E. (2004) *Phys. Rev. B*, **70**, 100201.

3
Anisotropic Physical Properties of Complex Metallic Alloys
Janez Dolinšek and Ana Smontara

3.1
Introduction

The anisotropic crystallographic structures of complex metallic alloys (CMAs) result in anisotropic magnetic, electrical and thermal transport properties, when measured along different crystallographic directions. Interesting classes of CMA compounds are the Al_4TM (TM = transition metal) and $Al_{13}TM_4$ families of intermetallics, which are periodic approximants to the decagonal quasicrystals (*d*-QCs). Their structures can be viewed as a stack of atomic planes and this structural anisotropy is at the origin of the anisotropic physical properties. While the stacked-layer crystallographic structure is a common property of the Al_4TM and $Al_{13}TM_4$ families, member compounds differ in the unit cell size and the number of atomic layers in the unit cell.

In the following we present a study of the anisotropic physical properties (magnetic susceptibility, electrical resistivity, thermoelectric power, Hall coefficient and thermal conductivity) of three stacked-layer CMAs of systematically increasing structural complexity comprising two, four and six atomic layers in the unit cell. The first is the $Al_{76}Co_{22}Ni_2$ compound [1, 2], known as the Y-phase of Al-Ni-Co, which belongs to the $Al_{13}TM_4$ class and is a monoclinic approximant to the decagonal phase with two atomic layers within one periodic unit of ≈ 0.4 nm along the stacking direction and a relatively small unit cell, comprising 32 atoms.

The second is the orthorhombic o-$Al_{13}Co_4$ compound [3], belonging to the $Al_{13}TM_4$ class of decagonal approximants with four atomic layers within one periodic unit of ≈ 0.8 nm along the stacking direction and a unit cell comprising 102 atoms. The third is the Al_4(Cr,Fe) complex metallic alloy with composition $Al_{80}Cr_{15}Fe_5$ [4, 5], belonging to the class of orthorhombic Al_4TM phases, which are approximants to the decagonal phase with six atomic layers in a periodic unit of 1.25 nm and 306 atoms in the giant unit cell. The above selection of samples allowed us to consider the evolution of anisotropic physical properties of the stacked-layer CMAs with increasing structural complexity and the unit cell size.

Complex Metallic Alloys: Fundamentals and Applications
Edited by Jean-Marie Dubois and Esther Belin-Ferré
Copyright © 2011 WILEY-VCH Verlag GmbH & Co. KGaA, Weinheim
ISBN: 978-3-527-32523-8

3.2
Structural Considerations and Sample Preparation

We describe here structural details of the investigated Y-Al-Ni-Co, o-Al$_{13}$Co$_4$ and Al$_4$(Cr,Fe) phases and the sample-preparation methods.

3.2.1
Y-Al-Ni-Co

The Y-Al-Ni-Co phase is described in the literature as the Al$_{13-x}$(Co$_{1-y}$Ni$_y$)$_4$ monoclinic phase [6], belonging to the Al$_{13}$TM$_4$ class of decagonal approximants. Other members of this class are monoclinic m-Al$_{13}$Co$_4$ [7], orthorhombic o-Al$_{13}$Co$_4$ [8], monoclinic Al$_{13}$Fe$_4$ [9], monoclinic Al$_{13}$Os$_4$ [10], Al$_{13}$Ru$_4$ (isotypical to Al$_{13}$Fe$_4$) [11] and Al$_{13}$Rh$_4$ (also isotypical to Al$_{13}$Fe$_4$) [12]. The structure of Al$_{13-x}$(Co$_{1-y}$Ni$_y$)$_4$ with $x = 0.9$ and $y = 0.12$, corresponding to composition Al$_{75}$Co$_{22}$Ni$_3$, was first described by Zhang et al. [6]. Lattice parameters of the monoclinic unit cell (space group C2/m (No. 12)) are $a = 1.7071(2)$ nm, $b = 0.40\,993(6)$ nm, $c = 0.74\,910(9)$ nm, $\beta = 116.17°$ and Pearson symbol mC34–1.8 with 32 atoms in the unit cell (8 Co/Ni and 24 Al), which are placed on 9 crystallographically inequivalent atomic positions (2 Co/Ni and 7 Al). Two of these are partially occupied (Al(6) by 90% and Al(6') by 10%). X-ray diffraction data revealed that the Al$_{13-x}$(Co$_{1-y}$Ni$_y$)$_4$ phase is identical to the previously reported Y–phase, found as predominant phase in samples with nominal compositions Al$_{75}$Co$_{20}$Ni$_5$ and Al$_{75}$Co$_{15}$Ni$_{10}$ [6, 13]. The structure of the Al$_{13-x}$(Co$_{1-y}$Ni$_y$)$_4$ is built up of one type of flat atomic layers, which are related to each other by a 2$_1$ axis, giving ≈ 0.4 nm period along the [010] stacking direction (corresponding to the periodic direction in the related decagonal d-Al-Ni-Co quasicrystal) and two atomic layers within one periodicity unit. Locally, the structure shows close resemblance to the d-Al$_{70}$Co$_{15}$Ni$_5$ quasicrystal [14], which also consists of only one type of a quasiperiodic layer, repeated by a 10$_5$-axis and giving the same ≈ 0.4 nm period.

The single crystal used in our study was grown from an incongruent Al-rich melt of initial composition Al$_{81.9}$Co$_{14.5}$Ni$_{3.6}$ by the Czochralski method using a native seed. The composition of the crystal (rounded to the closest integers) was Al$_{76}$Co$_{22}$Ni$_2$ and its structure matched well to the monoclinic unit cell of the Zhang et al. model [6] (who studied the composition Al$_{75}$Co$_{22}$Ni$_3$). In order to perform crystallographic-direction-dependent studies, we have cut from the ingot three bar-shaped samples of dimensions $2 \times 2 \times 6$ mm^3, with their long axes along three orthogonal directions. The long axis of the first sample was along the [010] stacking direction (designated in the following as b), which corresponds to the periodic direction in the related d-Al-Ni-Co quasicrystal. The (a,c) monoclinic plane corresponds to the quasiperiodic plane in d-QCs and the second sample was cut with its long axis along the [001] (c) direction, whereas the third one was cut along the direction perpendicular to the (b,c) plane. This direction is designated as a^* (it lies in the monoclinic plane at an angle 26° with respect to a and perpendicular to c). For each sample, the orientation of the other two crystallographic directions was also known. The so-prepared samples enabled us to determine the anisotropic physical properties along the three orthogonal directions of the investigated monoclinic $Al_{76}Co_{22}Ni_2$, abbreviated as Y-Al-Ni-Co in the following.

3.2.2
o-Al$_{13}$Co$_4$

The orthorhombic o-$Al_{13}Co_4$ phase is another member of the $Al_{13}TM_4$ class of decagonal approximants. According to the structural model by Grin et al. [8], lattice parameters of the o-$Al_{13}Co_4$ orthorhombic unit cell (space group $Pmn2_1$, Pearson symbol $oP102$) are $a = 0.8158$ nm, $b = 1.2342$ nm and $c = 1.4452$ nm with 102 atoms in the unit cell. The structure corresponds to a four-layer stacking along [100] [6, 8], with flat layers at $x = 0$ and $x = 1/2$ and two symmetrically equivalent puckered layers at $x = 1/4$ and $3/4$, giving ≈ 0.8 nm period along [100].

The o-$Al_{13}Co_4$ single crystal used in our study was grown by the Czochralski technique (the details are described elsewhere [15]) and its structure matched well to the orthorhombic unit cell of the Grin et al. model [8]. In order to perform crystallographic-direction-dependent studies, we have cut from the ingot three bar-shaped samples of dimensions $2 \times 2 \times 7$ mm^3, with their long axes along three orthogonal directions. The long axis of the first sample was along the [100] stacking direction (designated in the following as a), which corresponds to the pseudo-10-fold axis of the o-$Al_{13}Co_4$ structure and is equivalent to the periodic (10-fold) direction in the related d-QCs. The (b,c) orthorhombic plane corresponds to the quasiperiodic plane in the d-QCs and the second sample was cut with its long axis along the [010] (b) direction and the third one along the [001] (c) direction. For each sample, the orientation of the other two crystallographic directions was also known.

3.2.3
Al$_4$(Cr,Fe)

The Al_4(Cr,Fe) phase [16] belongs to the Al_4TM class of body-centered orthorhombic phases. The Al_4TM phase has been so far observed in six different Al-TM alloys, so that it must be a common structure to this class of alloys. The Al_4TM structure can be described as a periodic repetition of a sequence P'FPp'fp of six atomic layers stacked within one periodicity length of 1.25 nm along a, showing close structural relationship to the six-layer Al-TM d-QCs with the same periodicity. The block P'FP is composed of a flat layer F at $x = 0$ and a puckered layer P at $x \approx a/6$, whereas the puckered layer P' is in a mirror-reflecting position across the F layer. The block p'fp equals the block P'FP translated by ($a/2$, $b/2$, $c/2$).

The single crystal used in our study was grown from an incongruent Al-rich melt of initial composition $Al_{87}Cr_7Fe_6$ by the Czochralski method using a native seed. The composition of the sample (rounded to the closest integers) was $Al_{80}Cr_{15}Fe_5$ and its structure could be assigned to the orthorhombic phase, previously described by Deng et al. [16], with the following crystallographic parameters: Pearson's symbol $oI366$–59.56, space group $Immm$ (No. 71), unit cell parameters $a = 1.2500(6)$ nm, $b = 1.2617(2)$ nm, $c = 3.0651(8)$ nm and 306.44 atoms in the giant unit cell. Due to body centering, the primitive unit cell contains only half as many atoms. Cr and Fe atoms are not differentiated crystallographically. In order to perform crystallographic-direction-dependent studies, we prepared three bar-shaped samples of dimensions $2 \times 2 \times 8$ mm^3 with their long axes along the three crystallographic directions of the orthorhombic unit cell. The [100] stacking direction (designated in the following as a) corresponds to the periodic direction in d-QCs, whereas the [010] (b) and [001] (c) directions lie within the atomic planes (corresponding to the quasiperiodic directions in d-QCs).

3.3
Anisotropic Magnetic Properties

3.3.1
Y-Al-Ni-Co

The magnetization as a function of the magnetic field, $M(H)$, and the temperature-dependent magnetic susceptibility, $\chi(T)$, were investigated in the temperature interval between 300 and 2 K, using a Quantum Design SQUID magnetometer, equipped with a 50-kOe magnet. In the orientation-dependent measurements, magnetic field was directed along the long axis of each sample, thus along the a^*, b and c crystallographic directions. The $M(H)$ curves at $T = 5$ K are displayed in the upper panel of Figure 3.1, showing a linear dependence of the magnetization on the magnetic field in the whole investigated field range up to 50 kOe, except in the vicinity of $H = 0$, where small hysteresis loops are observed for all three directions due to a small ferromagnetic (FM) component in the magnetization. The slopes of the $M(H)$ curves appear in the order $(\partial M/\partial H)_b < (\partial M/\partial H)_c \approx (\partial M/\partial H)_{a^*}$ and show the following anisotropy: the slopes for the two inplane directions a^* and c are positive paramagnetic and there is little difference in magnitude between these two directions. In contrast, the $\partial M/\partial H$ slope for the field along the stacking b direction is much larger and negative diamagnetic.

The temperature-dependent anisotropic magnetic susceptibility $\chi = M/H$ along the three crystallographic directions, measured in a field 10 kOe and appearing in the order $\chi_b < \chi_c \approx \chi_{a^*}$, is displayed in the lower panel of Figure 3.1. The two inplane susceptibilities χ_{a^*} and χ_c are positive paramagnetic and there is not much difference in their magnitudes. In contrast, the susceptibility along the stacking b direction χ_b is negative diamagnetic, in agreement with the anisotropic behavior of the $M(H)$ curves shown in the upper panel of Figure 3.1. At temperatures below about 20 K, all three susceptibilities show a Curie upturn, typical of localized paramagnetic impurities.

3.3.2
o-Al$_{13}$Co$_4$

The magnetic field was again directed along the long axis of each sample, thus along the a, b and c crystallographic directions. The $M(H)$ curves at $T = 5$ K are displayed in

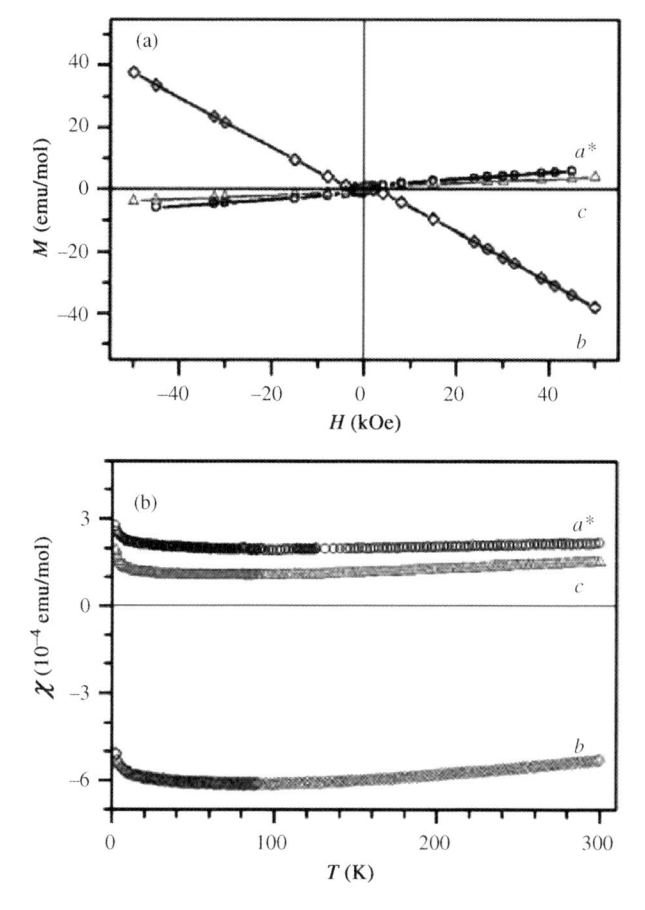

Figure 3.1 (a) Magnetization M of Y-Al-Ni-Co as a function of the magnetic field H at $T = 5$ K with the field oriented along three orthogonal crystallographic directions a^*, b and c. (b) Temperature-dependent magnetic susceptibility $\chi = M/H$ in the field $H = 10$ kOe applied along the three crystallographic directions.

the upper panel of Figure 3.2. For all three directions, the $M(H)$ dependence is linear and positive paramagnetic in the whole investigated field range up to 50 kOe, except in the close vicinity of $H = 0$, where a curvature typical of a small ferromagnetic (FM) component in the magnetization is observed. The $M(H)$ curves show anisotropy in the slopes in the following order: $(\partial M/\partial H)_a \ll (\partial M/\partial H)_b < (\partial M/\partial H)_c$. The anisotropy between the two inplane directions b and c is small, whereas the anisotropy to the stacking direction a is considerably larger.

The temperature-dependent magnetic susceptibility $\chi = M/H$ in the field of 1 kOe applied along the three crystallographic directions a, b and c is displayed in the lower panel of Figure 3.2. For each direction, both zero-field-cooled (zfc) and field-cooled (fc) temperature runs were performed. The existence of the FM component in the

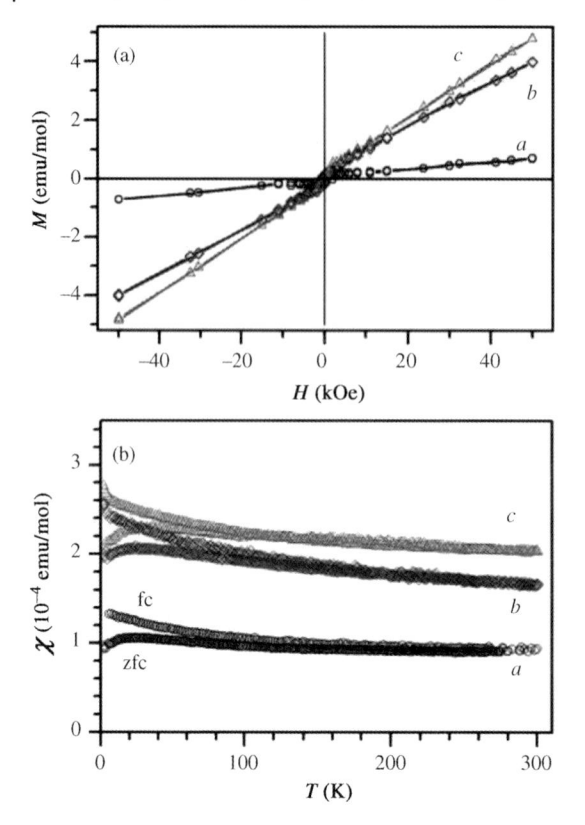

Figure 3.2 (a) Magnetization M of o-Al$_{13}$Co$_4$ as a function of the magnetic field H at $T = 5$ K with the field oriented along three orthogonal crystallographic directions a, b and c. (b) Temperature-dependent magnetic susceptibility $\chi = M/H$ in the field $H = 1$ kOe applied along the three crystallographic directions. Both zero-field-cooled (zfc) and field-cooled (fc) runs are shown.

susceptibility is manifested in the zfc–fc susceptibility splitting, observed below about 100 K. The presence of the zfc–fc splitting demonstrates remanence of the spins within the FM clusters. The anisotropic susceptibility appears in the order $\chi_a < \chi_b < \chi_c$, in agreement with the anisotropy determined from the $M(H)$ relation of Figure 3.2 (upper panel).

3.3.3
Al$_4$(Cr,Fe)

The $M(H)$ curves of Al$_{80}$Cr$_{15}$Fe$_5$ for the field up to 50 kOe along the three crystallographic directions are displayed in the upper panel of Figure 3.3. The curves are nonlinear and appear in the order $M_a < M_b \approx M_c$. There is almost no anisotropy

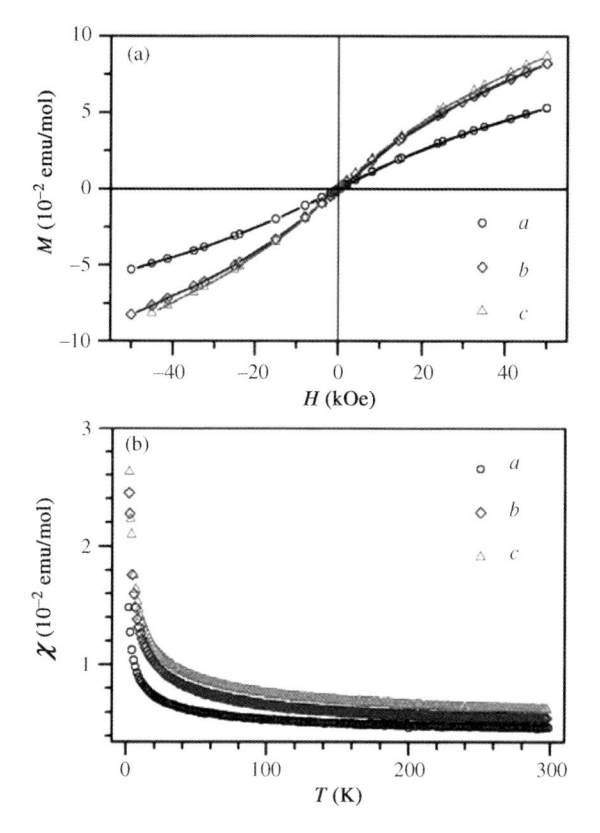

Figure 3.3 (a) Magnetization M of $Al_{80}Cr_{15}Fe_5$ as a function of the magnetic field at $T = 2$ K for the field oriented along three orthogonal crystallographic directions a, b and c of the orthorhombic unit cell. (b) Temperature-dependent magnetic susceptibility χ in a field $H = 1$ kOe applied along the three crystallographic directions.

for the two inplane crystallographic directions b and c, whereas the anisotropy to the stacking direction a is larger.

The susceptibility $\chi = M/H$ was investigated in a magnetic field $H = 1$ kOe (Figure 3.3, lower panel), which was directed along the a, b and c crystallographic directions. All three susceptibilities exhibit a Curie-type paramagnetic behavior of localized magnetic moments and show the anisotropy in the order $\chi_a < \chi_b < \chi_c$.

Summarizing the experimental anisotropic magnetic properties, we observe that all three compounds show weak inplane anisotropy of the susceptibility and the magnetization, whereas the anisotropy to the stacking direction is larger. The inplane magnetism is always considerably stronger than that along the stacking direction. Comparing the three compounds, the magnetization along the stacking direction shows the following regularity: for the Y-Al-Ni-Co it is diamagnetic; it becomes marginally paramagnetic for the o-$Al_{13}Co_4$ and strongly Curie-paramagnetic for the $Al_4(Cr,Fe)$.

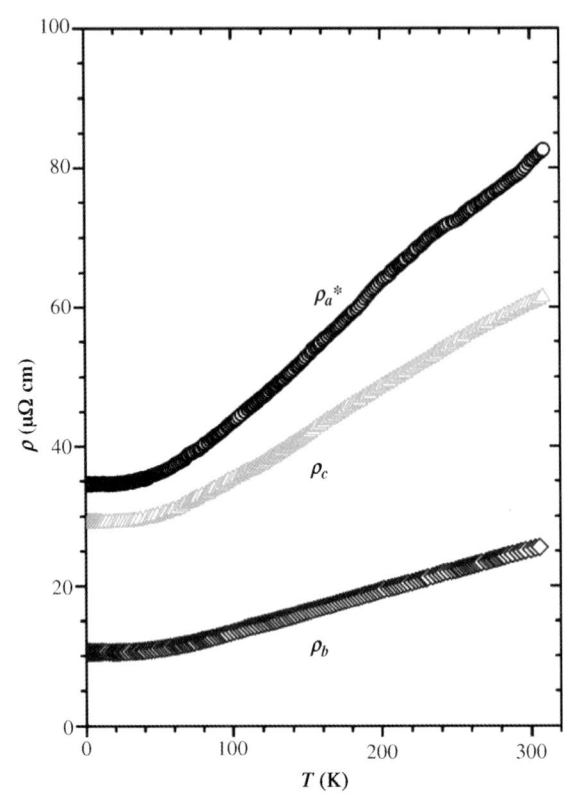

Figure 3.4 Temperature-dependent electrical resistivity of Y-Al-Ni-Co along three orthogonal crystallographic directions a^*, b and c.

3.4
Anisotropic Electrical Resistivity

3.4.1
Y-Al-Ni-Co

Electrical resistivity was measured between 300 and 2 K using the standard four-terminal technique. The $\varrho(T)$ data for the three crystallographic directions are displayed in Figure 3.4. The resistivity is the lowest along the stacking b direction perpendicular to the atomic planes, where its room temperature (RT) value amounts $\varrho_b^{300K} = 25\,\mu\Omega$ cm and the residual resistivity is $\varrho_b^{2K} = 10\,\mu\Omega$ cm. The two inplane resistivities are higher, amounting to $\varrho_c^{300K} = 60\,\mu\Omega$ cm and $\varrho_c^{2K} = 29\,\mu\Omega$ cm for the c direction and $\varrho_{a^*}^{300K} = 81\,\mu\Omega$ cm and $\varrho_{a^*}^{2K} = 34\,\mu\Omega$ cm for the a^* direction.

Whereas ϱ_b is considerably smaller than ϱ_{a^*} and ϱ_c by a factor of about 3, the two inplane resistivities are much closer, $\varrho_{a^*}/\varrho_c \approx 1.3$. The above resistivity values, appearing in the order $\varrho_b < \varrho_c < \varrho_{a^*}$ (even the inequality $\varrho_b \ll \varrho_c < \varrho_{a^*}$ may be considered to hold), reveal that Y-Al-Ni-Co is a good electrical conductor along all three crystallographic directions. The ratios of the resistivities along different crystallographic directions vary little over the whole investigated temperature range 300–2 K, amounting at RT $\varrho_{a^*}/\varrho_b \approx 3.2$, $\varrho_c/\varrho_b \approx 2.5$ and $\varrho_{a^*}/\varrho_c \approx 1.3$. The strong positive temperature coefficient (PTC) of the resistivity along all three crystallographic directions demonstrates the predominant role of the electron–phonon scattering mechanism, so that the resistivity is of the Boltzmann type.

3.4.2
o-Al$_{13}$Co$_4$

The $\varrho(T)$ data of o-Al$_{13}$Co$_4$ along the three crystallographic directions are displayed in Figure 3.5. The resistivity is the lowest along the stacking a direction perpendicular to the atomic planes, where its RT value amounts $\varrho_a^{300K} = 69\ \mu\Omega$ cm and the residual resistivity is $\varrho_a^{2K} = 47\ \mu\Omega$ cm. The two inplane resistivities are higher, amounting to $\varrho_b^{300K} = 169\ \mu\Omega$ cm and $\varrho_b^{2K} = 113\ \mu\Omega$ cm for the b direction and $\varrho_c^{300K} = 180\ \mu\Omega$ cm and $\varrho_c^{2K} = 129\ \mu\Omega$ cm for the c direction. The anisotropy of the two inplane

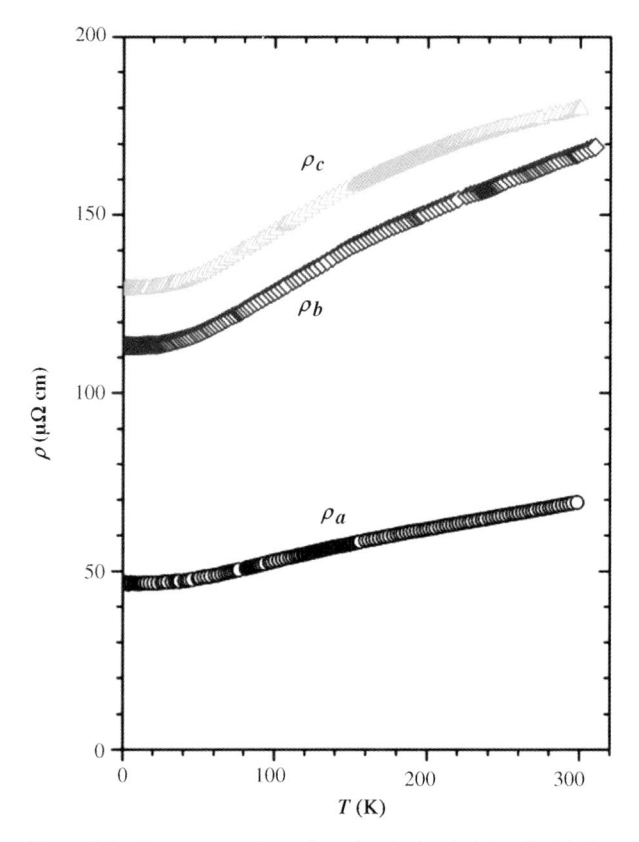

Figure 3.5 Temperature-dependent electrical resistivity of o-Al$_{13}$Co$_4$ along three orthogonal crystallographic directions a, b and c.

resistivities is small, amounting at RT to $\varrho_c^{300K}/\varrho_b^{300K} = 1.07$, whereas the anisotropy to the stacking direction is considerably larger, $\varrho_c^{300K}/\varrho_a^{300K} = 2.6$ and $\varrho_b^{300K}/\varrho_a^{300K} = 2.5$. The anisotropic resistivities thus appear in the order $\varrho_a < \varrho_b < \varrho_c$ (even the inequality $\varrho_a \ll \varrho_b < \varrho_c$ may be considered to hold), which is the same order as that of the anisotropic magnetic susceptibility shown in Figure 3.2. The PTC of the resistivity along all three crystallographic directions demonstrates the predominant role of the electron–phonon scattering mechanism and the resistivity is of Boltzmann type.

3.4.3
Al$_4$(Cr,Fe)

The anisotropic electrical resistivity of Al$_{80}$Cr$_{15}$Fe$_5$ is displayed in Figure 3.6. The resistivity is the lowest along the stacking a direction perpendicular to the atomic planes. ϱ_a shows a PTC in the whole investigated temperature interval and a RT value $\varrho_a^{300K} = 297$ μΩ cm. The resistivities within the atomic planes are higher and exhibit qualitatively different temperature dependencies with a broad maximum, where the temperature coefficient is reversed. ϱ_b exhibits a maximum at about 125 K with the peak value 375 μΩ cm and the RT value $\varrho_b^{300K} = 371$ μΩ cm. The resistivity ϱ_c is the highest; its maximum value 413 μΩ cm occurs at 100 K and the RT value is $\varrho_c^{300K} = 407$ μΩ cm. At RT, the ratios of the resistivities amount $\varrho_c/\varrho_a = 1.37$, $\varrho_b/\varrho_a = 1.25$ and $\varrho_c/\varrho_b = 1.10$. The resistivity of Al$_{80}Cr_{15}Fe_5$ is thus qualitatively different from the Boltzmann-type PTC resistivities of Y-Al-Ni-Co and o-Al$_{13}$Co$_4$. In the following we give some theoretical consideration of this non-Boltzmann behavior.

In an anisotropic crystal, the electrical conductivity $\sigma = \varrho^{-1}$ (the inverse resistivity) is generally a symmetric (and diagonalizable) tensor, relating the current density $\vec{j}$ to the electric field $\vec{E}$ via the relation $j_i = \sum_j \sigma_{ij} E_j$, where $i,j = x,y,z$ denote crystallographic directions. The tensorial ellipsoid exhibits the same symmetry axes as the crystallographic structure. For the orthorhombic Al$_{80}$Cr$_{15}$Fe$_5$ crystal this implies that the conductivity tensor is diagonal in the basis of the crystallographic directions a, b and c. The geometry of our samples (their long axes were along the three crystallographic directions) and the direction of the electric field applied along their long axes imply that diagonal elements $\sigma_{xx} = \sigma_a$, $\sigma_{yy} = \sigma_b$ and $\sigma_{zz} = \sigma_c$ were measured in our experiments. The temperature dependence of each of these elements was then analyzed using the theory of slow charge carriers by Trambly de Laissardière et al. [17], which applies to any diagonal element of the conductivity tensor (in [17], σ_{xx} is considered, but x may be any crystallographic direction).

According to the theory of slow charge carriers [17], the semiclassical (Bloch–Boltzmann) model of conduction breaks down when the mean free path of charge carriers is smaller than a typical extension of their wavefunction. This situation is realized for sufficiently slow charge carriers (where low electronic velocity is a consequence of weak dispersion of the electronic bands) and leads to a transition from a metallic to an insulating-like regime when scattering by defects or temperature effects increases. According to the Einstein relation, the conductivity σ depends

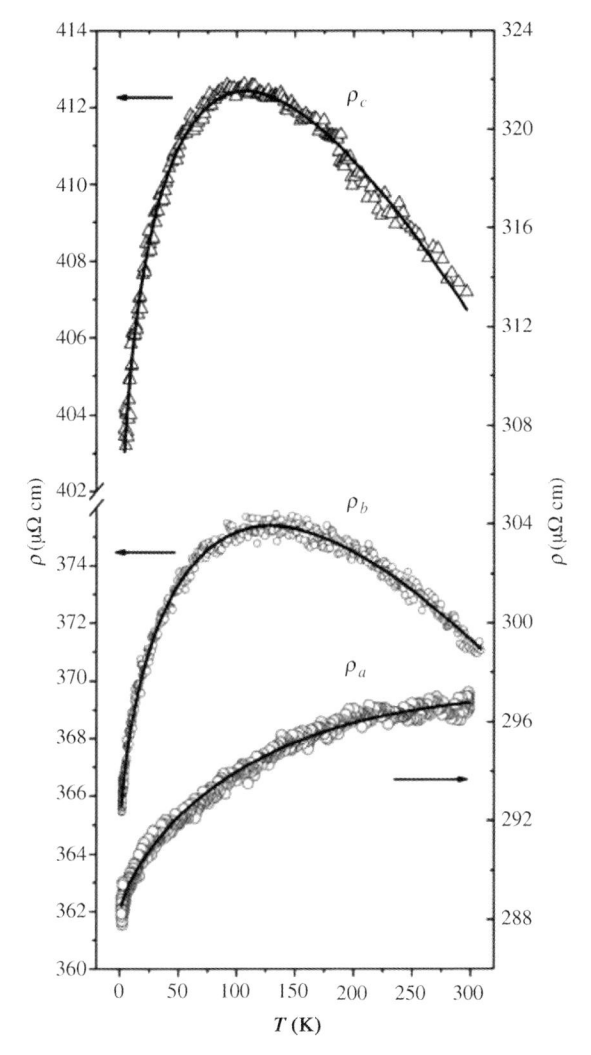

Figure 3.6 Temperature-dependent electrical resistivity of $Al_{80}Cr_{15}Fe_5$ along the three crystallographic directions a, b and c of the orthorhombic unit cell. Solid lines are fits with Equation 3.2 and the fit parameter values are given in Table 3.1.

on the electronic density of states (DOS) $g(\varepsilon)$ and the spectral diffusivity $D(\varepsilon)$ within the thermal interval of a few $k_B T$ around the Fermi level ε_F. In the case of slowly varying metallic DOS around ε_F it is permissible to replace $g(\varepsilon)$ by $g(\varepsilon_F)$. For the diffusion constant it was shown [17] that it can be written as $D = v^2\tau + L^2(\tau)/\tau$, where v is the electronic velocity, τ the scattering (relaxation) time and $L^2(\tau)$ is the nonballistic (non-Boltzmann) contribution to the square of spreading of the quantum state at energy ε due to diffusion, averaged on a time scale τ. $L(\tau)$ is bound by the unit cell length and saturates to a constant value already for short averaging time. The dc conductivity of the system in the crystallographic direction j can be written as

$$\sigma_j = \sigma_{Bj} + \sigma_{NBj} = e^2 g(\varepsilon_F) v_j^2 \tau_j + e^2 g(\varepsilon_F) \frac{L_j^2(\tau_j)}{\tau_j} \tag{3.1}$$

where σ_{Bj} is the Boltzmann contribution and σ_{NBj} is the non-Boltzmann contribution. The scattering rate τ^{-1} will generally be a sum of a temperature- and orientation-independent rate τ_0^{-1} due to scattering by quenched defects and a temperature-dependent term due to scattering by phonons τ_p^{-1}. The anisotropy of the atomic structure implies that the phonon spectrum will also be anisotropic, so that the scattering rate will generally depend on the crystallographic direction, $\tau_j^{-1} = \tau_0^{-1} + \tau_{pj}^{-1}$. In the simplest case, τ_{pj} can be phenomenologically written as a power-law of temperature, $\tau_{pj} = \beta_j / T^{\alpha_j}$. Assuming that $L^2(\tau_j)$ can be replaced by its limiting value, a constant L_j^2, Equation 3.1 yields a minimum in the conductivity σ_j as a function of τ or temperature (or equivalently, there is a maximum in the resistivity $\varrho_j = \sigma_j^{-1}$) at the condition $\tau_j = L_j / v_j$. Above the resistivity maximum, the non-Boltzmann contribution prevails and the resistivity exhibits a nonmetallic negative temperature coefficient (NTC), whereas below the maximum, the resistivity exhibits a metallic PTC due to dominant Boltzmann contribution. The resistivity maxima, as observed for ϱ_b and ϱ_c in Figure 3.6, can thus be considered as a consequence of a crossover from dominant ballistic conductivity at low-T to dominant nonballistic conductivity at high-T due to small velocities of the charge carriers.

Defining $A_j = e^2 g(\varepsilon_F) v_j^2 \tau_0$, $B_j = e^2 g(\varepsilon_F) L_j^2 / \tau_0$ and $C_j = \tau_0 / \beta_j$, Equation 3.1 can be rewritten as

$$\sigma_j = \frac{A_j}{1 + C_j T^{\alpha_j}} + B_j (1 + C_j T^{\alpha_j}) \tag{3.2}$$

that contains four crystallographic-direction-dependent fit parameters A_j, B_j, C_j and α_j (the last two always appear in a product $C_j T^{\alpha_j}$). The zero-temperature conductivity is obtained as $\sigma_j^0 = A_j + B_j$. In the regime of dominant scattering by quenched defects, $\tau_0 / \tau_{pj} = C_j T^{\alpha_j} \ll 1$, normally realized at low temperatures, expansion of Equation 3.2 yields the low-temperature form of the conductivity $\sigma_j = \sigma_j^0 - \sigma_1 T^{\alpha_j}$ (provided $A_j > B_j$). This can be viewed as a generalized Bloch–Grüneisen law that yields a metallic PTC resistivity. In the other extreme of dominant phonon scattering, $\tau_0 / \tau_{pj} = C_j T^{\alpha_j} \gg 1$, normally realized at high temperatures, Equation 3.2 yields the high-temperature form of the conductivity as $\sigma_j = \sigma_2 T^{\alpha_j}$, yielding an insulator-like NTC resistivity. The relative magnitudes of A_j, B_j and $C_j T^{\alpha_j}$ coefficients thus determine the temperature dependence of the resistivity within a given temperature range, which can either be in the metallic or insulating-like regimes, or at a crossover between these two regimes (in which case the resistivity exhibits a maximum). Since these coefficients depend on the electronic structure of the investigated compound ($g(\varepsilon_F)$ and v_j), its crystallographic details (L_j), defect concentration (τ_0) and phononic spectrum (τ_{pj}), they are specific to a given structure and sample purity.

Table 3.1 Fit parameters of the electrical resistivity (solid curves in Figure 3.6, as calculated from Equation 3.2).

Crystallographic direction a			
$A_a (\mu\Omega\,\text{cm})^{-1}$	$B_a (\mu\Omega\,\text{cm})^{-1}$	C_a	α_a
2.16×10^{-3}	1.32×10^{-3}	4.40×10^{-3}	0.70
Crystallographic direction b			
$A_b (\mu\Omega\,\text{cm})^{-1}$	$B_b (\mu\Omega\,\text{cm})^{-1}$	C_b	α_b
1.70×10^{-3}	1.04×10^{-3}	1.27×10^{-2}	0.64
Crystallographic direction c			
$A_c (\mu\Omega\,\text{cm})^{-1}$	$B_c (\mu\Omega\,\text{cm})^{-1}$	C_c	α_c
1.57×10^{-3}	9.34×10^{-4}	2.06×10^{-2}	0.57

The units of the coefficients C_j are chosen so that the temperature in the expression $C_j T^{\alpha_j}$ is dimensionless.

The fits of the resistivities with Equation 3.2 are displayed in Figure 3.6 as solid lines and the fit parameters are collected in Table 3.1. The fits are excellent for all three crystallographic directions. The A_j parameter values enable us to estimate the anisotropy of the electronic average velocities along the three crystallographic directions. We get $v_a/v_b = \sqrt{A_a/A_b} = 1.13$, $v_a/v_c = 1.17$ and $v_b/v_c = 1.04$, so that the velocity is the highest along the stacking a direction and there is also small anisotropy within the atomic planes. The parameters C_j and α_j describe the anisotropy of the electron–phonon scattering rate.

Here, it should be mentioned that a maximum in the resistivity at low temperatures is also predicted by the theory of weak localization [18], frequently used to analyze the temperature-dependent resistivity of icosahedral quasicrystals. Weak localization is considered to introduce small temperature-dependent correction to the Boltzmann conductivity due to spin-orbit and inelastic scattering processes of electrons. However, while the validity of the weak localization concept is restricted to low temperatures, the theory of slow charge carriers of Equation 3.2 is applicable at all temperatures and does not involve any electron localization.

Electrical resistivities of the investigated Y-Al-Ni-Co, o-$Al_{13}Co_4$ and Al_4(Cr,Fe) thus show similar anisotropy, being weak for the two inplane directions and stronger to the stacking direction. For all three compounds, the resistivity is the lowest along the stacking direction. The resistivity values increase with increasing complexity of the compounds, being lowest for the Y-Al-Ni-Co (RT values along the three orthogonal crystallographic directions in the range 25–81 $\mu\Omega\,\text{cm}$), significantly higher for the o-$Al_{13}Co_4$ (RT values in the range 69–180 $\mu\Omega\,\text{cm}$) and even higher for the Al_4(Cr,Fe) (RT values in the range 297–413 $\mu\Omega\,\text{cm}$). This overall increase of the resistivity is also accompanied by the change of the temperature coefficient: while the resistivities of Y-Al-Ni-Co and o-$Al_{13}Co_4$ show Boltzmann-type PTC for all crystallographic directions, the temperature-dependent inplane resistivity of Al_4(Cr,Fe) is non-Boltzmann, exhibiting a maximum by changing the slope from PTC to NTC, whereas the resistivity along the stacking direction still shows PTC in the investigated temperature range, but of the opposite curvature to those of the Y-Al-Ni-Co and o-$Al_{13}Co_4$ compounds.

3.5
Anisotropic Thermoelectric Power

3.5.1
Y-Al-Ni-Co

The thermoelectric power (the Seebeck coefficient S) was measured between 300 and 2 K by using a standard temperature-gradient technique. The sample was mounted on two small, electrically insulated copper thermal reservoirs. Electrical connections to the sample were made via annealed gold leads attached to its ends using silver paint. The temperature difference on the sample was measured by means of a differential Au(0.03% Fe)–chromel thermocouple with junctions glued by a General Electric varnish as close as possible to the electrical connections for measuring the thermopower. At each temperature, we have reversed the temperature difference on the sample several times, keeping it lower than 1 K. The experimental setup was checked by measuring a piece of lead wire. The anisotropic thermopower data of Y-Al-Ni-Co are displayed in Figure 3.7. Thermopower is negative for all three directions, suggesting that electron-type carriers dominate the thermoelectric transport. The RT values are in the range between -2 and $-20\,\mu$V/K in the order $S_{a^*} < S_c < S_b$. The $S(T)$ characteristics for all directions are qualitatively similar, except for the variation in magnitude. In all cases, a change of slope is observed at about 70 K, where the low-temperature slope is higher than the high-temperature one. Nonlinearities in the thermopower in this temperature range are often associated with electron–phonon effects, which typically reach their maximum value at a temperature that is some fraction of the Debye temperature θ_D. The importance of phonons

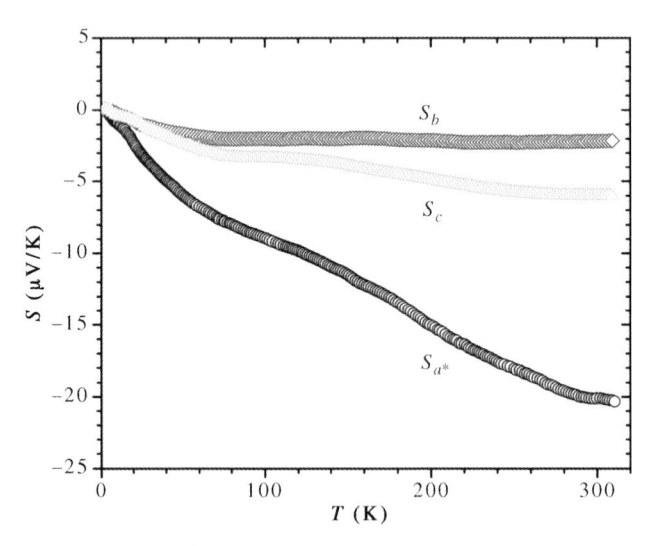

Figure 3.7 Temperature-dependent thermoelectric power (the Seebeck coefficient S) of Y-Al-Ni-Co along three orthogonal crystallographic directions a^*, b and c.

in the temperature dependence of the thermopower of Y-Al-Ni-Co is analogous to the temperature-dependent electrical resistivity of this compound, where electron–phonon interaction represents the main scattering mechanism, leading to the PTC of the resistivity. The thermopower in all three directions extrapolates approximately linearly to zero upon $T \rightarrow 0$, a feature that is usually associated with metallic diffusion thermopower.

3.5.2
o-Al$_{13}$Co$_4$

The thermopower data of o-Al$_{13}$Co$_4$, measured along the three crystallographic directions a, b and c, are displayed in Figure 3.8. The thermopower appears in the order $S_b < S_c < S_a$ and shows the following anisotropy: it is positive along the stacking a direction with the RT value $S_a^{300K} = 18.6\,\mu V/K$, it becomes almost symmetrically negative for the inplane b direction with $S_b^{300K} = -17.1\,\mu V/K$ and is close to zero for the second inplane c direction, amounting $S_c^{300K} = -2.9\,\mu V/K$. Whereas S_a and S_b exhibit a relatively strong linear-like temperature dependence (a small change of slope at about 50 K may be noticed, a feature that is often associated with electron–phonon effects), S_c shows almost no temperature dependence in the investigated temperature range. The observed anisotropy of the thermopower, ranging between positive and negative values, demonstrates that the Fermi surface is highly anisotropic, consisting of electron-like and hole-like parts, which may compensate each other for some crystallographic directions to yield a thermopower close to zero.

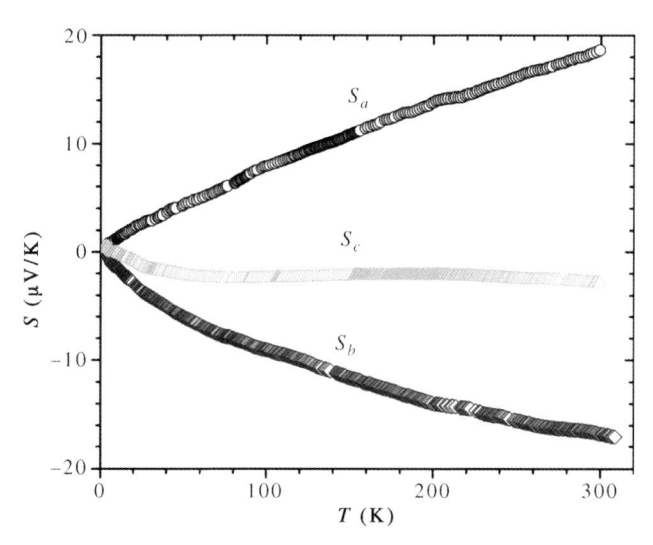

Figure 3.8 Temperature-dependent thermoelectric power (the Seebeck coefficient S) of o-Al$_{13}$Co$_4$ along three orthogonal crystallographic directions a, b and c.

3.5.3
Al4(Cr,Fe)

The anisotropic thermopower of $Al_{80}Cr_{15}Fe_5$ is displayed in Figure 3.9. Thermopower is small in all three directions and the values are in the range between $+1$ and $-2.5\,\mu V/K$. S_a, S_b and S_c exhibit slight quantitative differences in the details of their rather complicated temperature dependencies, but all of them exhibit qualitatively similar features of a pronounced minimum around 80 K and additional local minima and maxima. In the minimum at 80 K, the S_a value is the smallest, whereas at elevated temperatures, S_a exhibits stronger growth than the inplane coefficients S_b and S_c.

Based on the above-presented experimental data, no systematic differences between the anisotropic thermopowers of the investigated Y-Al-Ni-Co, o-$Al_{13}Co_4$ and Al_4(Cr,Fe) compounds can be inferred.

3.6
Anisotropic Hall Coefficient

3.6.1
Y-Al-Ni-Co

The Hall coefficient measurements were performed by the five-point method using a standard ac technique in magnetic fields up to 10 kOe. The current through the samples was in the range 10–50 mA. The measurements were performed in the

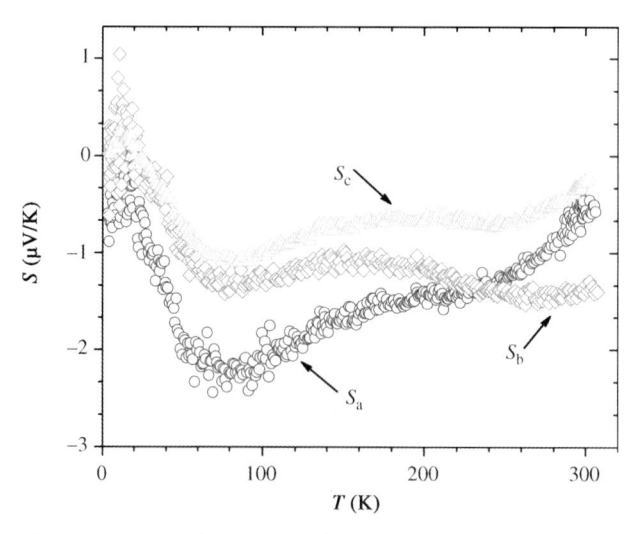

Figure 3.9 Thermoelectric power of $Al_{80}Cr_{15}Fe_5$ along three crystallographic directions a, b and c of the orthorhombic unit cell.

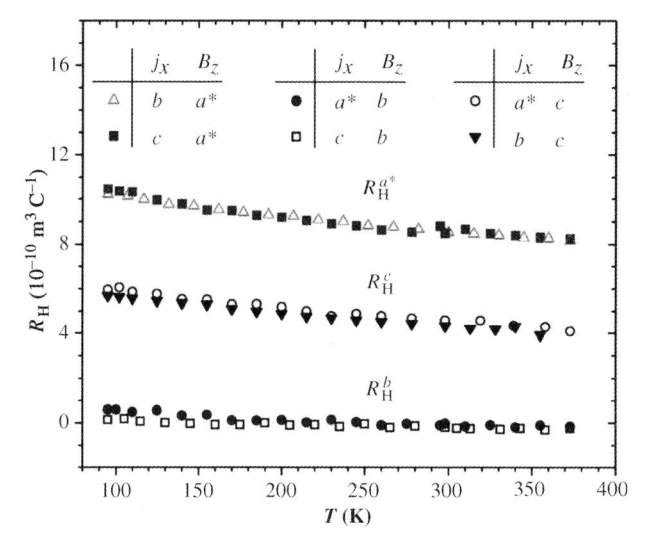

Figure 3.10 Anisotropic temperature-dependent Hall coefficient $R_H = E_y/j_x B_z$ of Y-Al-Ni-Co for different combinations of directions of the current j_x and magnetic field B_z (given in the legend). The superscript a^*, b or c on R_H denotes the direction of the magnetic field.

temperature interval from 90 to 370 K. The temperature-dependent Hall coefficient $R_H = E_y/j_x B_z$ of Y-Al-Ni-Co is displayed in Figure 3.10. In order to determine the anisotropy of R_H, three sets of experiments were performed with the current along the long axis of each sample (thus along a^*, b and c, respectively), whereas the magnetic field was directed along each of the other two orthogonal crystallographic directions, making six experiments altogether. For all combinations of directions, the R_H values are typical metallic in the range $10^{-10}\,\mathrm{m^3\,C^{-1}}$ (with the experimental uncertainty of $\pm 0.1 \times 10^{-10}\,\mathrm{m^3\,C^{-1}}$), showing a weak temperature dependence. This temperature dependence shows a tendency to disappear at higher temperatures. R_Hs exhibit pronounced anisotropy with the following regularity. The six R_H sets of data form three groups of two practically identical R_H curves, where the magnetic field in a given crystallographic direction yields the same R_H for the current along the other two crystallographic directions in the perpendicular plane. Thus, identical Hall coefficients were obtained for combinations $E_b/j_c B_{a^*} = E_c/j_b B_{a^*} = R_H^{a^*}$ (where the additional superscript on the Hall coefficient denotes the direction of the magnetic field), amounting $R_H^{a^*}(300\,\mathrm{K}) = 8.5 \times 10^{-10}\,\mathrm{m^3 C^{-1}}$, $E_{a^*}/j_c B_b = E_c/j_{a^*} B_b = R_H^b$ with $R_H^b(300\,\mathrm{K}) \approx 0$ and $E_b/j_{a^*} B_c = E_{a^*}/j_b B_c = R_H^c$ with $R_H^c(300\,\mathrm{K}) \approx 4.5 \times 10^{-10}\,\mathrm{m^3 C^{-1}}$. The two rather high positive values $R_H^{a^*}$ and R_H^c for the field lying in the (a,c) atomic plane and the almost zero value of R_H^b for the field along the perpendicular b direction (periodic direction in d-QCs) reflect strong anisotropy of the Fermi surface that consists mostly of hole-like parts, whereas electron-like and hole-like parts are of comparable importance for the field perpendicular to the (a,c) plane.

3.6.2
o-Al$_{13}$Co$_4$

The temperature-dependent anisotropic Hall coefficient R_H, of o-Al$_{13}$Co$_4$ is displayed in Figure 3.11. In order to determine the anisotropy of R_H, three sets of experiments were performed with the current along the long axis of each sample (thus along a, b and c, respectively), whereas the magnetic field was directed along each of the other two orthogonal crystallographic directions, making six experiments altogether. For all combinations of directions, the R_H values are typical metallic in the range 10^{-10} m^3 C^{-1} (with the experimental uncertainty of $\pm 0.1 \times 10^{-10}$ m^3 C^{-1}). R_H of o-Al$_{13}$Co$_4$ exhibits similar anisotropy as the previously described Y-Al-Ni-Co with the following regularity. The six R_H sets of data form three groups of two practically identical R_H curves, where the magnetic field in a given crystallographic direction yields the same R_H for the current along the other two crystallographic directions in the perpendicular plane. Thus, practically identical Hall coefficients are obtained for combinations $E_b/j_c B_a = E_c/j_b B_a = R_H^a$ (where the additional superscript on the Hall coefficient denotes the direction of the magnetic field), amounting $R_H^a(300\,\text{K}) = -6.5 \times 10^{-10}$ m^{3}C^{-1}, $E_a/j_c B_b = E_c/j_a B_b = R_H^b$ with $R_H^b(300\,\text{K}) = 3.5 \times 10^{-10}$ m^{3}C^{-1} and $E_b/j_a B_c = E_a/j_b B_c = R_H^c$ with $R_H^c(300\,\text{K}) = -0.6 \times 10^{-10}$ m^{3}C^{-1}. R_H^b and R_H^c are almost temperature independent within the investigated temperature range, whereas R_H^a shows moderate temperature dependence that tends to disappear at higher temperatures.

The observed R_H anisotropy reflects complicated structure of the Fermi surface. The negative $R_H^a < 0$ is electron-like for the magnetic field along the stacking a

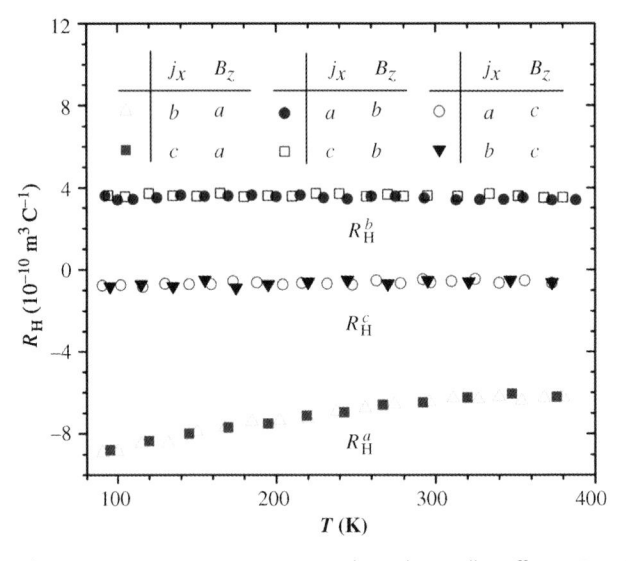

Figure 3.11 Anisotropic temperature-dependent Hall coefficient $R_H = E_y/j_x B_z$ of o-Al$_{13}$Co$_4$ for different combinations of directions of the current j_x and magnetic field B_z (given in the legend). The superscript a, b or c on R_H denotes the direction of the magnetic field.

direction, whereas the positive $R_H^b > 0$ behaves hole-like for the field along the inplane b direction. For the field along the second inplane direction c, $R_H^c \approx 0$ suggests that electron-like and hole-like contributions are of comparable importance.

3.6.3
Al$_4$(Cr,Fe)

The temperature-dependent anisotropic Hall coefficient $R_H = E_y/j_x B_z$ of $Al_{80}Cr_{15}Fe_5$ is displayed in Figure 3.12. In the first set of experiments, the current j_x was directed along the stacking a direction perpendicular to the atomic planes and the magnetic field B_z was applied along either of the two inplane crystallographic directions (b and c). In the second set, the current was flowing inplane along b, whereas the field pointed either along the second inplane direction c or perpendicular to the planes along a. In all cases, R_H shows a metallic behavior by being temperature independent at typical metallic values in the range about $\pm 1 \times 10^{-10}$ m^3 C^{-1}. R_H exhibits pronounced anisotropy. Whereas $R_H = E_c/j_a B_b$ is positive, the other three combinations of directions yield negative Hall coefficients.

The anisotropic Hall coefficient of the investigated Y-Al-Ni-Co, o-Al$_{13}$Co$_4$ and Al$_4$(Cr,Fe) shows the following regularity: the application of the field along the stacking direction always yields the lowest value of the Hall coefficient (for o-Al$_{13}$Co$_4$ and Al$_4$(Cr,Fe), the corresponding Hall coefficient is negative electron-like, whereas

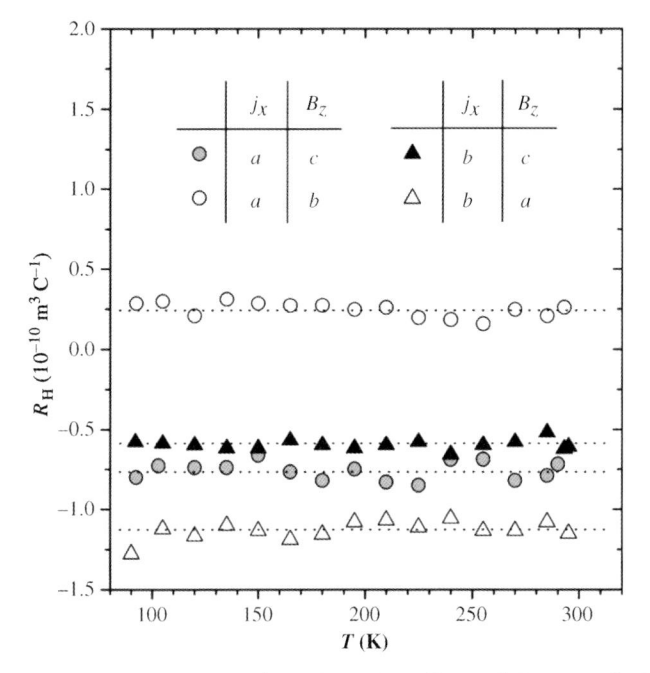

Figure 3.12 Temperature-dependent Hall coefficient of Al$_{80}$Cr$_{15}$Fe$_5$ for different combinations of directions of the current j_x and the magnetic field B_z (given in the legend).

for the Y-Al-Ni-Co, it is practically zero), whereas the application of the field inplane results in higher R_H values and a change of sign to positive hole-like for at least one of the inplane directions. No systematic change of R_H with increasing structural complexity can be claimed.

3.7
Anisotropic Thermal Conductivity

3.7.1
Y-Al-Ni-Co

The thermal conductivity κ of Y-Al-Ni-Co was measured along the a^*, b and c directions using an absolute steady-state heat-flow method. The thermal flux through the samples was generated by a 1-kΩ RuO_2 chip-resistor, glued to one end of the sample, while the other end was attached to a copper heat sink. The temperature gradient across the sample was monitored by a chromel-constantan differential thermocouple. The phononic contribution $κ_{ph} = κ - κ_{el}$ was estimated by subtracting the electronic contribution $κ_{el}$ from the total conductivity using the Wiedemann–Franz law, $κ_{el} = π^2 k_B^2 T σ(T)/3e^2$ and the measured electrical conductivity data $σ(T) = ϱ^{-1}(T)$ from Figure 3.4. Though the use of the Wiedemann–Franz law is a rough approximation, in this way determined $κ_{ph}$ gives an indication of the anisotropy of the phononic spectrum. The total thermal conductivity κ along the three crystallographic directions is displayed in the upper panel of Figure 3.13 and the electronic contribution $κ_{el}$ is shown by solid curves. At 300 K, we get the following anisotropy: $κ^{a^*} = 12.5$ W/mK, $κ_{el}^{a^*} = 9.1$ W/mK with their ratio $(κ_{el}^{a^*}/κ^{a^*})_{300 K} = 0.73$, $κ^b = 46.3$ W/mK, $κ_{el}^b = 29.2$ W/mK with $(κ_{el}^b/κ^b)_{300 K} = 0.63$ and $κ^c = 17.4$ W/mK with $(κ_{el}^c/κ^c)_{300 K} = 0.70$. Electrons are thus the majority heat carriers at RT for all three directions. The anisotropic thermal conductivities appear in the order $κ^{a^*} < κ^c < κ^b$ and similarly $κ_{el}^{a^*} < κ_{el}^c < κ_{el}^b$, which is identical to the order in which the anisotropic electrical conductivities of Y-Al-Ni-Co appear (Figure 3.4): $σ_{a^*} < σ_c < σ_b$. The phononic thermal conductivity is shown in the lower panel of Figure 3.13. We observe that anisotropic $κ_{ph}$s again appear in the same order, $κ_{ph}^{a^*} < κ_{ph}^c < κ_{ph}^b$, so that the phononic conductivity is the highest along the stacking b direction perpendicular to the (a,c) atomic planes, whereas the two inplane conductivities are lower and show smaller anisotropy. For all directions, $κ_{ph}$s show a typical phonon Umklapp maximum at about 40 K. The above results show that Y-Al-Ni-Co is the best conductor for both the electricity and heat along the stacking b direction, whereas both conductivities are smaller in the (a,c) plane.

3.7.2
o-Al$_{13}$Co$_4$

The thermal conductivity κ of o-$Al_{13}Co_4$ was measured along the a, b and c directions and the phononic contribution $κ_{ph} = κ - κ_{el}$ was estimated by subtracting the

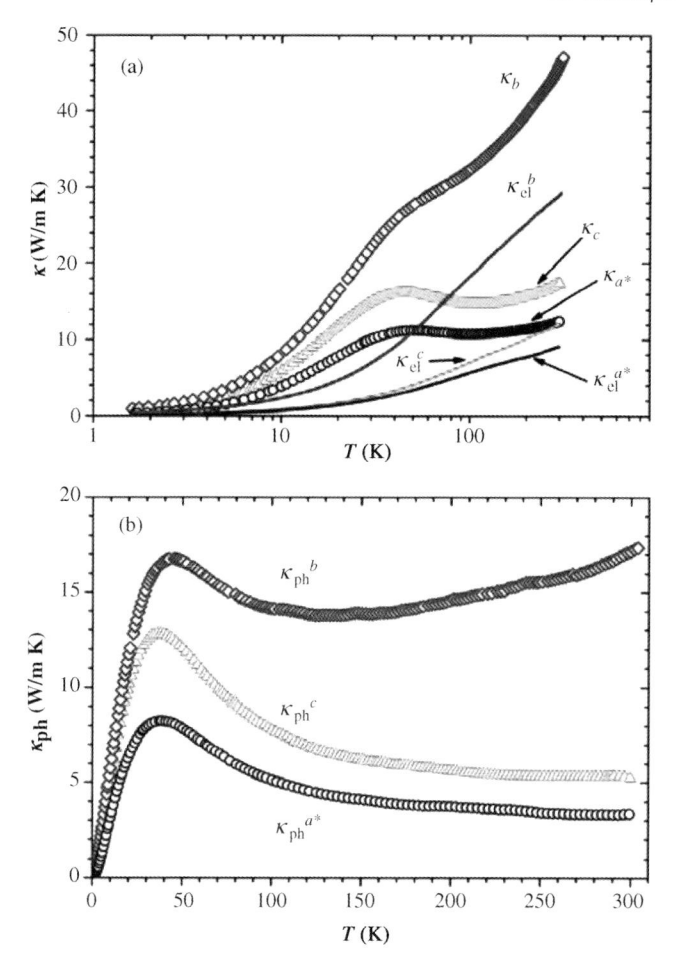

Figure 3.13 (a) Thermal conductivity κ of Y-Al-Ni-Co along the three crystallographic directions a^*, b and c. Electronic contributions κ_{el}, estimated from the Wiedemann–Franz law, are shown by solid curves. (b) Phononic thermal conductivity $\kappa_{ph} = \kappa - \kappa_{el}$ along the three crystallographic directions.

electronic contribution κ_{el} from the total conductivity κ using the Wiedemann–Franz law. The total thermal conductivity κ along the three crystallographic directions is displayed in the upper panel of Figure 3.14 and the electronic contribution κ_{el} is shown by solid curves. At 300 K, we get the following anisotropy: $\kappa^a = 12.5$ W/m K, $\kappa^a_{el} = 10.2$ W/m K with their ratio $(\kappa^a_{el}/\kappa^a)_{300\,K} = 0.82$, $\kappa^b = 6.1$ W/m K, $\kappa^b_{el} = 4.4$ W/m K with $(\kappa^b_{el}/\kappa^b)_{300K} = 0.72$, $\kappa^c = 6.2$ W/m K, $\kappa^c_{el} = 4.1$ W/m K with $(\kappa^c_{el}/\kappa^c)_{300K} = 0.66$. Electrons (and holes) are thus majority heat carriers at RT for all three directions. The anisotropic thermal conductivities appear in the order $\kappa^c \approx \kappa^b < \kappa^a$ and the same order applies to the electronic parts $\kappa^c_{el} \approx \kappa^b_{el} < \kappa^a_{el}$. The thermal conductivity is thus the highest along the stacking a direction, whereas the inplane conductivity is smaller with no noticeable anisotropy between the two inplane directions b and c. Since Figure 3.5 shows that the electrical conductivity

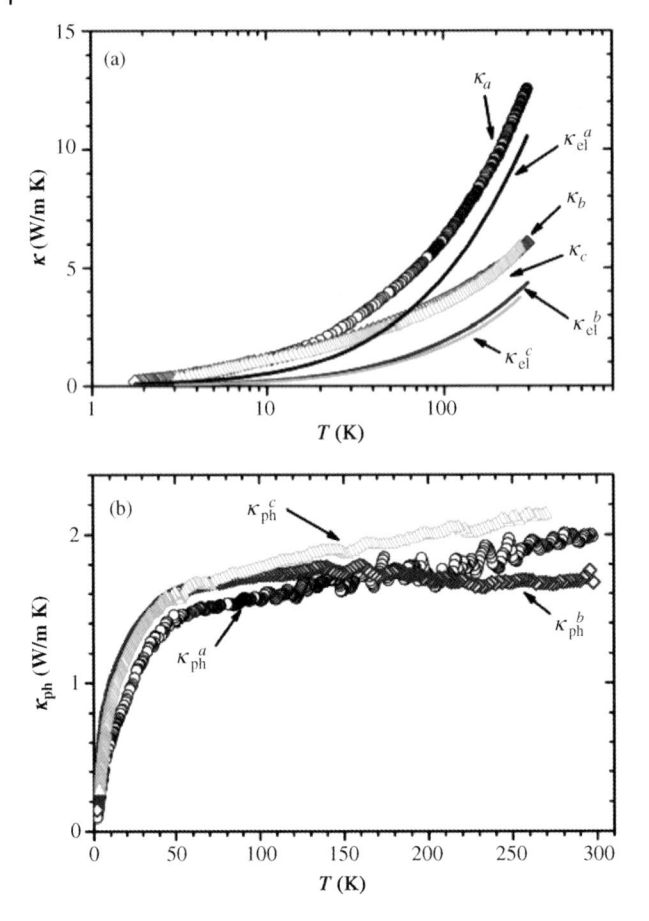

Figure 3.14 (a) Total thermal conductivity κ of o-Al$_{13}$Co$_4$ along the three crystallographic directions a, b and c. Electronic contributions κ_{el}, estimated from the Wiedemann–Franz law, are shown by solid curves. (b) Phononic thermal conductivity $\kappa_{ph} = \kappa - \kappa_{el}$ along the three crystallographic directions.

of o-Al$_{13}$Co$_4$ is also the highest along a (appearing in the order $\sigma_c < \sigma_b < \sigma_a$), this material is the best conductor for both the electricity and heat along the stacking a direction perpendicular to the (b,c) atomic planes. The phononic thermal conductivity is shown in the lower panel of Figure 3.14. We observe that the anisotropy of κ_{ph} is small and no systematic differences between the three directions can be claimed unambiguously.

3.7.3
Al$_4$(Cr,Fe)

The anisotropic thermal conductivity κ of Al$_{80}$Cr$_{15}$Fe$_5$ was measured along the a, b and c crystallographic directions. The total thermal conductivity κ and the electronic contribution κ_{el}, estimated from the Wiedemann–Franz law, are displayed in the

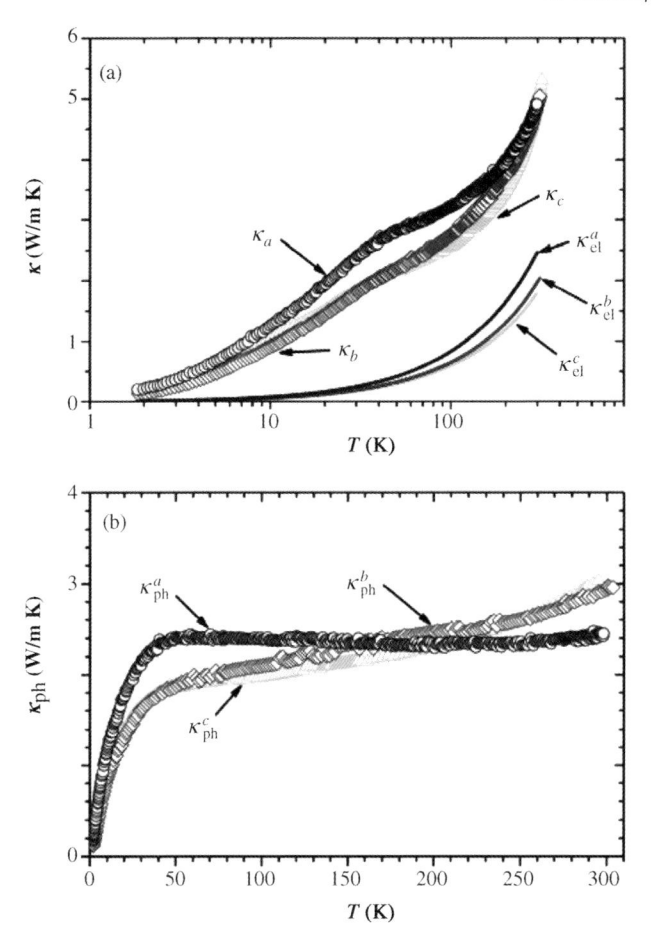

Figure 3.15 (a) Total thermal conductivity κ of $Al_{80}Cr_{15}Fe_5$ along the three crystallographic directions a, b and c of the orthorhombic unit cell. Electronic contributions κ_{el} estimated from the Wiedemann–Franz law, are shown by solid curves. (b) Phononic thermal conductivity $\kappa_{ph} = \kappa - \kappa_{el}$ along the three crystallographic directions.

upper panel of Figure 3.15. κ and κ_{el} are somewhat higher along the stacking a direction, whereas the anisotropy between the two inplane (b,c) conductivities is small. At elevated temperatures, the anisotropy of κ tends to disappear. The phononic contribution $\kappa_{ph} = \kappa - \kappa_{el}$ is shown in the lower panel of Figure 3.15, where we observe that the conductivity κ_{ph}^a along the a direction in the low-T regime below 50 K (which can be associated with the regime where Umklapp processes are still ineffective) is the highest, whereas the two inplane conductivities κ_{ph}^b and κ_{ph}^c are somewhat smaller and also show very weak inplane anisotropy.

The anisotropic thermal conductivity of the investigated Y-Al-Ni-Co, o-$Al_{13}Co_4$ and $Al_4(Cr,Fe)$ thus behaves in complete analogy to the electrical resistivity, by showing weak inplane anisotropy and considerable anisotropy to the stacking direction. For all three compounds, the thermal conductivity is the highest along the stacking direction, so that the investigated decagonal approximant phases are the best conductors for both electricity and heat along the stacking direction perpendicular to the atomic planes. At RT, Y-Al-Ni-Co shows the highest thermal conductivity, and $Al_4(Cr,Fe)$ the lowest one, indicating that increased complexity of the structure results in less efficient electronic and phononic transport of the heat.

3.8
Fermi Surface and the Electronic Density of States

The anisotropy of the electronic transport coefficients originates from the anisotropy of the electronic band structure and the associated anisotropic Fermi surface. Here, we present a theoretical *ab initio* calculation of the Fermi surface and the electronic DOS for the known crystal structural models of the Y-Al-Ni-Co, o-$Al_{13}Co_4$ and $Al_4(Cr,Fe)$ phases.

3.8.1
Y-Al-Ni-Co

The *ab initio* calculation of the band structure $\varepsilon_{\vec{k},n}$ (where n is the band index) of Y-Al-Ni-Co was performed within the framework of the density functional theory by applying Wien97 code [19], which adopts the full-potential linearized-augmented-plane-wave (FLAPW) method [20]. Calculations were based on the Y-Al-Ni-Co structural model by Zhang et al. [6], where we have replaced Ni atoms by Co (thus considering the composition $Al_{75}Co_{25}$ instead of the "Zhang" composition $Al_{75}Co_{22}Ni_3$). The partially occupied sites Al(6) and Al(6′) were taken with probabilities 1 and 0, respectively. The muffin-tin radii around the atoms were 1.16 Å that yielded the basis-set energy cutoff parameter 10 eV. The $\vec{k}$-space summation was performed in terms of the modified tetrahedron method [21]. We used 180 $\vec{k}$ points for the self-consistent cycle and 4320 $\vec{k}$ points in the full Brillouin zone (BZ) for the additional iteration, which was performed in order to obtain a dense mesh of the energy eigenvalues $\varepsilon_{\vec{k},n}$ required for the calculation of the temperature-dependent transport coefficients. The criterion for the self-consistency was the difference in the total energy after the last two iterations, being less than 1×10^{-4} Ry. Calculating the electronic DOS for the Zhang et al. model [6], we found that the forces on some of the atoms were as large as 170 mRy/Å, so that the structure is obviously not in equilibrium. To bring the atoms to their equilibrium positions, we performed structural relaxation as implemented in the ABINIT code [22], which was stopped after the forces on all atoms were less than 0.02 mRy/Å in the relaxed structure. The final atomic coordinates before and after the structural relaxation are published elsewhere [2]. The DOSs of the original and the relaxed model are presented in Figure 3.16. The DOS is strongly dominated by the transition-metal $3d$ states and exhibits a modest pseudogap close to the Fermi level ε_F without any spikes. The calculated Fermi surface in the first BZ of the original and the relaxed model is displayed in Figure 3.17, using the drawing program XCrysden [23]. The Fermi

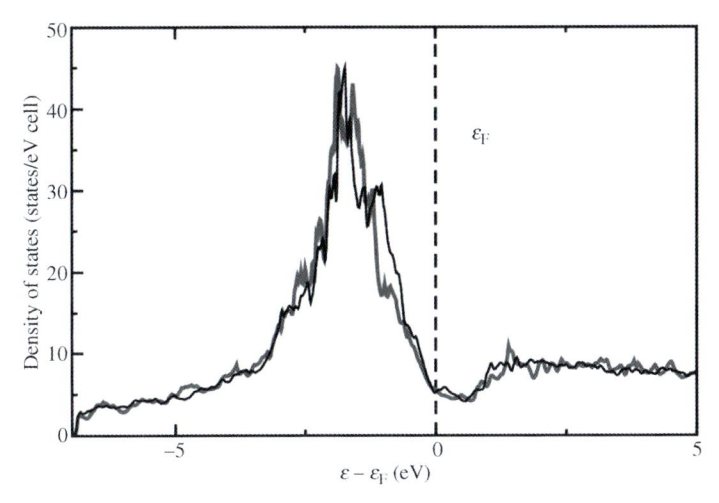

3.8 Fermi Surface and the Electronic Density of States

Figure 3.16 Theoretical electronic DOS of the Y-Al-Ni-Co phase, calculated *ab initio* for the original structural model of Zhang *et al.* [6] (thick gray curve) and the relaxed model [2] (thin black curve), assuming composition $Al_{75}Co_{25}$.

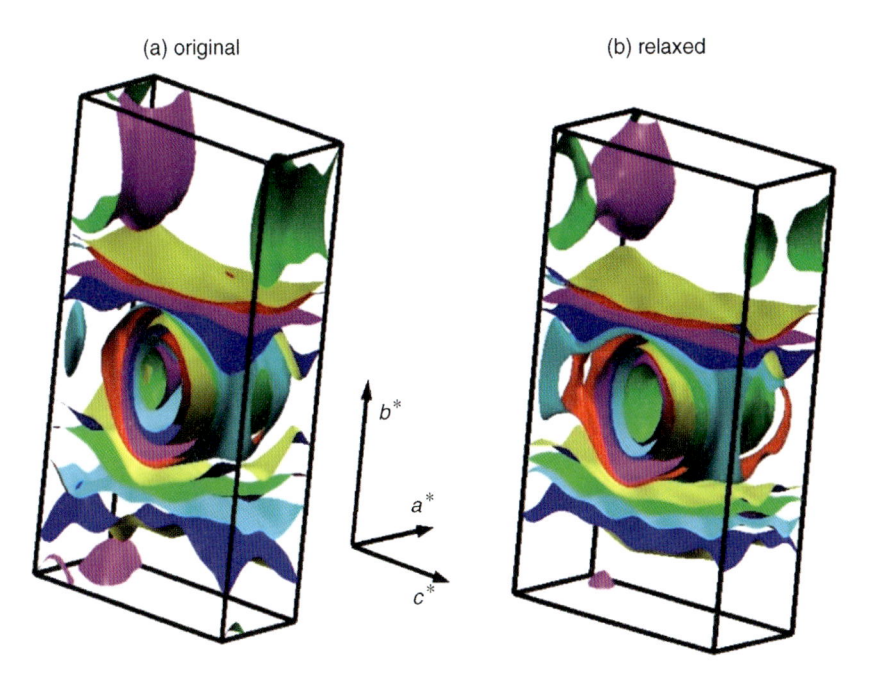

Figure 3.17 Fermi surface in the first Brillouin zone, calculated *ab initio* for (a) the original Y-Al-Ni-Co model (Zhang *et al.* [6]) of composition $Al_{75}Co_{25}$ and (b) its relaxed version [2]. Orientation of the reciprocal-space axes a^*, b^* and c^* is also shown. While a^* and c^* are perpendicular to b^*, the angle between a^* and c^* amounts to $63.83°$.

surface is contributed by eleven bands that cross ε_F, resulting in a significant complexity and highly anisotropic structure. This anisotropy is at the origin of the experimentally observed anisotropy of the Y-Al-Ni-Co electronic transport coefficients.

3.8.2
o-Al$_{13}$Co$_4$

The *ab initio* calculation of the electronic band structure $\varepsilon_{\vec{k},n}$ of o-Al$_{13}$Co$_4$ was performed for the structural model by Grin *et al.* [8]. This model contains no fractionally occupied sites. The atomic coordinates were taken as published in the literature and no additional structural relaxation was performed. We applied the ABINIT [22] code and the local-density approximation (LDA) [24] for the exchange-correlation potential. The electron–ion interactions were described with the norm-conserving pseudopotentials [25] of the Troullier–Martins [26] type. Due to a relatively large number of atoms (102) in the unit cell, the plane-wave cut-off parameter ε_{cut} was limited to 220 eV, whereas, according to the tests, $N_{\vec{k}} = 672\,\vec{k}$ points in the full Brillouin zone (BZ) ($N_{\vec{k},n} = 96$ in the irreducible BZ) were enough to obtain a dense mesh of the energy eigenvalues $\varepsilon_{\vec{k},n}$ required for the calculation of the temperature-dependent transport coefficients, assuring converged results. The *ab initio* calculated Fermi surface, visualized by using the XCrysden program [23], is presented in Figure 3.18. There are eight bands crossing the Fermi energy ε_F, resulting in a significant complexity and a highly anisotropic Fermi surface.

3.8.3
Al$_4$(Cr,Fe)

Containing 306 atoms in the giant unit cell, the *ab initio* calculations of the electronic structure of the Al$_4$(Cr,Fe) phase are more demanding than those for the other two compounds that contain considerably smaller number of atoms in the unit cell. The *ab initio* electronic structure study of Al$_4$(Cr,Fe) was performed with the full-potential linearized augmented plane wave (LAPW) computer program WIEN2k [27], which is based on density functional theory. For our calculations we used the local density approximation of Perdew and Wang [24]. Important details of the calculations are as follows. The radii of the muffin-tin spheres around the atoms were 2.06 Å for the Al atoms, 2.11 Å for the Cr and 2.35 Å for the Fe. The Cr-3p and Fe-3p semicore states, which are not well localized within the muffin-tin spheres (more than 0.01 electrons/atom "leak" out of the spheres), were treated as full valence states using local orbitals [28]. This treatment ensured the orthogonality of the higher-lying valence states with the semicore states. Apart from the local orbitals for the semicore states, the basis further included LAPWs with energies up to 127 eV and additional local orbitals for the Al-3d, Cr-3d and Fe-3d states using the formalism described elsewhere [29, 30]. The $\vec{k}$ space was sampled with 16 special $\vec{k}$ points (corresponding to 64 $\vec{k}$ points in the entire Brillouin zone) using temperature smearing (the electronic temperature was $T_e = 2$ mRy).

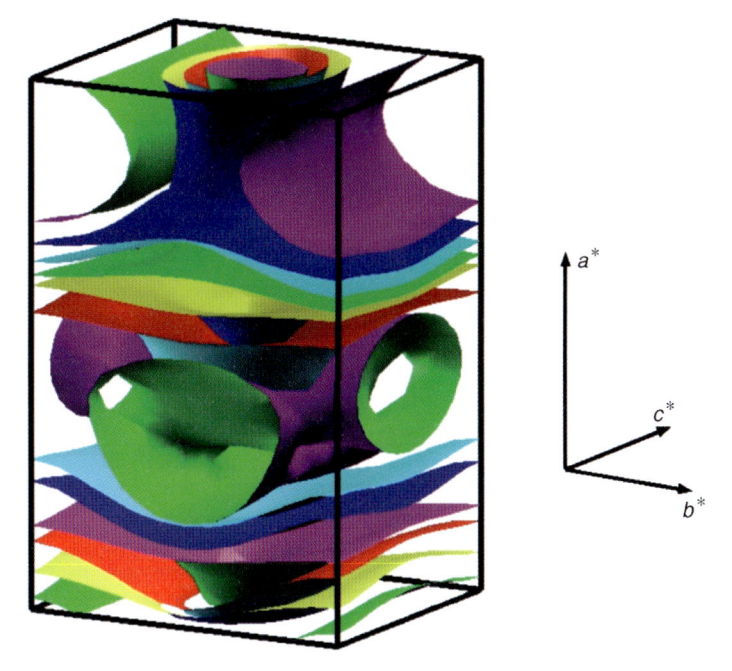

Figure 3.18 Fermi surface in the first Brillouin zone, calculated *ab initio* for the Grin et al. [8] structural model of o-Al$_{13}$Co$_4$. Orientation of the (orthogonal) reciprocal-space axes a^*, b^* and c^* is also shown.

In the Al$_4$(Cr,Fe) structural model of Deng et al. [16], no distinction is made between the Cr and Fe sites. In addition, there exist sites of mixed and partial occupancy. For our calculations, where each site must be either vacant or occupied by a specific atom, we made the following choices: (1) the sites of mixed occupancy 75% Al and 25% (Cr,Fe) were occupied with Al; (2) to avoid short distances, sites of partial occupancy were occupied in 50% of the cases (though in the structure refinement of the Deng model, the occupancy ranges from 50–67%) and (3) for the (Cr,Fe) sites we tried two models, one where neighboring Fe–Fe contacts were avoided as much as possible and another where we tried to avoid Cr–Cr contacts. We found that the first model was more stable, by approximately 2 eV per primitive unit cell, and we used this model as a basis for further *ab initio* calculations. This model had 148 atoms per primitive unit cell ($\approx$153 in the original Deng model) and its composition was Al$_{80.4}$Cr$_{10.8}$Fe$_{8.8}$ (compared to Al$_{80.6}$Cr$_{10.7}$Fe$_{8.7}$ in the original model).

Calculating the DOS for the Deng et al. [16] model with the sites of mixed and partial occupancy occupied as described above, we found that the forces on some of the atoms were as large as $\approx$ 94 mRy/Å. The reason for this is most probably the fact that the model gives atomic positions that are averaged over the different possible occupations of the mixed and partially occupied sites, whereas in our model we have made specific choices described in the previous paragraph. To bring the atoms to their equilibrium positions, we performed structural relaxation, which was stopped

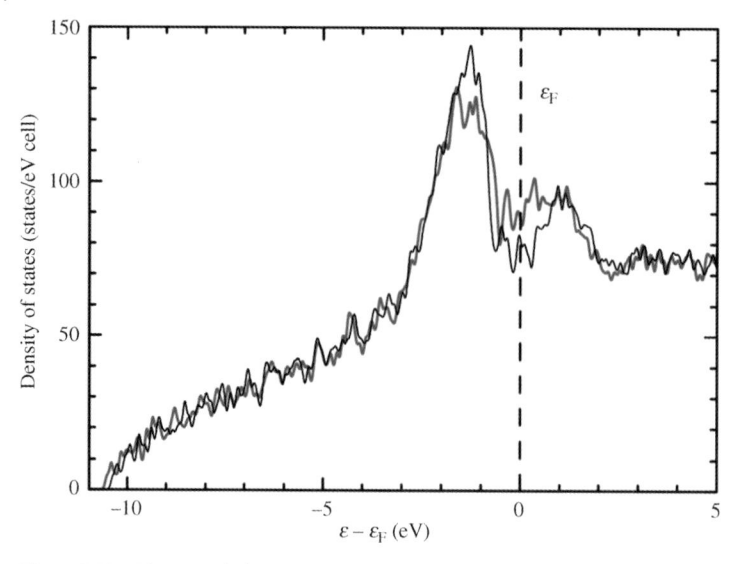

Figure 3.19 Theoretical electronic DOSs of the $Al_4(Cr,Fe)$ phase, calculated *ab initio* for the original (unrelaxed) Deng et al. [16] structural model (thick gray curve) and the relaxed model [4] (thin black curve).

after the forces on all atoms were less than 7 mRy/Å. The relaxed model with the atomic coordinates after the structural relaxation can be found elsewhere [4].

The electronic DOSs of the original Deng model of $Al_4(Cr,Fe)$ and the relaxed model are shown in Figure 3.19. At about 10 eV below the Fermi energy, the free-electron-like DOS of Al can be clearly recognized. Around 1–2 eV below ε_F, there is a peak due to the Cr-$3d$ and Fe-$3d$ states. Comparing the two DOSs, it can be seen that the structural relaxation has caused the DOS at ε_F to become somewhat lower, whereas the (Cr,Fe)-$3d$ peak has become more pronounced. The lowering of the DOS near ε_F indicates further electronic stabilization of the relaxed structure, as compared to the original model with unrelaxed atomic coordinates. For the electronic conduction it is important to note that the DOS at ε_F is high with no tendency to exhibit a pseudogap. The calculated DOS of $Al_4(Cr,Fe)$ is thus metal-like, so that the nonmetallic electrical resistivity displayed in Figure 3.6 occurs in the presence of a high density of charge carriers.

3.9
Theoretical *Ab Initio* Calculation of the Electronic Transport Coefficients

Knowing the Fermi surface, the anisotropic electronic transport coefficients can be evaluated by an *ab initio* calculation. We discuss here the transport coefficients of the Y-Al-Ni-Co and o-$Al_{13}Co_4$ compounds, which show PTC electrical resistivity typical of regular metallic alloys and compounds and can be described, to a good approximation, by the Boltzmann semiclassical transport theory. The non-Boltzmann resistivity of $Al_4(Cr,Fe)$, which exhibits a maximum in $\varrho(T)$ and the change of slope from NTC to PTC for some crystallographic directions, has already been discussed by the theory of slow charge carriers in Section 3.4.3.

The Boltzmann semiclassical theory in the form suitable for *ab initio* calculations is implemented in the BoltzTraP code [31]. The electrical conductivity tensor $\sigma_{ij}(T,\mu)$, as a function of the temperature T and the chemical potential μ, reads as

$$\sigma_{ij}(T,\mu) = \int \sigma_{ij}(\varepsilon)\left(-\frac{\partial f_\mu(T,\varepsilon)}{\partial \varepsilon}\right)d\varepsilon \qquad (3.3)$$

where $f_\mu(T,\varepsilon)$ is the Fermi–Dirac distribution and $i,j = x, y, z$ denote crystallographic directions. Since $-\partial f_\mu/\partial \varepsilon$ is a narrow bell-like function peaked around ε_F with the width of the order k_BT, this restricts the relevant energies entering Equation 3.3 to those in the close vicinity of the Fermi surface. The distribution $\sigma_{ij}(\varepsilon)$ is the sum over the $N_{\vec{k}}$ points $\vec{k}$ and bands n

$$\sigma_{ij}(\varepsilon) = \frac{1}{\Omega}\sum_{\vec{k},n}\sigma_{ij}(\vec{k},n)\delta(\varepsilon - \varepsilon_{\vec{k},n}) \qquad (3.4)$$

where Ω is the unit-cell volume. The tensor

$$\sigma_{ij}(\vec{k},n) = \frac{e^2\tau}{\hbar^2}\frac{\partial \varepsilon_{\vec{k},n}}{\partial k_i}\frac{\partial \varepsilon_{\vec{k},n}}{\partial k_j} \qquad (3.5)$$

where e is the elementary charge, and depends on the relaxation time τ. Neglecting its possible dependence on the band index n, τ varies with the temperature and crystallographic direction. This variation is generally not known, so that we are able to present only the product $\varrho_{ii}(T,\mu)\tau = \tau/\sigma_{ii}(T,\mu)$. In our calculations, the chemical potential μ equals the Fermi energy, obtained from the *ab initio* calculation. The temperature-dependence of the theoretical $\varrho_{ij}(T)\tau$ originates from the Fermi–Dirac function.

Similarly, the electronic contribution $\kappa_{el}^{ij}(T,\mu)$ to the total thermal conductivity κ_{ii} is τ dependent, too

$$\kappa_{el}^{ij}(T,\mu) = \frac{1}{e^2T}\int \sigma_{ij}(\varepsilon)(\varepsilon-\mu)^2\left(-\frac{\partial f_\mu(T,\varepsilon)}{\partial \varepsilon}\right)d\varepsilon \qquad (3.6)$$

so that we are able to present the quantity κ_{el}^{ij}/τ. On the other hand, the relaxation time drops out from the expressions for the Seebeck coefficient $S_{ij} = E_i^{ind}(\nabla_j T)^{-1}$, where E_i^{ind} is the thermoelectric field in the direction i and $\nabla_j T$ is the temperature gradient along j,

$$S_{ij}(T,\mu) = \sum_\alpha (\sigma)_{\alpha i}^{-1}(T,\mu)\frac{1}{eT}\int \sigma_{\alpha j}(\varepsilon)(\varepsilon-\mu)\left(-\frac{\partial f_\mu(T,\varepsilon)}{\partial \varepsilon}\right)d\varepsilon \qquad (3.7)$$

and the Hall coefficient

$$R_H^{ijk}(T,\mu) = \frac{E_j}{j_i B_k} = \sum_{\alpha,\beta}(\sigma)_{\alpha j}^{-1}(T,\mu)\sigma_{\alpha\beta k}(T,\mu)(\sigma)_{i\beta}^{-1}(T,\mu) \qquad (3.8)$$

where $\alpha, \beta = x, y, z$ and

$$\sigma_{\alpha\beta\gamma}(T,\mu) = \int \sigma_{\alpha\beta\gamma}(\varepsilon)\left(-\frac{\partial f_\mu(T,\varepsilon)}{\partial \varepsilon}\right)d\varepsilon \qquad (3.9)$$

The distribution $\sigma_{\alpha\beta\gamma}(\varepsilon)$ is the sum over the $N_{\vec{k}}$ points $\vec{k}$ and bands n

$$\sigma_{\alpha\beta\gamma}(\varepsilon) = \frac{1}{\Omega}\sum_{\vec{k},n} \sigma_{\alpha\beta\gamma}(\vec{k},n)\delta(\varepsilon-\varepsilon_{\vec{k},n}) \qquad (3.10)$$

with

$$\sigma_{\alpha\beta\gamma}(\vec{k},n) = \frac{e^3\tau^2}{\hbar^4}\varepsilon_{\gamma\mu\nu}\frac{\partial \varepsilon_{\vec{k},n}}{\partial k_\alpha}\frac{\partial \varepsilon_{\vec{k},n}}{\partial k_\nu}\frac{\partial^2 \varepsilon_{\vec{k},n}}{\partial k_\beta \partial k_\mu} \qquad (3.11)$$

where $\varepsilon_{\gamma\mu\nu}$ denotes the Levi-Civita tensor [32, 33]. Therefore, since the dependence of the relaxation time τ on the temperature and the crystallographic direction is not known, we are able to present the *ab initio* calculated products $\varrho_{ij}(T)\tau$ and $\kappa^{ij}_{el}(T)/\tau$, whereas $S_{ij}(T)$ and $R^{ijk}_H(T)$ can be calculated directly in absolute figures. It is worth mentioning that the electrical resistivity ϱ_{ii} and the electronic thermal conductivity κ^{ij}_{el} do not distinguish between the negative electron-type carriers and the positive hole-type carriers (the elementary charge appears in their expressions as e^2 and hence does not distinguish between the electrons $(-e)$ and holes $(+e)$, whereas the thermopower S_{ij} and the Hall coefficient R^{ijk}_H distinguish between the electrons and holes (the charge in their expressions appears as e or $1/e$, respectively, hence distinguishing its sign).

3.9.1
Anisotropic Hall Coefficient of Y-Al-Ni-Co

The theoretical anisotropic Hall coefficient of the original [6] and the relaxed [2] Y-Al-Ni-Co structural models (as discussed in Section 3.8.1), using the Fermi surface from Figure 3.17, and calculated for the same set of combinations of the current and field directions as the experimental ones in Figure 3.10 (reproduced here in panel (*i*) of Figure 3.20,), is shown in Figure 3.20, panels (*ii*) and (*iii*). For both models, the six theoretical R^{ilm}_H sets of data form three groups of two similar R^{ilm}_H curves, where the magnetic field in a given crystallographic direction yields similar values of the coefficient for the current along the other two crystallographic directions in the perpendicular plane. There exist some temperature-dependent differences within each group, where the temperature dependence originates from the Fermi–Dirac function. At a quantitative level, the relaxed model (Figure 3.20, panel (*ii*)) gives better matching of the theoretical R^{ilm}_H values to the experimental ones. At 300 K, the relaxed model yields $R^{cba^*}_H \approx R^{bca^*}_H = R^{a^*}_H \approx 8.8 \times 10^{-10}$ m^{3}C^{-1} which compares well to the experimental value $R^{a^*}_H = 8.5 \times 10^{-10}$ m^{3}C^{-1}. The figures for the other four combinations of the current and field directions are theoretical $R^{ba^*c}_H \approx R^{a^*bc}_H = R^c_H \approx 5.5 \times 10^{-10}$ m^{3}C^{-1} versus the experimental $R^c_H = 4.5 \times 10^{-10}$ m^{3}C^{-1} and (taking the average of $R^{ca^*b}_H$ and $R^{a^*cb}_H$) theoretical $R^{ca^*b}_H \approx R^{a^*cb}_H = R^b_H \approx$

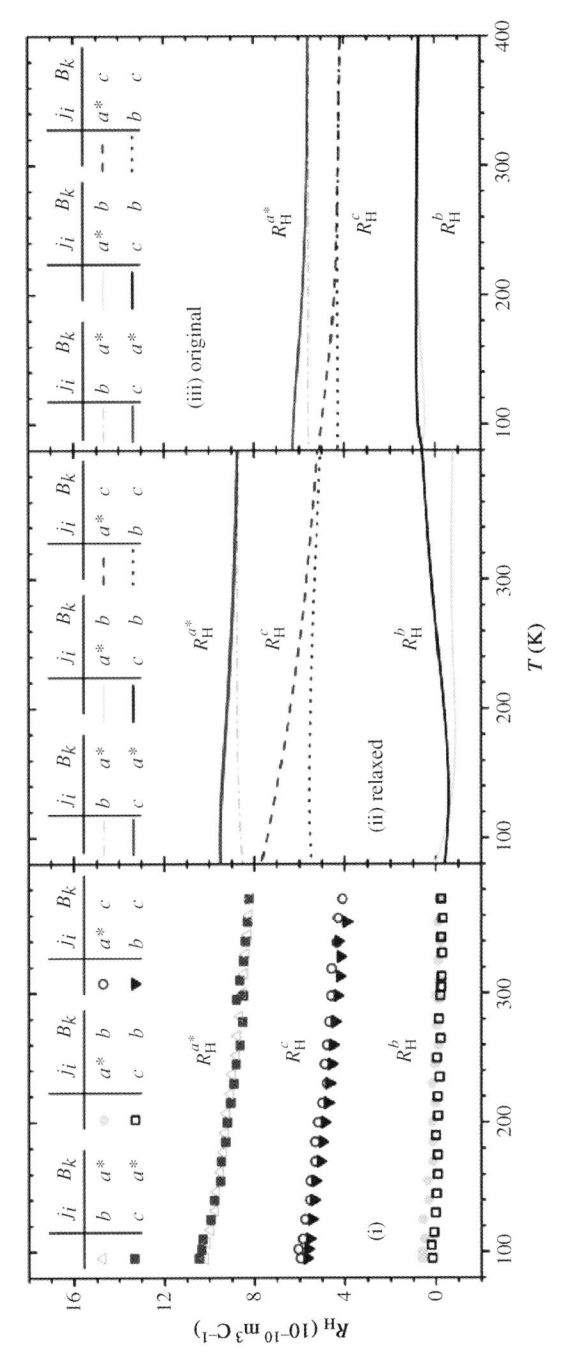

Figure 3.20 (i) Experimental anisotropic temperature-dependent Hall coefficient $R_H^{j\|k} = E_l/j_iB_k$ of Y-Al-Ni-Co, reproduced from Figure 3.10 The directions of the current (j_i) and field (B_k) are given in the legend. The superscript a^*, b or c on R_H denotes the direction of the magnetic field. (ii) Theoretical anisotropic Hall coefficient for the same set of current and field directions as in (i), calculated *ab initio* for the relaxed structural model [2] of the Y-Al-Ni-Co phase of composition Al$_{75}$Co$_{25}$. (iii) Theoretical anisotropic Hall coefficient calculated along the same lines as in (ii), by using the original Zhang *et al.* [6] structural model of Y-Al-Ni-Co.

-0.3×10^{-10} m^{3}C^{-1} versus the experimental $R_H^b \approx 0$. The relaxed model thus reproduces reasonably well the anisotropy of the Hall coefficient at a quantitative level. The small differences between the theoretical and the experimental values very likely originate from the fact that the experiment was performed on the sample of composition Al$_{76}$Co$_{22}$Ni$_2$, whereas the calculation was performed for the composition Al$_{75}$Co$_{25}$. Likewise, the fact that the two data sets within each of the three groups are experimentally practically indistinguishable (i.e. $R_H^{ilm} \approx R_H^{lim}$), whereas theoretically some temperature-dependent differences exist, can be attributed to the inevitable residual structural inhomogeneity of the macroscopic samples, which tends to average out the differences within each group. Considering the theoretical R_H^{ilm} values of the original model (Figure 3.20, panel (iii)), the agreement to the experiment is less satisfactory (the theoretical 300 K values $R_H^{a^*} = 5.6 \times 10^{-10}$ m^{3}C^{-1}, $R_H^c = 4.1 \times 10^{-10}$ m^{3}C^{-1} and $R_H^b = 0.8 \times 10^{-10}$ m^{3}C^{-1} versus the experimental $R_H^{a^*} = 8.5 \times 10^{-10}$ m^{3}C^{-1}, $R_H^c = 4.5 \times 10^{-10}$ m^{3}C^{-1} and $R_H^b \approx 0$). While the theoretical R_H^c and R_H^b values of the original model still match the experimental values well, this is not the case for the $R_H^{a^*}$ value, where the relaxed model gives much better matching to the experimental $R_H^{a^*}$.

The theoretical Hall coefficient, calculated *ab initio* for the Al$_{75}$Co$_{25}$ composition of the Y-Al-Ni-Co phase thus reproduces well the experimental features of the anisotropic Hall coefficient. The relaxed version [2] of the Zhang et al. [6] model reproduces the anisotropic Hall coefficient at a quantitative level for all six investigated combinations of directions of the electric current and magnetic field, whereas the original model gives less satisfactory matching to the experiment, the discrepancy being mainly in the $R_H^{a^*}$ element. The origin of the anisotropic Hall coefficient is the anisotropic Fermi surface, the anisotropy of which in turn originates from the specific stacked-layer structure of the Y-Al-Ni-Co compound and the chemical decoration of the lattice.

3.9.2
Anisotropic Transport Coefficients of o-Al$_{13}$Co$_4$

For the o-Al$_{13}$Co$_4$, the following electronic transport coefficients were calculated *ab initio*: the product of the electrical resistivity and the relaxation time, the ratio of the electronic thermal conductivity and the relaxation time and the Hall coefficient. For the thermopower, no agreement between the theory and experiment was obtained and the results are not presented. The calculations were performed for the o-Al$_{13}$Co$_4$ structural model by Grin et al. [8].

3.9.2.1 Electrical Resistivity

The theoretical $\varrho_a\tau$, $\varrho_b\tau$ and $\varrho_c\tau$ are displayed in Figure 3.21. In comparison to the experimental resistivity of o-Al$_{13}$Co$_4$ (Figure 3.5), the order of the anisotropic resistivity is correctly reproduced (the theoretical $\varrho_a\tau < \varrho_b\tau < \varrho_c\tau$ agrees with the experimental $\varrho_a < \varrho_b < \varrho_c$). In the approximation of a crystallographic-direction-independent τ (so that τ drops out of the resistivity ratios), we obtain the theoretical ratios at $T = 300$ K $\varrho_c/\varrho_b = 1.6$, $\varrho_c/\varrho_a = 7.7$ and $\varrho_b/\varrho_a = 4.8$, which compare at a

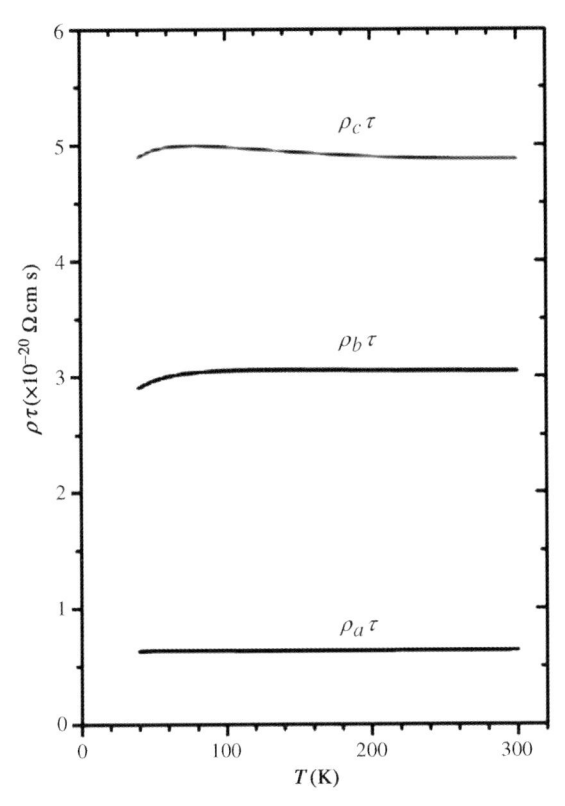

Figure 3.21 Theoretical anisotropic $\varrho_a\tau$, $\varrho_b\tau$ and $\varrho_c\tau$ (products of the electrical resistivity and the relaxation time), calculated ab-initio for the Grin et al. [8] structural model of o-Al$_{13}$Co$_4$.

qualitative level to the experimental 300 K ratios $\varrho_c/\varrho_b = 1.07$, $\varrho_c/\varrho_a = 2.6$ and $\varrho_b/\varrho_a = 2.5$.

3.9.2.2 Electronic Thermal Conductivity

The theoretical κ_{el}^a/τ, κ_{el}^b/τ and κ_{el}^c/τ are displayed in Figure 3.22. In comparison to the experimental electronic thermal conductivity of o-Al$_{13}$Co$_4$ (solid lines in Figure 3.14), the order of the anisotropic κ_{el} is correctly reproduced (the theoretical $\kappa_{el}^c/\tau < \kappa_{el}^b/\tau < \kappa_{el}^a/\tau$ agrees with the experimental $\kappa_{el}^c < \kappa_{el}^b < \kappa_{el}^a$). The theory also reproduces the experimental fact that the anisotropy between the two inplane conductivities κ_{el}^b and κ_{el}^c is small. In the approximation of a direction-independent τ, we obtain the theoretical ratios at $T = 300$ K $\kappa_{el}^a/\kappa_{el}^b = 4.7$, $\kappa_{el}^a/\kappa_{el}^c = 7.6$ and $\kappa_{el}^b/\kappa_{el}^c = 1.6$, whereas the experimental 300 K ratios are $\kappa_{el}^a/\kappa_{el}^b = 2.3$, $\kappa_{el}^a/\kappa_{el}^c = 2.5$ and $\kappa_{el}^b/\kappa_{el}^c = 1.07$, so that the agreement is qualitative only. One reason for the qualitative level of agreement between the theory and experiment is also the fact that the experimental κ_{el}^{ii} coefficients were derived from the total thermal conductivity κ^{ii} by using the Wiedemann–Franz law, which is by itself an approximation to the true κ_{el}^{ii}.

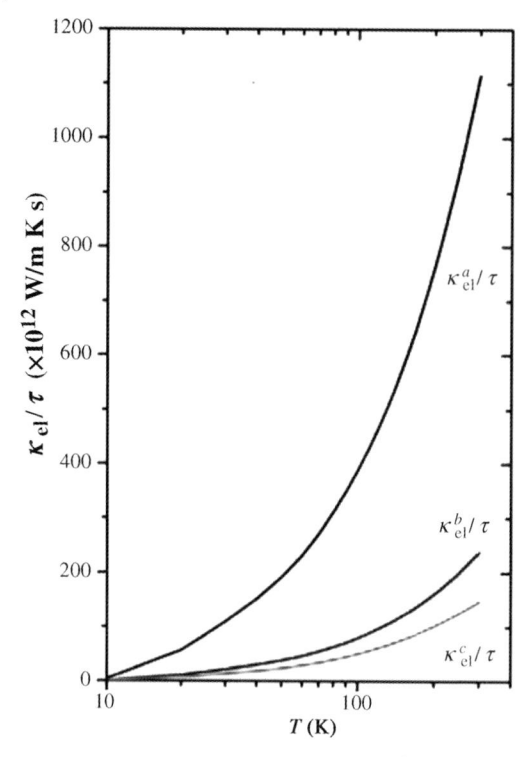

Figure 3.22 Theoretical anisotropic κ_{el}^a/τ, κ_{el}^b/τ and κ_{el}^c/τ (electronic thermal conductivities divided by the relaxation time), calculated *ab initio* for the Grin *et al.* [8] structural model of o-Al$_{13}$Co$_4$.

3.9.2.3 Hall Coefficient

The Hall coefficient R_H^{ijk} can be calculated in absolute figures. The theoretical anisotropic Hall coefficient of o-Al$_{13}$Co$_4$, calculated for the same set of combinations of the current and field directions as the experimental one in Figure 3.11, is shown in Figure 3.23. The theory correctly reproduces the experimental fact that the six theoretical R_H^{ijk} data sets form three groups of two R_H^{ijk} curves, where the magnetic field in a given crystallographic direction yields the same value of the Hall coefficient for the current along the other two crystallographic directions in the perpendicular plane. The theory also correctly reproduces the order of appearance of the anisotropic Hall coefficient (the theoretical order $R_H^a < R_H^c < R_H^b$ matches the experimental one from Figure 3.11) and the crystallographic-direction-dependent change of sign $R_H^b > 0$ and $R_H^a < 0$. The R_H^c coefficient, the value of which is intermediate to R_H^b and R_H^a, is theoretically less well reproduced. Experimentally we have $R_H^c \approx 0$ (even slightly negative), whereas the theory yields $R_H^c > 0$. At a quantitative level, the theory fails by giving too large values for all combinations of directions.

The lack of quantitative agreement between the *ab initio*-calculated theoretical transport coefficients and the experiments could indicate limited applicability of the Boltzmann transport theory to the complex electronic structure of the o-Al$_{13}$Co$_4$ giant-unit-cell intermetallic, where the possible electronic interband transitions

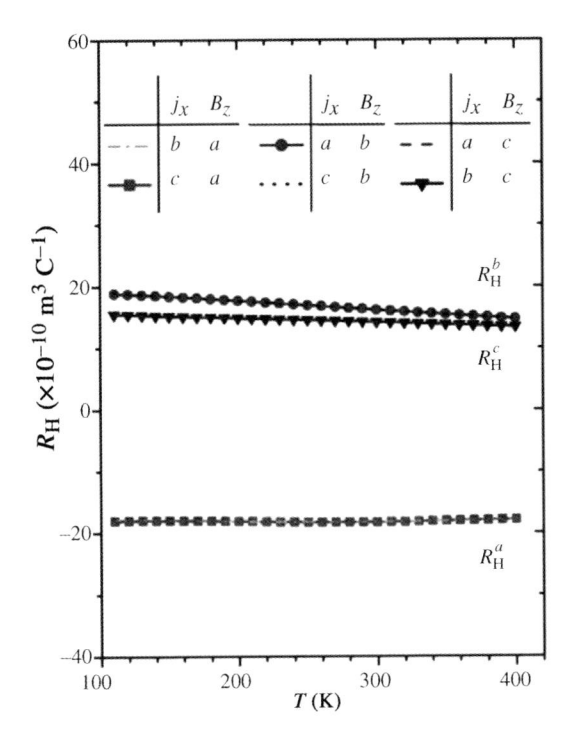

Figure 3.23 Theoretical anisotropic Hall coefficient, calculated *ab initio* for the Grin et al. [8] structural model of o-Al$_{13}$Co$_4$. R_H was calculated for the same set of current and field directions as for the experimental measurements shown in Figure 3.11 The superscript *a*, *b* or *c* on R_H denotes the direction of the magnetic field. There are three groups of curves, each consisting of two practically indistinguishable curves.

between relatively flat (weak dispersion) and closely spaced electronic bands are neglected. Structural relaxation of the Grin et al. [8] model of o-Al$_{13}$Co$_4$ could also yield better agreement with the experiment (for the calculations, the atomic coordinates were taken as published in the literature [8]). The inevitable residual structural disorder in macroscopic real samples may add to the qualitative, but not quantitative agreement between the theory and the experiment.

3.10
Conclusion

We have investigated magnetic susceptibility, electrical resistivity, thermoelectric power, Hall coefficient and thermal conductivity of the Y-Al-Ni-Co, o-Al$_{13}$Co$_4$ and Al$_4$(Cr,Fe) complex metallic alloys of different structural complexity, where these phases are approximants to the decagonal quasicrystal. The main objective was to determine crystallographic-direction-dependent anisotropy of the physical parameters when measured within the atomic planes, corresponding to the quasiperiodic planes in the related decagonal quasicrystals, and along the stacking direction perpendicular to the planes, corresponding to the periodic direction of decagonal quasicrystals. For all three phases, the stacking direction was found to be the most conducting direction for electricity and heat. The origin of the anisotropic physical properties is the anisotropy of the Fermi surface that reflects structural and chemical anisotropy of the investigated stacked-layer compounds. The anisotropic Hall coefficient R_H reveals that the complicated Fermi surface contains electron-like and hole-like contributions. Depending on the combination of directions of the current and magnetic field, electron-like ($R_H < 0$) or hole-like ($R_H > 0$) contributions may dominate, or the two contributions compensate each other, yielding $R_H \approx 0$. Similar complicated anisotropic behavior was observed also in the thermopower that may be either positive, negative or close to zero, depending on the crystallographic direction.

The investigated decagonal approximant phases exhibit anisotropic physical properties qualitatively similar to the d-Al-Ni-Co–type decagonal quasicrystals. The approximants and the decagonal quasicrystals both have in common the structural detail that atomic planes are stacked periodically. This stacked-layer structure and the chemical details of the lattice appear to be at the origin of the anisotropic physical properties, whereas the inplane structural details (either infinite quasiperiodic order in decagonal quasicrystals or periodic crystalline order in the approximant phases) seem to be of marginal importance (if they are of any at all) for the anisotropic magnetic and electronic transport properties of these stacked-layer intermetallic compounds.

Acknowledgement

This work was done within the activities of the 6th Framework EU Network of Excellence "Complex Metallic Alloys" (Contract No. NMP3-CT-2005-500140). We thank Peter Gille and Birgitta Bauer for provision of the samples. We thank Peter Jeglič, Stanislav Vrtnik, Matej Bobnar, Marko Jagodič, Zvonko Jagličić, Igor Smiljanić, Ante Bilušić, Jovica Ivkov, Osor S. Barišić, Denis Stanić and Petar Popčević for their help in the experimental measurements and Matej Komelj and Eeuwe S. Zijlstra for performing the theoretical calculations. A.S. acknowledges support of the Ministry of Science, Education and Sports of the Republic of Croatia through the Research Project No. 035-0352826-2848.

References

1 Smontara, A., Smiljanić, I., Ivkov, J., Stanić, D., Barišić, O.S., Jagličić, Z., Gille, P., Komelj, M., Jeglič, P., Bobnar, M., and Dolinšek, J. (2008) *Phys. Rev. B*, **78**, 104204.

2 Komelj, M., Ivkov, J., Smontara, A., Gille, P., Jeglič, P., and Dolinšek, J. (2009) *Solid State Commun.*, **149**, 515.

3 Dolinšek, J., Komelj, M., Jeglič, P., Vrtnik, S., Stanić, D., Popčević, P., Ivkov, J., Smontara, A., Jagličić, Z., Gille, P. and Grin, Yu. (2009) *Phys. Rev. B* **79**, 184201.

4 Dolinšek, J., Jeglič, P., Komelj, M., Vrtnik, S., Smontara, A., Smiljanić, I., Bilušić, A., Ivkov, J., Stanić, D., Zijlstra, E.S., Bauer, B.,

and Gille, P. (2007) *Phys. Rev. B*, **76**, 174207.
5. Dolinšek, J., Vrtnik, S., Smontara, A., Jagodič, M., Jagličić, Z., Bauer, B., and Gille, P. (2008) *Philos. Mag.*, **88**, 2145.
6. Zhang, B., Gramlich, V., and Steurer, W. (1995) *Z. Kristallogr.*, **210**, 498.
7. Hudd, R.C. and Taylor, W.H. (1962) *Acta Crystallogr.*, **15**, 441.
8. Grin, J., Burkhardt, U., Ellner, M., and Peters, K. (1994) *J. Alloys Compd.*, **206**, 243.
9. Grin, J., Burkhardt, U., Ellner, M., and Peters, K. (1994) *Z. Kristallogr.*, **209**, 479.
10. Edshammar, L.-E. (1964) *Acta Chem. Scand.*, **18**, 2294.
11. Edshammar, L.-E. (1965) *Acta Chem. Scand.*, **19**, 2124.
12. Chaudhury, Z.A. and Suryanarayana, C. (1983) *J. Less-Common Met.*, **91**, 181.
13. Kek, S. (1991) PhD thesis, University Stuttgart, Germany.
14. Steurer, W., Haibach, T., Zhang, B., Kek, S., and Lück, R. (1993) *Acta Crystallogr. B*, **49**, 661.
15. Gille, P. and Bauer, B. (2008) *Cryst. Res. Technol.*, **43**, 1161.
16. Deng, D.W., Mo, Z.M., and Kuo, K.H. (2004) *J. Phys.: Condens. Matter*, **16**, 2283.
17. Trambly de Laissardière, G., Julien, J.-P., and Mayou, D. (2006) *Phys. Rev. Lett.*, **97**, 026601.
18. Fukuyama, H. and Hoshino, K. (1981) *J. Phys. Soc. Jpn.*, **50**, 2131.
19. Blaha, P., Schwarz, K., Sorantin, P., and Trickey, S.B. (1990) *Comput. Phys. Commun.*, **59**, 399.
20. Wimmer, E., Krakauer, H., Weinert, M., and Freeman, A.J. (1981) *Phys. Rev. B* **24**, 864.
21. Blöchl, P.E., Jepsen, O., and Andersen, O.K. (1994) *Phys. Rev. B*, **49**, 16223.
22. Gonze, X., Beuken, J.-M., Caracas, R., Detraux, F., Fuchs, M., Rignanese, G.-M., Sindic, L., Verstraete, M., Zerah, G., Jollet, F. *et al.* (2002) *Comput. Mater. Sci.*, **25**, 478; the ABINIT computer program is a common project of the Université Catholique de Louvain, Corning Incorporated, and other contributors (URL http://www.abinit.org).
23. Kokalj, A. (2003) *Comp. Mater. Sci.*, **28**, 155; code available from http://www.xcrysden.org.
24. Perdew, J.P. and Wang, Y. (1992) *Phys. Rev. B*, **45**, 13244.
25. Fuchs, M. and Scheffler, M. (1999) *Comput. Phys. Commun.*, **119**, 67.
26. Troullier, N. and Martins, J.L. (1991) *Phys. Rev. B*, **43**, 1993.
27. Blaha, P., Schwarz, K., Madsen, G.K.H., Kvasnicka, D., and Luitz, J. (2001) *WIEN2k, An Augmented Plane Wave + Local Orbitals Program for Calculating Crystal Properties*, Karlheinz Schwarz, Techn. Universitaet Wien, Austria.
28. Singh, D. (1991) *Phys. Rev. B*, **43**, 6388.
29. Sjöstedt, E., Nordström, L., and Singh, D.J. (2000) *Solid State Commun.*, **114**, 15.
30. Madsen, G.K.H., Blaha, P., Schwarz, K., Sjöstedt, E., and Nordström, L. (2001) *Phys. Rev. B*, **64**, 195134.
31. Madsen, G.K.H. and Singh, J. (2006) *Comput. Phys. Commun.*, **175**, 67.
32. Hurd, C.M. (1972) *The Hall Effect in Metals and Alloys*, Plenum Press, New York, London.
33. Weber, H.J. (2004) *Essential Mathematical Methods for Physicists*, Elsevier Academic Press, San Diego, USA.

4
Surface Science of Complex Metallic Alloys
Vincent Fournée, Julian Ledieu, and Jeong Y. Park

4.1
Introduction

Studies of solid surfaces are motivated by both fundamental and practical reasons. At a fundamental level, surfaces are of great interest as they break the periodicity of the three-dimensional solid. This can lead to structural changes compared to a simple bulk truncation, as well as the introduction of localized electronic states at the surface. In turn, the basic knowledge of the atomic and electronic structures of solid surfaces is a necessary step towards the understanding of surface properties in the real world. One of the most cited example is probably heterogeneous catalysis, in which the rate of a specific chemical reaction is enhanced by the presence of a catalyst surface. Knowledge of the surface atomic and electronic structures and identification of active sites on the surface are key elements to uncovering mechanisms that govern catalytic activity and selectivity. Corrosion and oxidation resistance, friction and adhesion are additional examples of some of the surface properties that have motivated the development of surface science over the last fifty years. Surface studies of complex metallic alloys started in the 1990s, following the discovery of stable quasicrystalline phases and motivated by the observation reported by Dubois that such materials had potentially useful properties [1]. These include low friction coefficients, poor wetting behavior and high resistance against corrosion, which could combine to form efficient new coating materials. In this chapter, we give an overview of the progress achieved over the last 25 years in the field of surface science of complex metallic alloys (CMA). We mainly focus on clean model surfaces studied in an ultrahigh vacuum environment. All aspects related to the interaction of CMA surfaces with the environment, like oxidation and corrosion, will be treated in Chapter 6. The chapter is organized as follows. In Section 4.2, we summarize our current knowledge of the surface atomic structures of CMA, determined using real and reciprocal space methods, or by using *ab initio* calculations. Section 4.3 deals with the electronic

Complex Metallic Alloys: Fundamentals and Applications
Edited by Jean-Marie Dubois and Esther Belin-Ferré
Copyright © 2011 WILEY-VCH Verlag GmbH & Co. KGaA, Weinheim
ISBN: 978-3-527-32523-8

structure of CMA surfaces and the nature of electronic states in these materials. Nucleation and growth studies are summarized in Section 4.4. This covers experiments as well as simulations that have been developed to identify the nucleation mechanisms and delineate the growth conditions under which the complex structure of the substrate can be enforced in the metal film. Other phenomena, like intermixing and alloying, rotational epitaxy and electron-confinement effects are also discussed in this section. Finally, in Section 4.5 we present our current knowledge of the origin of some of the unusual properties of CMA surfaces, such as wetting, adhesion and friction.

4.2
Surface-Structure Determination

4.2.1
Surface Preparation

Surface-science studies require the use of single crystals, which present a well-defined orientation and no defects such as grain boundaries over macroscopic scales, because the surface structure and properties of crystalline materials strongly depend on their crystallographic orientation. The investigated surface should thus retain the same orientation over distances compatible with the size of the probe, which can be of several mm^2, depending on the technique used. Although a large number of complex metallic alloy phases have been discovered, over the last 25 years following the discovery of quasicrystals, the successful growth of single crystals has been achieved only for a limited number of systems. This explains why most surface studies of CMAs concern the stable F-type icosahedral (i-) quasicrystals (like the i-Al-Pd-Mn, i-Al-Cu-Fe and i-Al-Cu-Ru phases) and the decagonal (d-) quasicrystals (like the d-Al-Ni-Co and d-Al-Cu-Co phases). The number of surface studies of periodic CMAs is even more limited but this is a blooming field thanks to recent achievements in single-crystal growth [2].

Investigations of the intrinsic surface of a material must be performed in an ultrahigh vacuum (UHV) environment (i.e. pressure range of 10^{-9} to 10^{-11} mbar). This is required to keep the surface free of contamination during the time scale of the experiment. As a rule of thumb, the equivalent of one monolayer of residual gas particles (10^{15} atoms per cm^2) will impinge on the surface in one second for a base pressure of 10^{-6} torr (1.33×10^{-6} mbar). If all impinging molecules stick on the surface, then one monolayer (ML) of contaminants will be formed every second. Therefore, even in UHV conditions, a clean surface will remain clean only for a few hours, depending on the surface reactivity. Another reason that justifies the use of the UHV environment is the short inelastic mean-free path of particles interacting with the surface, which amounts to only a few atomic layers in photoemission or low-energy electron diffraction (LEED) experiments.

In this chapter, we are interested in the structure and properties of *intrinsic* surfaces, obtained by truncating the bulk structure of a perfect single crystal. There are several ways to prepare such intrinsic surfaces in UHV, starting from a well-oriented monocrystal. The main methods are (i) cleavage (to expose a fresh surface), (ii) ion bombardment following by annealing, (iii) heating (to desorb contamination layers) or (iv) chemical treatment, all of them being performed *in situ*. Mainly the first two methods have been used for CMA surfaces. In the following, we describe how surface preparation can affect the resulting surface morphology.

The cleaved surface of an *i*-Al-Pd-Mn quasicrystal oriented perpendicular to a high-symmetry direction has been investigated by Ebert *et al.* using scanning tunneling microscopy (STM) [3]. The surface is found to be rough, presenting a hill and valley structure with a typical lateral size of several nanometers. Smaller features can be observed at higher magnification inside these domains, with a diameter of 0.6–1 nm. This fine structure has been interpreted as reminiscent of the bulk structure, which is described by a network of highly symmetric interpenetrated clusters (pseudo-Bergman and Mackay clusters) [4]. It was argued that these clusters are more than just geometrical objects used to describe atomic positions in the aperiodic solid, i.e. a special stability can be associated to them. If this is the case, then one would expect that the fracture plane would try to circumvent the clusters in order to break only the weakest chemical bonds located in-between the clusters. However, Ponson *et al.* recently proposed a different interpretation of the same STM images based on an analysis of the scaling properties of the surface roughness [5]. They concluded that the measured roughness can also be interpreted as the signature of mechanical damage occurring within a 2–3 nm wide zone at the crack tip. Beside this uncertainty in the origin of the roughness of cleaved surfaces, the main drawback of this preparation route is that it offers only a limited control over the surface orientation of the cleavage plane. It is also unreliable in the sense that two cleaved samples will not necessarily expose the same fracture surface. In addition, it is a sample-consuming technique as only a limited number of cleavages can be realized on the same sample, each time offering a fresh surface to be investigated during the next few hours only. However, this destructive technique is sometimes the only solution to prepare a clean surface. It has been used, for example, for photoemission studies of CMAs containing low vapor pressure elements like Mg or Zn, which preclude the use of the sputter-annealing technique.

This latter method is the most widely used for surface preparation of CMA. Typically, a flux of noble gas ions accelerated to 0.5–5 keV is directed toward the sample surface in order to remove the contaminants. The sputtering rate depends on the mass of the noble gas but it is usually larger than unity. One drawback of this method when applied to multicomponent alloys is the preferential sputtering rate of some elements [6]. The structural and chemical disorder induced by sputtering can be recovered by annealing at a temperature that needs to be optimized for each alloy. The optimized annealing temperature must be high enough to promote diffusion, but low enough to avoid preferential evaporation of any alloy component. This surface preparation can lead to a surface terminated by wide terraces (>100 nm) of a very low corrugation (40 pm) separated by steps. This is comparable to the usual terrace and step morphology observed on simple metals and alloys. The method is highly reproducible and is preferred over cleavage. It is important to note that the annealing temperature and the thermal history of the sample are critical and can influence the surface characteristics.

A sputtered surface of an Al-TM quasicrystal leaves an Al-depleted overlayer resulting from the preferential sputtering of Al. For low annealing temperatures (lower than the optimized annealing temperature) the diffusion is limited and the surface composition is not recovered for kinetic reasons. As a result, and in agreement with the phase diagram, cubic overlayers are formed with composition close to $Al_{50}TM_{50}$ and a CsCl type structure [7]. Overheating the sample in UHV can lead to different structures as overlayers. Thus, hexagonal, orthorhombic as well as decagonal structures can coexist with the structure of the i-Al-Pd-Mn phase. These nonicosahedral overlayers may be induced by preferential evaporation. More generally, they are extra phases in thermodynamic equilibrium which coexist with the icosahedral phase [8]. As shown in Figure 4.1 for the i-Al-Pd-Mn quasicrystal, the single phase region shifts in composition with temperature, especially above 900 K. Therefore, depending on the initial composition of the solidified alloy and the annealing temperature used for the surface preparation, high-temperature annealing can lead to a stable two phase surface structure. This has been observed in a number of studies.

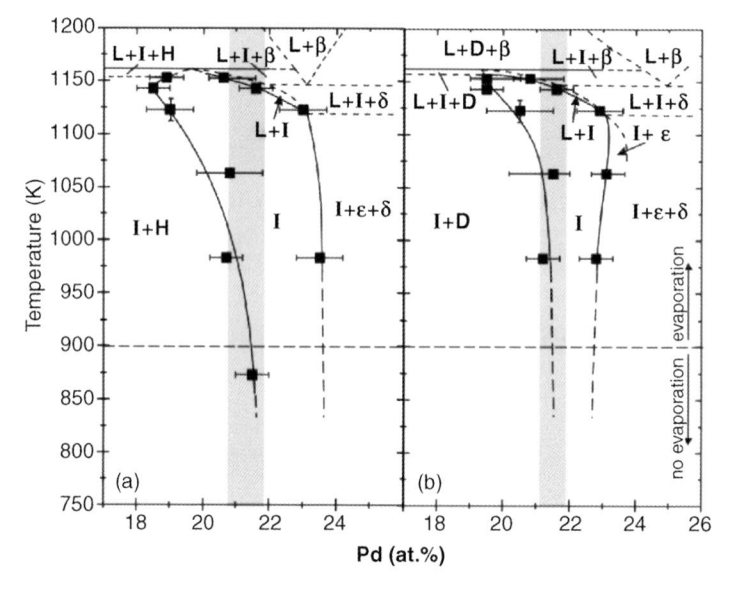

Figure 4.1 Polythermal cross sections through the Al-Pd-Mn phase diagram in the vicinity of the icosahedral phase area. On the Pd-poor sides of the cross sections, the icosahedral phase I is in equilibrium with the (a) orthorhombic H and (b) decagonal D phases. The solid lines should guide the eye to highlight the position of the boundary of the icosahedral single-phase area. The shaded vertical bars highlight the measured composition range of as-grown Al-Pd-Mn single quasicrystals. From reference [8].

The annealing treatment also promotes the diffusion of thermal vacancies, which condense at the surface. It must be noted that single grains of quasicrystals usually contain a high density of thermal vacancies, larger than what is usually observed for other intermetallics. The density of thermal vacancies depends on the thermal history of the sample. As they condense at the surface during annealing, they form faceted voids of different length scales. The surface can become matt as a result of microvoids observable by optical micrograph. At a much smaller length scale, faceted pits have been observed by STM on flat terraces on the five-fold surface of the i-Al-Cu-Fe quasicrystal [9] and on the pseudo-ten-fold surface of the ξ'-Al-Pd-Mn orthorhombic approximant. In this latter case, the size of the pits was found to range from a few nm to hundreds of nm in diameter and their depth is mostly uniform, equal to a single-step height (0.80 nm) [10].

The annealing treatment can also have more subtle effects on the surface characteristics. Annealing the same sample to a slightly higher temperature can reduce the root-mean-square roughness measured on terraces. The step height distribution can also be modified along with the average terrace size on quasicrystal surfaces [11]. In conclusion, surface preparation of CMAs requires fine tuning of several parameters and is slightly more complex compared to simple metal surfaces in general. However, clean and well-ordered surfaces can now be prepared for investigations of their surface structure and properties.

4.2.2
Structure from Real-Space Methods

Prior to reviewing the different aspects of CMA surfaces, it is important to succinctly describe the bulk structure of both icosahedral and decagonal quasicrystals. The bulk structure of the icosahedral sample can be described as a geometric cluster substructure consisting of 33-atom Bergman clusters and 51-atom pseudo-Mackay clusters. The structure along the five-fold axis can be viewed as quasiperiodically spaced atomic layers. The density, the chemical decoration, and the structure of individual planes can differ considerably. Perpendicular to the 5-fold axis, closely spaced layers are grouped into thin and thick blocks, each separated by empty regions (gaps) of various widths.

Decagonal quasicrystals possess a 10-fold symmetry axis along the periodic direction, and two sets of five equivalent 2-fold symmetry axes rotated by 36° in the quasiperiodic plane. The structure of the decagonal Al-Ni-Co phase consists of an ordered arrangement of columnar atom clusters of diameter equal to 20 Å along the ten-fold direction. Several types of phases exist in the Al-Ni-Co system ranging from Ni/Co-rich alloys to type-I superstructures. Depending on the decagonal Al-Ni-Co phase considered, the atomic distribution within the clusters will differ. The 20 Å clusters can be arranged in several manners such as to either decorate a rhombic Penrose tiling as in the case of the type-I superstructure or a pentagonal Penrose tiling for normal phases [12, 13]. The periodicity between the planes stacked along the 10-fold axis is phase dependent and each of these planes exhibits a 2D quasiperiodic structure.

We now concentrate on the works carried out on CMA surfaces. Because of the difficult task in determining the surface structures of aperiodic samples using diffraction methods, much of the progress made in recent years has resulted from the use of real-space methods like scanning tunneling microscopy (STM). Section 4.5 is entirely dedicated to the works and properties measured using atomic force microscopy (AFM).

Following the discovery of the first stable decagonal $Al_{65}Co_{20}Cu_{15}$ quasicrystal [14], the growth of millimeter-sized single grains marked the start of surface-science studies of CMAs. In 1990, the first STM experiment was carried out on a CMA sample by Kortan et al. [15] From their analysis, the surface structure of the decagonal $Al_{65}Co_{20}Cu_{15}$ quasicrystal could be described using a pentagonal quasilattice tiling, hence in good agreement with the high-resolution X-ray-diffraction measurements of the bulk. These STM results further confirmed that the atomic structure could not be understood by multiple twinning as initially suggested by double Nobel Prize winner Linus Pauling [16] to explain the diffraction pattern observed by Shechtman et al. [17].

Soon after these observations, the 5-fold surface of the icosahedral $Al_{70}Pd_{21}Mn_9$ sample was investigated by Schaub et al. [18, 19]. Today, it remains the most studied CMA surface. Atomically, at terraces separated by two main step heights (L and M) appeared at the surface according to a Fibonacci sequence (...LMLLMLML...). Their mutual ratio (L/M) was measured close to the irrational number $\tau = (1 + \sqrt{5})/2$, known as the golden mean. As the surface preparation was not yet optimized, the STM resolution was at this stage quite limited. Hence, only pentagonal depressions and five-fold stars arranged on a Fibonacci pentagrid could be distinguished. Simultaneously, real-space imaging of pentagonal symmetry elements was achieved for the first time using the secondary electron imaging technique (SEI) on the $i\text{-}Al_{70}Pd_{21}Mn_9$ surface [20]. The SEI patterns recorded upon annealing the sample to 700 K showed a high degree of orientational order consistent with icosahedral point-group symmetry [20]. These results were later confirmed by X-ray photoelectron diffraction (XPD) measurements that indicate that the local real-space environments of Al, Pd and Mn atoms possess icosahedral symmetry [21].

While looking for an improved surface preparation, STM measurements have shown that two different surface topographies, rough or flat, can exist at the surface of the $Al_{70}Pd_{21}Mn_9$ sample. When annealing the sample between 700 and 900 K, the surface of nominal bulk composition [22–24] is rough and reveals cluster-like protrusions. The LEED pattern is sharp and five-fold symmetric [25, 26]. At this stage, the SEI patterns can be well reproduced from a structural model based on a pseudo-Mackay cluster (PMI) [24]. For samples annealed briefly up to 925–970 K, the XPD patterns obtained experimentally match the single-scattering cluster calculations also performed for a PMI cluster [21]. Going one step further by using multiple scattering calculations to simulate their experimental XPD patterns, Zheng et al. have shown that the topmost surface is not reconstructed compared to the bulk structure. The planes that fit best their data consist of Al-rich surface planes followed by a dense mixed Al/Pd/Mn layer [27]. The structural relaxations perpendicular to the 5-fold axis are also consistent with the previous values reported using dynamical LEED [28].

Annealing to higher temperatures (between 830 and 930 K) for several hours leads to atomically flat terraces [25, 29]. From STM measurements, it has to be noted that void-rich terraces can appear at the surface, a sign of an intermediate stage of structural equilibration [30]. Following such preparation and complementary to the work reported by Schaub *et al.* [18], an additional step height (S) of $L-M = 2.40 \pm 0.2$ Å has been observed on the $Al_{70}Pd_{21}Mn_9$ surface [29]. From the analysis of the step morphologies, it has been shown that the step stiffness on the 5-fold Al-Pd-Mn surfaces increases as the step height increases. However, the step stiffness is lower than on Al (111) and on the ξ'-Al-Pd-Mn approximant [31].

To interpret the surface structure present on individual terraces, Shen *et al.* have compared autocorrelation functions calculated from experimental STM images and structural models available; experimental data could not distinguish between pseudo-Bergman and pseudo-Mackay clusters as possible building blocks of the quasicrystalline structure [29]. A different approach has been employed by Ledieu *et al.* to analyze atomically resolved STM images obtained from the $Al_{70}Pd_{21}Mn9$ surface: a tiling of the surface (see Figure 4.2) has been constructed using pentagonal prototiles of edge length 8 Å [32]. This dimension is consistent with predominant distances measured by low-energy ion scattering (LEIS) [33]. The resulting Penrose P1 tiling is found to be consistent with tiling derived from slices through the bulk geometric model [11] and the experimentally derived model of Boudard *et al.* [32, 34].

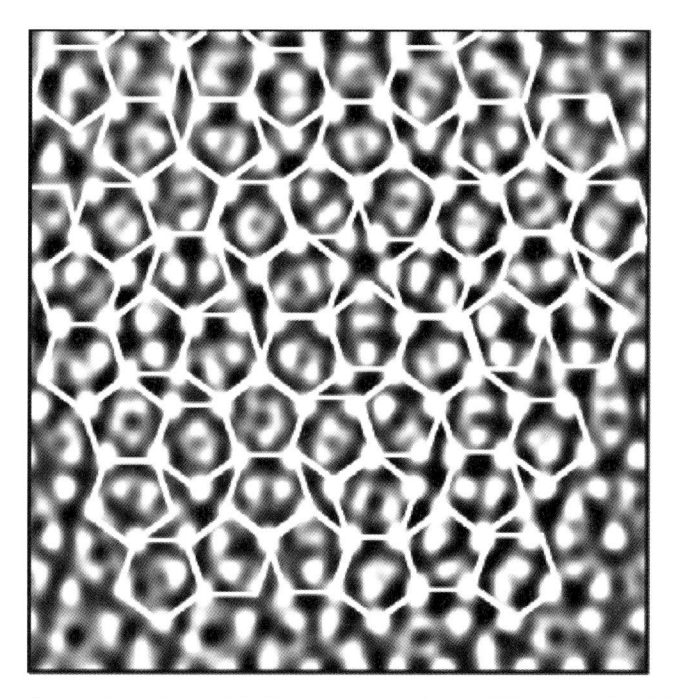

Figure 4.2 A Penrose P1 tiling superimposed on an STM image of the 5-fold Al-Pd-Mn surface (100 Å × 100 Å). From reference [11].

This good agreement confirmed earlier conclusions by Gierer *et al.* based on dynamical LEED analysis that quasicrystalline surfaces prepared by sputtering and annealing correspond to bulk truncation [28]. Using STM, Barbier *et al.* have also concluded a bulk-terminated surface and indicated that the succession of step heights separating terraces agree with the distances between dense Al planes [35]. Thus, both Bergman and pseudo-Mackay clusters must be truncated to generate the topmost surface layers. For crystals, the Bravais rule states that the surface usually forms at dense planes and at large interlayer spacings. Following these observations, Papadopolos *et al.* proposed a modified Bravais rule for CMA based on atomic density of terminations to predict the bulk planes expected at the surface [36].

Within terraces, several characteristic local features can be recognized and have been labeled white flowers (WF), and dark stars (DS) [11]. The first motif corresponds to an equatorially truncated pseudo-Mackay cluster surrounded by Bergman clusters resembling flower petals. The second pattern was first observed by Schaub *et al.* and corresponds to pentagonal depressions with the topmost surface planes. The origin of the DS motifs is still a subject of debate as it could be understood as originating either from truncated Bergman clusters [11, 37] or from the low-coordinated irregular first atomic shell of the pseudo-Mackay clusters [38, 39]. As will be shown in Section 4.4.1.2, these sites play a crucial role in adsorption studies. For the entire temperature range described above, the surface chemical composition of the 5-fold quasicrystal surface has been monitored using LEIS, medium-energy ion scattering (MEIS) and X-ray photoelectron spectroscopy (XPS). No chemical segregation has been reported so far on the quasicrystal surface. However, as mentioned in Section 4.2.1, upon annealing the $Al_{70}Pd_{21}Mn_9$ surface above 970 K, selective evaporation that obeys the vapor pressure of the three elements occurs [22, 23, 33, 40, 41]. Thus, the surface composition is shifted away from the narrow icosahedral region of the phase diagram and new crystalline phases are formed [41–43].

The 5-fold surface of the Al-Cu-Fe quasicrystal is another icosahedral phase that has also been investigated using real-space methods. The first STM images obtained on atomically flat terraces (corrugation of 0.25 Å) have been reported by Becker *et al.* [44]. The most dominant motifs of the surface structure are daisy-like ten-fold features. With an increased resolution, these ten petals flowers have been described by Cai *et al.* [9] as two concentric rings of ten atoms each with a filled center. Using two-dimensional cuts of bulk models derived from X-ray and neutron diffraction experiments, it has been possible to associate these flowers motifs and their distribution with features on Al-rich dense atomic planes. These results are also consistent with previous dynamic LEED calculations [45]. As for the *i*-Al-Pd-Mn surface, it has been concluded that the 5-fold *i*-Al-Cu-Fe surface is bulk-terminated [9]. Through the inspection of the step height distribution and atomically resolved STM images at the Al-Cu-Fe surface, Sharma *et al.* [46] have shown that the surface terminates preferentially at large gaps with the topmost layers having a large atomic density and a high Al content.

We now briefly review the works reported on decagonal Al-Ni-Co surfaces [46–50]. Perpendicular to ten-fold axis, terraces of width on the order of 100 Å are separated by 2 Å monoatomic step height [46, 48, 50, 51], consistent with the bulk interlayer spacing. A tendency towards step doubling has been reported by Sharma et al. on the Ni-rich (d-$Al_{71.7}Ni_{18.7}Co_{9.6}$) surface as well as for the type-I superstructure phase (d-$Al_{71.8}Ni_{4.8}Co_{13.4}$) [46, 52]. Regardless of the decagonal surfaces investigated several common features can be outlined. For instance, protrusions of atomic height sitting on top of pentagonal motifs have often been observed on terraces [46–48, 50, 51]. High-resolution STM images have shown that similar pentagonal features point in the same direction and are quasiperiodically distributed within each terrace [48]. Because adjacent terraces are related by inversion symmetry (pentagonal features rotated by $2\pi/10$ on successive terraces), the overall surface remains ten-fold symmetric [47, 48, 53]. For the Co-rich phase, connecting pentagonal features on individual terraces leads to a tiling of the quasicrystal surface described as a random pentagonal tiling [50]. This pentagonal tiling is in agreement with the tiling observed in the bulk structure of the Co-rich phase. The good agreement between the maxima observed in STM images and the density of Al surface atoms suggests that transition-metal atoms may not be observed at this decagonal quasicrystal surface [50], in contradiction with the recent calculations reported on the W-(Al-Co-Ni) surface [54]. At the surface of the type-I superstructure, two different terminations labeled "coarse" and "fine" coexist on the same single terrace. At high temperature, only the fine structure is maintained at the surface. Contrary to the Co-rich phase, the fine structure observed by STM is described using a random rhombic tiling of 20 Å edge length [46]. From autocorrelation analysis, this structure is found to be consistent with bulk truncations.

As mentioned above, the 2-fold surfaces of the d-Al-Ni-Co quasicrystal provide, within the same surface, a periodic direction along the 10-fold axis, and an aperiodic direction perpendicular to it. Surface properties that might depend on the existence of periodicity, such as friction, can thus be uniquely studied [55]. Park et al. have shown that the atomic structure of the 2-fold, (10000) surface of Al-Ni-Co mainly consists of atomic rows of atoms with 4 Å periodicity along the 10-fold direction and aperiodically spaced in the 2-fold direction [56]. Figure 4.3a shows a high-resolution scanning tunneling microscopy (STM) image of a terrace. Except for a defect in the form of a missing row (visible as a dark band), it consists of atomic rows of close, but not exactly the same, apparent height, with variations of 0.3 Å. Two different lengths, $S = 7.7 \pm 0.3$ Å and $L = 12.5 \pm 0.4$ Å, separate the rows and define the sides of pseudo unit cells. Secondary rows of lower apparent height are visible inside the cells, two within L and one within S, as shown in Figure 4.3b. The spacing between these secondary rows are $L_2 = 4.9 \pm 0.3$ Å, and $S_2 = 2.8 \pm 0.2$ Å. As shown in the figure, an intermediate-level partition can be considered with L_1 and S_1 separations, where $L_1 = L_2 + S_2$ and $S_1 = L_2$. The ratios L/S, L_1/S_1 or L_2/S_2 are all close to the Golden Mean. The L and S distances form an $LSLSLLSL$ sequence, which corresponds to a Fibonacci sequence (a Fibonacci sequence is a progression of numbers that are sums of the previous two terms: $f_{(n+1)} = f_{(n)} + f_{(n-1)}$).

Park et al. have found that this surface structure is reasonably consistent with the model of Takakura et al. [57] for $Al_{72}Ni_{20}Co_8$ in several aspects. First, its interlayer periodicity matches the 4 Å observed in the STM images. Secondly, among the many possible planes that can be generated by cuts parallel to the 2-fold plane, the planes

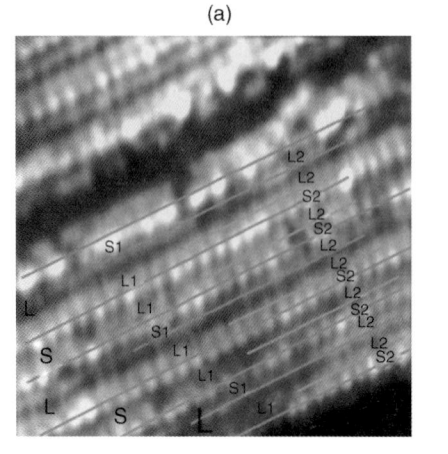

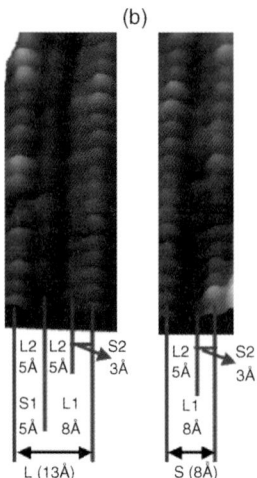

Figure 4.3 (a) STM images (90 Å × 90 Å) in the 2-fold Al-Ni-Co surface. The terraces are made of rows of periodically arranged atoms 4 Å along the 10-fold direction and separated by distances L and S. (b) Expanded view showing the interior in the L and S sections. L contains two atomic rows, separated by L_2 and S_2 distances. S contains one row, at distances of L_2 and S_2 from the boundary. The sequence of L and S spacings between rows follows a Fibonacci sequence. From reference [56].

made of pure Al are the only ones that: (a) contain rows with spacings matching the observed L and S sequences and also the daughter segments L_2 and S_2; (b) contain three compositionally different rows consistent with the observed bias dependence of the STM images. The Al termination is presumably driven by the lower surface energy of Al relative to the two transition metals, consistent with the observation that icosahedral quasicrystals exhibit Al-rich terminations, selected from among the possible bulk layers [56].

Finally, the surface of the ξ'-Al-Pd-Mn approximant phase has been investigated by Fournée et al. [43] Exceptionally large flat terraces can be prepared on this sample. Terraces of relatively smooth step edges compared to parent quasicrystals are separated by a unique height equal to half the period along the pseudo-10-fold axis. This step-height selection implies that the most favorable atomic layers for surface termination are pairs of layers related by a mirror plane. Within terraces, protrusions sitting on top of a flat layer have been understood as groups of atoms belonging to incomplete bulk planes ($y=0.25$ in the bulk model [58]). The dot-like features correspond to a decagonal ring of Al atoms that are part of the 3D cluster units of the phase, preferentially regrown upon annealing. The layer beneath these rings originates from closely spaced bulk planes [43].

From all the studies presented above, it appears that surface topmost layers are bulk terminated. For the selection of the plane as surface termination, a general trend has emerged: the densest layers containing the highest concentration of low surface energy elements are preferred surface terminations. A layer can be defined here as a combination of closely spaced atomic layers separated by less than 1 Å. An additional factor has to be taken into account for Al-based quasicrystal surfaces. It has been demonstrated that the surface terminates preferentially at large gaps with the topmost layers having a large atomic density and a high Al content.

4.2.3
Structure from Reciprocal-Space Methods

Low-energy electron diffraction (LEED) has been of great use to optimize the surface preparation of CMA samples and to discuss qualitatively the symmetry of the CMA surfaces and the spacing of the diffraction spots [7, 59–61]. Additional information on the atomic-scale composition of the surface structure can be gained by using the LEED optic in a dynamic mode. However, an initial structural model is required for such analysis. In 1997, Gierer *et al.* have exploited the full strength of dynamical LEED analysis (LEED I–V) to investigate atomic structure at the Al-Pd-Mn quasicrystal surface [28, 62]. Using approximations in LEED theory to accommodate for the aperiodicity of the sample, it has been demonstrated that the 5-fold surface of the Al-Pd-Mn quasicrystal consists of a mix of several relaxed bulk-like terminations. A contraction of 0.1 Å compared to the bulk value of 0.48 Å has been measured between the two topmost layers. The chemical composition for the outermost layer has been estimated at 93% Al and 7% Mn, whereas the second layer has a composition estimated at 49% Al, 42% Pd and 9% Mn. Hence, the overall composition for the two topmost layers is $Al_{77}Pd_{15}Mn_8$ with a 2D density calculated around 0.136 atoms/$Å^2$ [28, 62]. This lateral surface density is similar to that of the Al (111) surface. The spacing between adjacent terminations found through this LEED I–V analysis are in good agreement with the step structure measured by STM [18, 62]. Local atomic arrangements like pentagonal holes observed by real-space methods have also been correlated to structural features present on the identified bulk-like terminations.

By means of surface X-ray diffraction, Alvarez *et al.* [63] concluded that the annealed 5-fold Al-Pd-Mn sample exhibits an almost perfectly flat surface. From the angular width of the traverse scan, the lateral domain length is estimated at 380 Å, a value comparable to other perfect surfaces of single crystals [63]. The structural model used [64] has reproduced nicely the measured Bragg peaks and the surface signal. From the parameters obtained from the fit, the outermost layer tends to be contracted and Al and Pd rich compared to the bulk value, a composition slightly different from the LEED I–V analysis above [28].

Additional studies using complementary probes have demonstrated that the Al-Pd-Mn quasicrystal surface annealed above 900 K is indeed atomically flat and bulk terminated [22, 65]. The symmetry and the corrugation of the large terraces have also been analyzed by Barbier *et al.* [35] and Sharma *et al.* [66] using He diffraction patterns (HAS), an extremely surface sensitive technique that directly probes the reciprocal space of the surface corrugation.

The structural stabilities of the three high-symmetry surfaces (2-, 3- and 5-fold) of the Al-Pd-Mn sample have been reported by Shen *et al.* [67] using LEED. The diffraction patterns recorded for the different surfaces have been compared to X-ray diffraction data converted to the conditions of the LEED experiment. Interestingly, faceting occurs and has only been observed on LEED patterns measured on the 2-fold and 3-fold surfaces, which indicate qualitatively that they are less stable than the 5-fold surface. For surfaces annealed between 700 and 900 K, no massive lateral surface reconstruction has been noticed. The comparison with bulk diffraction data demonstrates that the LEED patterns obtained on the 2-fold, 3-fold and 5-fold surface are again consistent with expectations from bulk quasicrystalline structure [67].

A similar approach has been taken to investigate the five-fold i-Al-Cu-Fe quasicrystal surfaces, sample isostructural to the i-Al-Pd-Mn quasicrystal. Dynamical LEED analysis performed on the five-fold surface [45] has demonstrated that the outermost surface layer is Al-rich and the distance between the first two planes is contracted by 0.1 Å from the bulk value, similar to the 5-fold Al-Pd-Mn surface [28]. The compositions of the two topmost atomic planes are estimated at 90% Al and 10% Fe, and at 45% Al, 45% Cu, and 10% Fe atoms, respectively. The average lateral density of 0.14 atoms/Å^2 is similar to that of a single plane of Al (111). In the quasicrystal bulk structure, several atomic planes can be grouped in recurring patterns. From the LEED I–V analysis, the surfaces are found to form between these groups of closely spaced planes. Good agreement is found between the step heights measured on STM images and the thickness of the grouped layers favored in LEED [45].

In parallel to the works performed on icosahedral samples, several groups have focused their attention on the characterization of the d-Al-Ni-Co quasicrystal surface [46, 49, 68, 69]. Using spot profile analysis LEED (SPA-LEED), it appears that a single step height of 2.04 Å separates terraces on the 10-fold Al-Ni-Co surface. This value is in agreement with the bulk distance separating 5-fold quasiperiodic planes. The kinematic model calculations of the diffraction patterns show that the cluster structure present in the bulk is maintained within the surface region. Whereas the surface long-range order is consistent with ideally terminated bulk surface, a certain degree of lateral disorder is present within terraces [68]. On the Co-rich phase of Al-Ni-Co, dynamical LEED I–V analysis has indicated a 10% contraction between the first two layers ($d_{12} = 1.84 \pm 0.13$ Å) and a 5% expansion between the second and third layers ($d_{23} = 2.14 \pm 0.14$ Å) compared to the bulk spacing of 2.04 Å. The good match between the LEED structure and STM images suggests that lateral distortions are minimal at the surface [49].

Dynamical LEED analysis and HAS measurements [46] have also been carried out on the d-Al$_{71.8}$Ni$_{4.8}$Co$_{13.4}$ sample that corresponds to the type-I superstructure phase. As shown on Figure 4.4, the SPA-LEED pattern exhibits more than 500 diffraction spots, most of which can be indexed by the basic vectors (b_j) that correspond to the inplane bulk basis vectors. Additional spots appearing around strong diffraction spots and forming a decagon are identified as superstructure spots. These superstructure spots are expected in the diffraction pattern of the bulk truncated surface [46, 52]. The superstructures diffraction spots are much broader than the strong normal structure peaks. The widths of these spots are found to vary between different diffraction orders, an effect explained by the random nature of the tiling of the type-I structure. Investigation of the surface by HAS suggests an extremely low density of defects and a low corrugation of the surface. Because the peak positions are identical

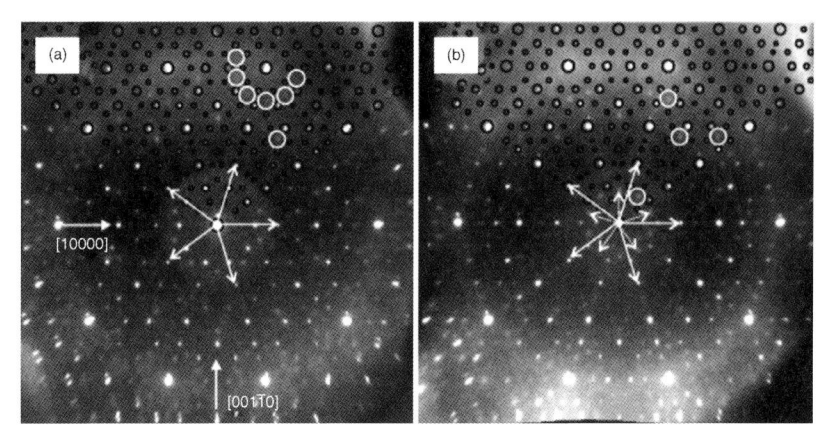

Figure 4.4 SPA-LEED images recorded at 65 and 75 eV electron energy from the ten-fold Å (00001) surface. From reference [46].

in HAS and SPA-LEED spectra, the topmost surface layer has a reciprocal lattice structure consistent with that of a bulk truncated surface [46, 52].

Regarding the 2-fold surfaces ((10000) and (00110)) of the d-$Al_{71.8}Ni_{14.8}Co_{13.4}$ sample, an 8 Å periodicity is measured along the periodic [00001] direction and this doubling of the lowest possible periodicity in decagonal samples is expected for the type-I phase [70, 71]. The HAS analysis reveals that both surfaces have a similar corrugation that remains much higher than on the ten-fold surface. Facets have been found to develop along the (10000) orientations, which suggests a lower surface energy of the (10000) surface compared to the (00110) surface [52]. The diffraction patterns for both surfaces are more consistent with surfaces generated by bulk truncations.

4.2.4
Structure from *Ab Initio* Methods

The structure, the stability, and the electronic properties of quasicrystal surfaces have been investigated using *ab intio* density functional methods [54, 72]. To perform the density functional theory (DFT) calculations, the Vienna *ab initio* simulation package (VASP) has been used [73, 74]. The structure model of the icosahedral quasicrystal is derived from the Katz–Gratias–Boudard bulk model. The atomic structure of the 5-fold Al-Pd-Mn surface is obtained from the icosahedral approximant model. This surface is cut perpendicular to one of its pseudo-5-fold axis and the position of the cleavage is chosen so as to create surface layers of high density consistent with experimental observations. The resulting surface structure is described by a Penrose P1 tiling that consists of pentagons, rhombi, crowns and pentagonal stars [11, 72]. Most of the tiling vertices coincide with the center of Bergman clusters.

From *ab initio* calculations performed on slabs models of the surface based on 2/1 and 3/2 approximant structures, surface charge-density minima coincide with P1 vertices and strong charge depletion is sometimes present within pentagonal tiles.

Upon structural relaxations, the skeleton of the P1 tiling fixed by TM atoms represent a stable surface termination, whereas considerable rearrangement of the Al atoms and relaxations of the interlayer distances occur [72]. Contrary to the XPD and LEED I–V analysis, the calculated surface relaxation leads to an increase of the interlayer spacing between the two topmost layers. The calculations suggest also that the reduction of the density of states (called the pseudogap) at the Fermi level is vanishing at the 5-fold Al-Pd-Mn surface. Although this is verified experimentally for cleaved surfaces, it has been shown that the pseudogap is restored for surfaces sputtered and annealed to sufficiently high temperatures [75]. From the calculated electronic charge-density distributions, simulated STM images have been generated. They reproduce the fine details observed on STM images obtained experimentally (see Figure 4.5). The dark star motifs discussed in Section 4.2.2 are now understood as resulting from deep charge minima associated with surface vacancies.

The brightness of the white flower center is associated with Mn atoms of reduced coordination decorating the center of equatorially truncated pseudo-Mackay clusters. Thus, Al atoms contribute to most of the STM signal while Pd atoms (0.4 Å beneath the topmost layer) are seen as dark spots [38]. The good agreement between experimental and simulated STM images supports the approach of using approximants to model quasicrystalline surfaces.

Similar calculations have been performed for the 10-fold Al-Ni-Co quasicrystal surface using the structure of the W-(Al-Co-Ni) approximant phase resolved by Sugiyama *et al.* [76]. Along the *c*-axis, the W structure can be described as a stacking of $ABA'B$ layers where A' correspond to A plane only shifted by half a period along the *a*-axis and B are puckered layers related to each other by mirror planes. The period along the *c*-axis is 8 Å. Upon structural relaxation, the changes of atomic positions are quite minimal and the change of positions of Al atoms is below 0.15 Å [54]. There is no surface reconstruction and both A and B surfaces are stable. Contrary to the work reported by Ferralis *et al.* [49], no interlayer relaxation appears within the calculations, a discrepancy probably linked to the reduced size of the model used here. The electronic structure calculated for the bulk indicates the presence of a pseudogap at

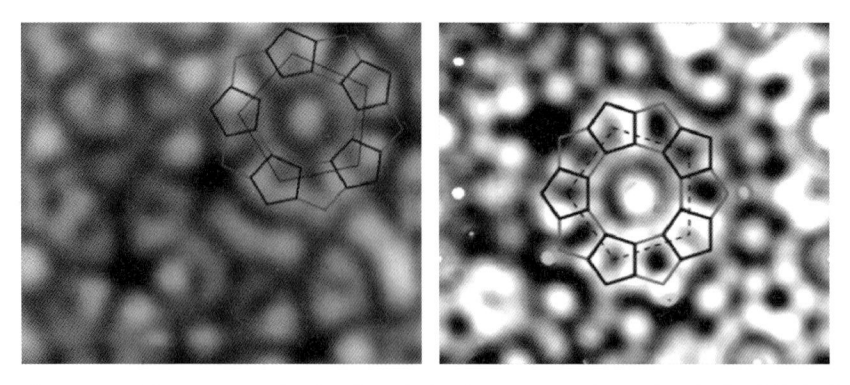

Figure 4.5 (left) Experimental and (right) calculated STM images obtained on the five-fold Al-Pd-Mn surface. The white flower motif is outlined on both images. From reference [38].

the Fermi level. This reduction of the DOS at E_F seems insensitive to the relative content of Ni and Co. However, the pseudogap is almost covered at the surface for both type of surface terminations (A or B). From electronic charge-density distribution, maxima are found on both terminations at transition metal positions. On termination A, enhance charge density have been calculated between Al atoms that indicate a significant degree of covalency in the bonding between two Al atoms [54]. As this bonding configuration is also observed in the electronic charge-density distribution of the bulk, it is thought that it should contribute to the stabilization of the quasicrystal structure. Simulated STM images for both types of terminations have been generated and compared to experimental ones. As for the 5-fold Al-Pd-Mn surface, Al atoms dominate the STM signal but TM atoms contribute significantly to the STM current too. The comparison with experimental images shows that only the smallest structural elements are reproduced in the calculated images [54].

4.2.5
Stability of Alloy Surfaces

We have seen in the previous section that specific planes of the bulk structure are selected as surface terminations. These planes are characterized by the fact that they have a high atomic density, contain mainly the element of the lowest surface energy, and are separated from neighboring planes by the largest gaps. We now discuss the possible mechanisms leading to the selection of these specific planes as surface terminations.

The driving force is the minimization of the surface free energy F_s. For any materials, there is a variety of possible mechanisms by which the surface free energy can be minimized and in general, the surface structure is not simply truncation of the bulk solid, as the local environment of surface atoms is different from that in the bulk. In the simple broken-bond approximation, the surface energy equals half the energy of the missing bonds at the surface and therefore will depend on the surface orientation. For an fcc crystal, the number of missing bonds is 3, 4 and 5 for (111), (100) and (110) surfaces, respectively. Therefore, the system may lower its surface free energy by developing preferential surface orientation. Surface atoms can also adjust their position to minimize F_s. In some cases, there is some oscillating behavior in the distance separating two adjacent planes parallel to the surface around the bulk value d_0, whose magnitude typically amounts to a few 10^{-2} Å. This is called surface relaxation. In other cases, the crystallography of the surface plane is modified, due to the buckling of the surface plane (some atoms will be displaced inwards and others outwards from the surface plane) or due to the absence of some of the atoms compared to the bulk termination. In this case, the surface unit cell is different from what is expected from a perfect bulk truncation. This is called surface reconstruction.

Surface orientation, interlayer relaxation and surface reconstruction can occur even for simple metals. For alloys, the introduction of different chemical elements offers additional mechanisms to minimize F_s, through chemical segregation or selection of planes with specific chemical composition. Surface segregation in alloys is driven by two main factors. One is the tendency for the element with the lowest surface energy to segregate at the surface. The other is the ordering tendency of the alloy, which depends on the sign and magnitude of the interatomic interactions. Taking the example of a binary AB alloy, ordered phases are expected if AB bonds are preferred compared to AA and BB bonds, as atoms of one element tend to be surrounded by atoms of the other element. In this case, the minority element should diffuse in the bulk, where the coordination is larger. In the opposite situation, a tendency to phase separation is expected that will reinforce the segregation phenomena driven by elemental surface energies. Thus, surface segregation of an alloy constituent is moderated by the ordering tendency (bond strength) and is expected to be manifested to a reduced extent in well ordered metallic alloys compared to solid solutions. A related phenomenon is the observation of chemically driven surface morphology in ordered binary A_3B alloys like Cu_3Au, Fe_3Al or $Cu_{83}Pd_{17}$ with $L1_2$-type structure. In the cubic unit cell, the majority atoms A occupy the center of the faces and the minority atoms occupy the edge position. Therefore, along [100, 110] directions, mixed A-B planes and pure A planes alternate, while along [111] directions, all planes are mixed A-B. STM images on vicinal surfaces, like $Cu_{83}Pd_{17}$ (110) and $Cu_{83}Pd_{17}$ (100), show evidence of step pairing, with pure Cu terraces separated by double steps [77]. This is consistent with the lower surface energy of pure Cu planes compared to that of mixed Cu-Pd planes. Therefore, the surface free energy is minimized by selecting only specific planes as surface termination, without reconstruction, thus satisfying both requirements (ordering tendency and element with the lowest γ at the surface) simultaneously.

We can make a parallel between quasicrystalline surfaces and surfaces of ordered $L1_2$ alloys mentioned above. In both cases, the strategy to minimize the surface free energy is to select planes containing the highest concentration of low surface energy elements while preserving the bulk structure and associated benefits in terms of interaction energies between the different species (alloying effect). Because of the aperiodic nature of quasicrystals, the selection of preferred terminations leads to a more complex step sequence compared to the step-pairing mechanism observed for $L1_2$ alloys. Recent studies of periodic CMAs suggest that the same principles dictate the formation of their surfaces. Like for aperiodic CMA, the surface corresponds to a bulk termination, with neither reconstruction nor chemical segregation effects, at the price of a step height selection mechanism in order to select only low-energy planes as surface termination [39, 62, 78]. This is also true for several pseudo-ten-fold surfaces of orthorhombic crystals like ξ'-$Al_{77.5}Pd_{19}Mn_{3.5}$ (Pnma, 140 atoms per unit cell) [43], T-Al_3(Mn, Pd) (Pnma, 156 atoms per unit cell) and o-$Al_{13}Co_4$ (Pmn21, 102 atoms per unit cell).

4.3
Electronic Structure

4.3.1
The Pseudo-Gap Feature

Initially, the Hume-Rothery rules established a correlation between the crystalline structure and the valence electron concentration in binary alloys for which atomic size ratios and electronegativity differences are small. These empirical rules were explained latter by Jones within the framework of the nearly free electron model. For alloys with small atomic size ratios and electronegativity differences, the internal energy U mainly depends on the band term E_b:

$$E_b = \int^{E_F} E.N(E).dE \tag{4.1}$$

In Hume-Rothery alloys, the band term (and thus internal energy) is lowered by the interaction of the Fermi sphere and Bragg planes. The mechanism is the following. In the nearly free electron approximation, the scattering of the electronic states by a Bragg plane defined by the reciprocal lattice vector K opens a gap in the dispersion relation $E(k)$ at $k = K/2$. The magnitude of the gap is a function of the Fourier component of the potential $V(K)$. Integrating over all directions in k space, Bragg scattering produces modulation of the density of states $N(E)$ compared to the free electron parabola. If the Fermi sphere is tangent to the Bragg plane, i.e. if $k_F \sim K/2$, then E_F falls within a minima of $N(E)$ and the band term is lowered as compared to the free electron case. The dip produced in the density of states at E_F is called a pseudogap. This implies a condition on the number of valence electrons per atom e/a:

$$(e/a) = \int_{-\infty}^{E_F} N(E).dE \tag{4.2}$$

Hume-Rothery rules were discovered first for simple *sp* alloys containing noble metals and polyvalent metals. They were later extended to alloys containing transition metals by Raynor (see reference 79 for recent review). A negative valence was ascribed to the transition metal (TM) element, to account for the coupling between *sp* and *d* states. More recent work on transition-metal aluminides by Trambly et al. has shown that the *sp–d* coupling contribute to the formation of the pseudogap and leads to a reinforcement of the number of *sp* states below the Fermi level [79]. This provides a natural interpretation of the negative valence associated with transition metals.

A Hume-Rothery-like stabilization mechanism also occurs even for amorphous alloys. In this case, no reciprocal lattice and no Brillouin zone can be defined. However, the diffraction pattern of such amorphous systems exhibit relatively sharp peak in the structure factor $S(q)$ at $q = K_p$, related to the short and medium range order. It can be used to define an isotropic pseudo-Brillouin zone of radius $K_p/2$. In real space, this corresponds to ions trying to arrange locally on concentric shells with sphere distances $\lambda_F = \pi/k_F$ and $k_F = (3\pi^2 n)^{1/3}$, with n the valence electron concentration.

Quasicrystals and periodic CMAS are also considered as a special kind of Hume-Rothery alloy [80]. Although nonperiodic, their diffraction pattern shows sharp peaks and a pseudo-Jones zone can be constructed from strongly scattering Bragg planes. Due to the high symmetry of the icosahedral point group, the multiplicity of Bragg planes is high and the pseudo-Jones zone is almost spherical. This allows a larger interaction with the Fermi sphere and thus a stronger depletion of $N(E)$ at E_F in quasicrystals compared to simple crystals of lower symmetry.

The density of states at E_F has been measured in many different CMA by using various techniques, like specific heat measurements or soft X-ray spectroscopies. A typical value of $N(E_F)$ amounts to about 1/3 of the free electron value for simple sp metal systems and to 1/10 of the free electron value in sp-d Al-TM systems [81]. The small values of $N(E_F)$ in CMAs have obvious consequences on the physical and chemical properties of CMA.

In the following, we will discuss the problem of the symmetry break introduced by the 2D surface on the persistence of the pseudogap up to the top layers.

The surface electronic structure of CMAs has been investigated mainly by using photoemission spectroscopy. Due to the small inelastic mean free path (*imfp*) of the photoemitted electrons (~1 nm), this technique provides information on the valence band structure in the near-surface region. A clear decrease of the spectral intensity has been measured in the vicinity of the Fermi cut-off, consistent with the expected pseudogap feature in $N(E)$. The near Fermi edge spectra can be fitted to quantify the width and depth of the pseudogap. The experimental intensity is expressed as:

$$I(E) = \int S(x).f(x,T); G_{exp}(x,E).dx \qquad (4.3)$$

where $S(E)$ is the spectral function, $f(E,T)$ is the Fermi function and $G_{exp}(E)$ is a gaussian instrumental function. Mori et al. [82] first suggested to approximate the DOS near E_F by a linear function modulated by a lorentzian dip centered at E_F to account for the pseudogap:

$$S(E) = N(ax+b).\left(1 - \frac{C.\Gamma_L^2}{x^2 + \Gamma_L^2}\right) \qquad (4.4)$$

In this approximation, the pseudogap is thus characterized by a half-width L and a depth C relative to the normal (linear) DOS. Typical values for Γ_L obtained for icosahedral and decagonal quasicrystals range from 0.2 to 1 eV, whereas the intensity at E_F represents a 30 to 90% reduction of the estimated normal DOS, in agreement with specific heat measurements mentioned above.

Valence-band photoemission spectra of Al-TM quasicrystals are usually dominated by d electrons of the transition metals, and the assumption that the DOS near E_F can be approximated by a linear function is not realistic in some cases, like for the d-Al-Ni-Co system. This is not the case for i-Mg-Zn-Re (RE: rare earth, e.g. Y, Ho or Er) quasicrystals, for which the valence band shows a simple sp metal behavior, at least down to 5 eV below E_F. The photoemission spectra show a clear metallic Fermi edge and a pseudogap feature (Figure 4.6), which were analyzed within the framework of the nearly free electron approximation by Suchodolskis et al. [83]. The experimental intensity was fitted using the above expression of $I(E)$ with $S(E)$ expressed as:

$$S(E) = S_0(E)[1+P(E)]; P(E) = P_g(E) = N\left(\frac{x_{max}-x_{min}}{2\sqrt{E/E_0}} - 1\right) \qquad (4.5)$$

where $S_0(E)$ is the normal free-electron DOS, N is the multiplicity of Bragg planes, $E_0 = h^2 (g/2)^2/2m$ the free0electron energy at $k = (g/2)$, g the reciprocal lattice vector

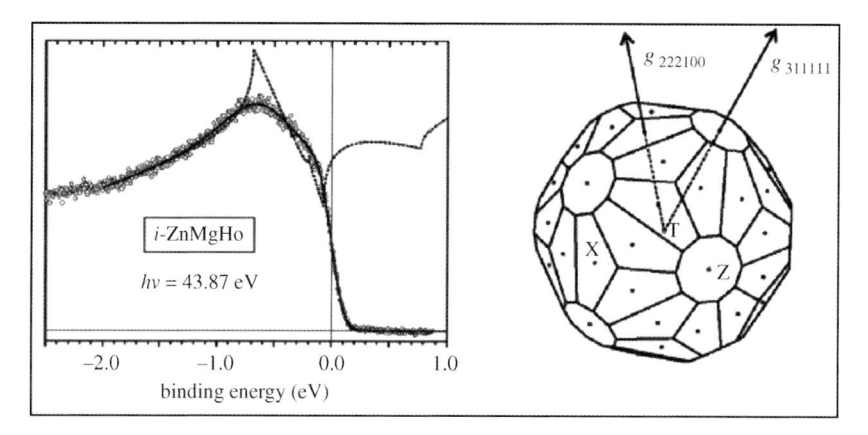

Figure 4.6 (Left) Near Fermi edge photoemission spectra of *i*-Zn-Mg-Ho recorded at 86.4 K. The dotted curve represents the experimental data points. The dashed curve represents the calculated DOS S(E) corresponding to a linear DOS modulated by van Hove singularities produced by the [222100] and [311111] families of Bragg planes. The full curve represents the calculated spectra. (Right) The [222100] and [311111] Jones zone of the icosahedral phase. From reference [38].

corresponding to the Bragg planes, and x_{min} and x_{max} are defined by $x_{min} = 1 - \left(1 + \left[E/E_0 + \sqrt{4E/E_0 + (\Delta/2E_0)^2}\right]^{1/2}\right)$; $\Delta = 2|V_g|$ is the pseudogap width and V_g is the pseudopotential, $x_{max} = 1 - \left(1 + \left[E/E_0 - \sqrt{4E/E_0 + (\Delta/2E_0)^2}\right]^{1/2}\right)$ for $E < E_0 - \Delta/2$; $x_{max} = 1$ for $E_0 - \Delta/2 < E < E_0 + \Delta/2$ and $x_{max} = 1 + [1 + E/E_0 - \left[\sqrt{4E/E_0 + (\Delta/2E_0)^2}\right]^{1/2}$ for $E > E_0 + \Delta/2$. An excellent fit of the photoemitted intensity could be obtained using this nearly free electron approximation with a pseudo-Jones zone constructed from the 60 equivalent [222100] Bragg planes and additional twelve [311111] zones. The Bragg planes are selected based on the matching condition $k_F = (3\pi 2n)^{1/3} = 1.49 \times 10^8$ cm$^{-1} \approx g_{222100}/2 = 1.53 \times 10^8$ cm^{-1} $\approx g_{31111}/2 = 1.59 \times 10^8$ cm^{-1} for *i*-Mg-Zn-Ho. The Jones zone is depicted in Figure 4.6 and the excellent fit indicates that the pseudogap in this system is due to van Hove singularities produced by the [222100] and [311111] families of Bragg planes.

As mentioned earlier, this type of analysis is not as straightforward for CMA containing transition metals, as the spectral intensity is dominated by TM *d* states in the vicinity of E_F. Another method used to probe the DOS at E_F is core-level XPS. It is well known that core-level lines in X-ray excited photoemission spectra of metallic compounds show an asymmetric tail. The asymmetry of the line shape results from intrinsic energy losses through the excitation of electron–hole pairs, which are created simultaneously with the core hole. The probability of these processes decreases rapidly with the electron–hole pair energy. Therefore, the main contribution comes from low-energy electron excitations across the Fermi level. This implies that the asymmetry of the XPS core-level lines depends on the local DOS at the Fermi level. Experimentally, the asymmetry parameter is derived by fitting the experimental function with a gaussian experimental function convoluted with a Doniach–Sunjic line shape, and a Shirley background is used to account for the emission of secondary electrons. Experimentally, the asymmetry parameter α is derived by fitting the experimental function with a Gaussian experimental function convoluted with a Doniach–Sunjic lineshape, and a Shirley background is used to account for the emission of secondary electrons:

$$f(E) = \frac{\Gamma(1-\alpha)}{\left((E-E_0)^2 + \gamma^2\right)^{(1-\alpha)/2}} \cos\left[\frac{\pi\alpha}{2} + (1-\alpha)\tan^{-1}\left(\frac{E-E_0}{\gamma}\right)\right] \quad (4.6)$$

A systematic loss of the asymmetric tail of TM 2p core levels has been measured in CMA like i-Al-Pd-Mn, i-Al-Cu-Fe, ξ'-Al-Pd-Mn and o-Al$_6$Mn, consistent with the pseudogap in these Hume-Rothery alloys [75, 84–86].

The Doniach–Sunjic line shape analysis thus provides an indirect method to probe $N(E_F)$. It has been used to investigate the metallic character as a function of the photoelectron escape depth. By using a tuneable synchrotron light source, it is possible to adjust the *imfp* of the photoelectrons which varies according to the universal curve. Neuhold et al. have used this method and reported an increase in the Al *2p* line asymmetry on the cleaved five-fold surface of the i-Al-Pd-Mn for most surface-sensitive experimental conditions [87]. This indicates an enhanced metallic character (reduction of the magnitude of the pseudogap at E_F) at the fracture surface of the icosahedral quasicrystal. It is consistent with a previous report by Ebert et al. on the two-fold fracture surface of the same alloy. Here, the surface DOS was investigated by scanning tunneling spectroscopy (STS) [88]. Current–voltage $I(V)$ curves were found to be linear, typical of a metallic behavior, without any indication of an irregularity in the DOS.

On the other hand, a Doniach–Sunjic line-shape analysis was performed later using the same i-Al-Pd-Mn quasicrystalline phase but the surface was prepared either by sputtering and annealing or by fracture in UHV [75]. The asymmetry parameter was derived from fitting of XPS core levels as a function of the emission angle in order to vary the surface sensitivity. The enhanced metallic character was confirmed for cleaved surfaces. However, it was found that the asymmetry remains as low as in the bulk even for the most surface-sensitive conditions when the surface has been prepared by sputter annealing. From these measurements, it follows that the pseudogap is maintained up to the extreme surface for the thermodynamically stable terrace/step surface but is smeared out for a rough cleavage surface.

In this section, we have summarized the various experimental approaches that have been used to identify the pseudogap at the Fermi level and the effect of surface preparation on the surface electronic structure. Finally, we would like to mention a related study of the electronic structure of incommensurate 2D superstructures by angle-resolved photoemission spectroscopy [89]. When Ag is deposited on the Cu(111) surface, it forms different moiré superstructures, depending on coverage. Moiré structures are frequently observed in metal-on-metal systems presenting a weak coupling between adsorbate and substrate and presenting a lattice mismatch. In the

Ag on Cu(111) system (13% lattice mismatch), the Ag monolayer forms a compressed, out-of-registry $\sim(9.5 \times 9.5)$ incommensurate moire superstructure. By contrast, the 2 monolayers thick film adopt a less compressed commensurate (9×9) moiré superstructure. Thus, the question arises why the system favors this incommensurate structure at low coverage in spite of this lattice compression. Photoemission data provided evidence that both structures display a filled surface-state band. However, it is found that a gap is formed at the Fermi energy by the interaction with the surface Brillion zone boundaries only in the case of the incommensurate structure. A crude estimate of the electronic energy gain showed that this effect can counterbalance the associated elastic energy. Therefore, the incommensurate structure appears to be electronically stabilized via a 2D Hume-Rothery like effect, which shares some similarities with CMA.

4.3.2
Nature of the Electronic States

The electronic structure of 1D quasiperiodic systems has been calculated by using the Fibonacci chain for example. The energy spectrum of such a model Hamiltonian is found to be singular continuous and the eigenstates are critical, decaying according to a power law $1/r^\alpha$, i.e. neither extended over the whole lattice as are the Bloch states in crystals, nor exponentially localized as in random systems [90]. Numerical studies of 2D quasiperiodic systems using approximants of the Penrose tiling also show the existence of critical wavefunctions [91]. An argument to explain the existence of critical states is as follows. We suppose that a wavefunction ψ_L is mainly localized in a region of radius L of the Penrose lattice. According to Conway's theorem for quasiperiodic lattices, this local environment of size L has some duplicates within a distance $2L$. The wavefunction ψ_L should thus have a probability to exist on this identical environment [92]. Defining a tunneling factor z between the two local environments, then $\psi_{2L} = z\, \psi_L$ and $\psi_{2^n L} = \psi_{z^n L}$. Introducing $L' = 2^n L$, we have $\psi_{L'} = (L/L')^\alpha \psi_L$, explaining qualitatively the power-law decay of the wavefunctions in a quasicrystal. The electronic structure of realistic quasicrystals has been calculated using *ab initio* methods. Since these calculations rely on the Bloch theorem being valid for periodic systems, only approximants can be treated in this way. Due to computational limitations, the size of the approximant must be kept below a few hundred atoms per unit cell. These calculations usually confirm the existence of the pseudogap at the Fermi level. In addition, they usually predict at electronic bands that produce a fine structure (spikes) in the density of states (see Figure 4.7) [93]. The spiky peaks in the DOS have a width of 10–20 meV and it has been suggested that the spiky DOS is a consequence of the quasiperiodicity and relates to the existence of critical states localized by the cluster building blocks.

High-resolution photoemission spectroscopy and point-contact tunneling spectroscopy have been performed in an attempt to confirm the spikiness of the DOS. However, no evidence for spikes in the DOS of quasicrystals and CMAs has ever been obtained, even using state-of-the-art resolution (5 meV at low temperature). It was suggested later that the spikiness could be an artifact of the calculations, due to the absence of convergence with respect to the number of k points in the Brillouin

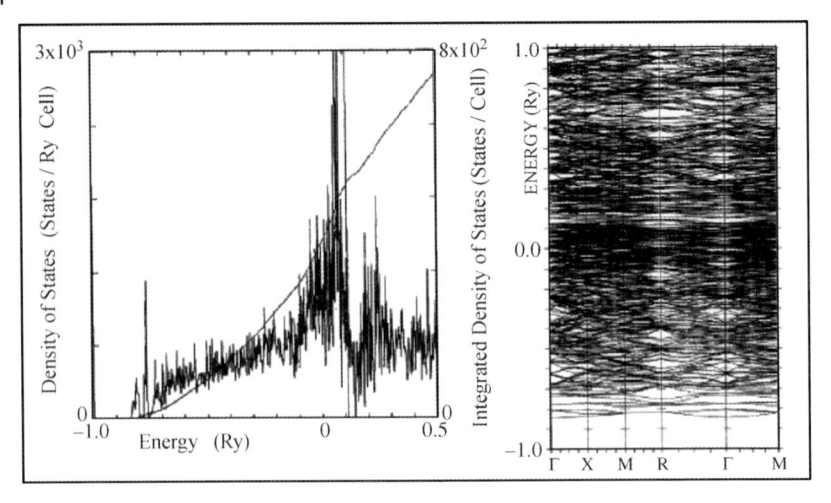

Figure 4.7 (Left) Total and integrated calculated DOS of α-Al$_{114}$Mn$_{24}$ approximant. (Right) Corresponding energy bands calculated along high-symmetry lines. From reference [94].

zone [95]. However, Widmer *et al.* recently reported a STS study of the clean five-fold surface of a *i*-Al-Pd-Mn quasicrystal at 5.3 K [96]. They showed that when the tunneling spectra are averaged over surfaces larger than a few nm^2, the measured DOS is smooth and reproduces the wide parabolic Hume-Rothery pseudogap. However, individual spectra representing the local DOS with subnanometer resolution exhibit a large number of energetically as well as spatially localized peaks. Localized states have a spatial extent of only 0.5–2 nm, i.e. of the same order of magnitude as the size of the cluster building blocks. These sharp features in the local DOS measured by low-temperature STS could be the first evidence of localized or critical states in a narrow energy window around E_F.

Providing full maps of the energy *vs.* momentum distribution of the valence levels of solids, angle-resolved photoemission spectroscopy is extremely useful to investigate the nature of the electronic states in solids. Delocalized electronic states in quasicrystals have been evidenced in the *d*-Al-Ni-Co phase by using this method [97]. In the energy range comprised between −2 to −8 eV below the Fermi level, angle-resolved photoemission spectra recorded on the ten-fold sputter annealed surface indicate the parabolic character of *sp*-like states (Figure 4.8). These extended states form bands that exhibit the same ten-fold symmetry observed in the LEED patterns. This is an important result. However, it does not rule out the existence of critical eigenstates and is therefore not contradictory with STS experiments mentioned above. Clearly, more work is necessary, in particular to establish the nature of the states at the Fermi level that dominate the transport properties of the alloy. Further investigation of the nature of the electronic states by angle-resolved photoemission spectroscopy in periodic CMA presenting local environments similar to those found in quasicrystals will therefore be of interest in the near future.

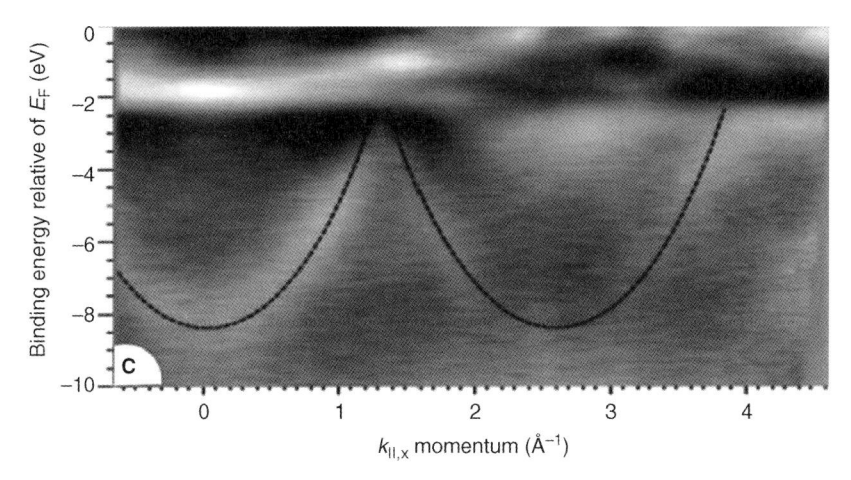

Figure 4.8 Intensity plot of angle-resolved photoemission spectra for d-Al-Ni-Co indicating the parabolic band-like character of the s-p-derived states. From reference [97].

Other related systems have been studied recently to shed light on the effect of quasicrystalline order on the electronic states. For example, Moras *et al.* have investigated Ag films on GaAs(110) by using angle-resolved photoemission [98]. The grown Ag film exhibits a one-dimensional quasiperiodic modulation, resulting in a Fibonacci sequence of parallel stripes with two different widths along the GaAs [001] direction. This system presents the advantage that it is made of a single element with a simple sp band structure, whereas quasicrystals are most often ternary alloys. It is found that the quasiperiodic modulation gives rise to a complex band structure along this direction, which displays aperiodically spaced replicas in reciprocal space, in contrast to the repeated-zone scheme of the electronic bands in crystalline materials.

4.4
Thin-Film Growth on CMA Surfaces

Thin film growth on quasicrystalline substrates is a very active topic, partly due to the prospect of forming single-element 1D, 2D or 3D quasiperiodic systems, and partly due to the possibility of using quasicrystals as templates for building novel nanostructures [52, 99, 100]. Pseudomorphic growth has now been achieved in a limited number of cases that will be described in Section 4.4.1.3. Many other interesting phenomena associated with solid film growth have also been observed. This includes intermixing, surface alloying, twinning of nanocrystals and the quantum size effect (QSE). All these phenomena will be briefly described in Section 4.4.2. In the following, we describe the first step of metal thin-film growth on CMA, in the submonolayer regime, which is characterized by a heterogeneous nucleation mechanism.

4.4.1
Low-Coverage Regime

4.4.1.1 Nucleation Mechanism

All thin-film growth experiments described in this chapter start with a clean well-ordered CMA surface prepared by sputtering and annealing as described previously. The films are grown by condensation on the substrate of metal vapor produced by either Knudsen cells or electron-beam sources, with a base pressure during deposition of typically 2×10^{-10} mbar. The sources are calibrated and the flux usually ranges from 10^{-3} to 10^{-2} monolayer (ML) per second, allowing precise control of the film thickness from submonolayer coverage up to a few tens of ML.

Precise characterization of the island formation during the initial stage of the film growth has been performed by the Ames group for Ag deposited on the five-fold surface of the i-Al-Pd-Mn quasicrystal [101, 102]. The initial stage of the growth is an important step as it can influence the subsequent development of film morphology. STM images of the surface exposed to 0.2 ML of Ag show the formation of isolated islands that are one atom high (2 Å) above the surface. From these images, it is possible to extract the average island density N_{av} by using image analysis tools. N_{av} (in nm^{-2}) at 0.2 ML does not vary significantly with deposition flux, at least for room-temperature deposition. Classic analysis for deposition on perfect crystalline surfaces shows that if (homogeneous) nucleation of stable island requires aggregation of $i + 1 \geq 2$ diffusing adatoms, then $N_{av} \sim F^{i/(i+2)}$. The flux independence of N_{av} rules out the possibility of homogeneous nucleation with critical size $i \geq 1$ on the quasicrystal surface. This indicates that nucleation occurred preferentially at specific trap sites intrinsic to the quasicrystalline surface. The temperature dependence of the Ag-island density at 0.2 ML has been further derived from variable temperature STM data. It appears that N_{av} is also independent of the temperature up to 300 K, but then decreases abruptly. This indicates that nucleation is dominated by trap sites up to 300 K.

A model has been developed by Unal et al. [102] to understand the dependence of island density on temperature and flux. First, the potential energy surface of an Ag adatom was determined by calculating the binding energy of Ag as a function of its position on the substrate. It was found that this potential-energy landscape is complex and several local configurations frequently identified in STM images correspond to strong bonding sites. Then, a rate-equation analysis is performed that incorporates among other parameters the density of strong bonding trap sites, the trap energy reflecting the additional binding energy at trap sites, and the critical nucleus size (i) above which islands are stable at trap sites. The temperature and flux dependence of the island density can be recovered by the model for parameters corresponding to binding energy of adatoms in the critical clusters significantly higher at trap sites than at free terrace sites. It was also found that the critical size i must be significantly greater than 1 and the best fit is found for $i = 5$, corresponding to stable 6 atom islands at trap sites. This value precisely corresponds to the number of atoms forming the so-called starfish islands that have been identified by STM for Al on i-Al-Cu-Fe [103, 104]. The identification of the trap sites for different adsorbates on CMA surfaces will be discussed in the next section.

For Ag deposition above 300 K on five-fold i-Al-Pd-Mn, the trap sites are not so efficient and they are not saturated under these conditions. Experimentally, it is observed that the island density varies across terraces [105]. This feature could have several origins, such as a variation in terrace width, variation of the Erlich–Schwoebel barrier (which is the additional energy barrier associated with diffusion of an adatom down a step) that could depend of the step height bordering a specific terrace, or the density of trap sites across terraces. However, rate-equation analysis suggests that the island density is most strongly influenced by the effective diffusion barrier in this case. It is indeed reasonable to expect some variation in terrace diffusion on icosahedral quasicrystals, because adjacent terraces display small differences in atomic density and chemistry. Similar terrace-dependent nucleation has been reported for example in Al, Sn, or Bi thin films grown on the five-fold surface of i-Al-Cu-Fe, but the origin of the dependence could not be ascertained. In general, a combined experimental and computational approach provides a much deeper understanding of the nucleation and growth of metal thin films on CMA surfaces. However, the latter has been possible only in recent years, thanks to the development of specific models (like the disordered bond network model) which are able to take into account the structural complexity of CMA surfaces.

4.4.1.2 Identification of the Trap Sites

At submonolayer (ML) coverages, the use of scanning tunneling microscopy has been crucial to identify preferential nucleation sites on CMA surfaces. Adsorption of Buckminsterfullerene molecules (C_{60}) has been carried out on the five-fold surface of the Al-Pd-Mn quasicrystal surface [106]. The choice of this molecule is motivated by its geometry, which displays hexagonal, and more importantly, pentagonal facets that matches the symmetry of the aperiodic substrate. In addition, the cage diameter of the molecule is in good agreement with the height of pentagonal hollows present on the substrate and referenced as dark stars in Section 4.2.2. Upon adsorption, there is no evidence for islands formation or for step-edge decoration by C_{60} molecules. The formation of hexagonal over layers observed, while dosing C_{60} on Al(111) is also not present in this case [107]. At 0.065 ML, the C_{60} molecules can be grouped into three classes depending on their apparent heights. These height differences could be explained by molecules bonding in hollow sites (topographical origin) and/or to different atomic species (Al/Mn atoms) in the surface. As demonstrated by Schaub *et al.* [18], pentagonal hollow sites are aligned along a Fibonacci pentagrid. Hence, if one multiplies the distance between two dark stars by the golden number τ or a multiple of τ, then the resulting distance should locate another such pentagonal depression. Using high-resolution STM images, it has been possible to locate several such pentagonal hollow sites on the dosed surface. From the local-scale relationships evident between adsorbate–pentagonal hollow site and adsorbate–adsorbate, the nucleation site has been tentatively assigned as the dark stars present on the five-fold Al-Pd-Mn quasicrystal surface [106].

The same adsorption sites are populated by Al adatoms upon adsorption on the five-fold Al-Cu-Fe surface, sample isostructural to the icosahedral Al-Pd-Mn quasicrystal. When adsorbed on the surface, Al atoms diffuse across terraces and drop into pentagonal hollow sites where they become trapped. Five additional adsorbates are stabilized in the periphery of this central trapped atom. The positions of the five surrounding Al adatoms are very near to Al quasilattice sites, i.e. the site that would be occupied by Al atoms present in the upper plane in the bulk model [103]. The island density being independent of the deposition flux also points towards an inhomogeneous nucleation. At 0.04 ML, these 6-atom clusters form pentagonal island "dubbed" starfish [103]. These pseudomorphic islands of identical orientation and dimension across terraces are monoatomic in height with an edge length equal to 5.1 ± 0.2 Å, which is consistent with the edge lengths of the dark stars. These experimental observations and the identication of the dark stars as preferential sites are also supported by kinetic Monte Carlo simulations [104] presented in Section 4.4.1.1.

A second preferential adsorption site has been identified on the five-fold icosahedral surface. It corresponds to the so-called "white flower" motifs discussed in Section 4.2.2. At 0.25 ML, a remarkable degree of alignment is observed between individual Si adatoms and aperiodically spaced lines derived from a Fibonacci pentagrid [108]. The FFT and autocorrelation have been calculated from STM images where only adsorbates have been selected by thresholding the images. The ten-fold FFT displays rings of spots whose radii are related to each other by powers of the golden ratio. These observations indicate the formation of an ordered quasiperiodic array of Si adatoms. The 2D autocorrelation function calculated from the thresholded images shows a spatial correlation of adsorbates over distances of at least 130 Å. The atomically resolved substrate around the adsorbates reveals that Si adatoms sit in the center of equatorially truncated pseudo-Mackay clusters, i.e. in the center of the "white flower." This conclusion is further supported by comparing simulated and experimentally derived radial distribution functions [108].

Recently, Smerdon et al. have shown that Bi adatoms preferentially decorate "white flower" sites [109]. Upon deposition at low submonolayer coverages (0.13 ML), the formation of Bi pentagonal island (resembling those reported on the i-Al-Cu-Fe/Al system) is observed. Starfish made of 5 Bi adatoms point in the same direction on the surface and exhibit an edge length equal to 4.9 ± 0.2 Å. This Bi arrangement is qualitatively comparable with the Pb starfish formed on the same substrate [110]. However, as shown elsewhere [111], 10 atoms are required to stabilize Pb pentagonal island. From FFT calculations, the pentagonal Bi island are shown to be quasiperiodically distributed on the surface. This suggests that adatoms nucleate on sites that are themselves quasiperiodically distributed on the substrate. From a comparison between experimental and simulated STM images, it has been possible to superimpose a Penrose P1 tiling on images of the dosed surface. While all dark stars fall in pentagons pointing down on the tiling, almost all Bi clusters occupy pentagonal tiles pointing in the opposite direction. Thus, adatoms decorate tiles with Mn atoms in the surface plane. From a thorough inspection of STM images (where both the underlying surface and the Bi adatoms are resolved) and from the above analysis, truncated pseudo-Mackay clusters are proposed as nucleation sites [109].

As has been presented above, two main adsorption sites related to truncated clusters have been identified and their preferential decoration is adsorbate specific. This unique feature to CMA surfaces offers the opportunity for surface patterning.

4.4.1.3 Pseudomorphic Layers

This section presents the pseudomorphic growth of single element on CMA surfaces. The term pseudomorphic defines a situation where the adlayer deposited adopts the structure of the CMA substrate [9]. The first pseudomorphic layers were reported by Franke et al. [112] while depositing Bi and Sb on the five-fold Al-Pd-Mn and ten-fold Al-Ni-Co quasicrystal surfaces. The growth mode of both adsorbates has been monitored using a HAS apparatus and the deposition has been calibrated to the known coverage of the Bi monolayer on GaAs (110). The adatom densities on the Al-Pd-Mn and Al-Ni-Co surfaces are measured as $(0.9 \pm 0.2) \times 10^{15}$ cm^{-2} and $(0.8 \pm 0.2) \times 10^{15}$ cm^{-2}, respectively, and match the density of Al atoms in the two quasicrystal surfaces [112]. To investigate their atomic structure, high-quality Bi and Sb monolayers have been prepared by deposition at 573 K (above the multilayer desorption temperature) followed by subsequent annealing to 823 K for Sb monolayers. The LEED and HAS measurements reveal that Bi and Sb monolayers are highly ordered with a very low defect density and both adopt a quasicrystalline structure. From the HAS analysis, it was concluded that the monolayers have a corrugation higher than the clean surfaces and this has been recently confirmed by STM measurements [109]. Similarly, Bi adsorption has also been carried out on the icosahedral Al-Cu-Fe sample [113]. At submonolayer coverage and for room-temperature deposition, the diffusion across terraces is inhibited due to a strong Bi-adsorbate interaction. Local pentagonal motifs formed by five Bi adatoms have been observed by STM. In addition, the ten-fold FFT calculated from STM images and the unmodified RHEED patterns suggest that Bi grows quasiperiodically up to one monolayer on the Al-Cu-Fe surface [113]. At 523 K, terrace diffusion is activated and the formation of Bi island is observed by STM. The sticking coefficient is drastically reduced at this dosing temperature. In the submonolayer regime, Bi coverage is terrace dependent. This was interpreted as a direct consequence of an inhomogeneous density of trap sites on terraces. Upon further deposition, a smooth wetting layer is obtained with an intrinsic roughness higher than for the clean surface. The RHEED patterns recorded on the monolayer are consistent with a pseudomorphic adlayer [114]. Similarly, a smooth film of height equal to one half of the Sn bulk lattice constant has been obtained when dosing Sn on the same surface. From STM images, calculated FFT and autocorrelation functions, it has been suggested that Sn grows in a pseudomorphic manner; hence the monolayer exhibits a quasicrystalline structure [115].

The Al-Pd-Mn and Al-Ni-Co surfaces have been successfully used as templates to grow quasiperiodic lead monolayers [110, 116]. In the first case, the Pb monolayer self-assembles via a network of pentagonal island/starfish in the temperature range of 57–653 K (see Figure 4.9a). The density of the film is estimated at 0.09 atom/Å^2 from XPS measurements. Contrary to the previous works reported [109, 112], the monolayer roughness is comparable to that of the clean quasicrystal surface. Atomically resolved STM images and LEED patterns show that the structure of the Pb adlayer is quasiperiodically ordered. Here, the Penrose P1 tiling derived from the monolayer is τ inflated compared to the tiling observed on the clean surface [11]. Upon further deposition, the sticking coefficient is drastically reduced and only one monolayer can be adsorbed on the quasicrystal surface. UPS and STS measurements

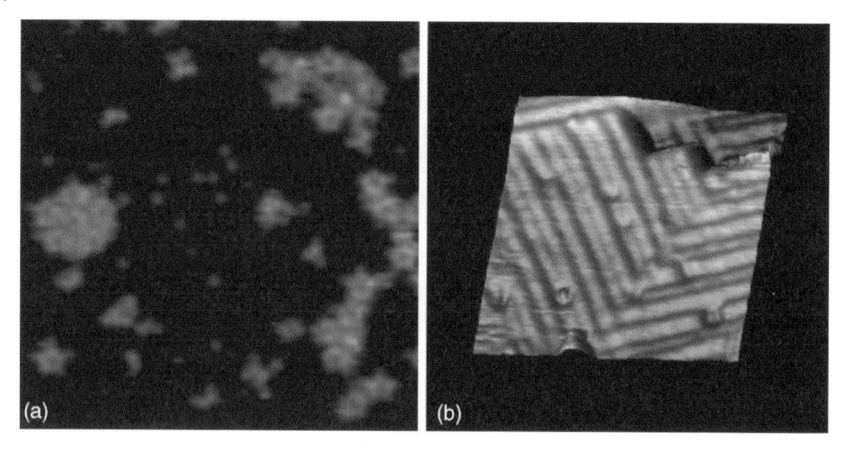

Figure 4.9 (a) STM image of the 5-fold surface of the Al-Pd-Mn after deposition of 0.2 ML of Pb (180 Å × 180 Å). (b) STM image of the 5-fold surface of the Al-Pd-Mn after deposition of 5.5 ML of Cu (100 Å × 100 Å). From references [110] and [117].

reveal that the Pb monolayer exhibits a pseudogap in the density of states at the Fermi level [110]. Due to the reduced chemical complexity, this study makes it possible to correlate the quasiperiodic structure of the film with the formation of the pseudogap at E_F. In a similar manner, only one Pb monolayer can be deposited on the ten-fold Al-Ni-Co surface due to the sticking coefficient of Pb vanishing after completion of the first wetting layer. The adlayer adopts a quasiperiodic structure, as is evident from LEED and STM measurements. Upon annealing the Pb film to 600 K, its structural quality is improved but pores develop on the terraces ($12 \pm 2\%$ of the surface) [116]. Adsorption of C_{60} molecules on the surface has demonstrated that these pores are Pb containing.

The growth of rare gases has been investigated using LEED and the grand canonical Monte Carlo method (GCMC) [118–120]. The adsorption of Xe on the ten-fold Al-Ni-Co surface has been monitored and adsorption isobars have been obtained by maintaining a fixed Xe pressure while changing the temperature of the sample and recording LEED frames [118]. These isobars indicate that Xe adatoms grow in a layer-by-layer fashion for at least the first two layers in the temperature range 60–80 K. The half-monolayer heat of adsorption has been measured at 250 ± 10 meV, consistent with Xe adsorption on other metal surfaces. Using dynamic LEED analysis and FFT calculated on a model structure of hexagonal islands, it has been demonstrated that the second layer consists of five different rotational domains of a hexagonal structure. The location of the diffraction spots is consistent with the Xe–Xe nearest-neighbor distance. Hence, the structure of the second layer is consistent with Xe(111) and the (111) plane is parallel to the surface plane [118]. From the background intensity and the absence of new diffraction spots in the LEED patterns, it has been suggested that the first monolayer might be quasiperiodically ordered. GCMC simulations have shown that, at low coverage, atoms are localized preferentially in the deepest parts of the computed Xe surface potential, which reflects the substrate symmetry. Upon completion of the first monolayer, the Fourier transform of the film remains ten-fold symmetric. With the second layer formed, the inplane structure is now six-fold symmetric [120], in agreement with the LEED analysis [118]. In addition to experimental observations, the calculations indicate that the transition from epitaxial 5-fold to 6-fold ordering is temperature dependent (development of the bulk Xe earlier at higher temperature). It is also mentioned that stacking faults occur in the multilayer films at all temperatures [120].

The growth of a pseudomorphic single element multilayer film has been achieved by deposition of Cu on the 5-fold Al-Pd-Mn quasicrystal surface [117]. Using LEED and STM, the initial growth up to 8 ML proceeds in a layer-by-layer manner. Above 4 ML, the structure observed by STM consists of rows aligned in five directions (see Figure 4.9b). As shown by height-profile measurements, the inter-row distances in each domain form quasiperiodic sequences having short ($S = 4.5 \pm 0.2$ Å) and long ($L = 7.3 \pm 0.3$ Å) separations. The mutual ratio of these two separations (L/S) approaches the golden mean. Because atomic resolution has been limited along the Cu rows, structural information has been extracted from LEED measurements carried out at 85 K [121]. The sharp LEED patterns with a low background exhibit streaks that are periodically spaced and present in five directions. Within the streaks, diffraction spots are quasiperiodically spaced and commensurate with those observed on the clean surface. The presence of the streaks indicates a periodic structure along the Cu rows, i.e. in the direction perpendicular to the quasiperiodic ordering. From momentum-transfer measurements combined with STM observations and Fourier transform simulations, it has been found that Cu atoms are organized periodically with a nearest-neighbor distance of 2.5 ± 0.1 Å along the aperiodically spaced rows [121]. A similar structure type has been obtained while dosing Co on the 10-fold Al-Ni-Co and the 5-fold Al-Pd-Mn surfaces [122]. STM and LEED analysis point towards a quasiperiodically modulated row structure composed of domains of aperiodically spaced Co rows having a periodic lattice parameter along the rows of 2.5 ± 0.1 Å [122].

In parallel, *ab initio* calculations have been performed to investigate the structure of Sn, Bi, and Sb monolayers on a five-fold Al-Pd-Mn surface [123]. These calculations have shown that unsupported quasiperiodic monolayers are unstable and triangular or square arrangement of atoms in a plane are preferred. As explained in Section 4.2.4, the surface structure can be understood in terms of a Penrose P1 tiling with vertices fixed by transition-metal atoms. Potential-energy mapping obtained for the three adsorbates shows the importance of the P1 skeleton for the stabilization of a quasiperiodic monolayer. Apart from surface vacancies present in the topmost plane [72], adatoms preferentially decorate vertices of the P1 tiling (binding energies of ~ 4 eV/atom) then midedge positions of the tiling and Mn atoms exposed at the surface. The decoration of these sites leads to a quasiperiodic layer with a density equal to ~ 0.09 atoms/Å^2, which is in agreement with experimental results [112]. The diffraction pattern calculated exhibits a pseudodecagonal symmetry. Upon relaxation, while the monolayer skeleton based on the P1 tiling is found to be stable, the atomic decoration inside the tiles differs from adsorbates and appears partially distorted. Due to the relatively large size of Bi atoms, partial occupancy of the center of the pentagonal tiles, for instance, are requested to improve the structural model [123].

An identical approach has been used to study the adsorption of atoms from groups one to three of the Periodic Table [124, 125]. It has been found that the adsorption energies scale with the valence of the adatoms. While binding on top of Al substrate atoms appears relatively weak, adsorbates bind more strongly in quasiperiodically distributed charge density minima and surface vacancies. As in the case of a Na monolayer, adatoms trapped in surface vacancies are not considered in the quasiperiodic overlayer. In addition, the site-dependent differences in the binding energies tend to decrease with increasing adatom size [125]. These preferential adsorption sites derived from potential-energy landscape are located at the vertices of a DHBS tiling (decagons (D), hexagons (H), boats (B) and stars (S)). Based on this tiling, an idealized quasiperiodic adlayer would have a density close to 0.066 atoms/Å^2. Starting with a quasiperiodic structure based on the P1 tiling, upon relaxation adatoms within the monolayers spontaneously rearrange in a quasiperiodic structure based on the DHBS tiling. It is interesting to note that the decagonal tiles coincide with the white-flower motif. In addition to the strong binding of the adatoms to the substrate, Krajčí et al. have shown that the size of the adatoms and the overlayer density play a crucial role in determining the stability of the quasiperiodic ordering of the adsorbed monolayer [125]. From the above criteria, it is predicted that adatoms with an atomic size of 3.7 ± 0.4 Å can form a dense quasiperiodic overlayer of density equal to 0.066 atoms/Å^2, equivalent to 0.5 ML coverage. Hence, Na, Ca, Y, La and most rare-earth elements appear to be good candidates for the formation of highly regular quasiperiodic monolayers based on the DHBS tiling [124, 125].

4.4.2
Multilayer Regime

4.4.2.1 Twinning of Nanocrystals

With increasing coverage, most of the elements adsorbed on quasicrystalline surfaces tend to recover their own structure. However, thick films retain some of the symmetry elements of the substrate on which the growth started, though at a mesoscopic scale. This is evidenced by the formation of n-fold twinning of nanocrystals, where the index n depends on the substrate symmetry [126]. For example, a thick film (10 ML) of Ag deposited on the five-fold surface of the i-Al-Pd-Mn consists of fcc nanocrystals with dense (111) planes parallel to the surface plane. The lateral dimension of the Ag crystallites is of a few tens of nanometers, and they have five different possible orientations with respect to the substrate, rotated by $2\pi/5$ from each other. In the case of the p-10f surface of the approximant, the LEED patterns indicate that the Ag nanocrystallites are formed according to two distinct orientations relative to the substrate, rotated by $\sim 12°$ around their [111] growth axis.

The orientation relationship between the metal and the CMA substrate can be explained by the coincidence of densely packed atomic rows common to the substrate and the adsorbate. Within the 5-fold plane of the quasicrystalline surface, the most densely packed directions are defined by a set of five 2-fold axis separated by $36°$. In this case, the most densely packed rows ($[1\bar{1}0]$) in the fcc(111) plane cannot coincide with the dense directions of the 5-fold substrate because a $36°$ rotation is not commensurate with a 60° rotation. Therefore, the Ag islands should occur in five different domains, rotated by 36°, again in agreement with experiments. Within the (a, c) plane perpendicular to the pseudo-10-fold axis of the orthorhombic phase, high-density atomic rows can be found along the $[10\bar{1}]$ and $[\bar{1}0\bar{1}]$ directions. The angle between these two directions is 108.4°. Dense atomic rows of Ag fcc (111) planes are aligned along these dense rows of the pseudo-10-fold plane, thus Ag nanocrystals appear according to two different orientations rotated by 11.6°, consistent with the experiment. Note that the Ag(111)/p-10f-ξ' interface looks very similar to that observed for *fcc* (111)/*bcc* (110) interfaces with the Kurdjumov–Sachs orientation relationship ($fcc[1\bar{1}0]//bcc[\bar{1}11]$ or $bcc[1\bar{1}1]$). This is actually not surprising considering the close similarity between the angle formed by high-density atomic rows in $bcc(110)$ and p-10f-ξ' substrates, i.e. 109.47° and 108.4°, respectively.

The formation of *n*-fold twinning of nanocrystals has been observed in many other systems, like Al, Ag, Bi or Xe on *d*-Al-Ni-Co; Ag, Al, Co, Fe, Ni or Bi on *i*-Al-Pd-Mn; Bi or Sn on *i*-Al-Cu-Fe. Bilki *et al.* have investigated the crystal/quasicrystal interface using molecular dynamics, taking Al on the ten-fold surface of *d*-Al-Ni-Co as an example [127]. This model considers the dynamics of adatoms constrained to move in a plane located at some distances above the substrate surface. Adatom–adatom and adatom–substrate interactions are accounted for by a simple Lennard-Jones potential and the equations of motion are solved numerically to reach the lowest-energy configurations. It is found that the structural mist between adsorbate and substrate leads to the formation of nanoscale crystalline domains that are aligned along the distinct symmetry direction of the quasicrystal. Increasing the strength of adatom-adatom interactions with respect to adatom–substrate interactions promotes the formation of six-fold symmetric crystalline patches. The optimal domain size of the crystals appears to be dominated by a competition between lattice strain and substrate energy. Indeed, lattice strain builds up with increasing domain size due to the structural mismatch. Therefore, there is a competition between lattice strain and substrate energy that determines the optimal domain size. Experimentally, the typical size of the domains deduced from LEED patterns or STM images ranges from a few nm to a few tens of nm.

4.4.2.2 Intermixing and Alloying

In the above, we have considered a sharp interface between the quasicrystalline substrate and the film. However, intermixing and/or alloying sometimes occur upon metal deposition on CMA surfaces. Shimoda *et al.* first reported the formation of an $AuAl_2$ or $PtAl_2$ surface alloy after annealing Au or Pt thin films (500 K) on the ten-fold surface of *d*-AlNiCo [128]. The surface composition and the film structure were investigated by XPS, RHEED and XPD and were found to be consistent with ten different domains of $AuAl_2$ or $PtAl_2$ alloy (*fcc*, CaF_2 structure type) exposing their (110) surface. Each domain appeared to be aligned along the high-symmetry directions of the substrate, resulting in a ten-fold twinning of crystalline alloys. Intermixing occurs already at room temperature in the case of Ni or Co deposited on the five-fold surface of *i*-Al-Pd-Mn, based on SEI, LEED and Auger spectroscopy measurements. Weisskopf *et al.* proposed a model in which Al from the substrate diffuses toward the surface to form an Al-TM (TM = Ni or Co) alloy with CsCl-type structure [129, 130]. This leads to an Al-depleted interface and a consequent phase transformation at the interface. Five cubic domains exposing their (110) faces parallel to the surface are formed. This is similar to the phase transformation induced by preferential sputtering mentioned earlier. Again, these domains are azimuthally rotated by 72° with respect to each other and are aligned with high symmetry directions of the icosahedral substrates. Upon further Co deposition, bcc Co grows epitaxially on the Al-Co domains. Therefore, the five-fold twinning in the Co film is mediated from the substrate through the formation of a twinned interfacial alloy. More complex structure appears upon low-temperature annealing of Cu thin films grown on the five-fold surface of i-Al-Pd-Mn or i-Al-Cu-Fe [131, 132]. In both cases, a five-fold twinning of γ-Al$_4$Cu$_9$ (110) domains is observed. This phase is a Hume-Rothery alloy with physical properties intermediate between a simple metal and a quasicrystal, and as such it could be used as an interface buffer layer to enhance adhesion between a quasicrystal coating and a metal substrate.

More generally, metal deposition on CMA surfaces and subsequent low-temperature annealing can lead to the formation of alloys at the surface. These alloys are neighboring phases of the initial CMA in the phase diagram and they usually present rotational epitaxy within the surface plane with the CMA substrate.

Interfaces are regions of high energy compared to the bulk, where atomic positions need to be adjusted on both sides of the interface to accommodate the two different lattices. How to describe interfaces and how nature minimizes the interface energy between a periodic and a quasiperiodic lattice is important in many aspects and can be described in terms of coincidence of reciprocal lattice site or locking into registry of the two half-crystal atomic structures. A review on some of these aspects concerning interfaces between quasicrystals and crystals has been published recently [133] and we will not discuss these matters further in this chapter.

4.4.2.3 Electron Confinement

In this section, we describe the influence of electronic confinement (resulting from an interface effect) on the growth mode of metal thin films on CMA substrates. Electron confinement in systems having at least one dimension smaller than the electron coherence length leads to the formation of electron standing waves, which are not observed in macroscopic systems. These quantum-well states (QWS) were first identified by Janklevic et al. in tunneling experiments through Pb thin films of various thicknesses ranging from 20 to 100 nm, sandwiched between oxide layers [134]. The tunneling conductance exhibited a set of oscillations, with an energy spacing between peaks that decreases with increasing thickness. They interpreted their result in terms of standing-wave states obtained by requiring nodes of the wavefunction at the boundaries of the metal film. This is equivalent to saying that the electron momentum k normal to the film surface should satisfy $k = n\pi/Nd$, with n and N integers and d the lattice spacing, thus making a connection with the period of the oscillations and the film thickness. Later, Schulte et al. predicted that the work function, Fermi energy, charge spilling, and electron density should oscillate as a function of film thickness [135]. They have simply used a jellium model in these calculations. More-refined models have been proposed since then, in order to take into account the finite potential barrier at both the film/vacuum and film/substrate interfaces, but the general picture remains the same [136, 137]. Electron confinement in thin films, in the direction perpendicular to the surface, produces QWS whose number and position with respect to the Fermi level oscillate with film thickness. This is clearly seen in photoemission spectra, and is well documented for a number of systems including metal on metal and metal on semiconductor systems. Because most properties depend directly on the electron density of states at the Fermi level, they will also oscillate with film thickness. Some of the most visual evidence for such a quantum size effect is the formation of "magic" or preferred island heights and the self-organization of the metal layer into these heights, reported for example in Pb on Si(111).

The formation of magic island heights has been reported recently for several metal thin films grown on CMA substrates, like Ag or Bi on i-Al-Pd-Mn, i-Al-Cu-Fe or d-Al-Ni-Co [114, 138, 139]. For Ag on i-Al-Pd-Mn, the roughness of the film at 1 ML coverage was found to be extremely high at room temperature, with Ag island forming needle-like structures 4 to 5 atoms high. Slightly increasing the deposition temperature (365 K) drastically affects the film morphology (Figure 4.10). Under these conditions, most of the Ag islands have a height corresponding to the stacking of 4 atomic layers. Similar observations were made for Bi thin films grown on i-Al-Cu-Fe, i-Al-Pd-Mn and d-Al-Ni-Co at room temperature. STM images show the formation of Bi crystallites with sharp edges and flat tops with a thickness of 13 Å or a multiple of this height, corresponding to the stacking of 4 atomic planes. Immediately after deposition, one sometimes observe islands of irregular shapes with a thickness of 6.5 Å (2 atomic planes), but these islands quickly reshape spontaneously into more stable 13 Å thick islands.

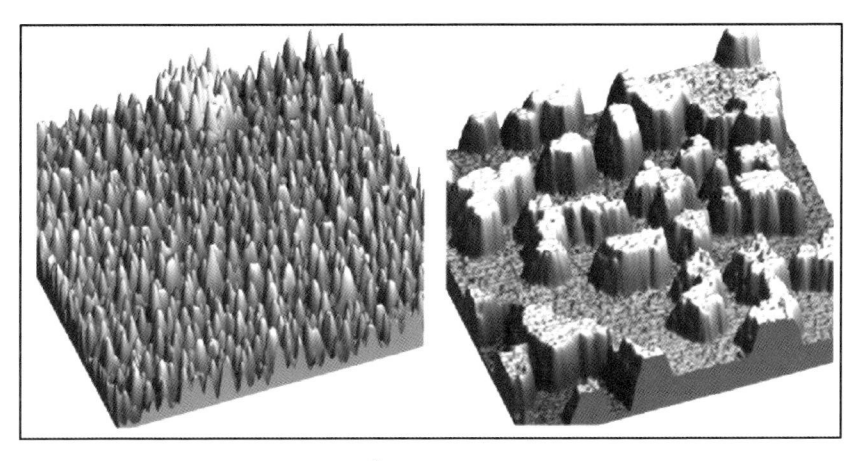

Figure 4.10 STM images (243 × 243 nm^2) of a 1 ML Ag film on the five-fold surface of i-Al-Pd-Mn, deposited at 300 K (top) and 365 K (bottom). Most of the islands are 4 atoms high. From reference [99].

In metal on metal or metal on semiconductor systems, the confinement barrier for the electron is ensured by the image potential at the vacuum side, and by a relative gap in the electronic structure at the interface side. While the existence of a gap is obvious for semiconductor substrates, the gap for metal substrates is symmetry dependent (i.e. noble metals usually exhibit a relative gap in their electronic structure along the [111] direction). In the case of the CMA substrate, the origin of the confinement has been ascribed to either the pseudogap in their electronic structure or to an incompatibility of the point-group symmetries that the wavefunction should satisfy across the boundary. In any case, photoemission studies of Ag films grown on quasicrystalline surfaces indeed reveal the existence of quantum-well states and confirm the interpretation of magic island in terms of quantum size effects.

4.5
Adhesion, Friction and Wetting Properties of CMA Surfaces

Quasicrystals are complex metallic alloys exhibiting long-range atomic order but no translational periodicity, which reveal remarkable mechanical properties, such as low friction, high hardness, low surface energy, and high wear resistance [1]. The discovery of these remarkable mechanical properties has led to several applications for these materials as thin films, coatings, sinters, and fillers for composites. Commercialization of these materials, of course, has additional requirements, such as reliability and low cost. From the point of view of fundamental studies, the role of periodicity and aperiodicity in friction and surface energy is particularly interesting, and has attracted broad interest in the field of surface chemistry and tribology. In this section, we focus on the fundamental aspects of mechanical properties of CMA surfaces, including quasicrystals and particularly the intrinsic relationship between these remarkable mechanical properties and the unique aperiodic atomic structure. We outline recent experimental results on the wetting, adhesion and friction properties of CMA surfaces.

4.5.1
Wetting Properties

Wetting refers to the contact between a liquid and a solid surface caused by the intermolecular interactions of the two contacting media. Wetting phenomena are quantified by measuring the contact angle – the angle at which the liquid/vapor interface meets the solid/liquid interface. If wetting is favorable, the contact angle will be low, and the fluid will spread to cover a larger area of the surface. It is known that the equilibrium shape of the droplet results from the balance of the surface tensions (as shown in the inset of Figure 4.11), which obey Young's equation [140]:

$$\gamma_{SV} = \gamma_L \cos\theta + \gamma_{SL} \tag{4.7}$$

where the terms γ_L, γ_{SV}, γ_{SL} are the surface tension of the liquid, the surface tension of the solid (in the presence of the liquid vapor) and the interfacial tension between solid and liquid, respectively.

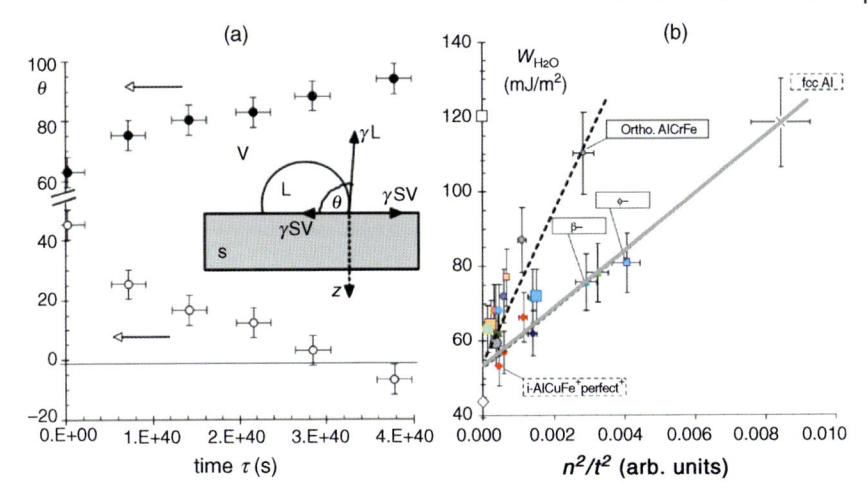

Figure 4.11 (a) The plot of the contact angle θ (top part) and cos θ (bottom part) versus time elapsed in ambient air measured on orthorhombic Al-Cu-Fe-Cr approximant of the decagonal phase. The inset shows the contact angle of a liquid droplet wetted to a rigid solid surface [141]. (b) Variation of the reversible adhesion energy of water, W_{H2O}, as a function of the Al concentration in the samples (inset) or $(n/t)^2$, where n is the Al 3p partial density of states at Fermi energy and t is the oxide thickness. The two straight lines are for specimens with different 3d states contributions at the Fermi energy. The line with the largest slope corresponds to Al-Cr-Fe samples whereas the other is for Al-(Cu, Fe) specimens. From reference [141].

Anomalously large values of the contact angle of water droplets were observed on quasicrystal surfaces. In ambient air, contact angles of water droplets for a quasicrystal of high lattice perfection, like an annealed single-domain icosahedral Al-Pd-Mn quasicrystal, are in the range $90° < h < 100°$, whereas pure aluminum metal shows values around $70°$ [141].

It was found that the contact angle increases as the quasicrystalline surface is exposed to air. Figure 4.11a shows the contact angle θ (top part) and cosθ (bottom part) versus time elapsed in ambient air after polishing a sintered sample of orthorhombic Al-Cu-Fe-Cr approximant of the decagonal phase in pure water, showing that surface energy decreases as the oxide layer gets thicker. It was also shown that the surface energy is associated with the electronic partial density of states based on X-ray photoelectron spectroscopy results. Figure 4.11b shows the plot of surface energy as a function of [Al 3p density of states (n)/thickness of oxide (t)]² measured on the various specimens prepared with different compositions, revealing the linear relation between W_{H2O} (adhesion energy of water) and $(n_{Al3p}/t)^2$. Within experimental accuracy, two classes of W_{H2O} data are shown. One data set corresponds to Al-Cu-Fe specimens while the other set relates to the Al-Cr-Fe(-Cu) system. This trend can be associated with a weak contribution of Fe 3d states at the Fermi energy in Al-Cu-Fe, but a stronger contribution of Cr and Fe 3d states in Al-Cr-Fe(-Cu).

Unlike water droplets, the contact-angle measurement with liquid metal droplets on thin polycrystalline films of decagonal $Al_{13}Co_4$ showed little difference in the contact angles as compared to the crystalline aluminum films, within experimental error [142]. This discrepancy is probably associated with the polar nature of water dipoles that give rise to electrostatic image forces induced by the dipoles of the water molecules. Experimental results on wetting behavior of water are in agreement with the model based on the electrostatic force suggested by Dubois [143].

4.5.2
Atomic-Scale Adhesion Properties of Complex Metallic Alloys

Quasicrystal surfaces also exhibit low adhesion properties that are associated with low surface energy. In this section, we show recent results of work on the adhesion of various complex metallic alloys under different conditions (surface modification, oxidation, nature of contact) that were obtained with atomic force microscopy and the mechanical continuum model.

Many factors can influence surface energy and surface adhesion, making the understanding of adhesion surface phenomena such as friction and adhesion quite complex [144]. Thus, it is important to characterize these mechanical properties at a well-defined interface in a well-controlled environment, such as an oxide- or hydrocarbon-free interface, obtainable in ultrahigh vacuum (UHV). The nature of the interface formed by the mechanical contact also plays a crucial role in determining atomic-scale mechanical properties. Depending on whether the bonds formed across an interface are weak or strong, the volume properties or the interface chemistry may dominate the tribological behavior. Thus, while in the elastic (reversible) regime the interaction between the tip and surface can be described by continuum models that predict the relationship between friction, adhesion and applied load. On chemically active surfaces strong bonds can be formed such that displacements of one surface relative to the other produce damage (plastic, or irreversible deformation). In these circumstances it is difficult to separate plastic effects originating from the creation of slippage in subsurface planes when critical loads are reached, from effects associated with bond rupture at the surface. Hence, the elastic regime offers a more straightforward interpretation of experimental data. One way to make the elastic regime more accessible, experimentally, is to weaken surface adhesion by passivating either the surface or the probe with an adsorbed molecular layer that exposes relatively inert chemical groups. Modifying the surface or tip with an adsorbate allows us to delineate the influence of chemical modification on adhesion properties.

4.5.2.1 Continuum-Mechanics Models
Previous experiments with well-characterized surfaces carried out in ultrahigh vacuum (UHV) have shown that the relation between the contact area and the applied load can be described by the Derjaguin–Muller–Toporov (DMT) [145, 146] or the Johnson–Kendall–Roberts (JKR) model [147], depending on the adhesion energy and on the hardness of the contacting materials. These two models have been developed as approximations for the elastic behavior in two opposite extremes, one for soft and adhesive materials, and the other for hard and poorly adhesive ones. Any real situation is of course intermediate between these two extremes [148, 149].

4.5 Adhesion, Friction and Wetting Properties of CMA Surfaces

Friction appears to scale with load in proportion to the area of contact as predicted for a continuous, elastic, single asperity contact. That is, $F_f = \tau \cdot A$, where, F_f is the frictional force, A the contact area, and τ the shear strength (shear force/area). In DMT and JKR models, the area (A) can be described as:

$$A_{JKR} = \pi \left(\frac{3R}{4E^*}\right)^{2/3} \left[L + 3\pi\gamma R + \sqrt{6\pi\gamma RL + 3\pi\gamma R^2}\right]^{2/3}$$

$$A_{DMT} = \pi \left(\frac{3R}{4E^*}\right)^{2/3} [L + 2\pi\gamma R]^{2/3}$$

(4.8)

where R is the tip radius, L the applied load, γ the interfacial energy per unit area (also known as the work of adhesion), and E^*, the combined elastic modulus of the two materials, given by $E^* = [(1-v_1^2)/E_1 + (1-v_2^2)/E_2]^{-1}$, where E_1 and E_2 are their Young's module and v_1 and v_2 are the Poisson ratios.

The pull-off force is related to the work of adhesion. In the JKR and DMT models, the effective value of γ is given by $L_c/(1.5\pi R)$ or $L_c/(2\pi R)$, respectively, where L_c is the adhesion force. To decide whether the behavior is closer to that predicted by DMT or JKR, an empirical nondimensional Tabor parameter $\tau = (16R\gamma^2/9E^2 z_0^3)$ [13], can be used. In this formula, z_0 is the equilibrium spacing of two surfaces (roughly an atomic distance), and empirically, it is found that the JKR model is a good approximation when $\tau > 5$, while DMT is more appropriate when τ is less than 0.1.

4.5.2.2 Adhesion on Clean and *In-Situ* Oxidized Quasicrystal Surfaces

AFM is a good probe of adhesion. The adhesion force can be obtained by measuring the force–distance curve or approach–retraction curve as shown in Figure 4.12a [150]. The approach curve is the plot of the vertical cantilever bending versus the displacement of the rear end of the cantilever base. Figure 4.12b shows examples of force–distance curves measured on the clean or oxidized quasicrystal surfaces. During retraction of the AFM probe, at point A, the probe snaps out of contact with the surface. At this point, the tensile load equals the adhesion force of the tip–sample junction. Thus, the difference in force between A and B (free position) is attributed to the adhesion force.

Force–distance curves were measured in the middle of the terraces with TiN or W_2C-coated cantilevers. The adhesion force was found to be $1.02 \pm 0.12\,\mu N$ (Figure 4.12b). The error corresponds to the standard deviation of multiple measurements [151]. Using W_2C-coated cantilevers of the same spring constant (48 N/m) the adhesion force was found to be $1.07 \pm 0.15\,\mu N$. In another series of measurements with W_2C-coated cantilevers with a spring constant of 11.5 N/m, a value of $0.85 \pm 0.09\,\mu N$ was obtained. Variations in adhesion force are likely caused by differences in parameters such as cleanliness and tip radius, which are difficult to control. The radius of the tip was determined to be ~150 nm from field-emission scanning electron microscopy carried out after a friction measurement. For a purely elastic tip–sample contact, the measured pull-off forces would correspond to a work of adhesion of 1.14 J/m² and 1.51 J/m² with the tip radius of 150 nm based upon the DMT and JKR models, respectively. As we will discuss later, however, this value is values with increasing oxide thickness is also consistent with wetting experiments on quasicrystalline surfaces (Figure 4.11a).

The adhesion forces of various complex metal alloys with air oxides were measured. Five single-grain CMA samples were examined: 2-fold and 5-fold surfaces of i-Al-Pd-Mn quasicrystals, a 2-fold surface of the ξ'-Al-Pd-Mn approximant, and 2-fold and 10-fold surfaces of d-Al-Ni-Co [155]. A sixth sample, polycrystalline Al, was obtained by depositing an aluminum film (thickness of 500 nm) onto a Si(100) wafer. Prior to AFM/FFM measurements, the samples were oxidized in ambient air at room temperature.

For quantitative measurements of friction and adhesion forces, it is crucial to have constant cantilever parameters, such as spring constant and tip radius. For this reason, the same cantilever was used for the whole series of friction measurements. To check whether or not the radius of the tip remained constant, friction was measured as a function of load on a reference sample before and after each set of friction measurements. The same friction values and lateral resolution were measured on the reference sample, confirming that the spring constant and tip radius remained constant throughout the experiments. The adhesion forces of the samples were ~17–25 nN. In Table 4.1, the values of adhesion force, and work of adhesion (using a tip radius of 150 nm) for different samples are listed. The results indicate that the adhesion mainly depends on the nature of aluminum oxide formed on the complex metallic alloys or aluminum surface.

4.5.3
Atomic-Scale Friction Properties

Anomalously low coefficients of friction were revealed on quasicrystalline materials sliding against diamond and steel by Dubois et al. [156]. An explanation of this effect has been sought ever since. Understanding it is important in order to unravel the basic physics of friction and to facilitate practical applications. The most intriguing possibility is that the low friction is related to the exotic atomic structure of the bulk material [157]. In general, frictional energy dissipation measured under such conditions includes contributions from such diverse factors as the breaking of chemical bonds, generation of point defects, interactions with wear debris, phase transformations near the sliding track, and in crystalline materials by creation of dislocations and propagation of slip planes in specific crystallographic directions. It has been suggested that some of these factors play a role in the observed low friction coefficients of quasicrystals. In this section, we review the recent studies on atomic-scale frictional properties on quasicrystal surfaces.

4.5.3.1 Friction-Measurement Apparatus – FFM and Tribometer
Various techniques have been used to elucidate the friction properties of quasicrystalline materials. Here, we discuss two tribological techniques, pin-on-disk and AFM/FFM, which yield tribological data at the macroscale and nanoscale, respectively. In earlier studies, macroscopic techniques such as the pin-on-disk tribometer was utilized to reveal tribological properties on quasicrystalline materials [158, 159]. The

Table 4.1 Adhesion forces between 2-fold and 10-fold Al-Ni-Co surfaces and a TiN-coated tip, measured in various states, such as clean, *in-situ* oxidized, air-oxidized, and ethylene-passivated. The table also shows adhesions of other CMA surfaces (2-fold and 5-fold surfaces of *i*-Al-Pd-Mn quasicrystals, a 2-fold surface of the ξ'-Al-Pd-Mn approximant) and that on Al film for comparison. Work of adhesion is estimated with DMT or JKR model, and tip radius of 150 nm. From references [150] and [155].

Types of surfaces	Surface conditions	Adhesion force (μN)	Work of adhesion (J/m²)	Mechanical regime
Pt(111)	clean	10	12(DMT) ~ 16 (JKR)	Inelastic
10-fold d-Al-Ni-Co	clean	0.7 ± 0.2	0.7 (DMT) ~ 0.9 (JKR)	Inelastic
	200 L oxygen *in-situ*	0.4 ± 0.1	0.4 (DMT) ~ 0.5 (JKR)	Inelastic
	ethylene passivated (high load)	0.07 ± 0.01	0.07 (DMT) ~ 0.09 (JKR)	Inelastic
	ethylene passivated (low load)	0.013 ± 0.002	~0.013 (DMT)	Elastic
	short air oxidized	0.04 ± 0.012	~0.04 (DMT)	Elastic
	long air oxidized	0.02 ± 0.004	~0.02 (DMT)	Elastic
2-fold d-Al-Ni-Co	clean	0.35 ± 0.08	0.35 (DMT) ~ 0.5 (JKR)	Inelastic
	clean surface-with passivated probe (high load)	0.4 ± 0.05	0.4 (DMT) ~ 0.5 (JKR)	Inelastic
	clean surface-with passivated probe (low load)	0.17 ± 0.03	0.18(DMT) ~ 0.22 (JKR)	Elastic
	short air oxidized	0.045 ± 0.01	~0.045(DMT)	Elastic
	long air oxidized	0.02 ± 0.004	~0.02(DMT)	Elastic
Approximant Al-Pd-Mn	long air oxidized	0.025 ± 0.01	~0.025(DMT)	Elastic
5-fold i-Al-Pd-Mn	long air oxidized	0.02 ± 0.005	~0.02(DMT)	Elastic
2-fold i-Al-Pd-Mn	long air oxidized	0.02 ± 0.005	~0.02(DMT)	Elastic
Polycrystalline Al	long air oxidized	0.02 ± 0.004	~0.02(DMT)	Elastic

pin-on-disk tribometer apparatus consists of a "pin" in contact with a rotating disc. In a typical pin-on-disc experiment, the coefficient of friction is continuously monitored by measuring the friction force with force sensors while the fixed load is applied to the pin–sample contact. As wear occurs, the friction coefficient changes due to chemical interactions between the pin and the surface.

The atomic and friction force microscope (AFM/FFM) has been used more recently to address the atomic scale origin of friction due to the nanometer size of the tip–sample contact area. A schematic of AFM/FFM is shown in Figure 4.14. In

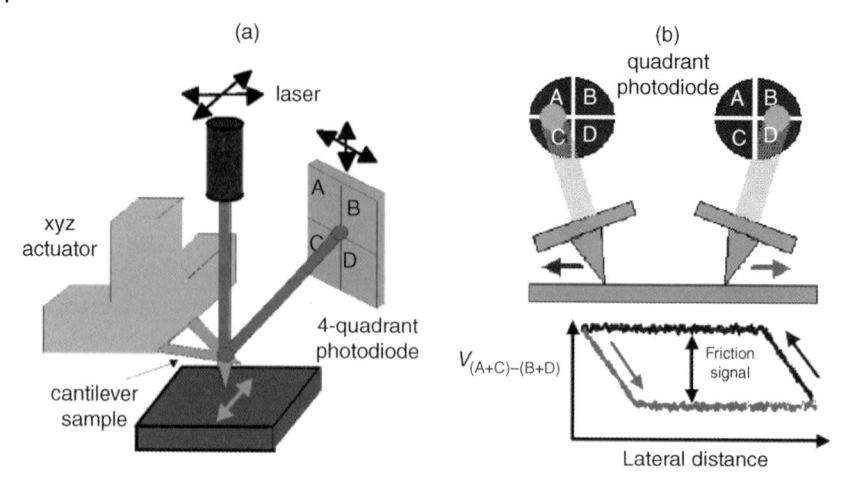

Figure 4.14 Schematic of (a) atomic force microscopy (AFM) and (b) scheme of friction measurement.

AFM, a sharp tip is brought into contact with a surface, which causes normal bending (z-deflection) of the cantilever supporting the tip (Figure 4.14a). If the tip is then shifted with respect to the sample (or *vice versa*), the cantilever is also twisted. The two deformations (lever bending and lever twisting) can be detected by a laser beam, which is reflected from the rear of the cantilever into a four-quadrant photodetector. The normal force acting on the cantilever can be deduced from the normal signals acquired with the photodetector ((A + B)–(C + D) in Figure 4.14a), provided that the spring constants of the cantilever and the sensitivity of the photodetector are known. Figure 4.14b shows a friction loop for a uniform surface. At the beginning of the scanning, the tip sticks on the surface because of stiction (static friction), and the lateral signal changes linearly with the lateral displacement (x) until the tip slides over the surface. As the tip starts sliding (in the regime of dynamic friction), the lateral signal becomes constant. The friction signal is simply the "gap" between the lateral signals from tracing and retracing, as shown in Figure 4.14b.

4.5.3.2 Friction on Atomically Clean and *In-Situ* Oxidized Quasicrystal Surfaces

The first measurements of tribological properties in UHV, where the oxide could be truly circumvented, were carried out in the laboratory of Andrew Gellman. His group employed a tribometer in which two clean surfaces of identical structure, composition, and history could be brought into contact. They found that the friction coefficient between a pair of clean, single-grain 5-fold Al-Pd-Mn surfaces was half of that between a pair of crystalline Al-Pd-Mn approximant surfaces (as shown in Figure 4.15a). Here, the Al-Pd-Mn approximant had a cubic-based structure that was a clear contrast to the icosahedral Al-Pd-Mn surface, but the chemical compositions were reasonably similar. It was found that adsorption of oxygen or water on the 5-fold surface of *i*-Al-Pd-Mn resulted in a decrease of friction force by about a factor of 2 [158]. Mancinelli *et al.* found that adsorption of oxygen or water on a crystalline

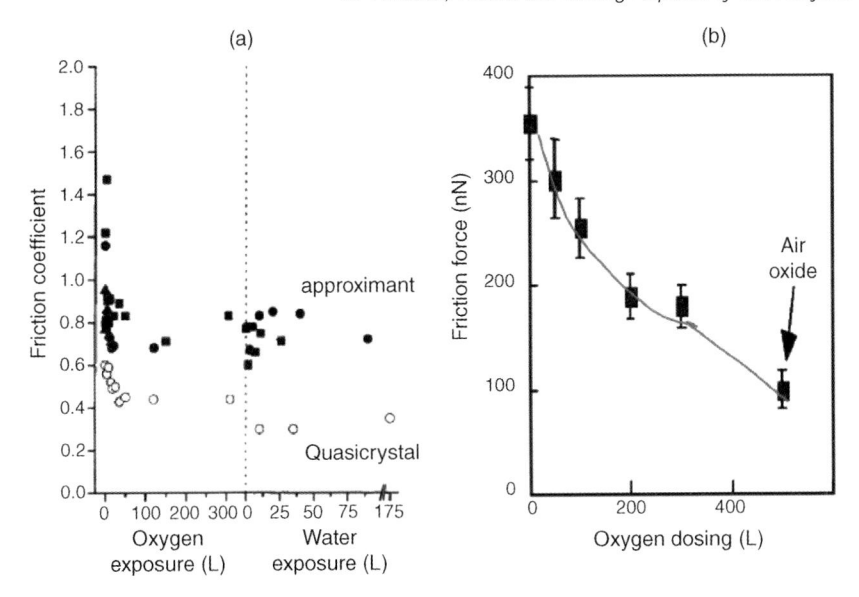

Figure 4.15 (a) Friction coefficients measured between the surfaces of pairs of $Al_{48}Pd_{42}Mn_{10}$ approximants (filled symbols) and between pairs of $Al_{70}Pd_{21}Mn_9$ quasicrystals (open circles) as a function of the exposure of the surfaces to O_2 and then H_2O. (b) Plot of friction forces as a function of oxygen dosing measured on 10-fold Al-Ni-Co quasicrystal surface, showing the rapid decrease in friction that saturates at high exposures in vacuum. Only after air oxidation is a substantial decrease in friction observed. From references [151] and [160].

approximant caused a decrease of the same magnitude (Figure 4.15a) [160]. In these experiments contacts were probably inelastic, although that was not determined. These results showed that surface oxidation inhibits friction, but that this inhibition is not unique to quasicrystals.

Similar frictional behaviors at the nanoscale contact were revealed by Park et al. who carried out AFM/FFM measurement on the atomically clean 10-fold decagonal Al-Ni-Co quasicrystal surfaces [151]. Figure 4.15b also shows the friction force as a function of oxygen exposure at an applied load of 1000 nN. Friction forces decreased rapidly in the early stages of oxygen adsorption (100 L) and became saturated after a 200 L dose. This is entirely consistent with previous tribological measurements on the 5-fold i-Al-Pd-Mn quasicrystal surfaces.

4.5.4
Friction Anisotropy

Several observations of friction anisotropy in elastic and reversible contacts have been attributed to the degree of registry commensurability between two solids as a function of sliding angle [161, 162]. Efficient energy dissipation through phonon generation normally requires atomic-scale instabilities, where atoms suddenly "jump" or "slip" from metastable to stable configurations. In the case of wearless friction, the forces between atoms across the contact are normally relatively weak compared to the interatomic forces within each solid. If this is not the case, then irreversible contacts with inelastic deformation and atom transfer are likely, as we see for unpassivated contacts between an AFM tip and a clean quasicrystal surface. These weak forces normally result in gradual, adiabatic displacements of contact atoms from their equilibrium positions and little phonon generation. Even weak interaction potentials, however, can exhibit nonadiabatic "stick-slip" behavior for periodic contacts, while incommensurability inhibits phonon excitation at the sliding interface, and hence can lead to low friction forces.

4.5.4.1 Friction Anisotropy of Clean 2-Fold Al-Ni-Co Surface

In order to investigate the role of surface structure, it was desirable to probe periodic and aperiodic atomic arrangements nearly simultaneously. To this end, Park et al. used a 2-fold surface of the decagonal Al-Ni-Co phase, since it presents both periodic and aperiodic atomic arrangements. The atomically resolved STM image of the 2-fold Al-Ni-Co surface revealed the presence of atomic rows along the 10-fold direction with an internal periodicity of 0.4 nm, as shown in Figure 4.16a. Along the orthogonal axis in the surface plane, the spacing between the rows followed a Fibonacci sequence with inflation symmetry [56]. As shown in Figure 4.16b, a series of experiments were carried out to explore friction anisotropy as a function of scanning direction. In a

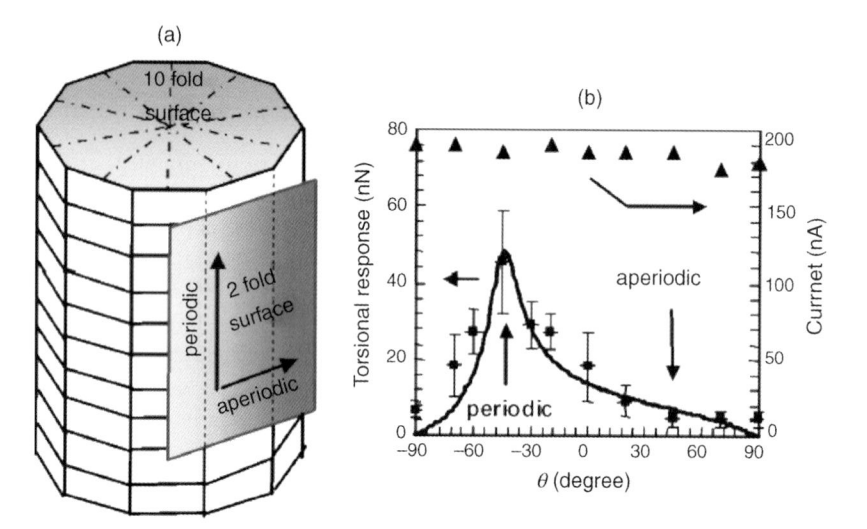

Figure 4.16 (a) Schematic model of a decagonal Al-Ni-Co quasicrystal, showing the orientation of decagonal and two-fold planes. The 2-fold plane is periodic along the 10-fold direction and aperiodic along the 2-fold direction (b) Torsional response of the cantilever measured as a function of scanning angle on the two-fold surface of the Al-Ni-Co decagonal quasicrystal at zero external load. The torsional response was higher along the periodic direction than along the aperiodic direction. The solid line shows the calculated torsional response with scanning angle for an elliptical anisotropy factor (ratio of torsional response) of 8. The current at the sample bias of 1.0 V was measured at the same time as torsional response. From reference [154].

conventional scanning force microscope, friction is measured by scanning the tip along the surface perpendicular to the cantilever axis. The torsional response of the cantilever is then proportional to tip–sample friction, and the normal load applied to the tip–sample contact is not affected by friction. In an optical deflection AFM, a laser beam reacted off the cantilever surface close to the tip measures changes in the cantilever slope. Torsional and normal slope changes are proportional to frictional and normal forces acting on the tip, respectively. Rotating the scan angle relative to the cantilever axis introduces complications, since frictional forces now modulate the normal load applied to the contact, and the optical deflection signals show a mixed response to normal and frictional forces.

Friction measurement on the two-fold Al-Ni-Co surface revealed high friction anisotropy, with friction being 8 times higher along the periodic direction than along the aperiodic direction [55]. Figure 4.16b shows the torsional response of the cantilever as a function of scanning direction at an applied load of 0 nN. The overlaid curve shows the calculated torsional response as a function of rotation angle assuming an elliptical friction anisotropy ratio of 8, consistent with the measured variation of the friction in a previous, more limited experiment. Figure 4.16b shows the current measured simultaneously with torsional response. In the elastic regime, conductance is a convenient way to check for constancy of the contact area. In this experiment, the conductance was constant within 5%, indicating that the contact area was invariant with scanning angle.

Why is the friction force higher in the periodic than the quasiperiodic direction? Ever since low friction was first discovered in quasicrystals, incommensurability has been considered as a possible cause. In this experiment, however, the TiN tip is of a different material and probably amorphous, and hence should be incommensurate in both periodic and aperiodic directions. Registry is therefore unlikely in any scanning direction. However, periodic stick-slip has frequently been reported for AFM contacts between a periodic surface and an amorphous tip. The two-fold decagonal quasicrystal surface does allow us the unique opportunity to slide the same tip across a chemically identical surface, changing the degree of order by changing direction. The observation of significant friction anisotropy adds weight to the notion that order plays a significant role in wearless friction.

The two strongest options are electronic and phononic friction, i.e. energy dissipation via excitation of electron–hole pairs or phonons, respectively. Direct creation of electron–hole pairs has been invoked as a mechanism of frictional energy dissipation, and both experimental and theoretical efforts have been made to address the importance of this dissipation channel. The other mechanism is phononic friction, in which vibrations of the surface atoms are excited and subsequently damped by energy transfer to the bulk material through the propagation of phonon modes and in metals also by electronic excitations. Phononic friction is a stronger candidate than electronic friction, since studies show that it generally dominates electronic friction. The excitation and propagation of phonons along the aperiodic direction could be inhibited by phonon gaps, leading to low energy dissipation. Such gaps are predicted theoretically [163], but have not been observed experimentally for quasicrystals.

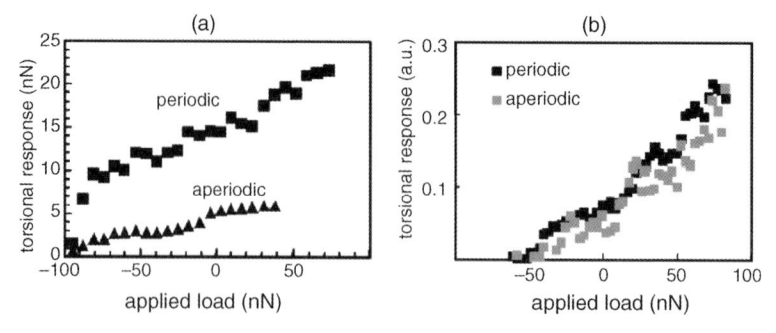

Figure 4.17 (a) Torsional response versus applied load measured on 2-fold Al-Ni-Co quasicrystal surface with the molecule-passivated AFM probe (a) after exposing the surface to 100 L of ethylene, and (b) after short air oxidation. From reference [154].

4.5.4.2 Friction Anisotropy After Surface Modification

The torsional anisotropy discussed in the previous section could be influenced by surface modifications. For example, it was found that the friction anisotropy measured with a passivated AFM probe dropped by half to 3 to 4 after exposing the surface to 100 L of ethylene gas, as shown in Figure 4.17a [154]. In this experiment, ethylene exposure was performed by backfilling the chamber to a pressure of 4.0×10^{-7} Torr at room temperature. No ordered structures were formed on this surface, as shown by LEED. The decrease in the friction anisotropy suggests that the anisotropy of the clean surface arises from short-range interactions between tip and surface. In a different experiment it was shown that the friction anisotropy disappeared completely when the surface was oxidized by exposure to air. Figure 4.17b shows the torsional response measured on a surface oxidized by several months exposure to air. The thicker oxide layer is more effective at screening the interaction responsible for the anisotropy than the ethylene monolayer.

4.5.4.3 Low Friction of Quasicrystals and Its Relation with Wetting and Adhesion

Quasicrystal surfaces appear to have low adhesion based on AFM in UHV. Wetting experiments with polar solutions also show that the contact angle is high, compared to the crystalline materials, and it depends on the electronic nature of quasicrystalline materials and oxide thickness. There are evidences that the low surface energy of quasicrystalline material is related to the quasiperiodic atomic structure. Low friction, as revealed with a tribometer or AFM/FFM, can be attributed (at least in part) to low adhesion because the lower adhesion decreases the effective applied load, and therefore decreases the contact area and friction forces.

AFM/FFM studies on two-fold decagonal quasicrystal surfaces revealed the strong friction anisotropy. The degree of friction anisotropy depends on the ordering of the surface structure. We argue that this friction anisotropy cannot be due to adhesion, but rather is probably due to anisotropic phonon excitation cross sections at the sliding interface. Hence, it would appear that two different fundamental factors contribute to low friction at quasicrystal surfaces: low adhesion (mainly an electronic effect) and inefficient phonon excitation (a dynamic effect).

The relative importance of these two factors will depend upon the conditions of sliding and the type of comparison that is being made. In comparisons among clean metals, or between metals where the clean surface is exposed by wear, the dynamic factor will contribute most in elastic contacts; adhesion must play an increasing role in inelastic contacts.

4.6 Conclusion

From the general overview on the surface structure of quasicrystals and periodic CMAs presented in this chapter, it is clear that they are structurally similar to what could be expected from simple bulk terminations, at least when the surface is prepared by sputtering and annealing in UHV. No indication of chemical segregation or surface reconstruction are present in the available dataset. This result is *a priori* surprising and is achieved via the selection of specific planes as surface terminations. These planes are characterized by their high atomic density and their high content in the lowest surface energy element.

Nucleation and growth on CMA are strongly affected by the complex potential energy surface experienced by an adsorbate. Strong adsorption sites have been identified on the surface, which act as traps for diffusing adatoms. Heterogeneous nucleation of island at these specific sites appears to be quite general on CMA surfaces. Trap sites are quasilattice sites corresponding to truncated 3D clusters and are quasiperiodically distributed on the surface. Therefore, there is a possibility that CMA surfaces could be used as templates to grow self-organized arrays of nanometer-sized islands under specific growth conditions. For some specific adsorbates like rare gases of low melting point metals, a quasiperiodic order develops in the first monolayer. These films are model systems for studies of the structure–property relationships, independently of the chemical complexity usually associated with CMA. Such opportunities have not been fully investigated and represent a new area for future research. Other interesting phenomena in thin-film growth on CMA have been reported in this overview, including the occurrence of a self-selection of particle size in crystalline thin films or the existence of quantum size effects manifesting at room temperature or even above.

The complex atomic ordering of CMA has important consequences on their surface properties. One characteristic of quasicrystals and periodic CMAs is that they exhibit a pseudogap in their electronic density of states at the Fermi level. We have seen that the pseudogap feature persists up to the top surface layers. In addition, bandgaps in the phonon density of states have also been predicted theoretically, although this has not been confirmed experimentally. These peculiarities of CMA surfaces have some influence on their adhesion and friction properties. In particular, we have argued that the strong friction anisotropy measured by AFM on the two-fold decagonal quasicrystal surface is probably due to anisotropic phonon excitation at the sliding interface, while their low adhesion is related to the electronic structure. Friction in the elastic regime is a matter of energy dissipation and we have seen that both the electronic and phononic structure contribute to the low friction measured on CMA surfaces. Other surface mechanisms relying on energy dissipation, like chemical reactions in heterogeneous catalysis, might also be unusual compared to normal metals.

Acknowledgments

We acknowledge the European Network of Excellence on Complex Metallic Alloys (CMA) contract NMP3-CT-2005-500145 for financial support. J. Y. P. acknowledges the support by WCU program through the National Research Foundation of Korea Funded by the Ministry of Education, Science and Technology (31-2008-000-10055-0).

References

1 Dubois, J.M. (2005) *Useful Quasicrystals*, World Scientific, Singapore.
2 Gille, P. and Bauer, B. (2008) *Cryst. Res. Technol.*, **43**, 1161.
3 Ebert, Ph., Feuerbacher, M., Tamura, N., Wollgarten, M., and Urban, K. (1996) *Phys. Rev. Lett.*, **77**, 3827.
4 Gratias, D., Puyraimond, F., Quiquandon, M., and Katz, A. (2001) *Phys. Rev. B*, **63**, 024202.
5 Ponson, L., Bonamy, D., and Barbier, L. (2006) *Phys. Rev. B*, **74**, 184205.
6 Jenks, C.J., Burnett, J.W., Delaney, D.W., Lograsso, T.A., and Thiel, P.A. (2000) *Appl. Surf. Sci.*, **157**, 23.
7 Shen, Z., Kramer, M.J., Jenks, C.J., Goldman, A.I., Lograsso, T.A., Delaney, D.W., Heinzig, M., Raberg, W., and Thiel, P.A. (1998) *Phys. Rev. B*, **58**, 9961.
8 Yureschko, M., Grushko, B., and Ebert, P. (2005) *Phys. Rev. B*, **96**, 256105.
9 Cai, T., Fournée, V., Lograsso, T.A., Ross, A.R., and Thiel, P.A. (2002) *Phys. Rev. B*, **65**, 140202.
10 Sharma, H.R., Shimoda, M., Fournée, V., Ross, A.R., Lograsso, T.A., and Tsai, A.P. (2005) *Phys. Rev. B*, **71**, 224201.
11 Papadopolos, Z., Kasner, G., Ledieu, J., Cox, E.J., Richardson, N.V., Chen, Q., Diehl, R.D., Lograsso, T.A., Ross, A.R., and McGrath, R. (2002) *Phys. Rev. B*, **66**, 184207.
12 Steurer, W., Haibach, T., Zhang, B., Kek, S., and Lück, R. (1993) *Acta. Crystallogr. B*, **49**, 661.
13 Yamamoto, A. and Weber, S. (1997) *Phys. Rev. Lett.*, **78**, 4430.
14 Kortan, A.R., Thiel, F.A., Chen, H.S., Tsai, A.P., Inoue, A., and Masumoto, T. (1989) *Phys. Rev. B*, **40**, 9397.
15 Kortan, A.R., Becker, R.S., Thiel, F.A., and Chen, H.S. (1990) *Phys. Rev. Lett.*, **64**, 200.
16 Pauling, L. (1985) *Nature*, **317**, 512.
17 Shechtman, D., Blech, I., Gratias, D., and Cahn, J.W. (1984) *Phys. Rev. Lett.*, **53**, 1951.
18 Schaub, T.M., Burgler, D.E., Guntherödt, H.-J., and Suck, J.-B. (1994) *Phys. Rev. Lett.*, **73**, 1255.
19 Schaub, T.M., Burgler, D.E., Guntherödt, H.-J., and Suck, J.-B. (1994) *Z. Phys. B*, **96**, 93.
20 Erbudak, M., Nissen, H.-U., Wetli, E., Hochstrasser, M., and Ritsch, S. (1994) *Phys. Rev. Lett.*, **72**, 3037.
21 Naumovic, D., Aebi, P., Schlapbach, L., and Beeli, C. (1997) *New Horizons in Quasicrystals: Research and Applications* (eds A.I. Goldman, D.J. Sordelet, P.A. Thiel, and J.M. Dubois) World Scientific, Singapore.
22 Cappello G., Dchelette, A., Schmithusen, F., Chevrier, J., Comin, F., Stierle, A., Formoso, V., De Boissieu, M., Lograsso, T., Jenks, C., and Delaney, D. (1999) *Mater. Res. Soc. Symp. Proc.*, **553**, 243.
23 Schmithusen, F., Cappello, G., de Boissieu, M., Comin, F., and Chevrier, J. (2000) *Surf. Sci.*, **444**, 113.
24 Bolliger, B., Erbudak, M., Vvedensky, D.D., Zurkirch, M., and Kortan, A.R. (1998) *Phys. Rev. Lett.*, **80**, 5369.

25 Ledieu, J., Munz, A.W., Parker, T.M., McGrath, R., Diehl, R.D., Delaney, D.W., and Lograsso, T.A. (1999) *Surf. Sci.*, **433/435**, 665.

26 Ledieu, J., Munz, A.W., Parker, T.M., McGrath, R., Diehl, R.D., Delaney, D.W., and Lograsso, T.A. (1999) *Mater. Res. Soc. Symp. Proc.*, **553**, 237.

27 Zheng, J.C., Huan, C.H.A., Wee, A.T.S., Van Hove, M.A., Fadley, C.S., Shi, F.J., Rotenberg, E., Barman, S.R., Paggel, J.J., Horn, K., Ebert, Ph., and Urban, K. (2004) *Phys. Rev. B*, **69**, 134107.

28 Gierer, M., Van Hove, M.A., Goldman, A.I., Shen, Z., Chang, S.-L., Jenks, C.J., Zhang, C.-M., and Thiel, P.A. (1997) *Phys. Rev. Lett.*, **78**, 467.

29 Shen, Z., Stoldt, C.R., Jenks, C.J., Lograsso, T.A., and Thiel, P.A. (1999) *Phys. Rev. B*, **60**, 14688.

30 Unal, B., Lograsso, T.A., Ross, A.R., Jenks, C.J., and Thiel, P.A. (2005) *Phys. Rev. B*, **71**, 165411.

31 Ledieu, J., Cox, E.J., McGrath, R., Richardson, N.V., Chen, Q., Fournée, V., Lograsso, T.A., Ross, A.R., Caspersen, K.J., Unal, B., Evans, J.W., and Thiel, P.A. (2005) *Surf. Sci.*, **583**, 4.

32 Ledieu, J., McGrath, R., Diehl, R.D., Lograsso, T.A., Ross, A.R., Papadopolos, Z., and Kasner, G. (2001) *Surf. Sci. Lett.*, **492**, L729.

33 Jenks, C.J., Ross, A.R., Lograsso, T.A., Whaley, J.A., and Bastasz, R. (2002) *Surf. Sci.*, **34**, 521.

34 Boudard, M., de Boissieu, M., Janot, C., Heger, G., Beeli, C., Nissen, H.-U., Vincent, H., Ibberson, R., Audier, M., and Dubois, J.M. (1992) *J. Phys.: Condens. Matter*, **4**, 10149.

35 Barbier, L., Le Floch, D., Calvayrac, Y., and Gratias, D. (2002) *Phys. Rev. Lett.*, **88**, 85506.

36 Papadopolos, Z., Pleasants, P., Kasner, G., Fournée, V., Jenks, C.J., Ledieu, J., and McGrath, R. (2004) *Phys. Rev. B*, **69**, 224201.

37 Ledieu, J. and McGrath, R. (2003) *J. Phys.: Condens. Matter*, **15**, S3113.

38 Krajci, M., Hafner, J., Ledieu, J., and McGrath, R. (2006) *Phys. Rev. B*, **73**, 024202.

39 Unal, B., Jenks, C.J., and Thiel, P.A. (2008) *Phys. Rev. B*, **77**, 195419.

40 Noakes, T.C.Q., Bailey, P., McConville, C.F., Parkinson, C.R., Draxler, M., Smerdon, J., Ledieu, J., McGrath, R., Ross, A.R., and Lograsso, T.A. (2005) *Surf. Sci.*, **583**, 139.

41 Naumovic, D., Aebi, P., Beeli, C., and Schlapbach, L. (1999) *Surf. Sci.*, **433–435**, 302.

42 Ledieu, J., Muryn, C.A., Thornton, G., Cappello, G., Chevrier, J., Diehl, R.D., Lograsso, T.A., Delaney, D., and McGrath, R. (2000) *Mater. Sci. Eng. A*, **294**, 871.

43 Fournée, V., Ross, A.R., Lograsso, T.A., Anderegg, J.W., Dong, C., Kramer, M., Fisher, I.R., Canfield, P.C., and Thiel, P.A. (2002) *Phys. Rev. B*, **66**, 165423.

44 Becker, R.S., Kortan, A.R., Thiel, F.A., and Chen, H.S. (1991) *J. Vac. Sci. Technol. B*, **9**, 867.

45 Cai, T., Shi, F., Shen, Z., Gierer, M., Goldman, A.I., Kramer, M.J., Jenks, C.J., Lograsso, T.A., Delaney, D.W., Thiel, P.A., and Van Hove, M.A. (2001) *Surf. Sci.*, **495**, 19.

46 Sharma, H.R., Franke, K.J., Theis, W., Riemann, A., Folsch, S., Gille, P., and Rieder, K.H. (2004) *Phys. Rev. B*, **70**, 235409.

47 Cox, E.J., Ledieu, J., McGrath, R., Diehl, R.D., Jenks, C.J., and Fisher, I. (2001) *Mater. Res. Soc. Symp. Proc.*, **643**, K11.3.1.

48 Kishida, M., Kamimura, Y., Tamura, R., Edagawa, K., Takeuchi, S., Sato, T., Yokoyama, Y., Guo, J.Q., and Tsai, A.P. (2002) *Phys. Rev. B*, **65**, 094208.

49 Ferralis, N., Pussi, K., Cox, E.J., Gierer, M., Ledieu, J., Fisher, I.R., Jenks, C.J., Lindroos, M., McGrath, R., and Diehl, R.D. (2004) *Phys. Rev. B*, **69**, 153404.

50 Yuhara, J., Klikovits, J., Schmid, M., Varga, P., Yokoyama, Y., Shishido, T., and Soda, K. (2004) *Phys. Rev. B*, **70**, 24203.

51 McGrath, R., Ledieu, J., Cox, E.J., and Diehl, R.D. (2002) *J. Phys.: Condens. Matter*, **14**, R119.

52 Sharma, H.R., Shimoda, M., and Tsai, A.P. (2007) *Adv. Phys.*, **56**, 403.

53 Diehl, R.D., Ledieu, J., Ferralis, N., Szmodis, A.W., and McGrath, R. (2003) *J. Phys.: Condens. Matter*, **15**, R1.

54 Krajci, M., Hafner, J., and Mihalkovic, M. (2006) *Phys. Rev. B*, **73**, 134203.

55 Park, J.Y., Ogletree, D.F., Salmeron, M., Ribeiro, R.A., Canfield, P.C., Jenks, C.J., and Thiel, P.A. (2005) *Science*, **309**, 1354.
56 Park, J.Y., Ogletree, D.F., Salmeron, M., Ribeiro, R.A., Canfield, P.C., Jenks, C.J., and Thiel, P.A. (2005) *Phys. Rev. B*, **72**, 220201.
57 Takakura, H., Yamamoto, A., and Tsai, A.P. (2001) *Acta Crystallogr. A*, **57**, 576.
58 Boudard, M., Klein, H., de Boissieu, M., Audier, M., and Vincent, H. (1996) *Philos. Mag. A*, **74**, 939.
59 McRae, E.G., Malic, R.A., Lalonde, T.H., Thiel, F.A., Chen, H.S., and Kortan, A.R. (1990) *Phys. Rev. Lett.*, **65**, 883.
60 Shen, Z., Pinhero, P.J., Lograsso, T.A., Delaney, D.W., Jenks, C.J., and Thiel, P.A. (1997) *Surf. Sci.*, **385**, L923.
61 Jenks, C., Delaney, D.W., Bloomer, T.E., Chang, S.-L., Lograsso, T.A., Shen, Z., Zhang, C.-M., and Thiel, P.A. (1996) *Appl. Surf. Sci.*, **103**, 485.
62 Gierer, M., Van Hove, M.A., Goldman, A.I., Shen, Z., Chang, S.-L., Pinhero, P.J., Jenks, C.J., Anderegg, J.W., Zhang, C.-M., and Thiel, P.A. (1998) *Phys. Rev. B*, **57**, 7628.
63 Alvarez, J., Calvayrac, Y., Joulaud, J.L., and Capitan, M.J. (1999) *Surf. Sci. Lett.*, **423**, L251.
64 Katz, A. and Gratias, D. (1993) *J. Non-Cryst. Solids*, **153–154**, 187.
65 Cappello, G., Schmithusen, F., Chevrier, J., Comin, F., Stierle, A., Formoso, V., De Boissieu, M., Boudard, M., Lograsso, T., Jenks, C., and Delaney, D. (2000) *Mater. Sci. Eng. A*, **294**, 822.
66 Sharma, H.R., Franke, K.J., Theis, W., Gille, P., Ebert, Ph., and Rieder, K.H. (2003) *Phys. Rev. B*, **68**, 054205.
67 Shen, Z., Raberg, W., Heinzig, M., Jenks, C.J., Fournée, V., Van Hove, M.A., Lograsso, T., Delaney, D., Cai, T., Canfield, P.C., Fischer, I.R., Goldman, A.I., Kramer, M.J., and Thiel, P.A. (2000) *Surf. Sci.*, **450**, 1.
68 Gierer, M., Mikkelsen, A., Graer, M., Gille, P., and Moritz, W. (2000) *Surf. Sci.*, **463**, L654.
69 Shimoda, M. and Sharma, H.R. (2008) *J. Phys.: Condens. Matter*, **20**, 314008.
70 Sharma, H.R., Franke, K.J., Theis, W., Riemann, A., Folsch, S., Rieder, K.H., and Gille, P. (2004) *Surf. Sci.*, **561**, 121.

71 Sharma, H.R., Theis, W., Gille, P., and Rieder, K.H. (2002) *Surf. Sci.*, **511**, 387.
72 Krajči, M. and Hafner, J. (2005) *Phys. Rev. B*, **71**, 054202.
73 Kresse, G. and Furthmuller, J. (1996) *Comput. Mater. Sci.*, **6**, 15.
74 Kresse, G. and Joubert, D. (1999) *Phys. Rev. B*, **59**, 1758.
75 Fournée, V., Pinhero, P.J., Anderegg, J.W., Lograsso, T.A., Ross, A.R., Canfield, P.C., Fisher, I.R., and Thiel, P.A. (2000) *Phys. Rev. B*, **62**, 14049.
76 Sugiyama, K., Nishimura, S., and Hiraga, K. (2002) *J. Alloys Compd.*, **342**, 65.
77 Barbier, L., Salanon, B., and Loiseau, A. (1994) *Phys. Rev. B*, **50**, 4929.
78 Sharma, H.R., Fournée, V., Shimoda, M., Ross, A.R., Lograsso, T.A., Tsai, A.P., and Yamamoto, A. (2004) *Phys. Rev. Lett.*, **93**, 165502.
79 Trambly de Laissardière, G., Nguyen-Manh, D., and Mayou, D. (2005) *Prog. Mater. Sci.*, **50**, 679.
80 Friedel, J. (1988) *Helv. Phys. Acta*, **61**, 538.
81 Belin-Ferré, E. and Dubois, J.-M. (2006) *Int. J. Mater. Res.*, **97–7**, 985.
82 Mori, M., Matsuo, S., Ishimasa, T., Matsuura, T., Kamiya, K., Inokuchi, H., and Matsukawa, T. (1991) *J. Phys.: Condens. Matter*, **3**, 767.
83 Suchodolskis, A., Assmus, W., Giovanelli, L., Karlsson, U.O., Karpus, V., Le Lay, G., Sterzel, R., and Uhrig, E. (2003) *Phys. Rev. B*, **68**, 542071.
84 Jenks, C.J., Chang, S.-L., Anderegg, J.W., Thiel, P.A., and Lynch, D.W. (1996) *Phys. Rev. B*, **54**, 6301.
85 Fournée, V., Anderegg, J.W., Ross, A.R., Lograsso, T.A., and Thiel, P.A. (2002) *J. Phys.:Condens. Matter*, **14**, 2691.
86 Horn, K., Theis, W., Paggel, J.J., Barman, S.R., Rotenberg, E., Ebert, Ph., and Urban, K. (2006) *J. Phys.:Condens. Matter*, **18**, 435.
87 Neuhold, G., Barman, S.R., Horn, K., Theis, W., Ebert, P., and Urban, K. (1998) *Phys. Rev. B*, **58**, 734.
88 Ebert, Ph., Yue, F., and Urban, K. (1998) *Phys. Rev. B*, **57**, 2821.
89 Schiller, F., Cordon, J., Vyalikh, D., Rubio, A., and Ortega, J.E. (2005) *Phys. Rev. Lett.*, **94**, 016103.
90 Kohmoto, M., Kadano, L.P., and Tang, C. (1983) *Phys. Rev. Lett.*, **50**, 1870.

References

91 Tsunetsugu, H., Fujiwara, T., Ueda, K., and Tokihiro, T. (1991) *Phys. Rev. B*, **43**, 8879.

92 Sire, C. (1995) *Proceedings of the 5th International Conference on Quasi crystals* (eds C. Janot and R. Mosseri), World Scientific, Singapore, p. 415.

93 Fujiwara, T. (1999) *Physical Properties of Quasicrystals*, (ed. Z.M. Stadnik), Springer-Verlag, Berlin, p. 169.

94 Fujiwara, T. (1989) *Phys. Rev. B*, **40**, 942.

95 Zijlstra, E.S. and Janssen, T. (2000) *Europhys. Lett.*, **52**, 578.

96 Widmer, R., Groning, O., Rueux, P., and Groning, P. (2006) *Philos. Mag.*, **86**, 781.

97 Rotenberg, E., Theis, W., Horn, K., and Gille, P. (2000) *Nature*, **406**, 602.

98 Moras, P., Theis, W., Ferrari, L., Gardnio, S., Fujii, J., Horn, K., and Carbone, C. (2006) *Phys. Rev. Lett.*, **96**, 156401.

99 Fournée, V. and Thiel, P.A. (2005) *J. Phys. D: Appl. Phys.*, **38**, R83.

100 Thiel, P.A. (2008) *Annu. Rev. Phys. Chem.*, **59**, 129.

101 Fournée, V., Cai, T.C., Ross, A.R., Lograsso, T.A., Evans, J.W., and Thiel, P.A. (2003) *Phys. Rev. B*, **67**, 033406.

102 Unal, B., Fournée, V., Schnitzenbaumer, K.J., Ghosh, C., Jenks, C.J., Ross, A.R., Lograsso, T.A., Evans, J.W., and Thiel, P.A. (2007) *Phys. Rev. B*, **75**, 064205.

103 Cai, T., Ledieu, J., McGrath, R., Fournée, V., Lograsso, T.A., Ross, A.R., and Thiel, P.A. (2003) *Surf. Sci.*, **526**, 115.

104 Ghosh, C., Liu, Da-Jiang, Schnitzenbaumer, K.J., Jenks, C.J., Thiel, P.A., and Evans, J.W. (2006) *Surf. Sci.*, **600**, 2220.

105 Unal, B., Evans, J.W., Lograsso, T.A., Ross, A.R., Jenks, C.J., and Thiel, P.A. (2007) *Philos. Mag.*, **87**, 2995.

106 Ledieu, J., Muryn, C.A., Thornton, G., Diehl, R.D., Delaney, D.W., Lograsso, T.A., and McGrath, R. (2000) *Surf. Sci.*, **472**, 89.

107 Johansson, M.K.-J., Maxwell, A.J., Gray, S.M., Brühwiler, P.A., Mancini, D.C., Johansson, L.S.O., and Martensson, N. (1996) *Phys. Rev. B*, **54**, 13472.

108 Ledieu, J., Unsworth, P., Lograsso, T.A., Ross, A.R., and McGrath, R. (2006) *Phys. Rev. B*, **73**, 012204.

109 Smerdon, J.A., Parle, J.K., Wearing, L.H., Lograsso, T.A., Ross, A.R., and McGrath, R. (2008) *Phys. Rev. B*, **78**, 075407.

110 Ledieu, J., Leung, L., Wearing, L.H., McGrath, R., Lograsso, T.A., Wu, D., and Fournée, V. (2008) *Phys. Rev. B*, **77**, 073409.

111 Ledieu, J., Krajci, M., Hafner, J., Leung, L., Wearing, L.H., McGrath, R., Lograsso, T.A., Wu, D., and Fournée, V. (2008) *Phys. Rev. B*, **79**, 165430.

112 Franke, K.J., Sharma, H.R., Theis, W., Gille, P., Ebert, P., and Rieder, K.H. (2002) *Phys. Rev. Lett.*, **89**, 156104.

113 Fournée, V., Sharma, H.R., Shimoda, M., Tsai, A.P., Unal, B., Ross, A.R., Lograsso, T.A., and Thiel, P.A. (2005) *Phys. Rev. B*, **95**, 155504.

114 Sharma, H.R., Fournée, V., Shimoda, M., Ross, A.R., Lograsso, T.A., Gille, P., and Tsai, A.P. (2008) *Phys. Rev. B*, **78**, 155416.

115 Sharma, H.R., Shimoda, M., Ross, A.R., Lograsso, T.A., and Tsai, A.P. (2005) *Phys. Rev. B*, **72**, 045428.

116 Smerdon, J.A., Leung, L., Parle, J.K., Jenks, C.J., McGrath, R., Fournée, V., and Ledieu, J. (2008) *Surf. Sci.*, **602**, 2496.

117 Ledieu, J., Hoeft, J.-T., Reid, D.E., Smerdon, J.A., Diehl, R.D., Lograsso, T.A., Ross, A.R., and McGrath, R. (2004) *Phys. Rev. Lett.*, **92**, 135507.

118 Ferralis, N., Diehl, R.D., Pussi, K., Lindroos, M., Fisher, I.R., and Jenks, C.J. (2004) *Phys. Rev. B*, **69**, 075410.

119 Trasca, R.A., Ferralis, N., Diehl, R.D., and Cole, M.W. (2004) *J. Phys.:Condens. Matter*, **16**, S2911.

120 Curtarolo, S., Setyaman, W., Ferralis, N., Diehl, R.D., and Cole, M.W. (2005) *Phys. Rev. Lett.*, **95**, 136104.

121 Ledieu, J., Hoeft, J.T., Reid, D.E., Smerdon, J.A., Diehl, R.D., Ferralis, N., Lograsso, T.A., Ross, A.R., and McGrath, R. (2005) *Phys. Rev. B*, **72**, 035420.

122 Smerdon, J.A., Ledieu, J., Hoeft, J.T., Reid, D.E., Wearing, L.H., Diehl, R.D., Lograsso, T.A., Ross, A.R., and McGrath, R. (2006) *Philos. Mag.*, **86**, 841.

123 Krajči, M. and Hafner, J. (2005) *Phys. Rev. B*, **71**, 184207.

124 Krajči, M. and Hafner, J. (2007) *Phys. Rev. B*, **75**, 224205.

125 Krajčiand, M. and Hafner, J. (2008) *Phys. Rev. B*, **77**, 134202.

126 Fournée, V., Ross, A.R., Lograsso, T.A., Evans, J.W., and Thiel, P.A. (2003) *Surf. Sci.*, **537**, 5.

127 Bilki, B., Erbudak, M., Mungan, M., and Weisskopf, Y. (2007) *Phys. Rev. B*, **75**, 045437.
128 Shimoda, M., Guo, J.Q., Sato, T.J., and Tsai, A.P. (2000) *Surf. Sci.*, **454**, 11.
129 Weisskopf, Y., Erbudak, M., Longchamp, J.N., and Michlmayr, T. (2006) *Surf. Sci.*, **600**, 2594.
130 Weisskopf, Y., Burkardt, S., Erbudak, M., and Longchamp, J.N. (2007) *Surf. Sci.*, **601**, 544.
131 Bielmann, M., Barranco, A., Rueux, P., Groning, O., Fasel, R., Widmer, R., and Groning, P. (2005) *Adv. Eng. Mater.*, **7**, 392.
132 Duguet, T., Ledieu, J., Dubois, J.M., and Fournée, V. (2008) *J. Phys.: Condens. Matter*, **20**, 314009.
133 Fournée, V., Ledieu, J., and Thiel, P. (2008) *J. Phys.: Condens. Matter*, **20**, 310301.
134 Jaklevic, R.C., Lambe, J., Mikkor, M., and Vassell, W.C. (1971) *Phys. Rev. Lett.*, **26**, 88.
135 Schulte, F.K. (1976) *Surf. Sci.*, **55**, 427.
136 Chiang, T.C. (2000) *Surf. Sci. Rep.*, **39**, 181.
137 Milun, M., Pervan, P., and Woodruff, D.P. (2002) *Rep. Prog. Phys.*, **65**, 99.
138 Fournée, V., Sharma, H.R., Shimoda, M., Tsai, A.P., Unal, B., Ross, A.R., Lograsso, T.A., and Thiel, P.A. (2005) *Phys. Rev. Lett.*, **95**, 1.
139 Moras, P., Weisskopf, Y., Longchamp, J.N., Erbudak, M., Zhou, P.H., Ferrari, L., and Carbone, C. (2006) *Phys. Rev. B*, **74**, 121405.
140 Young, T. (1805) *Philos. Trans. R. Soc. Lond.*, **65**, 95.
141 Dubois, J.M., Fournée, V., Thiel, P.A., and Belin-Ferré, E. (2008) *J. Phys.: Condens. Matter*, **20**, 314011.
142 Bergman, C., Girardeaux, C., Perrin-Pellegrino, C., Gas, P., Dubois, J.-M., and Rivier, N. (2008) *J. Phys.: Condens. Matter*, **20**, 314010.
143 Dubois, J.-M. (2004) *J. Non-Cryst. Solids*, **334/335**, 481.
144 Carpick, R.W. and Salmeron, M. (1997) *Chem. Rev.*, **97**, 1163.
145 Derjaguin, B.V., Muller, V.M., Yu., P., and Toporov, J. (1975) *Colloid Interface Sci.*, **53**, 314.
146 Enachescu, M., Van Den Oetelaar, R.J.A., Carpick, R.W., Ogletree, D.F., Flipse, C.F.J., and Salmeron, M. (1998) *Phys. Rev. Lett.*, **81**, 1877.
147 Johnson, K.L. and Kendall, K. (1971) *Proc. Roy. Soc. Lond. A Mater.*, **324**, 301.
148 Carpick, R.W., Ogletree, D.F., and Salmeron, M. (1999) *J. Colloid Interface Sci.*, **211**, 395.
149 Schwarz, U.D. (2003) *J. Colloid Interface Sci.*, **261**, 99.
150 Park, J.Y., Ogletree, D.F., Salmeron, M., Ribeiro, R.A., Canfield, P.C., Jenks, C.J., and Thiel, P.A. (2006) *Philos. Mag.*, **86**, 945.
151 Park, J.Y., Ogletree, D.F., Salmeron, M., Jenks, C.J., and Thiel, P.A. (2004) *Tribol. Lett.*, **17**, 629.
152 Enachescu, M., Carpick, R.W., Ogletree, D.F., and Salmeron, M. (2004) *J. Appl. Phys.*, **95**, 7694.
153 Park, J.Y., Ogletree, D.F., Salmeron, M., Ribeiro, R.A., Canfield, P.C., Jenks, C.J., and Thiel, P.A. (2005) *Phys. Rev. B*, **71**, 144203.
154 Park, J.Y., Ogletree, D.F., Salmeron, M., Ribeiro, R.A., Canfield, P.C., Jenks, C.J., and Thiel, P.A. (2006) *Phys. Rev. B*, **74**, 024203.
155 Park, J.Y. and Thiel, P.A. (2008) *J. Phys.: Condens. Matter*, **20**, 314012.
156 Dubois, J.M., Kang, S.S., and Von Stebut, J. (1991) *J. Mater. Sci. Lett.*, **10**, 537.
157 Vanossi, A., Roder, J., Bishop, A.R., and Bortolani, V. (2001) *Phys. Rev. E – Statistical, Nonlinear and Soft Matter Physics*, **63**, 1.
158 Ko, J.S., Gellman, A.J., Lograsso, T.A., Jenks, C.J., and Thiel, P.A. (1999) *Surf. Sci.*, **423**, 243.
159 Dubois, J.-M., Brunet, P., Costin, W., and Merstallinger, A. (2004) *J. Non-Cryst. Solids*, **334**, 475.
160 Mancinelli, C., Jenks, C.J., Thiel, P.A., and Gellman, A.J. (2003) *J. Mater. Res.*, **18**, 1447.
161 Hirano, M., Shinjo, K., Kaneko, R., and Murata, Y. (1991) *Phys. Rev. Lett.*, **67**, 2642.
162 Dienwiebel, M., Verhoeven, G.S., Pradeep, N., Frenken, J.W.M., Heimberg, J.A., and Zandbergen, H.W. (2004) *Phys. Rev. Lett.*, **92**, 126101.
163 Theis, W., Sharma, H.R., Franke, K.J., and Rieder, K.H. (2004) *Prog. Surf. Sci.*, **75**, 227.

5
Metallurgy of Complex Metallic Alloys

Saskia Gottlieb-Schoenmeyer, Wolf Assmus, Nathalie Prud'homme, and Constantin Vahlas

5.1
Introduction

The availability of single crystals is very important for the understanding of the physical and chemical properties of materials. Measurements performed on single crystals in a specific orientation allow for a much better interpretation of the results. This holds especially for complex metallic alloys with their complicated structure. Therefore, availability of single crystals is the key for the understanding of these compounds and crystal growth is an essential topic for groups working in this research field. In this chapter we follow today's interpretation of a crystal: all materials that have sharp Bragg or Laue reflections are crystalline. This interpretation enables the inclusion of quasicrystals, whereas the old definition – crystals have a three-dimensional periodic ordering of building blocks (unit cells) – excludes this quasi-periodic ordered interesting group of materials.

On the other hand, the availability of reliable deposition processes to produce CMA coatings and films on top of various substrates opens up an avenue to a large variety of applications of CMAs: corrosion-resistant coatings, light absorbers, tribological applications, and so on. It is therefore critical to set up such processes, with the view that complex shapes are a key to many such applications. Besides thermal spray techniques, or physical vapor deposition methods, magnetron sputtering, and so on, which are hindered by unavoidable shadow effects, new chemical vapor deposition techniques are now available to perform deposition on complex surfaces. The second part of this chapter reports on the progress made along this line to obtain CMA thin films from metal gaseous precursors, a challenge that so far had not been met in the literature.

5.2
Basic Concepts of Crystal Growth

For the understanding of crystal-growth transport, surface kinetics and capillarity are important topics and should therefore be discussed. In a crystal, the atoms or molecules are incorporated at specific lattice sites, whereas the starting material – the constituents of the compound or the polycrystalline compound itself – do not have the specific well-ordered structure of the CMA. Most of the CMA single crystals are grown from a liquid state – a melt or a solution.

The constituents have to be transported to the growth front from the bulk melt, where they are incorporated into the crystal. In case of noncongruent melting behavior or solution growth the nonincorporated material must be transported away from the growth front to avoid a local stoichiometry change of the melt near the crystallization front. During incorporation, the latent heat of crystallization becomes free and this energy must be also transported away from the growth front, so that the interface of the growing crystal remains in "thermal equilibrium." The quotation marks show that at thermal equilibrium no phase change would happen and a crystal would never grow – but we have to stay near the thermal equilibrium.

After the transport of the atoms or molecules to the interface of the growing crystal they must be incorporated at the correct lattice position. Therefore, surface kinetics is the next process that controls crystal growth. The crystals grow preferentially at kink positions or dislocations, as the deposited atoms have more binding partners at such positions. This is the second transport process during the growth of crystals. The third topic, which is important for the growth, is named capillarity. This means that the melting/solidifying temperature of the melt depends on the curvature of the interface as described by the Gibbs–Thomson relation

$$\Delta T = T_m - T = \frac{2\gamma T_m}{L} \cdot \frac{1}{r} \tag{5.1}$$

T_m is the melting temperature of a flat interface, T is the real melting temperature, γ surface tension, L is the latent heat and r is the curvature of the interface.

For most measurements on CMA compounds large crystals are necessary. Large means at least large in comparison to lattice distances, or, more concrete, several mm in size. For the growth of large crystals homogenous growth conditions are necessary. Therefore, a cellular or dendritic interface must be avoided as crystals with good quality grow best from a planar growth front. Therefore, capillarity is of lower importance and transport plays the dominant role.

Far away from the interface of the growing crystal transport is performed by convection and diffusion, but at the interface within a boundary layer of thickness δ the transport is governed by diffusion only and is described by the diffusion equation

$$D\nabla^2 u = \frac{du}{dt} \tag{5.2}$$

This diffusion equation describes both thermal and compositional diffusion, where D is the appropriate diffusion coefficient, u the temperature or composition and t the time [1].

The boundary condition at the interface can be described by

$$D(\nabla u)_{\text{Interface}} = -kv \qquad (5.3)$$

where $(\Delta u)_{\text{Interface}}$ is the thermal or compositional gradient at the interface, v the local (interface) velocity and k the temperature rise associated with the latent heat from crystallizing a unit volume, or in the case of material diffusion the amount of the rejected – not incorporated – material when crystallizing the unit volume. This is of especially high importance in case of noncongruent melting behavior, solution growth or materials with high impurity content.

From this equation the high importance of the ratio D over v, which is named diffusion length, can be seen clearly: It is the length over which the temperature or chemical potential field can be made uniform in the vicinity of the growth front.

For the growth of large crystals the diffusion length should therefore be macroscopic. All crystal-growth experiments of CMAs with the aim of large crystals should be performed in such a way that the diffusion length during the whole growth process remains macroscopic.

Here, between the growth of metals with a high thermal diffusivity and insulators (e.g., oxides) with a low thermal diffusivity have to be distinguished. Metals have a thermal diffusivity $D \sim 0.1\text{–}1\,\text{cm}^2/\text{s}$. This means that at "normal" growth rates of 10^{-3} cm/s the thermal diffusion length is macroscopic. For rapid growth $v \sim 10$ cm/s D/v becomes small, the heat flow occurs on a local scale at the growth front and the result is a dendritic growth pattern that should be avoided for the growth of large CMA crystals.

Now, material diffusion should be discussed. The mass diffusivity in liquids and melts is considerably smaller (D approximately 10^{-5} cm/s) than the thermal diffusivity. This means that even at growth rates of $v = 10^{-3}$ cm/s the diffusion length is 10^{-2} cm. The result is that the material that is rejected by the growing crystal accumulates in the boundary layer at the growth front. The composition near the interface of the growing crystal changes during growth. This has to be avoided.

To summarize: for crystal growth, thermal diffusivity and material diffusivity must be compared. For the growth of CMAs the bottleneck is due to the high thermal diffusivity of metals and the low material diffusivity always the mass diffusivity. The next step is to reduce the growth velocity to values so that the diffusion length remains macroscopic.

The only way to solve the problem of the microscopic diffusion length is to reduce the growth velocity. This is necessary in such growth runs where a high amount of material must be transported away from the growth front. This is always the case for solution growth (growth rate approximately 10^{-6} cm/s or 1 mm/day). The compositional phase diagram of the CMA plays an important role. When the CMA has a congruent melting behavior – melt and crystal have the same composition – no material besides impurities have to be transported away from the interface and high growth rates, for example, 10 mm/h can be used even for CMAs. But in the case of noncongruent melting CMAs – unfortunately this is the case for a lot of compounds – much lower growth rates must be used. In principle a low growth rate is not disadvantageous except for the time taken, the crucible material, however, then often becomes a serious problem.

Crucible materials that are often used for the growth of CMA systems are the refractory metals Mo, Ta and W. As these materials are sensitive to oxygen at high temperatures – as are most of the CMAs too – crystal growth must be performed under controlled atmosphere or vacuum. As most of the compounds have a significant vapor pressure at the melting temperature, which results in stoichiometric deviations during long-lasting growth runs high-purity argon gas is recommended. Due to its high atomic weight argon reduces the evaporation much better than a light (e.g., He) atom. For other CMA materials, such as BaNiGe, glassy carbon or graphite is an acceptable crucible material, BN and refractory oxide crucibles are also used. For long-lasting growth experiments the crucible material is often a serious problem, especially when impurities at the ppm level must be avoided. Therefore, after growth an analysis for example, EPMA on the crucible material impurities should be performed.

Sometimes, no appropriate crucible material is available. Then, crucible-free methods like zone-melting or using a cold crucible (e.g., Hukin-type) or a cold boat is recommended where the melt levitates without crucible contact in the crucible or the boat. This levitation is induced by eddy currents from a high-frequency (RF) electromagnetic field that is also used for heating.

In the case of very high vapor pressures (like Yb compounds) closed crucibles must be used. This is described later in detail.

Besides working with starting materials of excellent quality the question of the growth method is always of crucial importance for retaining single crystals of good quality. The decision of which growth method is the most appropriate should be made on the basis of the phase diagram, if it is accessible. Naturally, arguments like the reactivity of the treated elements and their properties, like for example, a high vapor pressure, and so on, also have to be considered.

Within this chapter, it is not possible to describe all growth methods in detail. Instead, a short overview of the most relevant ones for growing CMAs is given. These are mainly growth methods from the melt like Bridgman, Czochralski, zone melting and flux growth [2].

5.2.1
Bridgman Method

The Bridgman technique is easy to handle but nevertheless a very powerful method. A crucible that is positioned vertically in a furnace is lowered slowly out of the hot zone. Solidification starts at the coldest part of the crucible (see Figure 5.1). The temperature field can be produced by a resistance or an inductive heating device, depending on the desired temperature range. Essential for the growth process is a

5.2 Basic Concepts of Crystal Growth

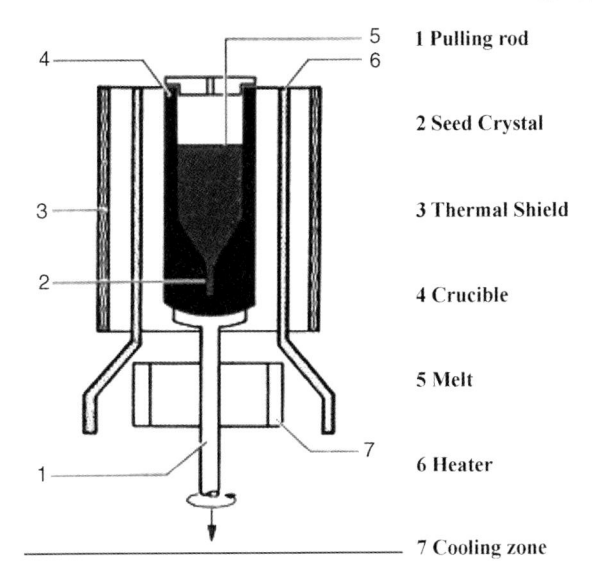

1 Pulling rod

2 Seed Crystal

3 Thermal Shield

4 Crucible

5 Melt

6 Heater

7 Cooling zone

Figure 5.1 Schematic Bridgman equipment.

well-defined temperature gradient. The crystallization starts when the melting temperature is reached at the bottom of the crucible and continues upwards with subsequent lowering of the melt out of the high-temperature region. Sometimes (e.g., when the growth process should occur in a vacuum) it can be useful to lift the furnace instead of lowering the crucible. Either way, the crystallization velocity is directly connected to the lowering velocity, although it is not the same. This depends on the heat conductivity of the crucible and the melt. Typical pulling velocities for metallic compounds range between 1 to 10 mm/h.

A possibly steep temperature gradient at the phase boundary between melt and solid is often required for (metallic) alloys. Such a gradient can be realized by using two temperature zones. The crystallization then starts at the transition interface from the hotter to the colder temperature zone. Another way to retain steep temperature gradients is cooling the bottom of the crucible, which is therefore mounted on a cooled pulling rod. Additionally, rotating the crucible avoids asymmetries in the temperature field.

To support the natural seed selection typical Bridgman crucibles are often designed with a conical- or tip-shaped bottom. The small diameter at the bottom supports the growth of only a small number of seeds. In an ideal situation, only one seed wins the competition and forms a huge single crystal at the end of the growth process. A more detailed inspection of the crucible form demands also the consideration of the growth behavior of the specific material. A growth of one preferred direction can be achieved by using an oriented seed crystal when the crystal direction is a stable growth direction. Certainly, this is an ambitious task due to the fact that in this case the seed crystal should not be melted. This cannot be checked visually like in the case of the Czochralski technique.

In the case of incongruently melting materials, which means that solid and liquid do not have the same composition, not all parts of the crystallized sample are of the desired phase. With ongoing crystallization additional secondary, ternary, and so on phases solidify. Thus, the amount of material that crystallizes in the desired phase depends only on the size of the temperature window of the primary solidification area.

A huge advantage of the Bridgman technique is that is does not have to be monitored all the time. The diameter of the crucible limits the diameter of the growing crystal. This means that the growth temperature can be chosen freely so that optimal growth conditions are fulfilled. Disadvantages of the technique are that the solidified ingot often sticks to the crucible after the growth. Removal of the ingot often leads to the destruction of the (expensive) crucibles and sometimes to fractures of the crystals.

A similar technique is the horizontal gradient freeze method.

5.2.2
Zone Melting

A single-crystal growth technique that is rather similar to the Bridgman growth method is the zone melting technique. Vertical and horizontal designs are possible. The difference from most other methods is the fact that only parts of the material, a well-defined zone, are molten. This can be realized by different heating techniques, such as for example, a very narrow inductive coil that can be driven slowly along the crucible. To realize a very-well defined melting zone more sophisticated techniques are required. Also, a laser or powerful lamp or an electron beam can be focused at one region of the sample. Often, mirror furnaces are used. Several elliptical formed mirrors focus the light of halogen bulbs at the sample.

This technique is mostly used for incongruent melting materials. Since only small regions of the sample are molten, the starting composition can be chosen in such a way that the growing crystal, except for the last section, is of constant composition.

A special arrangement of the zone melting technique is the floating zone melting. A rod of the prereacted material is fixed at both ends. Then, a small region is heated. The melt is hanging freely between both ends of the rod, only kept in place by its surface tension. The melt zone is slowly driven along the rod and often both parts are rotated in counter directions. This technique is often used when no appropriate crucible material can be found. Additionally, a purification effect can arise due to the evaporation of impurities or a distribution coefficient being not equal to one. Naturally the floating of the melt is an unstable process, especially when the melt zone gets too long (that is at least when the length of the melt zone exceeds its circumference) the surface tension is not strong enough and the liquid will be disrupted.

5.2.3
Czochralski Technique

Another single-crystal growth technique is the Czochralski method. The idea was first published in 1918 by Czochralski [1] and is today well known for the production

1 Crystal neck

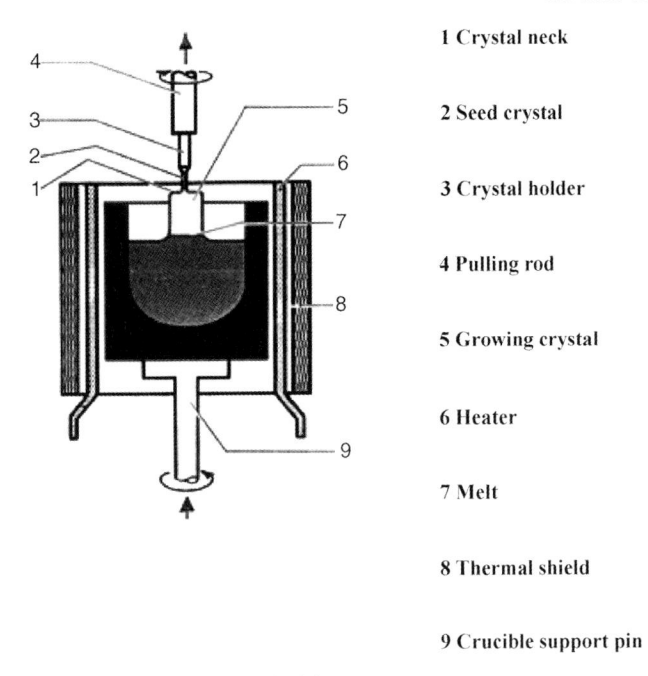

2 Seed crystal

3 Crystal holder

4 Pulling rod

5 Growing crystal

6 Heater

7 Melt

8 Thermal shield

9 Crucible support pin

Figure 5.2 Schematic Czochralski equipment.

of high-quality crystals for example, semiconductors like Si or crystals for nonlinear optics like borates.

A schematic setup is shown in Figure 5.2. The desired material is molten in an appropriate crucible at a temperature well above the melting temperature until a good homogenization of the melt is achieved. Then, the temperature is lowered and kept a little above the melting point. Now the critical part of the growth process starts. The cooled seed crystal that can be a piece of crystal from the desired phase, or if not available another higher melting material, is dipped into the melt and crystallization should start directly at the seed crystal that is continuously pulled upwards again. In an ideal situation the growth front is situated a little above the melting surface. The diameter of the growing crystal depends on the thermal balance of heating and cooling. A temperature rise results in a smaller cross section of the growing crystals whereas temperature reduction leads to an increase of the crystal diameter. Typical pulling velocities vary from 100 to 0.1mm/h. After dipping the seed into the melt usually a small neck is grown. Then, the diameter of the growing crystal is enlarged by reducing the temperature. Once the desired diameter is reached the temperature can be kept constant. Certainly in the case of incongruently melting systems the change of the liquidus temperature with changing composition has to be considered.

Although the basic idea of this growth method seems to be relatively simple, the system melt–crystal is very sensitive to small temperature fluctuations and good results can only be achieved when the main controlling parameters (temperature and pulling velocity) are monitored all the time. On the other hand, this is also a great advantage of this technique because the success of the growth experiment can be seen immediately. Another benefit is the quasicrucible freeness that provides an impurity free and strainless growth of the crystal, assuming that the melt is as clean as possible.

The choice of the seed crystal is of high importance. Its quality should be high, otherwise it is likely that crystal defects like grain boundaries and microcracks will spread into the bulk crystal. In principle, an orientational growth is possible if a preoriented seed is used. When no seeds of the phase to be grown exist, a material with a higher melting temperature can be used followed by a necking procedure, which leads to a seed selection. Afterwards, the growth process is continued as described above.

In the case of incongruently melting materials one has to keep in mind that the composition of the melt continuously changes during the crystallization process. Several arrangements have been designed to maintain a constant composition of the melt. Instead of working with only one big crucible it is advisable to work with virtually two crucibles that are connected in some way (e.g., one crucible that is divided into several parts). The growth process takes place in the smaller part and the loss of one component can be adjusted due to the connection to the second larger part where a constant melt composition can be easily provided.

Often, materials have very high vapor pressures, especially at high temperatures. This is generally not a problem if the evaporated material and melt have almost the same composition. But if only one component is volatile, working with an open melt is not possible. In this case, a modification of the Czochralski technique, the liquid encapsulation technique, can be used. Here, the melt is covered by another inert melt to avoid evaporation of one component. Often, boron trioxide (B_2O_3) is used, but eutectic mixtures of salts are also common. The layer thickness ranges from 5 to 10 mm. At the end of the growth process the crystal is pulled through the protective layer. To guarantee phase pureness of the growing crystal the solubility of the protective layer material in the melt must be negligible. Another disadvantage of this technique is that due to a very high temperature gradient between melt and protective coating, thermal stresses can occur. Besides, if the coating is not transparent the visual monitoring of the growth process is no longer possible. Nevertheless, this technique provides a successful method to handle highly volatile elements [2].

A special variant of the Czochralski technique is the Kyropoulus method (Figure 5.3). Instead of pulling the growing crystal out of the melt it grows into the melt. After the growth process the crystal is removed from the rest of the melt. This has to be performed before secondary, ternary phases, and so on, solidify.

5.2.4
Flux Growth Technique

The self-flux growth technique, also called flux growth technique, is a special case of solution growth. It has been applied for many different classes of materials such as oxide crystals, garnets, but also for the production of single quasicrystals [3]. It has the advantage that no new elements are added to the synthesis.

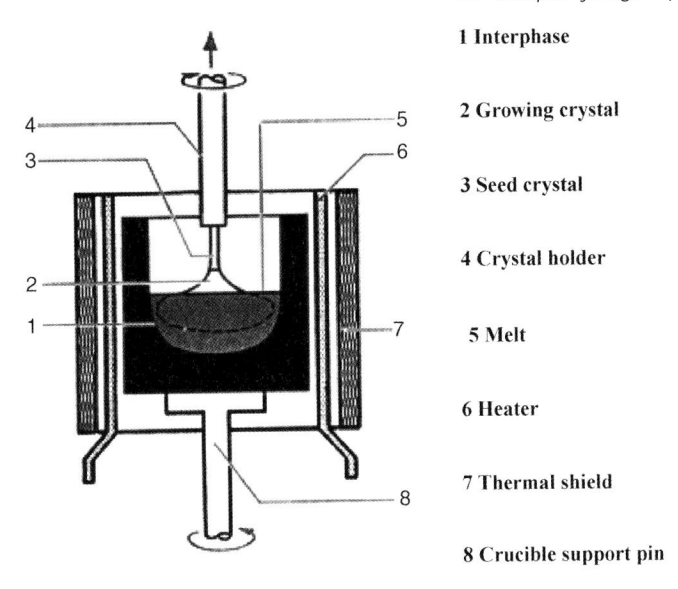

Figure 5.3 Schematic Kyropoulus equipment.

1 Interphase
2 Growing crystal
3 Seed crystal
4 Crystal holder
5 Melt
6 Heater
7 Thermal shield
8 Crucible support pin

An appropriate crucible containing the melt is placed in a furnace with a low-temperature gradient and a high-precision temperature regulation. The melt is first homogenized at a temperature well above the melting point of the alloy. A slow cooling process follows while the desired phase solidifies into the melt. Finally, the rest of the melt is decanted. The temperature program for the growth process and the starting composition of the melt has to be chosen according to the phase diagram. The decanting process has to be performed at temperatures that are still high enough. When the region of first solidification ends the crystallization of secondary, ternary phases is probable.

To reduce the number of grains a very slow cooling sequence is chosen and as in the case of the Bridgman technique tip-shaped crucibles and cold fingers are used. For decanting the crucible is often simply turned upside down. The residual melt is trapped in a second crucible that has been placed on top of the first one.

The more general description of some single-crystal growth techniques will be expanded in the following sections using the examples of some selected CMAs.

5.3
Examples of Single-Crystal Growth of CMAs

5.3.1
$Al_{13}Co_4$ and $Al_{13}Fe_4$ Using the Czochralski Technique

After Gille and Bauer who set up the growth parameters of these CMAs [4], $Al_{13}Co_4$ and $Al_{13}Fe_4$ can be seen as approximants of the decagonal quasicrystal found in the Al-Co-Ni system, which means both phases consist of clusters that are also part of the quasicrystalline structures. Thus, the comparison of physical properties of both quasicrystals and their periodic approximants is still a very active field of research. Therefore, single-crystalline material of appropriate size has to be available.

Both, orthorhombic $Al_{13}Co_4$ and monoclinic $Al_{13}Fe_4$ are incongruent melting phases that result from peritectic reactions at 1092 °C and 1160 °C, respectively. Regarding the phase diagrams [5, 6] $Al_{13}Co_4$ is stable within a temperature range from 974 °C to 1092 °C, whereas $Al_{13}Fe_4$ does not decompose until a temperature of 655 °C is reached. What makes the growth process an ambitious one has two main features: Only very low pulling rates can be used due to the fact that the growing crystal and melt differ widely in composition. Therefore, the excess component has to be carried away from the growth front. Besides, aluminum shows a huge affinity for oxygen, such that the growth chamber has to be gas tight and before starting the growth process the chamber was evacuated and baked for several days. Argon (5N grade) was used as a protective gas.

As starting compositions $Al_{86.5}Co_{13.5}$ and $Al_{89.5}Fe_{10.5}$ were chosen. Al (4N grade), Co (3N grade) and Fe (3N grade) in the form of metallic pieces (17g) that have been etched to avoid surface contaminations were used. In a first step a homogenous solution was prepared. Therefore, the elements were molten in an alumina crucible under argon atmosphere using a radio-frequency heating facility. After quenching and cooling the melt to room temperature it could be easily removed.

The presynthesized material again was placed into an alumina crucible and was homogenized in a second step. Therefore, for at least 12 h the material was kept at temperatures 100 K above the corresponding liquidus temperature that had to be found by direct observation. This was done in the following way: the melt was cooled until needles of $Al_{13}Co_4$ or $Al_{13}Fe_4$ form spontaneously. These needles were then redissolved by slowly heated again. The temperature at which only traces of the needles were found was the correct liquidus temperature (T_L).

When T_L was determined, the seed crystal that was fixed at a ceramic holder was wetted by the solution at temperature T_L. Afterwards, it was lowered by 1–2 mm into the solution and the growth process was started using pulling rates of 100–150 µm/h. For the first runs when no seed crystal was available a tapered aluminum rod was dipped into the melt such that one droplet of melt could crystallize at the top of the rod. Then, the crystallization process was continued as described before.

As in any Czochralski process the diameter was controlled by the temperature ramp. At the beginning of the growth process the temperature was lowered at less than 0.1 K/h to enlarge the diameter of the growing crystal to the favored size. With crystallization, and therefore decreasing T_L, the temperature ramp had to be adjusted and finally reached 0.5 K/h. After 2–3 weeks, when half of the mass of the initial melt was crystallized, the growth was stopped by very fast pulling of the crystal. Photographs of the grown single crystals can be seen in Figure 5.4.

The grown single crystals were used to redetermine the lattice constants and to analyze chemical bonding in a first step [7]. The experimental data are summarized in Table 5.1.

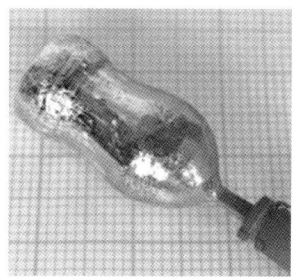

Figure 5.4 (left) $Al_{13}Co_4$ single crystal grown by the Czochralski technique using an [001]-oriented single-crystalline native seed. (right) $Al_{13}Fe_4$ single crystal grown by the Czochralski technique using an [010]-oriented single-crystalline native seed.

5.3.2
Single-Crystal Growth of β-Al-Mg

The growth of this material was revisited by Lipińska-Chwałek et al. [8] With lattice parameters of 2.8 nm and 1168 atoms per unit cell the phase β-Al_3Mg_2 (spacegroup Fd-3m) definitively belongs to the class of complex metallic alloys [9, 10]. Due to its low specific weight of 2.2 g/cm³ the material may offer some potential for technological exploration. Although the substance has been well known since the 1960s [11] its physical properties are nearly unexplored, because high-quality single-crystalline material has not been available.

For single-crystal growth of β-Al_3Mg_2 three techniques, the Bridgman method, the Czochralski technique and the flux-growth method, were tried out. For all experiments Al (5N grade) and Mg (3N) grade were used, only for the Czochralski growth experiments was Al (6N) used. For the first studies the phase diagram published by Murray was considered. According to that work β-Al_3Mg_2 melts congruently at 451 °C and its stability range extends from 38.5 at.% Mg to 39.9 at.% Mg at 400 °C. Later phase-diagram studies performed by the authors basically confirm these results. A slightly wider homogeneity range from 37.4 at.% Mg to 39.9 at.% Mg was found. Two additional high-temperature and one low-temperature phase transitions have been detected. The melting point was determined to 447 °C and as starting composition for crystal-growth experiments $Al_{61.5}Mg_{38.5}$ was chosen.

Table 5.1 Redetermined lattice constants of $Al_{13}Co_4$ and $Al_{13}Fe_4$.

	$Al_{13}Co_4$	$Al_{13}Fe_4$
space group	$Pmn2_1$	C2/m
a [Å]	8.158 (1)	15.488(1)
b [Å]	1.2342(1)	8.0866(5)
c [Å]	1.4452(2)	12.4769(8)
β [°]	90	107.669(4)
ϱ_{calc} [g/cm³]	3.966(1)	3.8415(8)
ϱ_{exp} [g/cm³]	3.962(1)	3.847(3)

For sample characterization light microscopy and scanning electron microscopy (SEM) were used. The alloy compositions were determined by energy-dispersive X-ray analysis (EDX), which was calibrated by inductively coupled plasma optical emission spectroscopy. Phase identification was done by powder X-ray diffraction.

5.3.2.1 Bridgman Growth

Both graphite and alumina tapered tip-shaped crucibles were used and the growth process was realized using a protective argon atmosphere of 260 mbar in a vertical tube furnace. While the furnace was kept constant at 520 °C during growth the crucible was slowly pulled out of the high-temperature region. Six growth runs were performed and the pulling rate was varied from 1 mm/h up to 20 mm/h. Although the crystallized material was single β-phase only relatively small grain sizes of 0.35 cm^3 could be produced using the graphite crucibles. By using alumina crucibles only smaller grains could be gained. Additionally, due to sticking it was difficult to remove the ingots from the crucibles. Nevertheless, the small grains could be used as seed crystals for the Czochralski growth runs.

5.3.2.2 Czochralski Growth

The growth was realized using alumina crucibles in a protective argon atmosphere of 400 mbar. A tungsten susceptor in a high-frequency field provided the desired temperature range. The seed was rotated at 25 turns per minute and a pulling rate of 15 mm/h was applied. Three growth runs were performed. For the first two runs a polycrystalline seed crystal cut from the Bridgman growth ingot was used. For grain selection a thin neck of 1 mm diameter was grown, then the diameter was increased up to 1 cm. The volumes of the achieved single crystals amounted to 3.6 cm^3. For the third run a [110]-oriented seed crystal was used, the other growth parameters were left unchanged. The grown single crystal showed clear facets and had a size of 4.2 cm^3.

5.3.2.3 Self-Flux Growth

Three growth runs were performed. The melt was placed into an alumina crucible. Both flat-bottom and tip-shaped crucibles were used. An inverted crucible was put on top and the whole assembly was sealed in a quartz ampoule under an argon atmosphere at 800 mbar. In a chamber furnace the ampoule was first heated up to 600 °C. After homogenization for one hour the temperature was lowered at 10 K/h to 540 °C, afterwards to 300 °C with a cooling rate of 1 K/h. By taking the ampoule out of the furnace at 300 °C the crystallization process was stopped. The grown samples were all single β-phase and showed very big single-crystal regions with volumes of 10 cm^3 up to 17 cm^3. The use of tip-shaped crucible forms did not decrease the number of grain boundaries.

Comparing all three growth methods the Bridgman technique seems to be the least adequate one, whereas both Czochralski and the flux technique have been very successful (see Figure 5.5). If oriented crystal growth is desired, the Czochralski technique should be the preferred method. If single crystals of possible large volume are favored the flux growth technique is the best choice according to the authors.

Figure 5.5 (a) (left): A sample grown by the self-flux growth technique – two grains are clearly visible. (b) (right): A sample grown by the Czochralski technique – fully single crystalline.

5.3.3
Single-Crystal Growth of $Mg_{32}(Al,Zn)_{49}$

The structure of $Mg_{32}(Al,Zn)_{49}$ was first solved by Bergman *et al.* [12] (1957). It crystallizes in the cubic body-centered space group $Im\bar{3}$ with 162 atoms per unit cell (lattice parameters of $a = 14.16$ Å). The structure often referred to as "Bergman phase" can be described as a body-centered arrangement of characteristic clusters.

The growth of a single crystal was solved recently by Feuerbacher *et al.* [13]. Experimental phase-diagram information can be found in Petrov *et al.* [14]. $Mg_{32}(Al,Zn)_{49}$ solidifies via a degraded peritectic reaction and has a large liquidus field. The stability range at 335 °C extends from $Mg_{38}Al_{18}Zn_{46}$ to $Mg_{40}Al_{48}Zn_{12}$ with a width of about 7 at.% Mg.

For crystal growth of $Mg_{32}(Al,Zn)_{49}$ both Bridgman and Czochralski growth methods were applied. In both cases single crystals were grown from a prealloyed melt. The grown crystals were analyzed by scanning electron microscopy (SEM). Some parts were also used for metallographic investigations. Compositions were checked by energy dispersive analysis (EDX) and single crystallinity was proved by Laue diffraction.

At a first preparation step, 32 at.% Al (5N grade), 37 at.% Mg (3N grade) and 31 at.% Zn (5N grade) were melted under protective argon atmosphere using a levitation induction furnace with a water-cooled copper crucible. To ensure a good homogeneity the melt was heated several times above the melting point. In this manner 20 g and 80 g melts were produced for Bridgman and Czochralski growth, respectively. Rod-shaped ingots of 9 cm length and a diameter of 8–10 mm for Bridgman growth were formed by casting the melt into a water-cooled tube mold.

5.3.3.1 Bridgman Growth
Bridgman growth was performed using a vertical tube furnace with a well-defined temperature gradient. The cylindrical crucibles, made from alumina and graphite, with a tip-shaped bottom (8 mm to 10 mm diameter) were put on a cold finger to ensure that solidification started at the lowermost part of the crucible. The growth was performed under a protective argon atmosphere and the furnace temperature was

maintained at 600 °C. Pulling velocities varied between 2 mm/h and 0.5 mm/h but the best results were obtained at speeds of 1 mm/h and below.

Due to the incongruent solidification behavior the last solidified part of all grown samples showed additional phases. Nevertheless, the largest part of the sample could be identified as single-phase $Mg_{36.8}Al_{28.9}Zn_{35.7}$ (average composition of all three performed growth runs). During the growth process the composition changes to lower Zn content, whereas the Mg rate stays more or less constant ($Mg_{37.9}Al_{35.1}Zn_{27.0}$, $Mg_{37.0}Al_{33.8}Zn_{29.2}$ and $Mg_{38.1}Al_{31.5}Zn_{30.4}$).

This is in agreement with published data. From the single-phase region single-crystalline samples of volumes up to 3 cm^3 could be extracted. The number of grains was reduced by using a tip-shaped crucible, in a way that two of three grown samples were fully single crystalline, whereas the third sample consisted of two grains that could be distinguished by the naked eye. The excellent structural quality of the samples could be verified by a neutron-diffraction study. Figure 5.6 shows one of the grown crystals.

5.3.3.2 Czochralski Growth

The melt was heated in an alumina crucible using a tungsten susceptor in a high-frequency field. A special wiper system was installed to remove the oxide layer that covered the melt after heating.

At a temperature of 534 °C a cylindrical seed crystal of 2.5 mm diameter cut from a Bridgman run was dipped into the melt. The seed was rotated at 30 turns/min and the crystal was pulled at 10 mm/h. By changing the heating power the diameter of the growing crystal was controlled. This resulted in temperature changes from 508 °C (thick diameter) in the beginning to 530 °C (small diameter) at the end of the growth process.

A fully single-crystalline sample of 1 cm^3 volume could be grown using the Czochralski technique. Compared to the Bridgman runs the composition was determined as $Mg_{36.4}Al_{33.4}Zn_{30.2}$. Figure 5.7 shows the grown crystal as it was mounted in the Czochralski apparatus. Due to using nonoriented seed crystals the growth was performed along an arbitrary direction.

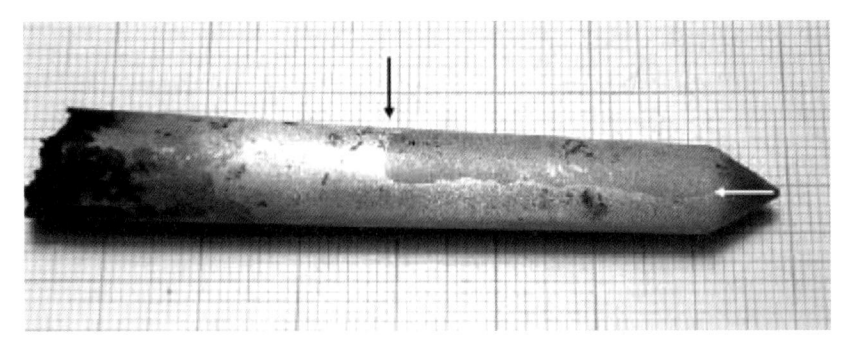

Figure 5.6 Crystal grown by the Bridgman technique.

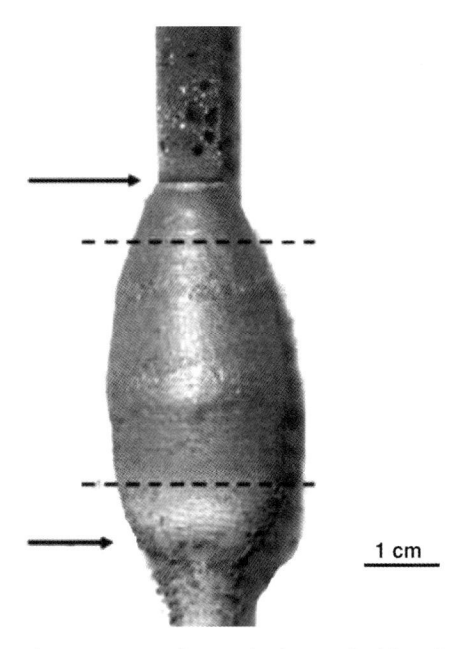

Figure 5.7 Crystal grown by the Czochralski technique.

5.3.4
Single-Crystal Growth of Al-Pd-Mn Approximants

In the Al-Pd-Mn system a family, the so-called ξ'-family, of complex phases exists. 320 atoms per unit cell form the basis phase of this family crystallizing in an orthorhombic structure-type (Pnma). The lattice can be described by alternatively arranged flattened hexagons ($a = 23.54$ Å, $b = 16.56$ Å, and $c = 12.34$ Å) [15] On the vertices of these hexagons lattice, pseudo-Mackay clusters are arranged. The other members of the ξ'-family are generated by linear defects, which consist of local atomic rearrangement of the ξ'-structure. The hexagon lattice is partly replaced by a combination of a banana shaped polygon and an attached pentagon. These so called phason lines tend to arrange closely neighbored along the [100] direction and form phason planes that in turn order periodic in a way that superstructures are built. One popular example is the ψ-Al-Pd-Mn or ε_{28}-phase ($c = 57$ Å) [16–18].

For growth of ξ'-Al-Pd-Mn the Bridgman technique was applied. There is still much discussion about the phase diagram [16–25], so details will not be discussed here.

A prealloyed melt of 79.0 at.% Al, 18.0 at.% Pd and 3.0 at.% Mn was synthesized using a levitation induction furnace. Afterwards, the melt was kept in a tip-shaped pyrolytic BN crucible mounted on a water-cooled molybdenum cold finger. The furnace temperature was kept constant at 900 °C and the crucible was pulled out of the hot zone at 1 mm/h. The grown sample was single phase with a composition of $Al_{73.9}Pd_{22.2}Mn_{3.9}$.

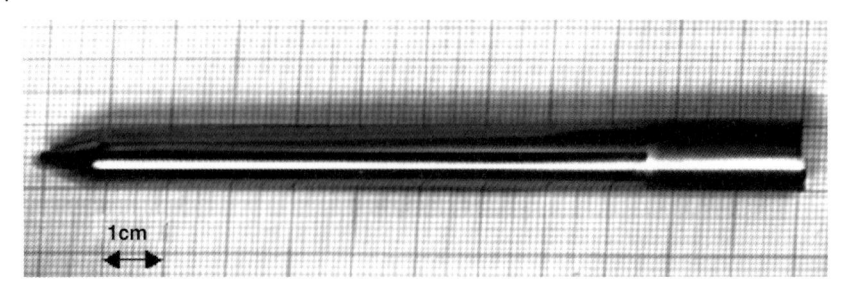

Figure 5.8 Single crystal of ξ'-Al-Pd-Mn grown by the Bridgman technique.

The last part (2.5 cm) showed a eutectic mixture of other phases. Large single-crystalline grains of 1 up to 9 cm^3 could be extracted from the first part (Figure 5.8). Analyzing these parts by selected-area electron diffraction reveals both regions of the ξ'-family and of the ψ-Al-Pd-Mn superstructure were found.

5.3.5
Crystal Growth of Yb-Cu Superstructural Phases

Gottlieb-Schönmeyer et al. [26] successfully grew the first single crystals of YbCu$_{4.4}$ and YbCu$_{4.25}$, which are both monoclinically distorted superstructural phases of the cubic AuBe$_5$-structure type. With several thousand atoms per unit cell and lattice parameters in the range of nanometers these are one of the most complex binary metallic alloys known to date. Along with YbCu$_{4.5}$ whose structure was solved by Černy et al. in 1996 [27] the two new detected superstructures follow the building principle that was proposed by Černy et al. in 2003 for three DyCu$_x$ phases (x = 4.5, 4.0 and 3.5) [28].

Before crystal-growth experiments were performed detailed phase diagram studies in the Yb-Cu system (from 16 at.% Yb to 24 at.% Yb) were done. These reveal a phase richness that has not been expected before. Besides the phases YbCu$_{4.5}$ (congruent melting at 937 °C) and YbCu$_{3.5}$ (peritectic formation at 825 °C) that have been reported in previous phase diagrams [29], two new phases YbCu$_{4.4}$ and YbCu$_{4.25}$ could be discovered. These two new phases form peritectically at 934 ± 2 °C and 931 ± 3 °C (see also Figure 5.9). Due to the high vapor pressure of ytterbium tantalum crucibles (Ø = 8 mm, height = 17 mm) that could be sealed with a specially designed press were used for phase-diagram studies. The measurements were carried out using a standard DSC device (Netsch, STA 409) that allows simultaneous thermogravimetric analysis (TGA). Heating and cooling rates of 2 °C/min up to 10 °C/min were used.

Crystal-growth experiments were done using the Bridgman technique. The pure elements (Yb 4N grade and Cu 5N grade) were sealed in a tantalum crucible (Ø = 9 mm, height = 85 mm), that was mounted on a cooled pulling rod. The starting composition was 19.21 at. % Yb. In addition due to the specially designed crucibles (small diameter at the bottom) the pulling rod acted as a cold finger. The sample was heated inductively under a protective argon atmosphere. The temperature was checked pyrometrically. After a homogenization process of half an hour, while the

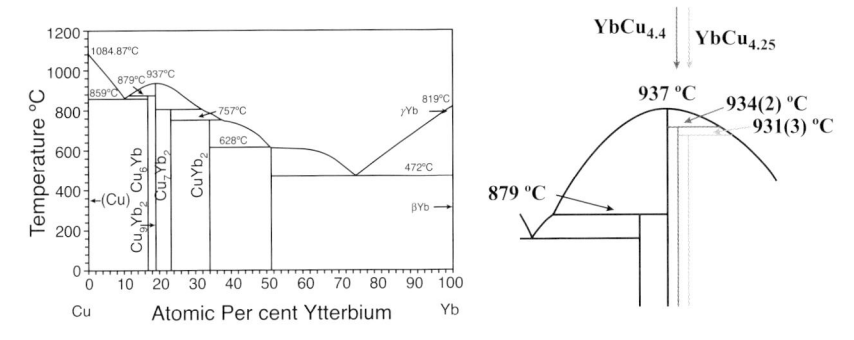

Figure 5.9 (left) Phase diagram of Yb-Cu after (Wilke and Bohm, 1988). (right) Modifications in the Cu-rich region.

whole sample was hold at a constant temperature of about 1000 °C, the crucible was pulled slowly out of the hot zone. Medial pulling velocities between 0.5 mm/h up and 3 mm/h), showed the best results.

All samples were cut by spark erosion and checked by EDX for phase homogeneity. Although no phase contrast could be detected (compositions range between 19.65 at. %-Yb and 20.4 at.%-Yb) the grown samples were not fully single phase, as analysis of different parts of the crucible by selected-area electron diffraction (SAED) and single-crystal X-ray diffraction (SC-XRD) show. Two superstructures could be detected that refer to two different phases $YbCu_{4.4}$ and $YbCu_{4.25}$ (see Figures 5.10a and b) [28]. The only slight compositional differences could not be resolved by EDX but due to the different number of satellite reflections the two phases could be distinguished by the diffraction techniques mentioned above. Instead of growing single crystals of $YbCu_{4.5}$ two new phases $YbCu_{4.4}$ and $YbCu_{4.25}$ have been discovered but since compositions as well as solidification temperatures are very close single-crystal

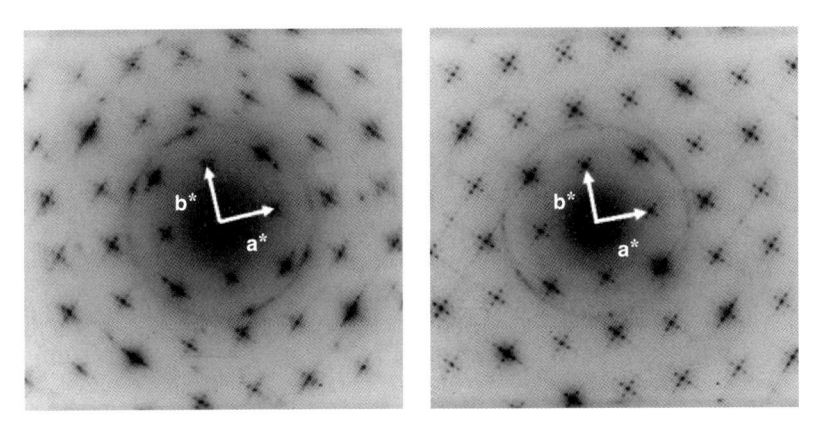

Figure 5.10 (left) SAED of $YbCu_{4.4}$ 5 × 5 satellite reflections in this plane. (right) SC-XRD of $YbCu_{4.25}$ 6 × 6 satellite reflections in this plane.

growth turns out to be a very challenging task. To realize the growth of only one defined superstructure more research work will have to be done.

5.3.6
Single-Crystal Growth of MgZn$_2$

The Laves phase MgZn$_2$ ($hP12$) is congruently melting at 600 °C [30, 31]. It was grown as a single crystal by Drescher (Diploma thesis, 2007). The lattice parameters of the hexagonal unit cell are: $a = 5.221$ Å and $c = 8.567$ Å, $\gamma = 120°$ [32, 33]. For single-crystal growth two techniques the Bridgman technique and the liquid-encapsulated Kyropoulus technique, were applied, where the latter technique seemed to be the more successful one. Centimeter-sized single crystals could be grown (Figure 5.11).

5.3.6.1 Bridgman Technique
The pure elements (Mg 4N and Zn 6N) were sealed in a tantalum crucible (Ø = 9 mm, height = 84 mm) that was mounted on a cooled pulling rod that acted as in the case of Yb-Cu as a cold finger. Due to the high vapor pressure of Zn sealing was indispensable. The starting composition was the stoichiometric one (66 at.% Zn). The crucible was heated inductively up to 800 °C under a protective argon atmosphere. The temperature was checked pyrometrically. After a homogenization process the sample was pulled out of the hot zone. Pulling velocities varied between 3.81 cm/h and 2.03 cm/h. After the growth process the samples were cut by spark erosion and checked by energy dispersive X-ray analysis (EDX) and powder X-ray diffraction for phase homogeneity. Unfortunately, the sample was not fully single phase.

5.3.6.2 Liquid-Encapsulated Kyropoulus Technique
Alumina crucibles (Ø = 5.9 cm, length = 4 cm) were charged with 80 g of a weighed sample of stoichiometric composition (Mg (4N) and Zn (6N) (66 at.% Zn)) and 20 g of an eutectic LiCl/KCl salt mixture. The salt mixture that remains liquid during the whole growth process prevents an evaporation of Zn and Mg. The crucible is positioned in a quartz glass cylinder that is affixed to a water-cooled high-grade

Figure 5.11 MgZn$_2$ single crystals grown by the liquid-encapsulated Kyropoulus method.

steel flange. The whole chamber is surrounded by a resistance furnace and the temperature can be measured by a NiCr-Ni thermocouple.

After repeated evacuation of the chamber the melt is homogenized for several hours at 700 °C under an argon atmosphere of 0.5 bar. To realize the temperature gradient that is necessary for the growth process the furnace is lowered some centimeters. Afterwards, a water-cooled tungsten seed crystal is dipped 5 mm into the melt (through the salt layer). The temperature is lowered from 620 °C to 545 °C at 2.5 °C/h. The solidified sample can be removed afterwards. The remaining salt layer can simply be removed by water.

Analyzing the samples by energy dispersive X-ray analysis and powder X-ray diffraction reveals phase-pure material with a composition of $MgZn_2$. Single crystallinity was checked by Laue diffraction. Figure 5.11 shows a single crystal of $MgZn_2$.

5.4
Introduction to Chemical Vapor Deposition of Coatings Containing CMAs

Chemical vapor deposition (CVD) is a process for the deposition of thin films on substrates. It can be schematized as follows: the part to be covered is positioned in the heated zone of a reactor; a gas phase, containing molecules with the elements to be deposited (named CVD precursors), is flowed in the reactor; the heat provided allows decomposition of the precursors in the vicinity of the piece or on its surface; the reaction produces the solid film and also volatile by-products that are evacuated from the reactor.

CVD is widely used in materials processing technology. This is mainly due to its high throughput and to its capacity to conformally cover complex-shaped parts. Versatility, cost effectiveness and environmental compatibility are additional advantages of CVD processes. Moreover, the use of molecular (organometallic and metalorganic) precursors in the last thirty years has allowed (MO)CVD processes to operate at low to moderate temperatures, thus extending the targeted applications spectrum so as to cover temperature-sensitive substrates. Both thermodynamically stable and metastable metallic, ceramic and polymeric films and coatings are nowadays processed by MOCVD.

The price to pay for this high potential is the need to manage the complex, gas-phase and surface chemistries. Such a delicate, "butterfly" chemistry, as it was pejoratively called by the past, imposes a series of challenges to be met. In addition to mastering the deposition reaction, these challenges also concern the design of the precursors upstream of the MOCVD process, the engineering of the MOCVD apparatus in terms of generation of precursor vapors and of energy providing means, the *in situ* and online diagnostics in order to obtain information on the reactions occurring in the vicinity of and on the growing surface. Multiscale modeling of the process, taking into account atomic interaction, molecular dynamics, chemical kinetics, fluid mechanics and phase equilibria is a necessary tool to meet these challenges.

The principle of CVD is illustrated in Figure 5.12. In this figure, the different steps of the process and the corresponding phenomena and tasks are shown. Observation

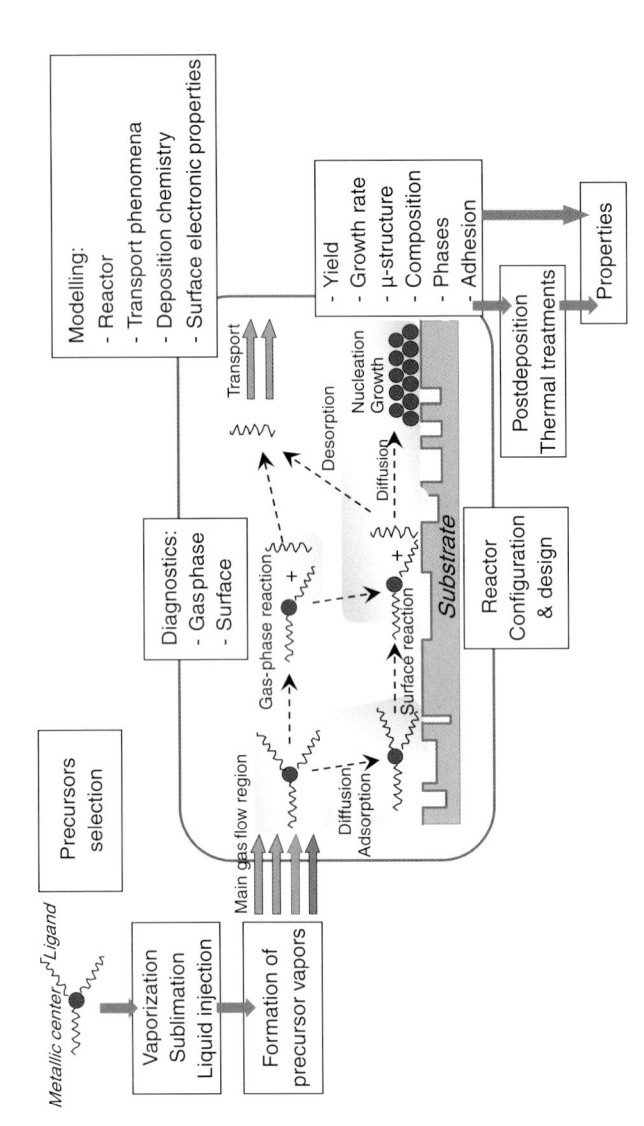

Figure 5.12 Schematic illustration of the CVD process. Molecular compounds (organometallic or metalorganic precursors) are used to identify the different steps and to illustrate the complexity of the process.

of this figure reveals that one point that is of importance in CVD concerns the design and the selection of the precursors, as well as the different ways of formation of precursor vapors and their introduction into the deposition chamber. A second point concerns the configuration and the design of the CVD reactor. Finally, a third one includes the different phenomena occurring during the deposition process: gas phase and surface reactions, diffusion, adsorption and desorption of molecules, and nucleation and growth of the film. The chemical, physical and technological options for each of these points influence the process performance, such as the yield and the growth rate. They also influence the characteristics of the obtained material, namely its microstructure, the elemental composition and the phases present in the film as well as its interaction with, and consequently the adhesion to the substrate.

Processing of CMA-containing thin films and coatings is expected to further extend the state-of-the-art of materials performance and to raise numerous bottlenecks in many application domains, including hard and barrier coatings, nonwetting surfaces, catalysis or thermoelectricity to name but a few. Such hopes are based on the excellent surface properties of CMAs. Films and coatings can be composed exclusively or in part of CMAs, the latter case allowing for the combination of the characteristics provided by the two (or more) components of the material. Due to its previously mentioned characteristics, CVD can play a major role in the implementation of "CMA solutions" to surface engineering. Parts containing nonline-of-sight surfaces can be coated by CVD processes. Examples are molds in the glass industry, turbine blades and vanes in aeronautic industry. Other possibilities concern porous peeforms whose internal surface must be functionalized, for example for the preparation of supported catalysts. However, the inherent difficulty to establish a robust CVD process is further amplified in the case of CMAs due to (a) the multielement nature of the CMAs, (b) their narrow stability domain. For these reasons, there are actually only a few CVD reports on the processing of multielement intermetallic compounds in general. Suhr *et al.* reported on the processing of thin metal alloy films of Fe/Co and Au/Pt/Pd by plasma-enhanced CVD [34]. They mention that "...*the formation of metal alloy films has... been considered a domain of sputtering since CVD processes using mixtures are faced with great difficulties resulting from differences in vapor pressure, as well as thermal and plasma stability of the components...*". To the best of our knowledge, CVD of CMA thin films has only been investigated in the frame of the CMA European Network of Excellence.

5.5
MOCVD Processing of Al-Cu-Fe Thin Films

This section resumes the strategy that has been adopted to investigate the MOCVD route for the processing of CMA films in the Al-Cu-Fe system, targeting especially the icosahedral $Al_{63}Cu_{25}Fe_{12}$ phase. The selection of this system was based on the thermodynamic stability of the icosahedral phase in a temperature range that includes ambient and expected processing and service conditions. The stability domain of this phase is illustrated in Figure 5.13 [35]. The authors demonstrated that

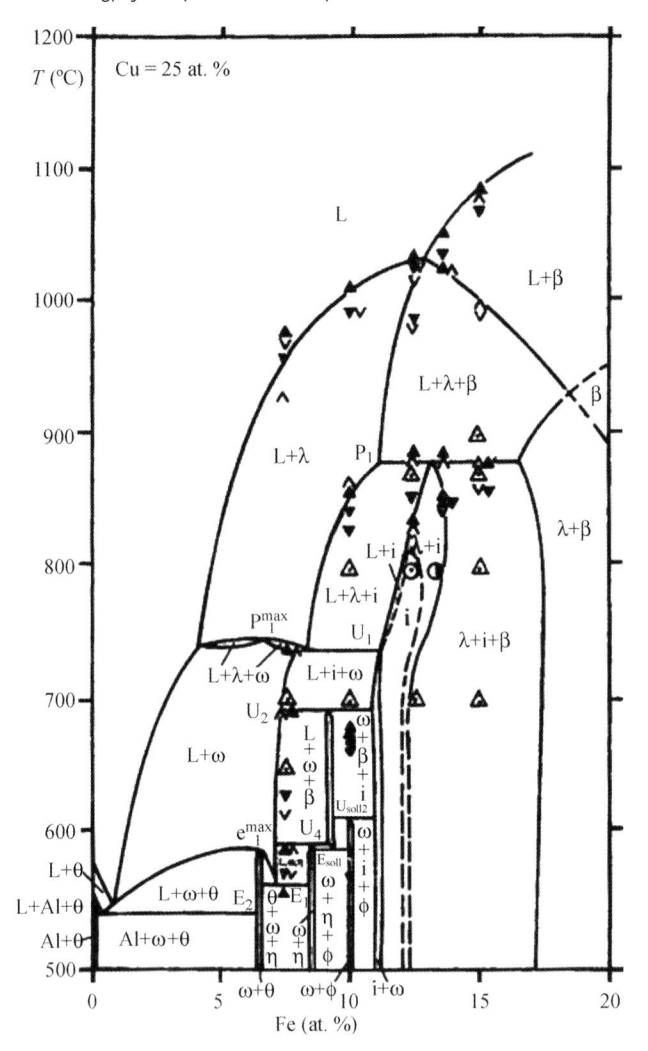

Figure 5.13 Isopleth of the Al-Cu-Fe phase diagram, with constant Cu concentration of 25 at.%. Adapted from reference 36.

the icosahedral phase is formed via a peritetic reaction at 882 °C (L + λ + β ↔ i) and that it is stable down to room temperature. However, it can be observed in Figure 5.13 that this phase has a very narrow compositional range not exceeding 2 at.%, which is also shifted to lower Fe content with decreasing temperature.

Another reason for the selection of the Al-Cu-Fe system is the promising properties of the targeted phase as have been compiled, for example by Huttunen-Saarivirta [36]. These include negative and linear temperature dependence of the electrical resistivity, large magnetoresistance at low temperature, linear increase of the optical conductivity with increasing frequency, with the possibility to provide high solar absorbance of 90% and a low thermal emittance.

Also, low thermal conductivity at low temperature, including those prevailing in cryogenic environment, low wetting angle by polar liquids such as water, low friction coefficient, relatively high microhardness in a temperature range corresponding to the brittle regime of the bulk material and also promising catalytic properties. Moreover, the binary Al-Cu and Al-Fe systems contain compounds and compositions, which also present promising properties. For example, Hsu *et al.* reported on the enhanced Young's modulus, good compressive strength and reasonably good compressive ductility of Al–Al$_2$Cu composites [37]. Jun *et al.* reported on the corrosion resistance of Al–Fe coatings [38]. In addition, the three alloying elements are reasonable in price, easily available and nontoxic.

Finally, a third reason for the selection of this system is the possibility to dispose of compatible metalorganic precursors of the three elements, Al, Cu and Fe. This point will be developed in the next section.

The adopted strategy to establish a robust CVD process for the deposition of intermetallic films and coatings in the Al-Cu-Fe system is schematically presented in Figure 5.14. According to this scheme, the work is initiated by the selection of the Al and Cu compounds. Deposition of unary Al and Cu films leads to the identification of parametric windows for the two processes. Matching of the two windows allows for the Al-Cu codeposition. This mainstreaming is consolidated if such codeposition provides coatings whose composition, phases and microstructure are the ones of interest and if such coatings are obtained with acceptable growth rate. In parallel, the selection of the appropriate precursors for the deposition of Fe is led and the corresponding parametric window is established. The final step is the matching of the deposition conditions for Al-Cu and Fe, followed by the determination of the material and process characteristics. Feedback loops at different steps of the approach allow for modification of the adopted solutions in order to satisfy the partial specifications and constraints.

In that which follows the criteria for the precursors selection and the different possibilities will be presented first. Then, deposition of aluminum and copper will be developed, followed by the presentation of the first results on the MOCVD of a CMA phase, namely the Al$_4$Cu$_9$ approximant. Finally, the deposition of iron films from original precursors will be presented before providing concluding remarks and perspectives for further research.

5.5.1
Precursors Selection

The general criteria qualifying an inorganic or molecular compound as precursor for CVD processes have been discussed in references [39] and [40]. In the case of CVD of intermetallic compounds including CMAs there are additional ones such as (a) similar transport behavior, (b) absence of heteroatoms in the ligands that may react with the other metal, (c) compatible decomposition schemes, and (d) belonging to a common family of compounds.

Taking into account the above constraints, the following molecular precursors were selected for the MOCVD of AlCuFe films: Dimethylethylamine alane

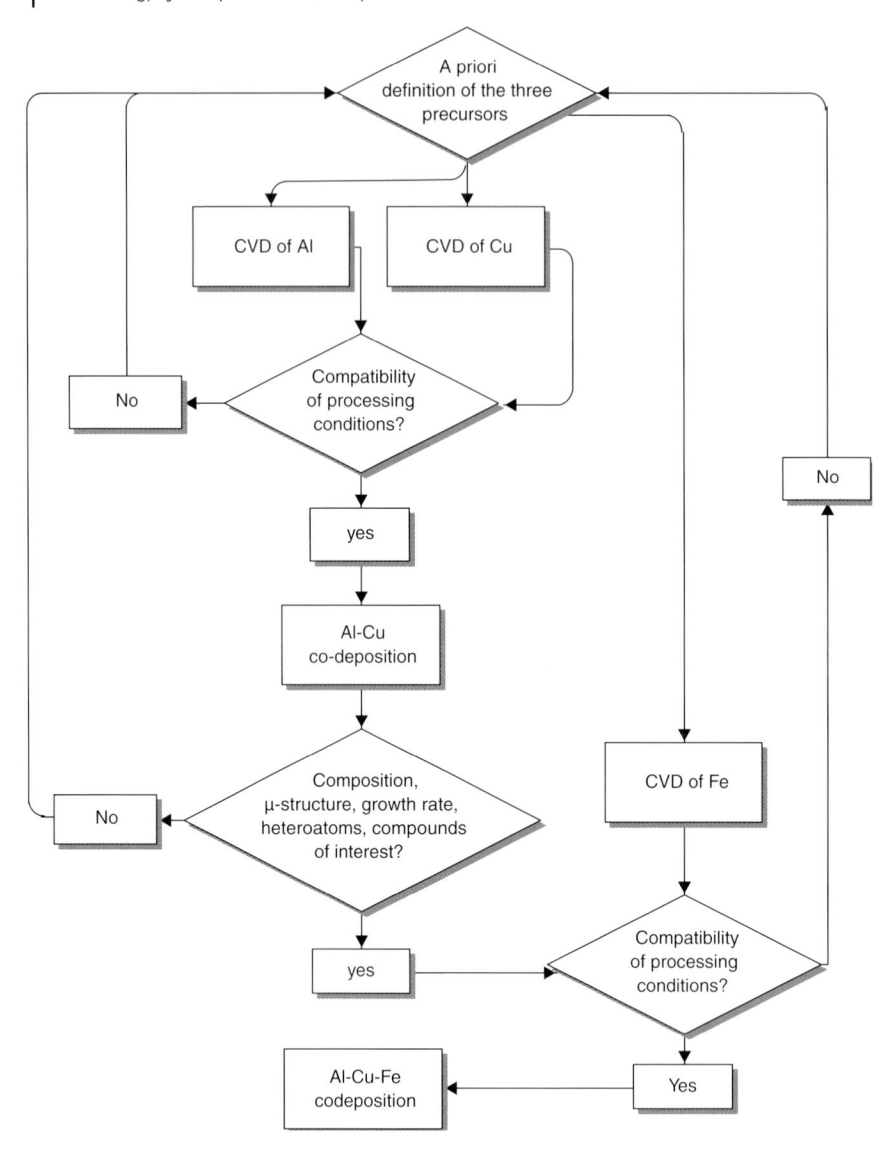

Figure 5.14 Flowchart illustrating the adopted strategy and the different steps for the MOCVD of coatings containing intermetallic phases in the Al-Cu-Fe system.

Me$_2$EtAlH$_3$ or DMEAA for aluminum, copper cyclopentadienyl triethyl phosphine (CpCuPEt$_3$) and iron bis(N,N'-di-tert-butylacetamidinate) (Fe2) for iron. The molecular structure of these three compounds is presented in Figure 5.15.

A variety of Al precursors has been successfully used for the MOCVD of high-purity Al films, including triisobutyl-alyminium Bu$^i_3$Al, tri-t-butyl-aluminum Bu$^t_3$Al, bis-trimethylamine alane AlH$_3$(NME$_3$)$_2$, dimethyl aluminum hydride Me$_2$AlH with

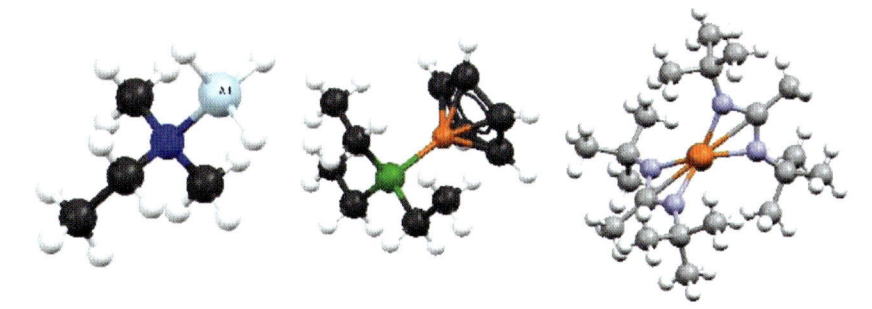

Figure 5.15 Molecular structure of the three precursors under investigation for the codeposition of Al-Cu-Fe films. From left: Dimethyl-ethyl-amine alane (DMEAA), copper cyclpentadieny triethyl phosphine (CpCuPEt₃) and iron bis(N,N'-di-tert-butylacetamidinate).

the related adducts, $Me_2AlH(NMe_3)$, $Me_2AlH(NMe_2Et)$ (see Jones *et al.* and references therein) [41] and DMEAA [42]. Among them, DMEAA presents attractive properties: it is liquid at ambient conditions and has a relatively high vapor pressure at room temperature (1.5 Torr) [43, 44]. However, it is unstable in ambient temperature and therefore it presents a limited shelf life unless it is stored at low temperature (~5 °C) DMEAA is thermally decomposed to DMEA ($[(CH3)_2C_2H_5]N$) and surface-adsorbed alane (AlH_3) that in turn is dissociated on the surface to aluminum and atomic hydrogen. Combination of two hydrogen atoms yields dihydrogen that is desorbed from the surface and is evacuated with the other gaseous by-products. Due to the absence of Al–C bonds in the precursor molecule, the obtained Al films are carbon-free [45, 46].

CpCuPEt₃ was synthesized and characterized in the late 1950s [47–49], but it had not been used as a MOCVD precursor before years [50, 51]. In the 1990s, it has been reported in Cu–Al codeposition in combination with alanes [52, 53]. However, in these reports copper concentration in the films was low, at the level of 1 wt.%. The physical and thermal properties of CpCuPEt₃ were investigated fifteen years later [54]. The conclusions of this work were that CpCuPEt₃ is stable below 70 °C and that its Clapeyron law in the 40–70 °C range is $\log P_{vap}$ (Torr) $= 9.671 - 3455/T(K)$.

A wide range of precursors has already been tested for MOCVD of iron, but an optimal iron compound for practical MOCVD of pure iron films is missing. Besides, the affinity of iron for carbon facilitates the formation of carbides and this strong trend must be avoided in the deposition of the pure metal. Compared with other metals, few open scientific publications refer to the thermal MOCVD of pure iron films. Basic molecular precursors have been studied, such as $Fe(Cp)_2$ [55], $Fe(CO)_5$ [56–61], $Fe_2Cp_2(CO)_4$ [62], $Fe(N(SiMe_3)_2)_3$ [63], $Fe(COT)CO_3$ [64], [(arene)(diene)Fe0] [65]. Except for $Fe((N(SiMe_3)_2)_3$ these compounds do not meet the prerequisites of oxygen-free ligands and metal–carbon-free bonds. Iron amidinates meet these prerequisites. Gordon's group showed that the monomeric bis(N, N'-di-*ter*-butylacetamidinato)iron(II) (**Fe2**) is volatile [66] and could be used as precursor for deposition of iron films by atomic layer deposition (ALD) [67]. **Fe2** is extremely sensitive to air and light. Upon storage in a metallic glove box under a continuously purified argon flow, the white powder turns gray. Due to its instability, the compound must be kept in sealed ampoules in a refrigerator. This compound was selected in the present work for testing as a precursor for the MOCVD of iron.

5.5.2
Deposition of Aluminum

Al deposition was experimentally investigated in a stagnant flow, cylindrical, vertical, stainless-steel MOCVD reactor. The reactor was equipped with a showerhead above a resistively heated susceptor whose diameter is 58 mm. Its base pressure was 10^{-6} Torr. Mass flow controllers delivered 99.999% pure nitrogen that was used both as carrier gas and as a dilution gas. Adduct-grade DMEAA vapor was delivered to the reactor chamber using a bubbler, maintained at 8–9 °C.

Figure 5.16 presents two scanning electron microscopy (SEM) micrographs of an aluminum film processed at 220 °C and 10 Torr on a silicon wafer. The film presents a continuous base and also surface characteristics of Al crystals. Surface roughness is about 0.3 μm. No preferential orientation could be noticed with X-ray diffraction analysis, and the carbon contamination is less than 1 wt.%.

Figure 5.17 presents the Arrhenius plot of Al deposition on silicon at 10 Torr. Despite the reduced number of investigated temperatures, the distinction among the

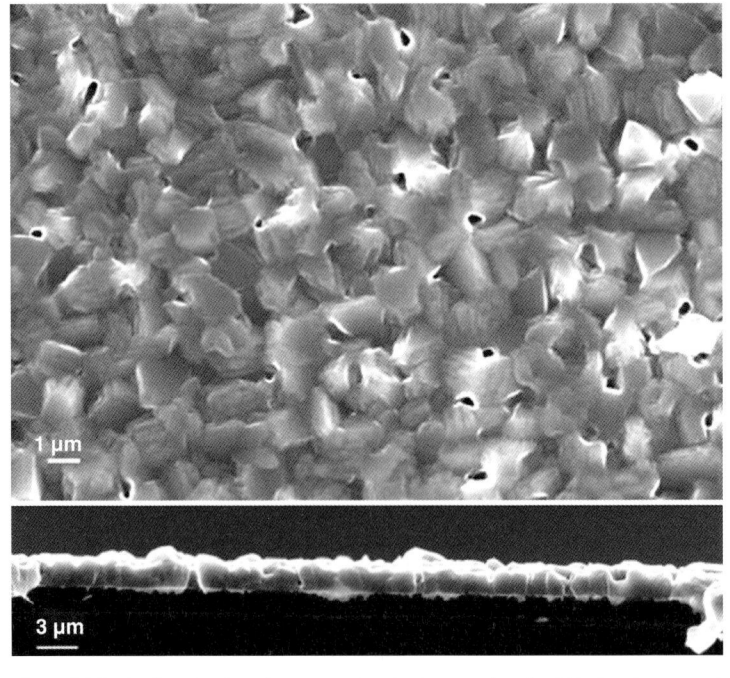

Figure 5.16 Surface (top) and cross-sectional micrographs of a CVD aluminum coating processed from DMEAA at 220 °C and 10 Torr.

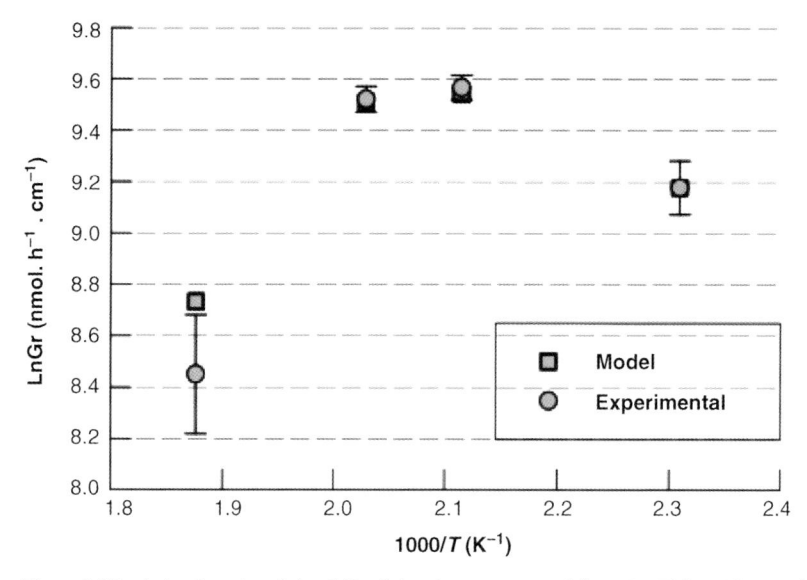

Figure 5.17 Arrhenius plot of the CVD of aluminum processed from DMEAA at 10 Torr. The experimental results are compared with those resulting from the modeling of the process.

typically met three domains is illustrated, namely the low-temperature one (up to 200 °C) where growth rate increases with increasing temperature, a relatively restricted plateau in the range 200 °C–220 °C and a third regime occurring at higher temperature where growth rate decreases with decreasing temperature. However, the influence of temperature to the growth rate is overall rather weak, leading to the conclusion that the aluminum growth is a process controlled by diffusion rather than kinetics. These results are in agreement with abundant literature information on the growth rate of MOCVD Al films from DMEAA [45, 46, 68–71].

This process was investigated theoretically in the frame of a computational analysis [72]. A computational fluid dynamics model was developed to describe the complex transport phenomena involved in the vertical cold-wall MOCVD reactor, also taking into account the decomposition chemistry of the DMEAA. The results of the model in the same processing conditions as the experimental ones are also presented in the diagram of Figure 5.17. It appears that the model predicts fairly well the experimentally measured Arrhenius plot. Similarly, the model is in satisfactory agreement with the experimentally determined film thickness profiles in the radial direction on the susceptor.

5.5.3
Deposition of Copper

Cu deposition was experimentally investigated in the same reactor that was used for Al deposition. $CpCuPEt_3$ was maintained in temperatures ranging between 60 °C and 90 °C, depending on the experiments. Hydrogen was used as a reactant gas and nitrogen was used both as carrier gas through the precursor sublimator and as dilution gas. Thermal decomposition of $CpCuPEt_3$ vapors was studied by *in situ* and online mass spectrometry [54]. This study revealed that the precursor is monomeric in the gas phase and there are no Cu-containing fragments in the gaseous decomposition products such as $CuPEt_3$, $CuCp$ or $CuCp_2$. $CuCpPEt_3$ is decomposed with the formation of a surface intermediate $\{CuCp\}_{surf}$ and departure of PEt_3 in the gas phase. $\{CuCp\}_{surf}$ is rapidly converted into $\{Cu\}_{surf}$ and $Cp_2(g)$. Identification of cyclopentadiene at temperature higher than 270 °C reveals a change of the decomposition mechanism in these conditions. The observed results allowed the following scheme to be proposed for thermal decomposition of the precursor on the growing surface:

$$T < 270\,°C: \quad CpCuPEt_3 \rightarrow \{CuCp\}_{surf} + PEt_{3\,gas}$$
$$\{CuCp\}_{surf} + CpCuPEt_3 \rightarrow 2\,Cu_{surf} + (Cp)_{2\,gas} + PEt_{3\,gas}\ \text{and/or}$$
$$2\{CuCp\}_{surf} \rightarrow 2\,Cu_{surf} + (Cp)_{2\,gas}$$
$$T > 270\,°C: \quad CpCuPEt_3 \rightarrow Cu_{surf} + HCp_{gas} + \text{other organics}$$

From the suggested mechanisms it can be concluded that decomposition proceeds with neither carbon nor phosphorous incorporation in the film.

Figure 5.18 presents two scanning electron microscopy (SEM) micrographs of a copper film deposited at 220 °C and 10 Torr on a silicon wafer. The growth is island-type, which is characteristic of the Volmer–Weber type of growth of Cu CVD. Film

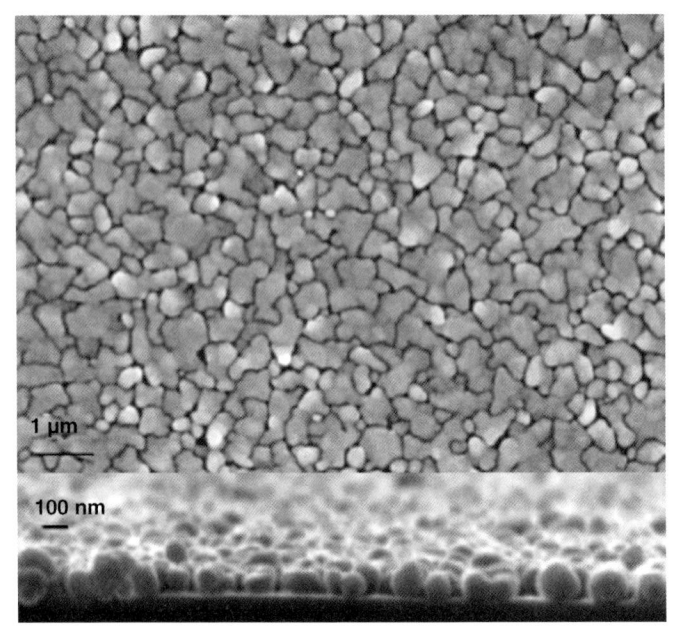

Figure 5.18 Surface (top) and cross-sectional micrographs of a CVD copper coating processed from $CpCuPEt_3$ at 220 °C and 10 Torr.

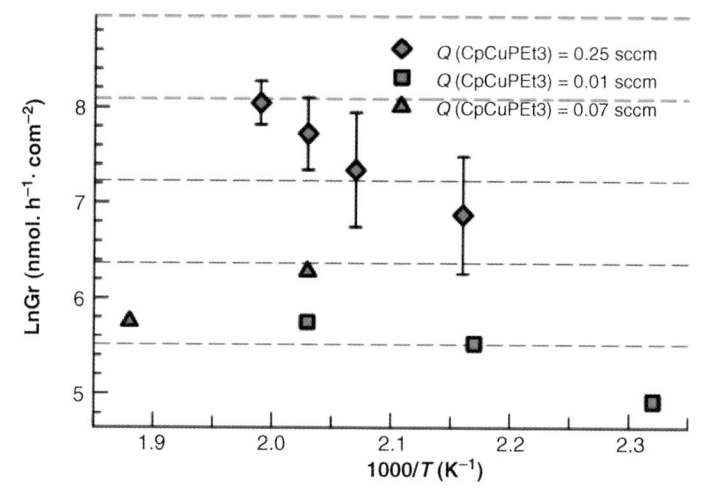

Figure 5.19 Arrhenius plot of the CVD of copper processed from CpCuPEt$_3$ at 10 Torr. Squares, triangles and diamonds correspond to deposition in precursor feeding rate of 0.01 sccm, 0.07 sccm and 0.25 sccm, respectively.

thickness and surface roughness are 100 nm and 40 µm, respectively. X-ray diffraction showed that the films exhibit a weak (111) texture. As expected, films are phosphorous- and carbon-free, within the detection limit of electron probe microanalysis (EPMA).

The Cu growth rate was studied for three different precursor flow rates in the temperature range between 158 °C and 260 °C. The obtained Arrhenius plot at 10 Torr on a silicon wafer is presented in Figure 5.19. Growth rate increases with increasing precursor concentration in the input gas. In the temperature range between 158 °C and 240 °C and for CpCuPEt$_3$ flow rates 0.01 sccm and 0.25 sccm, growth rate slightly increases with increasing temperature. In this temperature range, the activation energies were estimated to be 14 kJ/mol and 32 kJ/mol for $Q_{(CpCuPET3)} = 0.01$ sccm and 0.25 sccm, respectively. These values are comparable to, although slightly weaker than those previously reported [73]. The reported information indicates that in this temperature range, the growth of copper is controlled by the diffusion of CpCuPEt$_3$ to the surface through the boundary layer.

5.5.4
Deposition of the Al$_4$Cu$_9$ Approximant Phase

The results obtained from the investigation of the deposition of unary Al and Cu films provided guidance for the study of the codeposition of Al-Cu from the same precursors. Similar gas-phase composition (including H$_2$ for the hydrogenation of the cyclopentadienyl ligand) and sublimation conditions for the two precursors were adopted. Codeposition was investigated in the temperature range between 200 °C and 260 °C. The nature of the stable gaseous reactants and by-products was monitored by online mass spectrometry. Compared with the results obtained during the deposition of single Al and Cu, mass spectrometry revealed that there are simple additive effects in the gas phase during codeposition. Indeed, only fragments due to the decomposition of $CpCuPEt_3$ and of DMEAA were identified.

Al-Cu films with Cu content between 1 at.% and 93 at.% were deposited by varying the processing conditions. EPMA revealed that films are carbon-, nitrogen- and phosphorous-free. Al, Cu and various intermetallic phases, among which Al_2Cu and $Al_{6.108}Cu_{3.892}$, were identified by X-ray diffraction. Their formation depends on the processing conditions, namely the concentration of the input gas and the temperature of the substrate. The growth rate of the films decreases when their composition approaches 50 at.% Cu. This result is attributed to synergetic effects between adsorbed species from the two precursors, since only additive effects were observed in the gas phase in codeposition conditions.

Figure 5.20 shows a surface SEM micrograph of a coating, deposited at 200 °C. This film contains 60 at.% Cu. The X-ray diffractogram of this film is presented in the foreground of the micrograph. The only crystalline phase present is the Al_4Cu_9 approximant, whose JCPDS pattern is also presented in the figure for comparison. Interestingly, the Al_4Cu_9 approximant was obtained without postdeposition annealing. To the best of the authors' knowledge, this is the first time that a CMA phase has been deposited by CVD. This result is very promising since it provides the proof of principle for the use of MOCVD for the processing of this family of films. It paves the way towards the use of CMAs as coatings applied on complex-shaped pieces. However, a comprehensive investigation of the codeposition process is necessary

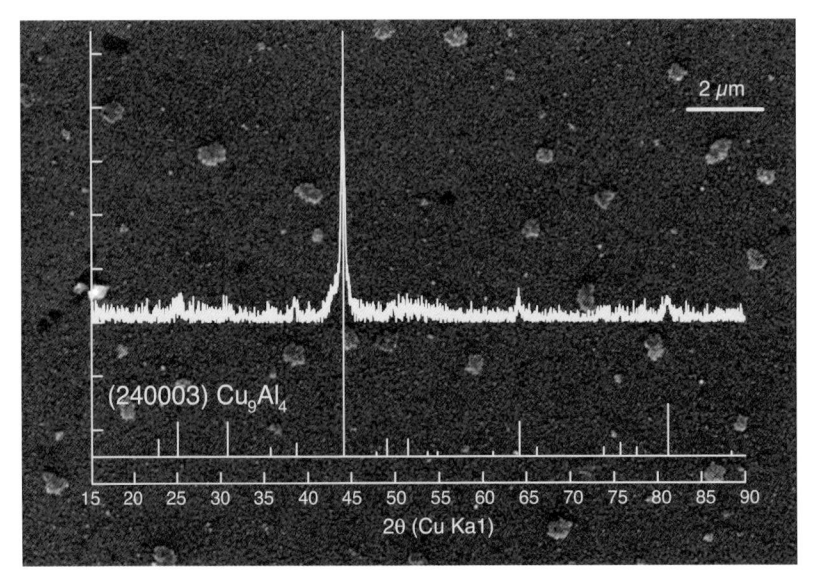

Figure 5.20 Surface micrograph of a CVD coating composed of pure Al_4Cu_9 approximant and corresponding X-ray diffractogram.

5.5.5
Deposition of Iron

Preliminary investigation of the thermal decomposition behavior of Fe2 was performed by *in situ*, time-of-flight mass spectrometry [74]. The feature of the mass spectra is the occurrence of relatively intense molecular peaks corresponding to the monomer $[FeL_2]^+$ and the free ligand $[HL]^+$ where L is $(C_4H_9)NC(CH_3)N(C_4H_9)$. There are no peaks at m/z higher than the value expected for FeL_2 that is, a value of mass over charge (m/z) equals 394 Fe2. Thus, the compound is likely to be monomeric in the gas phase. Decomposition is initiated at 210 °C, corresponding to a considerable decrease in the intensity of the initial compound ion peaks. Maximum decomposition occurs at 250 °C. Main gaseous by-products are the free ligand HL (L = $(C_4H_9)NC(CH_3)N(C_4H_9)$) ($m/z$ 170), CH_3CN (m/z 41), C_4H_8 (m/z 56), C_4H_9 (m/z 57), $(CH_3)_2CHNH$ (m/z 58). It is worth noting that the decrease in the intensity of the molecular peak on reaching the onset temperature is not sharp, and this is attributed to the weak vaporization stability of the precursor [75].

Deposition of iron films from Fe2 was performed at 10 Torr, in a temperature range varying between 200 °C and 450 °C and with the precursor container maintained at 85 °C. Films present a gray, mirror-like metallic surface. Figure 5.21 presents two, surface and cross-sectional, SEM micrographs of a film processed on silicon wafers at 280 °C. The film is composed of densely packed nanocrystallites whose size does not exceed 100 nm. This morphology prevails for deposition temperature up to 300 °C. Above this temperature, the films are well crystallized with grains of apparent cubic structure.

Elemental analysis of the films by EPMA revealed that they contain iron, nitrogen and carbon. According to X-ray diffraction the films are mainly composed of iron nitride, Fe_4N with <111> preferential orientation. Metallic Fe and cementite Fe_3C are also present in the films at a lesser extend. The chemical environment of Fe and of C was investigated by X-ray photoelectron spectroscopy (XPS) after sputtering of the surface of the films. Peak curve fittings in the Fe 2p region (706–714 eV) and the C 1s region (282–286 eV) confirmed the presence of Fe_3C. However, the Fe $2p_{3/2}$ binding energy in Fe_4N is too close to that of the bulk metal to be used to distinguish a Fe_4N from α-Fe. The evolution of the $Fe_3C/(Fe + Fe_4N)$ and of carbide/(total carbon) peak ratios in the investigated temperature domain show that the decrease of the deposition temperature allows decreasing the carbon content of the films.

Figure 5.22 presents the evolution of the growth rate of the films as a function of the inverse temperature at 10 Torr. Growth rate increases with increasing deposition temperature following a linear relation in the Arrhenius coordinates. This behavior reveals that deposition is limited by surface kinetics. The corresponding activation energy of the process is 226 kJ mol^{-1}.

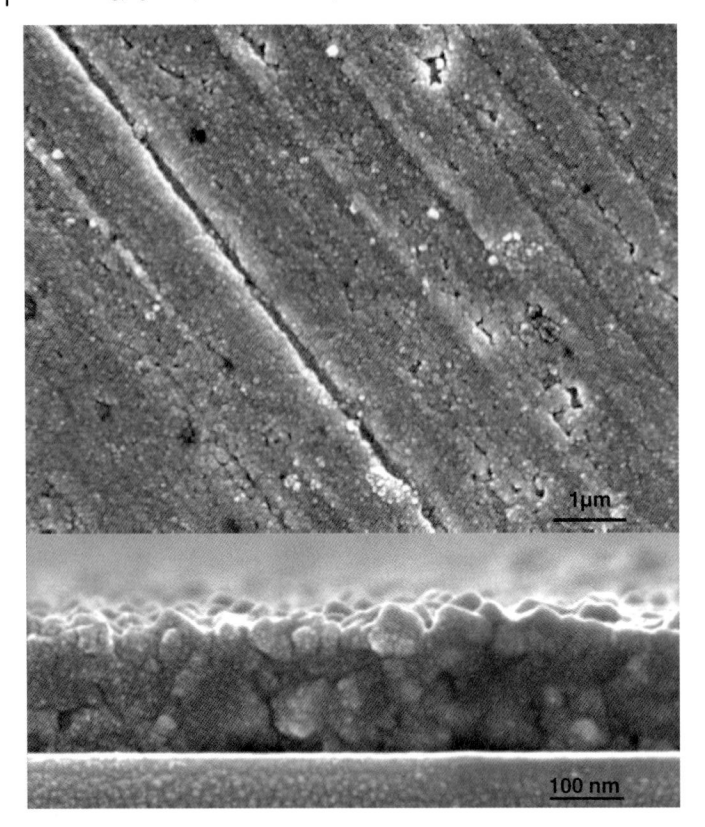

Figure 5.21 Surface (top) and cross-sectional SEM micrographs of a CVD iron film processed on silicon wafer at 280 °C.

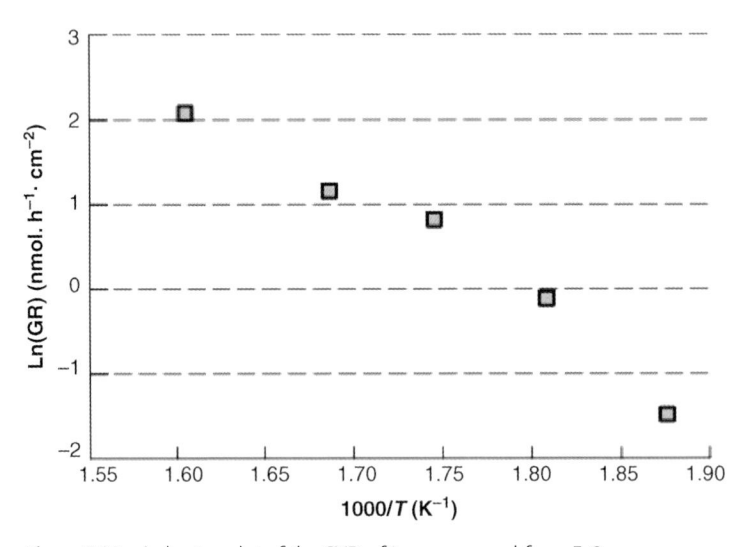

Figure 5.22 Arrhenius plot of the CVD of iron processed from Fe2.

5.6
Concluding Remarks

Initial screening of the metalorganic precursors of the elements of interest is a prerequisite for the convenient processing of CMA films and coatings by MOCVD. Tailoring of the metal coordination sphere by an appropriate set of ligands (alkoxides, β-diketonates, amides and thiolates, classical or functional) allows tuning properties and accessing well-defined precursors. Such a selection is the first step in the establishment of the CVD process and consequently it contains *a priori* decisions based on empirical criteria. However, it enters an optimization loop, since the use of molecular precursors in MOCVD conditions provides a way for establishing relationships between precursors and the final material. As a result, it allows insight into their decomposition pathways to be obtained and finally it helps in defining precursors with the appropriate architecture and consequently in optimizing this preliminary task.

The way to the MOCVD of CMAs is paved by a series of steps. The first one, after the definition of the molecular precursors, consists in tuning the processes for the deposition of the individual elements. The objective at this stage is to obtain unary films with smooth microstructure and acceptable purity. Another objective is to establish the Arrhenius plots for each element and through such diagrams, to identify a temperature window where the processes present growth rates with ratios similar to the ones among the elements in the targeted phases. The second step is codeposition of binary films. Precursor transport conditions and deposition parameters are defined based on the growth rates of the unary films as previously established. However, CVD is not a physical deposition process and consequently, the relation between the compositions of the gas phase and of the film is not straightforward. The key role played by the chemistry is materialized in the kinetically controlled regime of the Arrhenius plots. Therefore, an additional degree of freedom available in the frame of codeposition of intermetallic alloys is provided by the possibility to operate in a temperature window containing the diffusion-limited regime for one element and the kinetically limited one for the other. In such a way, the degree of incorporation of each element can be controlled by the concentration in the input gas of the precursor (diffusion-limited regime) and by the deposition temperature (surface-kinetics-limited regime).

Nevertheless, even in this case, the elemental composition of the films can be unexpectedly shifted with regard to the adopted strategy. Indeed, competitive phenomena occurring on the growing surface or between the surface and the gas phase, depending on reaction mechanisms such as the Langmuir–Hinshelwood and the Eley–Rideal ones, respectively, can be at the origin of particular surface reactions that control the growth.

Last but not least, ensuring appropriate elemental composition in the films does not guarantee the stabilization of the targeted phases. The activation energies for the nucleation of the different phases are not necessarily the same and corresponding differences may lead to a particular sequence of phase nucleation and ultimately to the formation of two or more undesired phases, even in the case when the global composition of the film meets that of the targeted phase.

Going back to the selection of the molecular precursors, it appears from the above discussion that an additional criterion in this stage should be the use of compounds with compatible, if not similar ligands. This condition reduces the probability of the occurrence of competitive phenomena on the surface and helps in establishing a more direct relation between gas-phase and film compositions. The ultimate step in this direction could be the use of single-source precursors, containing the elements to be deposited in the correct composition. Although this attractive solution may simplify the process, it presents inherent difficulties, especially on the formulation of such precursors. It should only be adopted at a final step, after the correlation between the characteristics of gas phase and those of the films has been established.

MOCVD of intermetallic films, including CMAs is actually in its infancy. Progress in numerous aspects must be made prior to consolidating the process. Modeling can strongly assist this evolution. Its efficiency will be higher if it is performed at the multiscale level, with continuous feedback between the different scales. Such different spatial scales range from the MOCVD reactor to the film feature and the film morphology progression. The former require continuum modeling, the latter discrete. The process modeling at the continuum level is based on, and the theoretical predictions are drawn from, first principles; that is, the conservation of mass/species, momentum and energy. The discrete models employ Monte-Carlo-type simulations, such as kinetic MC and ballistic techniques. Monte Carlo simulations are based on the results obtained from the investigation of the different aspects of surface reactivity as have been presented in Figure 5.12, obtained through *ab initio* methods.

Continuous and discrete models can be linked through flux-type boundary conditions at interfaces between the continuum and discrete transport regimes. Several reaction combinations should be explored for developing a comprehensive understanding of the studied processes, which may involve multiple chemical species, concentration ranges, and gas temperatures and pressures. Information obtained from the analysis of the thermodynamic equilibrium in the reactive gas phase and from past experimental and theoretical studies contributes to the determination of elementary reaction steps, which play important roles in the generation of the key intermediates in the gas phase and in the overall deposition characteristics (thickness, uniformity and film composition). These tools are the basis for the development of the surface models needed for the process simulation.

Acknowledgements

This work was supported by the European Commission under contracts no. NMP3-CT-2005-500140 and MC-IIF-39728, by the Agence Nationale de la Recherche in France under contract no. NT05-3-41834 and by the Deutsche Forschungs gemeinschaft (DFG grant "Physical Properties of Complex Metallic Alloys").

References

1. Czochralski, J. (1918) *Z. Phys. Chem.*, **92**, 219 (in German).
2. Langsdorf, A. and Assmus, W. (1999) *Cryst. Res. Technol.*, **34**, 261.
3. Langsdorf, A. and Assmus, W. (1998) *J. Cryst. Growth*, **192**, 152.
4. Gille, P. and Bauer, B. (2008) *Cryst. Res. Technol.*, **43**, 1161.
5. Gödecke, T. and Ellner, M. (1971) *Z. Metallkd.*, **62**, 842.
6. Kattner, U.R. (1990) *Binary Alloy Phase Diagrams*, vol. **1** 2nd edn (ed. T.B. Massalski), ASM International Materials Park, OH, p. 147.
7. Grin, Y., to be published.
8. Lipińska-Chwałek, M., Balanetskyy, S., Thomas, C., Roitsch, S., and Feuerbacher, M. (2007) *Intermetallics*, **15**, 1678.
9. Samson, S. (1965) *Acta Crystallogr.*, **19**, 401.
10. Urban, K. and Feuerbacher, M. (2004) *J. Non-Cryst. Solids*, **334–335**, 143.
11. Samson, S. (1969) *Developments in the Structural Chemistry of Alloy Phases* (ed. B.C. Giessen), Plenum Press, New York, London, p. 65.
12. Bergman, G., Waugh, J.L.T., and Pauling, L. (1957) *Acta Crystallogr.*, **10**, 254.
13. Feuerbacher, M., Thomas, C., and Roitsch, S. (2008) *Intermetallics*, **16**, 93.
14. Petrov, D., Watson, A., Gröbner, J., Rogl, P., Tedenac, J.C., Bulanova, M. et al. (2005) *Light Metal Ternary Systems*, vol. **11A3** (eds G. Effenberg and S. Ilyenko), Springer, Heidelberg, Berlin, p. 191.
15. Boudard, M., Klein, H., de-Boissieu, M., Audier, M., and Vincent, H. (1996) *Philos. Mag. A*, **74**, 309.
16. Matsuo, Y. and Hiraga, K. (1994) *Philos. Mag. Lett.*, **70**, 155.
17. Yurechko, M., Fattah, A., Velikanova, T., and Grushko, B. (2001) *J. Alloys Compd.*, **329**, 173.
18. Yurechko, M., Grushko, B., Velikanova, T., and Urban, K. (2002) *J. Alloys Compd.*, **337**, 172.
19. Klein, H., Audier, M., Boudard, M., Boissieu (de), M., Beraha, L., and Duneau, M. (1996) *Philos. Mag. A*, **73**, 309.
20. Klein, H., Feuerbacher, M., Schall, P., and Urban, K. (2000) *Philos. Mag. Lett.*, **80**, 11.
21. Klein, H. (1997) PhD thesis, Grenoble unpublished.
22. Beraha, L., Duneau, M., Klein, H., and Audier, M. (1997) *Philos. Mag. A*, **76**, 587.
23. Gähler, F., Kramer, P., Trebin, H.-R., and Urban, K. (2000) Proceedings of the 7th International Conference of Quasicrystals, *Mat. Sci. Eng. A.*, 294–296.
24. Yurechko, M., Velikanova, T., and Urban, K. (2009) Proc. PDMS 6 Kiew, in press.
25. Shramchenko, N. and Dénoyer, F. (2002) *Eur. Phys. J. B*, **29**, 51.
26. Gottlieb-Schönmeyer, S., Brühne, S., Ritter, F., Assmus, W., Balanetskyy, S., Feuerbacher, M., Weber, T., and Steurer, W. (2009) *Intermetallics*, **17**, 6.
27. Černý, R., François, M., Yvon, K., Jaccard, D., Walker, E., and Petříček, V. (1996) *J. Phys.: Condens. Matter*, **8**, 4485.
28. Černý, R., Guénée, L., and Wessicken, R. (2003) *J. Solid State Chem.*, **174**, 125.
29. Iandelli, A. and Palenzona, A. (1971) *J. Less-Common Met.*, **25**, 333.
30. Westbrook.
31. Massalski, T.B. (1992) *Binary Alloy Phase Diagrams*, 2nd edn. ASM International, Materials Park, OH.
32. Komura, Y. and Tokunaga, K. (1980) *Acta Crystallogr.*, **36**, 1548.
33. Villars, P. and Calvert, L.D. (1991) *Pearson's Handbook of Crystallographic Data for Intermetallic Phases*, vol. **4**, ASM Intern., Materials Park, Ohio.
34. Suhr, H., Etspüler, A., Feurer, E., and Kraus, S. (1989) *Plasma Chem. Plasma. P.*, **9**, 217.
35. Zhang, L. and Lück, R. (2003) *Z. Metallkd.*, **94**, 98.
36. Huttunen-Saarivirta, E. (2004) *J. Alloys Compd.*, **363**, 150.
37. Hsu, c.J., Kao, P.W., and Ho, N.J. (2005) *Scr. Mater.*, **53**, 341.
38. Jun, J.H., Jun, J.H., and Kim, K.Y. (2002) *J. Power Sources*, **112**, 153.
39. Maury, F. (1995) *J. de Physique IV*, **C5**, 449.
40. Maury, F., Gueroudji, L., and Vahlas, C. (1996) *Surf. Coat. Techn.*, **86–87**, 316.
41. Jones, A.C., Houlton, D.J., Rushworth, S.A., Flanagan, J.A., Brown, J.R., and Critchlow, G.W. (1995) *Chem. Vap. Dep.*, **1**, 24.

42 Gladfelter, W.L. and Phillips, E.C. (1993) US Patent US5191099 (A).
43 Delmas, M., Poquillon, D., Kihn, Y., and Vahlas, C. (2005) *Surf. Coat. Techn.*, **200**, 1413.
44 Yun, J.H., Kim, B.Y., and Rhee, S.W. (1998) *Thin Solid Films*, **312**, 259.
45 Jang, T.W., Rhee, H.S., and Ahn, B.T. (1998) *Mater. Res. Soc. Symp. Proc.*, **514**, 351.
46 Jang, T.W., Moon, W., Baek, J.T., and Ahn, B.T. (1998) *Thin Solid Films*, **333**, 137.
47 Wilkinson, G. and Piper, T.S. (1956) *J. Inorg. Nucl. Chem.*, **2**, 32.
48 Whitesides, G.M. and Fleming, J.S. (1967) *J. Am. Chem. Soc.*, **89**, 2855.
49 Cotton, F.A. and Marks, T.J. (1970) *J. Am. Chem. Soc.*, **92**, 5114.
50 Hara, K., Kojima, T., and Kukimoto, H. (1987) *Jpn. J. Appl. Phys.*, **26**, L1107.
51 Dupuy, C.G., Beach, D.B., Hurst, J.E., and Jasinski, J.M. (1989) *Chem. Mater.*, **1**, 16.
52 Katagiri, T., Kondoh, E., Takeyasu, N., Nakano, T., Yamamoto, H., and Otha, T. (1993) *Jpn. J. Appl. Phys.*, **32**, L1078.
53 Kondoh, E., Kawano, Y., Takeyasu, N., and Ohta, T. (1994) *J. Electrochem. Soc.*, **141**, 3494.
54 Senocq, F., Turgambaeva, A., Prud'homme, N., Patil, U., Krisyuk, V.V., Samelor, D., Gleizes, A., and Vahlas, C. (2007) *Surf. Coat. Technol.*, **201**, 9131.
55 Dormans, G.J.M. (1991) *J. Cryst. Growth*, **108**, 806.
56 Kaplan, R. and Bottka, N. (1982) *Appl. Phys. Lett.*, **41**, 972.
57 Walsh, P.J. and Bottka, N. (1984) *J. Electrochem. Soc.*, **131**, 444.
58 Stauf, G.T. and Dowben, P.A. (1988) *Thin Solid Films*, **156**, L31.
59 Lane, P.A., Wright, P.J., Oliver, P.E., Reeves, C.L., Pitt, A.D., and Keen, J.M. (1997) *Chem. Vap. Dep.*, **3**, 97.
60 Lane, P.A. and Wright, P.J. (1999) *J. Cryst. Growth*, **204**, 298.
61 Haugan, H.J., McCombe, B.D., and Mattocks, P.G. (2002) *J. Magn. Magn. Mater.*, **247**, 296.
62 Feurer, R., Larhrafi, M., Morancho, R., and Calsou, R. (1988) *Thin Solid Films*, **167**, 195.
63 Baxter, D.V., Chisholm, M.H., Gama, G.L., Hector, A.L., and Parking, I.P. (1995) *Chem. Vap. Dep.*, **1**, 49.
64 Luithardt, W. and Benndorf, C. (1996) *Thin Solid Films*, **291**, 200.
65 Michkova, K., Schneider, A., Gerhard, H., Popovska, N., Jipa, I., Hofmann, M., and Zenneck, U. (2006) *Appl. Catal. A*, **315**, 83.
66 Lim, B.S., Rahtu, A., Park, J.S., and Gordon, R.G. (2003) *Inorg. Chem.*, **42**, 7951.
67 Lim, B.S., Rahtu, A., and Gordon, R.G. (2003) *Nature Mater.*, **2**, 749.
68 Matsuhashi, H., Lee, C.H., Nishimura, T., Masu, K., and Tsubouchi, K. (1999) *Mater. Sci. Semicond. Proc.*, **2**, 303.
69 Neo, Y., Niwano, M., Mimura, H., and Yokoo, K. (1999) *Appl. Surf. Sci.*, **142**, 443.
70 Nakajima, T., Nakatomi, M., and Yamashita, K. (2003) *Molec. Phys.*, **101**, 267.
71 Delmas, M. and Vahlas, C. (2007) *J. Electrochem. Soc.*, **154**, D538.
72 Xenidou, T.C., Boudouvis, A.G., Markatos, N.C., Samélor, D., Senocq, F., Gleizes, A.N., and Vahlas, C. (2007) *Surf. Coat. Technol.*, **201**, 8868.
73 Beach, D.B., LeGoues, F.K., and Hu, C.K. (1990) *Chem. Mater.*, **2**, 216.
74 Turgambaeva, A.E., Krisyuk, V.V., Stabnikov, P.A., and Igumenov, I.K. (2007) *J. Organomet. Chem.*, **692**, 5001.
75 Krisyuk, V.V., Gleizes, A.N., Aloui, L., Turgambaeva, A.E., Sarapata, B., Prud'homme, N., Senocq, F., Samélor, D., Zielinska-Lipiec, A., Dumestre, F., and Vahlas, C. (2009) *J. Mater. Chem.*, submitted.

6
Surface Chemistry of CMAs
Marie-Geneviève Barthés-Labrousse, Alessandra Beni, and Patrik Schmutz

6.1
Introduction

Surface chemistry can be roughly defined as the study of the adsorption of gas or liquid molecules on the surface. It reflects the capacity of a material to form bonds with atomic and molecular species from the surrounding environment and is one of the most important properties in terms of applications in various fields such as corrosion protection, catalysis, adhesion, friction and wear... The interest in studying CMAs surface reactivity comes from their specific electronic structure, which is related to the existence of highly symmetric clusters that decorate the giant unit cells and could affect interactions of the surface atoms with surrounding atoms and molecules. In spite of this characteristic feature, a limited number of studies have been devoted so far to the chemical reactivity of CMA surfaces. Moreover, although the number of possible combinations of metal constituents should give rise to the formation of a huge variety of CMAs, most investigations have been performed so far on Al-based compounds.

The interest in studying oxidation of CMAs emerged from the excellent oxidation and corrosion resistance that was initially reported [1–3] and from the fact that many promising properties of these alloys, such as their low surface energy and friction coefficient, their optical emissivity can be affected by the nature and thickness of the oxide layer formed on the surface. Several studies were thus devoted to the oxidation characteristics of Al-rich complex alloys, but surprisingly enough very little work has been done in the field of aqueous (also called wet) corrosion.

Adsorption of simple molecules other than oxygen and water on CMAs surfaces under ultrahigh vacuum conditions has originally been performed both with the idea to form molecular ordered complex overlayers and, as a first step, to understanding the catalytic properties of these materials. In the present chapter, we will focus our attention on the adsorption step, but the reader can refer to Chapter 8, where it will be shown that CMAs are indeed very promising catalytic materials as they can present high activity and selectivity, they can be stable up to high temperature and, thanks to their brittleness, they can be easily crushed into powders at room temperature.

Complex Metallic Alloys: Fundamentals and Applications
Edited by Jean-Marie Dubois and Esther Belin-Ferré
Copyright © 2011 WILEY-VCH Verlag GmbH & Co. KGaA, Weinheim
ISBN: 978-3-527-32523-8

Our intention in this chapter is to point out the similarities and differences that CMAs can present with respect to more classical metallic alloys in terms of surface reactivity.

6.2
Surface Chemistry of CMAs Under UHV Environment

6.2.1
Interaction with Oxygen

The interest in studying oxidation of complex metallic alloys is twofold. On the one hand, it has been speculated that the persistence of a pseudogap at the Fermi level up to the surface could strongly influence the reactivity (hence the oxidability) of CMAs; on the other hand, as already mentioned in the introduction, many intriguing properties of these alloys can be affected by the nature and thickness of the surface oxide layer. Most investigations so far have been devoted to Al-based compounds. In particular, the first stages of oxidation of Al-Pd-Mn, Al-Cu-Fe and Al-Cr-Fe resulting from exposure to dry oxygen in an ultrahigh vacuum chamber have been extensively studied. The main questions raised were related to the existence of a measurable chemisorption step and to possible differences in the detailed mechanisms of oxide(s) nucleation between CMAs, related classical intermetallics and pure aluminum.

The first investigations related to surface oxidation of CMAs were undertaken by Thiel and coworkers on the fivefold surface of the icosahedral Al-Pd-Mn quasicrystal [4–7]. Upon exposure to dry oxygen in the temperature range 105–870 K, the oxidation characteristics of the fivefold i-Al$_{70}$Pd$_{21}$Mn$_9$ surface were found to be very similar to those of aluminum, with the existence of a chemisorbed phase that destroys the quasiperiodicity of the surface and is a precursor to the formation of a thin (4–8 Å) oxide layer consisting exclusively of amorphous Al oxide. While the inertness of Pd does not look surprising in view of the low enthalpy of formation of the bulk palladium oxide ($\Delta H = -85$ kJ mol^{-1} O$_2$ for PdO), oxidation of both Al and Mn would have been expected based upon thermodynamic considerations ($\Delta H = -1080$ kJ mol^{-1} O$_2$ for Al$_2$O$_3$ and $\Delta H = -1041$ kJ mol^{-1} O$_2$ for Mn$_3$O$_4$). In fact, Popovič et al. have shown that oxidation of Mn can be observed even at room temperature for extremely large exposure to oxygen (8400 L) [8]. The very slow oxidation rate is due to the specific layer-by-layer structure of the quasicrystalline i-Al-Pd-Mn surface, which is terminated by an Al-rich plane. As the diffusivities of the elements in a quasicrystal are much smaller than in conventional alloys, the initial oxidation process is governed by the oxidation of the outermost layer and leads to the formation of a passivating Al oxide that can act as a barrier and delay oxidation of the other elements.

Surprisingly, although most authors conclude there is formation of an amorphous aluminum oxide layer, Longchamp et al. [9] reported the formation of a well-ordered, 0.5-nm thick aluminum oxide film following oxidation in vacuum of the pentagonal surface of i-Al-Pd-Mn at 700 K and subsequent annealing at the same temperature.

The oxide layer consists of five pairs of domains similar to those of γ-Al_2O_3 aligning their nominal (111) face parallel to the substrate surface. The domains are of approximately 3.5 nm diameter, rotated by 72° with respect to each other and aligned along a two-fold symmetry direction of the pentagonal substrate surface. Similar ordered layers have been observed following high-temperature oxidation in vacuum of ordered binary alloys surfaces such as NiAl (110). However, the self-size selection of the domains is only observed on the pentagonal surface of i-Al-Pd-Mn and can be of great potential interest when using these oxidized surfaces as templates to grow nanostructures. The absence of any ordered layer reported in most papers has been ascribed to a critical dependence of the oxidation mechanisms on the Al-to-Pd concentration.

Further insight into the reactivity of CMAs has also been obtained by Popovič et al. [8] by comparing the kinetics of oxidation of quasicrystalline and crystalline surfaces of an icosahedral Al-Pd-Mn quasicrystal cut perpendicular to the fivefold axis and of Al (111). Figure 6.1 shows the results of their XPS study following sequential exposure to dry oxygen at room temperature.

The percentages of oxidized Al (Mn) in the total amount of Al (Mn) atoms have been deduced from the XPS Al $2p$ and Mn $2p_{3/2}$ core-level spectral lines monitored at normal emission (0°) and/or 45° off normal. It can be clearly seen from these results that Mn is strongly oxidized in the crystal even for low exposure to oxygen while very large exposure to oxygen was necessary to observe the formation of manganese oxide

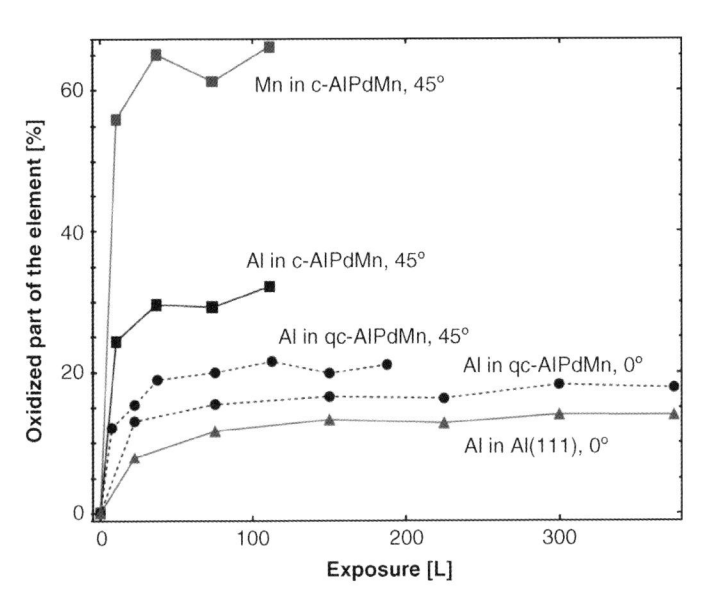

Figure 6.1 Percentages of oxidized Al (Mn) in total amount of Al (Mn) atoms following exposure of crystalline (c), quasicrystalline (qc) i-AlPdMn and Al(111) surfaces to oxygen at room temperature in ultrahigh vacuum. Values deduced from XPS data monitored at normal emission (0°) and/or 45° off normal. Reprinted from reference 8, Copyright © (2001), with permission from Elsevier.

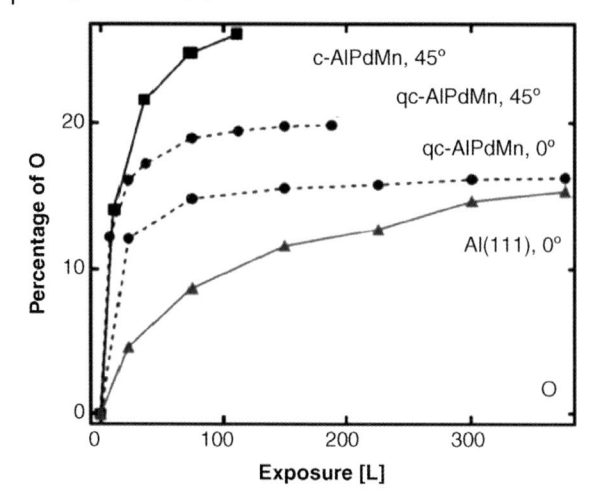

Figure 6.2 Variation of the percentage of O atoms during exposure of crystalline (c), quasicrystalline (qc) i-AlPdMn and Al(111) surfaces to oxygen at room temperature in ultrahigh vacuum. Values deduced from XPS data monitored at normal emission (0°) and/or 45° off normal. Reprinted from reference 8, Copyright © (2001), with permission from Elsevier.

on the surface of quasicrystalline Al-Pd-Mn. This can be related to differences in the composition of the outmost layer between the crystalline and the quasicrystalline phases as there is no Al-dense surface layer in the periodic phase and Mn is available at the surface for oxidation. In addition, the diffusivity of Mn is larger in the crystalline phase than in the quasicrystalline one, thus allowing Mn segregation in the former case. In the same way, oxidation of Al decreases in the order crystal > quasicrystal > Al(111). Al surface segregation is observed in both crystalline and quasicrystalline phases. Figure 6.2 compares the variations in the percentage of O during the oxidation process for the crystal, quasicrystal and Al(111).

It can be seen from Figure 6.2 that the crystalline phase of the Al-Pd-Mn alloy is more reactive towards oxygen than the quasicrystalline one, due to rapid penetration of oxygen in the rough surface created by sputtering treatment. However, the most striking feature in this figure is the larger reactivity of the quasicrystalline i-Al-Pd-Mn surface compared with Al (111). Although the plateau value looks very similar for both surfaces, the exposure time required to reach saturation is much longer for elemental aluminum. Combined with the stronger oxidation observed for quasi-crystalline i-Al-Pd-Mn in Figure 6.2, this suggests that the high reactivity of the quasicrystalline surface is related to the geometry of the surface that is more open than the densely packed Al(111) surface, rather than to the reduced density of states near the Fermi level.

Finally, Dubot et al. observed that the aluminum oxide layer that was formed following oxidation of i-Al-Pd-Mn by dry oxygen in vacuum exhibits specific electronic properties, close to those of α-alumina [10]. In particular, enhanced ionicity and stretching vibrational properties, compared to the amorphous layer grown on pure aluminum in similar conditions, have been evidenced and ascribed to the influence of the bulk quasicrystalline substrate dielectric function.

Some similarity with conventional aluminum-based alloys and pure aluminum has also been observed during oxidation of Al-Cu-Fe complex metallic alloys. In particular, oxidation of quasicrystalline i-Al$_{63}$Cu$_{24}$Fe$_{13}$ and ψ-Al$_{66}$Cu$_{22}$Fe$_{12}$ and of crystalline β-Al$_{51}$Cu$_{35}$Fe$_{14}$ and λ-Al$_{75}$Cu$_3$Fe$_{22}$ phases in vacuum conditions leads to the formation of a very thin passivating layer of aluminum oxide that protects the other elements from oxidation [11, 12]. Aluminum segregation is observed in all cases. No significant difference in the oxidation characteristics appears when comparing two crystalline phases (β-Al$_{51}$Cu$_{35}$Fe$_{14}$ and λ-Al$_{75}$Cu$_3$Fe$_{22}$) and one quasicrystalline phase (ψ-Al$_{66}$Cu$_{22}$Fe$_{12}$) saturated with oxygen at room temperature.

A detailed investigation of the oxidation mechanism of i-Al-Cu-Fe has been performed by Rouxel et al. [13, 14]. A three-stage oxidation process has been observed when the surface is exposed up to 7200 L of oxygen between room temperature and 700 °C. The first chemisorption stage is followed by nucleation and growth of an aluminum amorphous oxide until all aluminum surface atoms are oxidized and the surface is saturated in oxygen. Again, only aluminum is involved in the oxidation process and a thin passivating alumina layer is formed (less than 2 nm thick at room temperature). However, some peculiar behavior has been evidenced when measuring, using Auger electron spectroscopy, the variations in the slope at the origin of the plot of aluminum surface enrichment versus time of exposure to oxygen for various temperatures (see Figure 6.3). It has been shown that Al surface segregation is due to two distinct modes of atomic motion. Above 500 °C, diffusion is related to a classical vacancy diffusion mode with an activation energy equal to 2.2 eV at^{-1} similar to what

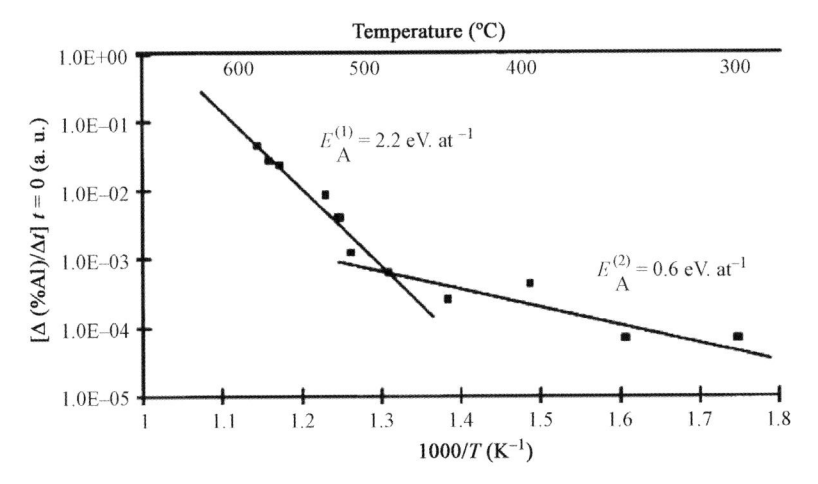

Figure 6.3 Variations in the slopes at the origin of the plot of aluminum surface enrichment versus time of exposure to oxygen as a function of temperature. The aluminum enrichment follows an Arrhenius law and the values of the activation energy can be deduced from the slope of the plots. Reprinted from reference 8, Copyright © (2006), with permission from Elsevier.

is observed for diffusion in metals and conventional intermetallics. Below 500 °C, the lower value of the activation energy (0.6 eV at^{-1}) can be ascribed to atomic mobility by a phason flip mechanism, a concept that is intrinsic to the quasiperiodic structure and independent of the formation and migration of vacancies.

In addition, some differences in the growth kinetics and oxidation of CMAs and classical intermetallics has been observed by Rouxel *et al.* [13, 15] when oxide layers are formed at elevated temperatures. These authors have compared the behavior of the quasicrystalline icosahedral *i*-Al$_{62}$Cu$_{25.5}$Fe$_{12.5}$ and the crystalline tetragonal ω-Al$_7$Cu$_2$Fe phases with pure aluminum following exposure up to 5000 L of oxygen vapor in vacuum at 600 °C. They observed that the thickness of the final oxide layer formed was increasing in the order Al $>$ ω-Al-Cu-Fe $>$ *i*-Al-Cu-Fe. This behavior has been ascribed to the difference in the substrate structures. When pure aluminum is oxidized at temperatures above 400 °C, nucleation and growth of γ-Al$_2$O$_3$ crystallites in epitaxy with the subjacent aluminum structure occurs at the interface between the amorphous oxide and the metal substrate. On ω-Al-Cu-Fe and *i*-Al-Cu-Fe the structural complexity of the surface may delay or even prevent the nucleation and epitaxial growth of the γ-Al$_2$O$_3$ crystallites.

Oxidation in vacuum of Al-Cr-Fe complex alloys of various compositions (Al$_{77}$Cr$_{16.5}$Fe$_6$; Al$_{72.5}$Cr$_{19.5}$Fe$_8$; Al$_{72.5}$Cr$_{21.5}$Fe$_6$; Al$_{67.6}$Cr$_{23.3}$Fe$_{9.1}$) has been studied by Demange *et al.* at room temperature and 450 °C [16, 17]. Here again, preferential oxidation of aluminum and aluminum segregation are observed, which lead to the formation of a thin passivating aluminum oxide layer (a few Å thick). It must be pointed out that, in such oxidizing conditions, the content in Cr has little influence on the thickness of the oxide layer.

In conclusion, as for more conventional intermetallics, oxidation of Al-based complex metallic alloys in vacuum environment leads to Al segregation and formation of a passivating aluminum oxide layer, a few Å thick. However, some major differences have also been observed in the growth mechanisms of this aluminum oxide layer, which can be related to the structural complexity of the CMAs or/and the existence of additional diffusion mechanisms. Moreover, the aluminum oxide layer that is formed can present specific electronic properties.

6.2.2
Interaction with Other Molecules

Only a limited number of studies are addressing the surface reactivity of CMAs with molecules other than oxygen or water. Most of them have been performed on clean quasicrystals surfaces under carefully controlled ultrahigh vacuum conditions in an attempt at creating artificial 2D aperiodic molecular overlayers, with a unit cell simply related to the substrate unit cell. In fact, atomic or molecular adsorption on well-defined surfaces can provide a simple route to produce nanoscale structures such as nanoclusters, quantum wires, 2-dimensional overlayers and 3-dimensional epitaxial layers. By using the surface as a template, atomic patterns can be rapidly generated over a macroscopic scale. For example, adsorption of simple metal atoms proved to be quite successful to obtain pseudomorphic adsorbed structures and to gain some

Table 6.1 Summary of the literature results for molecular adsorption onto i-AlPdMn and d-AlNiCo surfaces.

	i-AlPdMn	d-AlNiCo
Atomic sulfur	Surface reconstruction (disordered) [18–21])	
Methanol	Molecular adsorption and decomposition [22–24]	Dissociation [24]
Carbon monoxide	Molecular adsorption [22]	Molecular adsorption on Ni topsite [24]
	No adsorption [24]	
Nitrogen oxide	Dissociation [24]	Dissociation [24]
Formic acid	Dissociation [24]	Dissociation [24]
Benzene	Disordered molecular adsorption [24, 25]	
Aminobenzoic acid	Disordered overlayer [26]	
Iodoalkanes	Dissociation [22]	
Ethylene		Disordered adsorption [19, 27]
Xenon		Ordered monolayer [26, 28]
Buckminsterfullerene (C_{60})	Ordered overlayer at very low coverage [19, 20, 26, 29, 30]	Disordered adsorption on Al sites [31]

insight in the electronic and dynamic phenomena in such systems. However, we will focus here on studies of adsorption of simple molecules and readers are referred to Chapter 3 for more details on metal adsorption.

Most molecular adsorption experiments have been carried out on the fivefold surface of the icosahedral $Al_{70}Pd_{21}Mn_9$ (i-AlPdMn) and on the 10-fold surface of the decagonal $Al_{72}Ni_{11}Co_{17}$ (d-AlNiCo), which are the most intensively studied and the best understood aperiodic surfaces. The results from a literature survey are summarized in Table 6.1.

It can be seen from Table 6.1 that most systems produce dissociative adsorption or disordered overlayers, due to strong interactions between the molecular species and the surface, which result in multiple possible adsorption sites. However, some promising results have been obtained by McGrath and coworkers when using weakly interacting molecules such as xenon [26] or buckminsterfullerene (C_{60}) [19, 20, 26, 29–31]. Of particular interest is the low coverage (0.065 monolayer) adsorption of C_{60} onto the icosahedral Al-Pd-Mn surface. Following preparation by a series of sputtering and annealing cycles in vacuum, the fivefold i-AlPdMn surface consists of large ($\geq$150 nm) Al-rich atomically flat terraces. Although the detailed structure of the terraces remains unknown, several repeating geometric features have been identified. Among them, fivefold pentagonal hollows having a height of 0.7 ± 0.1 nm and displaying Fibonacci scaling relationships (i.e. with distances between the hollows proportional to the golden ratio $\tau = 1.618\ldots$) have been observed as dark fivefold stars in STM images and have been ascribed to the dissection of Bergman clusters during the annealing process [26, 29, 30]. During the first stages of

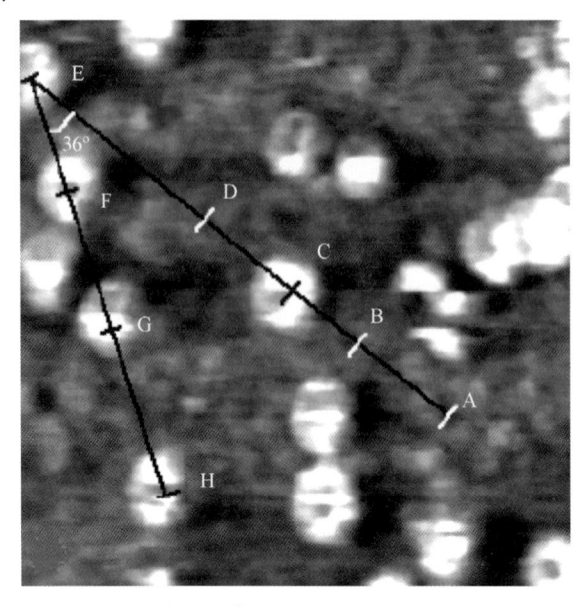

Figure 6.4 150 Å × 150 Å STM image of the *i*-AlPdMn surface covered with 0.065 monolayer of C_{60}. White spots correspond to adsorbed C_{60} molecules. Reprinted from reference 8, Copyright © (2003), with permission from IOP.

adsorption, these pentagonal hollows can act as preferential adsorption sites for C_{60} molecules that are therefore aligned along the same directions as the pentagonal hollows and that display the same Fibonacci scaling relationships. This is illustrated in Figure 6.4 where intermolecular distances are related by:

$$[AE] = \tau[AD] = \tau^2[AC] = \tau^3[AB]$$
$$[EH] = \tau[EG] = \tau^3[EF]$$

As can be seen, the attempt to use quasicrystalline surfaces as templates for aperiodic ordered molecular adsorption has not been very successful so far. As mentioned above, this can be ascribed to the strong bonding of the molecular species with the surface or/and to the presence of too many adsorption sites due to the chemical complexity of these multielement surfaces. A promising route to bypass this last problem has been recently suggested by using single-element metallic thin films that can form quasicrystalline intermediate overlayers when deposited on an aperiodic substrate [19].

It must be pointed out that, for some molecules, the adsorption behavior reported in Table 6.1 noticeably differs from what happens on metal or classical intermetallics surfaces. For example, sulfur is well known to form ordered self-organized overlayers on a number of single-crystal metal surfaces, whereas multiple-site adsorption and/or adsorbate-induced reconstruction is observed on the five-fold surface of *i*-Al-Pd-Mn. In the same way, *ab initio* DFT calculations performed by Krajčí and Hafner have evidenced some specificity in the CO adsorption and dissociation on the *i*-Al-Pd-Mn surface, due to a strong corrugation of the potential-energy surface for atoms and molecules associated with the complexity of the surface structure [32]. C−C bond breaking has also been observed for iodoalkanes adsorption on *i*-Al-Pd-Mn [22] while it is not observed for pure Al or Pd. All these results suggest that Complex Metallic Alloys surfaces can be more reactive than those of their pure metal constituents or classical intermetallics counterparts, thus opening up promising perspectives for applications in the field of catalysis (see Chapter 8).

6.3
Atmospheric Aging

In terms of atmospheric aging of CMA, the research field is again mostly focused on Al-based compounds. There is continuous transition between scientific investigations with controlled exposure to oxygen and water in ultrahigh vacuum conditions (UHV) and exposure for longer time to air in ambient conditions.

6.3.1
Atmospheric Oxidation

The oxidation experiments performed in controlled conditions, such as a UHV environment, were also developed in order to rationalize the behavior of both quasicrystals and quasicrystals approximants in atmospheric ambient [33]. The high interest resided primary on the evaluation of the possibility to exploit those materials from a technological point of view. This was surely determined by their multiple and promising properties that, in the case of Al-alloys in atmospheric environment, were thought to be correlated to the presence of a thin oxide (5–10 nm thick) that instantaneously formed on the metallic surface. As an example, we can consider the tribological properties associated to them [13].

Several studies were performed in order to assess the type of oxide, its thickness and its evolution with time. Up to now, the characterization was devoted to the atmospheric oxidation behavior at room temperature in order to mimic the real environment and the real behavior of the system in the environment.

In this kind of characterization, the sample was exposed to an "uncontrolled" environment constituted by "normal" air, where the sole parameter to be varied was humidity. According to that, several experiments were performed in order to characterize the types of oxide that were formed after the sample was stored in a desiccator with $CaSO_4$ or in an environment with increasing humidity, with the ultimate step being the direct dipping into ultrapure water. It is worth noticing that, the different types of characterizations that were developed in a controlled (UHV) or uncontrolled environment and reported here in various chapters, have been mainly conducted using the same type of samples and with the same preparation procedure. This allowed the characterization of the sample behavior in all types of environment and gave hints on the determination of the effect of the substrate composition (the metallic alloy) or of the substrate structure on the oxide formation.

The techniques that were used were various, since at the beginning, differently from the "controlled" characterization done in a UHV environment, there was no availability of techniques sensitive to the in-situ thin oxide growth in ambient and to its structure and composition. Several research groups started using the techniques exploited in surface science, mainly X-ray photoemission spectroscopy (XPS) or Auger spectroscopy, which were coupled with depth-profile analysis in order to gather information on the qualitative and quantitative composition of the oxide versus its thickness. One drawback was surely that of not being able to follow the evolution of the oxide with time.

The systems studied up to now, were mainly polycrystalline and polyphasic materials, due to the intrinsic difficulties residing in the growth of large-grain single crystals. The interest was initially focused on Al-Cu-Fe [12, 13] and then shifted to the Al-Cr-Fe [16, 17] alloys in order to enhance the oxidation stability, due to the presence of Cr oxide that was known to be thermodynamically stable in this condition (-1139.7 kJ/mol for Cr_2O_3).

The preparation of the sample was usually accomplished by the means of standard metallographic techniques, grinding using SiC down to 4000 grit and subsequent polishing with diamond paste up to 0.25 micrometer with ethanol as a solvent. Only in more recent times, and with the availability of single-grain and single-phase samples, has it been possible to prepare the surface in a well defined way, by the means of sputtering and annealing that were developed in the UHV chamber and that were followed by atmospheric oxidation [34].

Coming back to the first experiments performed by Demange and coworkers [16, 17] the comparison was made with the same alloys oxidized in UHV with pure oxygen. The samples were Al-Cr-Fe mixtures of approximants of composition $Al_{77}Cr_{16.5}Fe_6$, $Al_{72.5}Cr_{19.5}Fe_8$, $Al_{72.5}Cr_{21.5}Fe_6$ and $Al_{67.6}Cr_{21.5}Fe_{9.1}$ [16]. The XPS analysis revealed that the alloys were covered by Al oxide. The oxide thickness was evaluated using the Strohmeier formula [35] once the area of the XPS peak was known. In air and with a 6% humidity, the thickness of the oxide formed after 24 h was found to be higher than the one grown in a controlled atmosphere (in the 19–29 Å range for all the four samples, compared to 7–9 Å) and at the same time lower than the one grown after exposure to ultrapure water for 15 h. The latest was found to fall in the 49–78 Å range and to be composed of Al and Cr oxides. In the light of these results, a mechanism for ambient oxidation in humid air was proposed: oxidation of Al first, then segregation of Al with Al-depletion of the layer underneath and then Cr oxidation in that interfacial region. This was also in accordance with the results found by Pinhero et al. [36] in the angle-resolved XPS data of Al-Cu-Fe-Cr systems. It is worth noticing that, in an atmospheric environment, the outermost layer is always constituted by contamination of carbon.

The ultimate and most detailed experiment in air was done with humidity in the 54–67% range and was performed by Veys and coworkers, where the aging effect of the sintered γ-$Al_{65}Cr_{27}Fe_8$ phase with a γ-brass structure was followed with time [37]. X-ray reflectivity characterization gave the possibility to follow the characteristic of the oxide growth versus time. A freshly polished sample was in fact mounted on a diffractometer, carefully aligned and kept there for 15 days. Regularly repeated

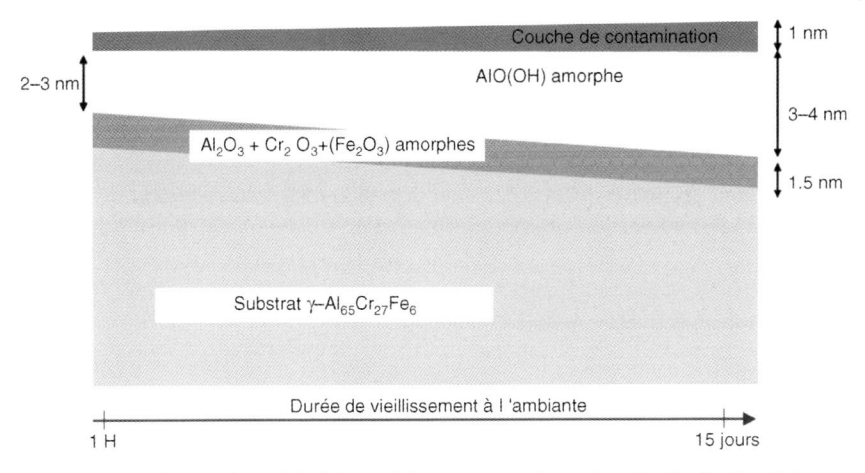

Figure 6.5 Oxide growth model of the multilayer structure formed on the $Al_{65}Cr_{27}Fe_8$ CMA as a result of aging in atmospheric conditions. Reproduced from reference [39].

measurements were done. From the interference effects of the X-ray beam with a film having different properties (such as electron density) with respect to the substrate, it was possible to gather information on the presence of multiple stacked layers, their thickness, the roughness of the interface and degree of crystallization. Additional secondary ion mass spectrometry (SIMS)-based surface-analytical techniques, the secondary neutral mass spectroscopy (SNMS) and XPS, allowed a model for the surface interaction of the $Al_{65}Cr_{27}Fe_8$ CMA to be proposed with atmospheric air during longer exposure time up to 15 days (Figure 6.5). The data were rationalized with the help of simulations and fitting procedures that included the former parameters. For this, the data relative to the electronic densities were calculated for several Al and Cr oxides, hydroxides and oxy-hydroxides, once the atomic structure and the number of electron per atom was known. Those data were compared with the angle-resolved XPS and depth-profile analysis performed on the same sample soon after polishing and after 15 days.

The evolution of the oxide with time followed the route that has already been hypothesized by Demange and coworkers [16, 17] and this time was qualitatively and quantitatively characterized: so that they could distinguish between the presence of an oxidized layer close to the metallic substrate, then a hydroxilated layer that was followed by a top layer of chemisorbed water. This was in accordance with the Simmons and Beard description [38]. Consequently, they proposed a surface structure consisting of a stacking of a mixed amorphous oxide layer of Al_2O_3, Cr_2O_3 and Fe_2O_3 of constant thickness close to 15 Å, followed by another amorphous oxidized layer containing only Al and oxygen (AlO(OH)) whose thickness increases upon aging time and was approximately 20–40 Å after 15 days of exposition and finally, a contamination layer of approximately 10 Å.

These aging conditions can further be used to investigate the influence of surface oxides (called passive film in the corrosion community) on the aqueous corrosion mechanisms. Aluminum itself is stable in "normal" atmospheric conditions, so that the topic of atmospheric corrosion of Al-based compounds has not attracted a large interest in the past. Some more specific studies based on the preliminary work mentioned previously would, however, be necessary to investigate the influence of humidity for example on the formation of the surface oxides and hydroxides and their structure and properties. This aspect, not directly relevant for corrosion resistance, could certainly influence other properties like surface energy and friction, for example.

6.3.2
Surface Properties in Atmospheric Conditions

Among the various surface properties where CMAs show interesting behavior, it is necessary to emphasize the very low surface energy and related low friction coefficient found for the Al-Cu-Fe system [40]. In the case of Al-based alloys, a direct relation has been found between the reversible adhesion energy and the ratio of the electron density at the Fermi level to the thickness of the oxide film (see Figure 6.6).

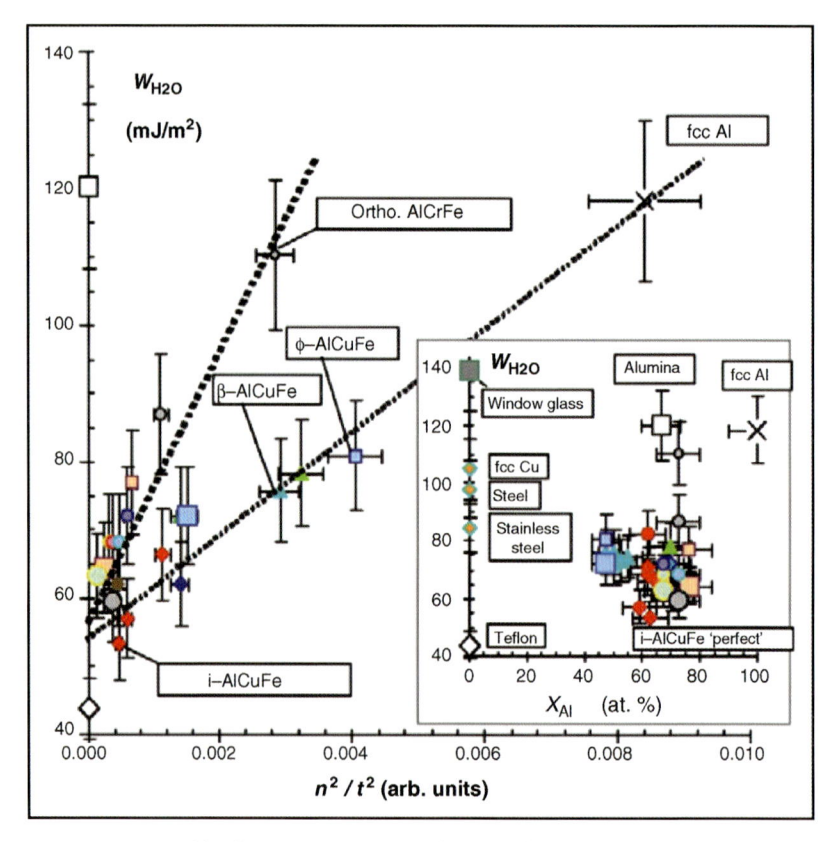

Figure 6.6 Reversible adhesion energy (W) as a function of the partial density of state of the Fermi level (n) divided by the oxide thickness (t). Reproduced from reference [40].

Furthermore, related very low friction coefficients have been measured, for example for the i-Al-Cu-Fe quasicrystalline compound. Following the comment on the environment of the previous section, the exact influence of the atmospheric aging and surface preparation on these properties has not yet been investigated in detail.

6.4
Surface Chemistry and Reactions in Aqueous Solutions

CMAs are materials with complex structures. Therefore, the interest in fundamental locally resolved oxidation investigations is clear. However, investigation of the corrosion behavior of CMAs or quasicrystals is in its very early stage and only very few studies have been published, mainly related to the electrochemical behavior of quasicrystals and their approximant phases for Al-Cu-Fe [41–43] and Al-Cr-(Cu)-Fe system [3, 44–47]. The corrosion resistance of Al-Cr-Cu-Fe and Al-Cu-Fe was investigated in 0.5M Na_2SO_4 in different pH during potentiodynamic polarization by Massiani et al. [44] However, the polarization curves have not been interpreted in detail in terms of electrochemical reactions. Veys et al. [45] also performed studies of the polarization resistance of Al-Cr-(Cu)-Fe alloys in a solution of a mixture of citric acid and sodium chloride. They found that the electrochemical properties of these alloys are affected by their chemical composition rather than by their crystallographic structure. The electrochemical impedance measurements of Al-Cr-Cu-Fe coatings carried out by Balbyshev et al. [46] demonstrated high corrosion resistance of the alloy during long immersion in dilute Harrison's solution. Veys [39] also investigated the localized corrosion susceptibility of Al-based CMA and found good corrosion resistance but with obvious signs of transient attacks especially when the surface was aged in air for longer times. The presence of a large amount of Al in most of the investigated Al-based CMA guarantees surface stability in its thermodynamic stability domain (approximately between pH 3 and pH 9). For this reason, the focus will be set in the following section on compounds with higher corrosion resistance in more severe environments. Although they show very interesting surface properties, the Al-Cu-Fe CMAs are not ideal compounds in terms of aqueous corrosion resistance.

6.4.1
Thermodynamic Stability

One of the reasons for the lack in corrosion investigations is certainly the small size of the available samples. Most of the investigations presented and mentioned have therefore been performed on powder metallurgically prepared samples. For the localized corrosion susceptibility assessment, these samples are not suitable because the oxide inclusions and defects induced by the production processes induce "weak" areas of the sample surface.

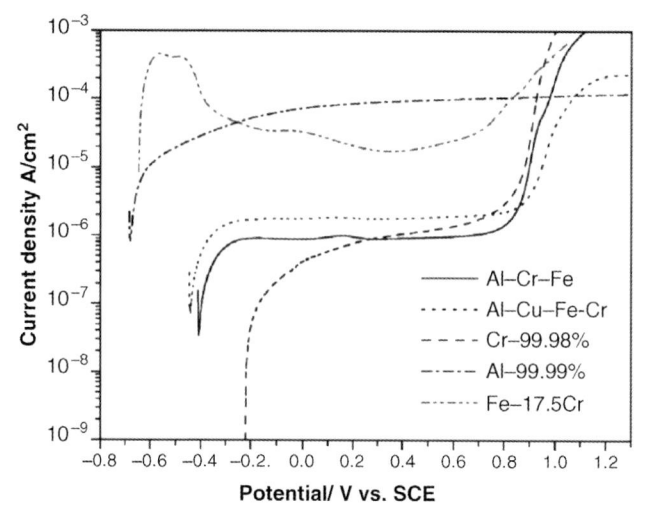

Figure 6.7 Electrochemical potentiodynamic polarization measurements on $Al_{65}Cr_{27}Fe_8$, $Al_{67.2}Cr_{10.4}Cu_{12}Fe_{10.4}$, Fe17.5Cr stainless steel and pure elements in 0.1M Na_2SO_4 (pH 2) [48].

The first aspect to be considered when addressing aqueous corrosion resistance is the thermodynamic stability of the surface oxide that forms in the presence of halide (Cl,...) -free solutions. This parameter defines the uniform corrosion resistance of a material. There are two classes of CMAs ($Al_{67.2}Cr_{10.4}Cu_{12}Fe_{10.4}$ and $Al_{65}Cr_{27}Fe_8$) that show very broad thermodynamic surface oxide stability in aqueous electrolytes [48]. A passive behavior (nm thick protecting oxide) is present from pH 0 to 14 during electrochemical potentiodynamic polarization experiments. Figure 6.7 shows the comparison of $Al_{65}Cr_{27}Fe_8$, $Al_{67.2}Cr_{10.4}Cu_{12}Fe_{10.4}$, stainless steel (Fe17.5Cr) and pure elements in a 0.1 M Na_2SO_4 solution adjusted to pH 2. Electrochemical current density in the microampere domain are measured for the CMAs, corresponding to the slow formation of a stable (hydr-)oxide. Aggressive chloride-containing electrolytes will be discussed in relation to localized corrosion susceptibility. For stainless steel, the threshold concentration in Cr for passivation is 12%, so that the mechanism is different in the case of $Al_{67.2}Cr_{10.4}Cu_{12}Fe_{10.4}$, for which the chromium content is below this concentration. The major advantage of these Al compounds is that they do not show the active dissolution observed in stainless steel (see Figure 6.7: current in mA range for steel) at low potential in acidic media before a stable oxide can form. In this sense, the presence of Al in the CMA improves the stability of the surface oxide in a pH domain where surface oxide on pure Al is not stable at all. The absence of active dissolution could be a key advantage in the corrosion resistance of these compounds. The problem of stainless steel is that, if the surface is activated by a tribocorrosion processes (local removal of the oxide) or in a localized corrosion attack (pit) inducing a very low pH, dissolution can proceed unhindered in an autocatalytic way. This is the most dangerous degradation mechanism that is suppressed in this case. One drawback

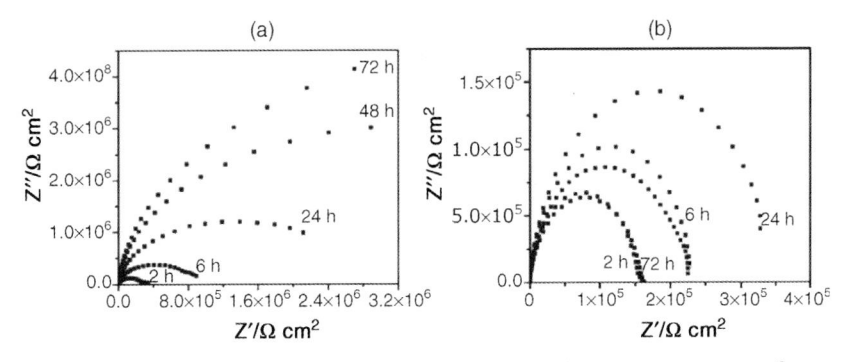

Figure 6.8 Electrochemical impedance spectroscopy (Nyquist plots) measurements as a function of time for: (a) $Al_{65}Cr_{27}Fe_8$, and (b) $Al_{67.2}Cr_{10.4}Cu_{12}Fe_{10.4}$ in 0.1M Na_2SO_4 (pH 2) [48].

(observed in Figure 6.7) of all the high Cr content materials is the transpassive dissolution (observed oxidation of Cr^{3+} in soluble Cr^{6+} at 0.8 V standard calomel electrode – SCE), but this problem only starts to be relevant at very high pH and is similar for Cr-containing stainless steel.

The electrochemical polarization investigations alone do not allow all the aspects of the uniform corrosion behavior to be described because they address the stability of a polarized sample surface. Oxide formation is forced by the electrochemical control and the stability of this oxide in a given solution is then assessed. Electrochemical impedance spectroscopy allows, by applying a very small (10 mV) potential perturbation around the corrosion or electrochemical open-circuit potential OCP and recording the current answer, to monitor the "freely" oxidizing surface stability. At the corrosion potential, the two considered compounds ($Al_{65}Cr_{27}Fe_8$, and $Al_{67.2}Cr_{10.4}Cu_{12}Fe_{10.4}$), show significant difference in stability (see Figure 6.8). During the first day of immersion in 0.1 M Na_2SO_4 (pH 2), the impedance value increase indicates a stabilization of the surface oxide, but afterwards the Cu-containing CMA started to show signs of dissolution. The impedance values in the range of 150 kΩ/cm^2 after 72 h is still not indicating fast uniform corrosion, but the stability of $Al_{65}Cr_{27}Fe_8$ is significantly better.

The surface protection is obtained because of the integration of Cr (hydr-)oxides in the nm-thick surface oxide. Figure 6.9 presents the XPS characterization of the sample surface after 48 h of exposure in 0.1 M Na_2SO_4 (pH 2). It is interesting to note that the passive layer is composed of a mixture of different Al (hydr-)oxides and with smaller amount of Cr (hydr-)oxides mainly at the surface in this case. Iron is not found in the oxidized state in the passive layer. The (hydr-)oxide film is very thin (3–4 nm), as evidenced by the presence of metallic components in the signals. The presence of a large amount of Al is a sign that a synergetic protecting effect with chromium is obtained. Chromium alone does not guarantee suppression of the active dissolution at very low pH.

To obtain detailed information about long-term stability and simultaneously a description of the dissolution mechanisms, ICP-MS (ion coupled plasma – mass spectroscopy) measurement of the dissolved species in 1 M H_2SO_4 and 1 M NaOH as

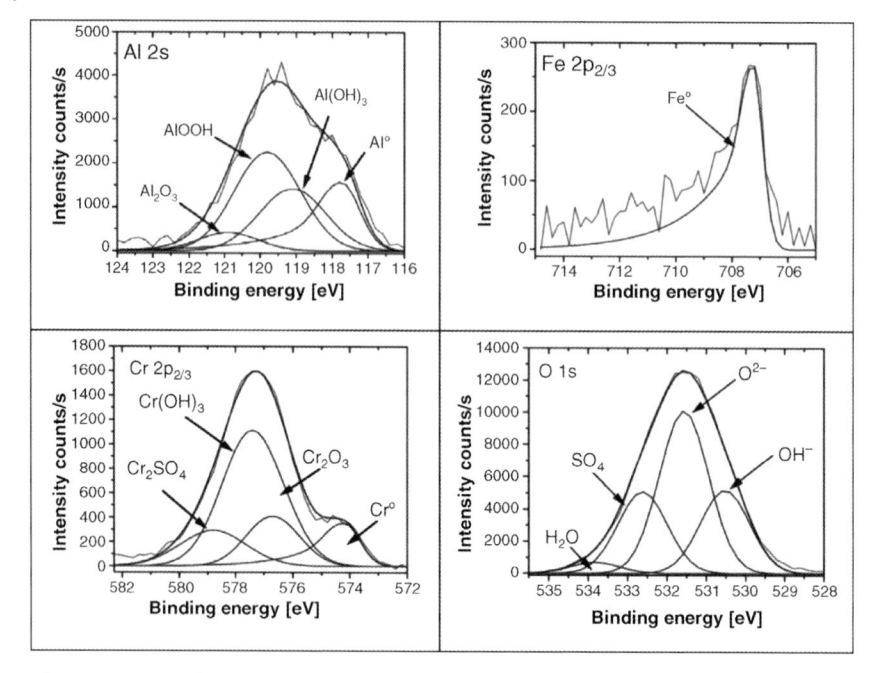

Figure 6.9 X-ray photoelectron spectroscopy spectra of the different elements present in the passive oxide formed after exposure of $Al_{65}Cr_{27}Fe_8$ to a 0.1 M Na_2SO_4 (pH 2) solution during 48 h [48].

a function of time was performed for a duration of 2 months for $Al_{65}Cr_{27}Fe_8$, and $Al_{67.2}Cr_{10.4}Cu_{12}Fe_{10.4}$. Figure 6.10 provides a clear picture of the differences in the extreme pH domain. The $Al_{65}Cr_{27}Fe_8$ compound is clearly very stable from pH 0 to pH 14. The maximal amount of 50 µg/g after 2 months for the element Al in acidic solution corresponds to passive film dissolution rates and even after 1 year of immersion, the sample surface does not show any visible signs of corrosion. This compound seems even more stable in very alkaline solution, most probably due to the high stability of Fe in alkaline solutions. For the Cu-containing $Al_{67.2}Cr_{10.4}Cu_{12}Fe_{10.4}$ the situation is completely different. The measured dissolved ions concentrations are more than 1000 times higher in sulfuric acid and 60 times higher in sodium hydroxide than for $Al_{65}Cr_{27}Fe_8$. In acidic media, this compound corrodes extremely rapidly. One observation is that all the elements are detected in large amount, except Cu. The exact mechanism is that all the element are dissolving, but due to the fact that the corrosion potential of the alloy is very low compared to the redox potential of the copper dissolution, all the copper is redeposited. This process is obviously very detrimental and is a concern for any copper-containing compounds that starts to corrode. In the alkaline domain, the difference is less extreme, but still significant corrosion occurs, the main difference is that in alkaline solutions, stable Fe-containing corrosion products can form. These last experiments allow between very corrosion-resistant materials ($Al_{65}Cr_{27}Fe_8$) and materials showing corrosion

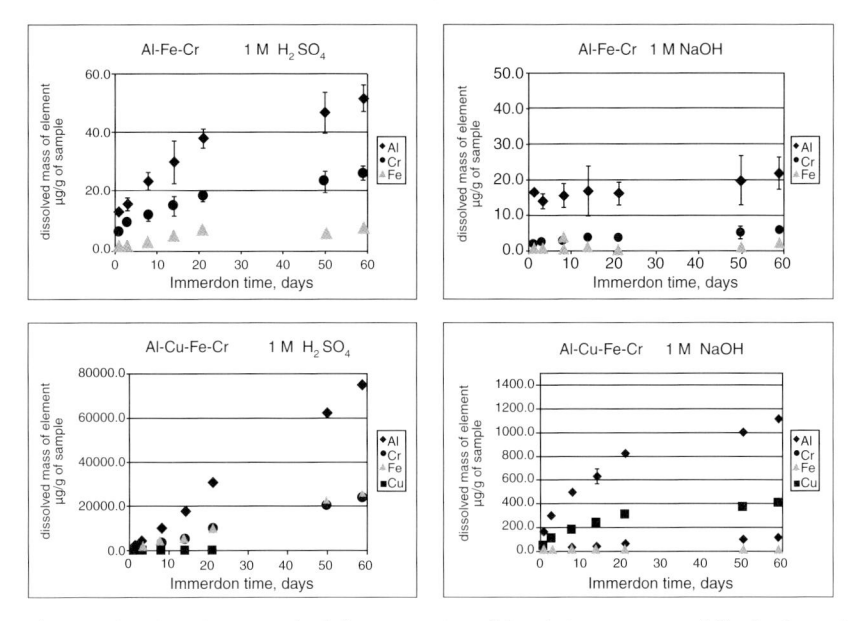

Figure 6.10 ICP-MS time-resolved characterization of dissolution processes of $Al_{65}Cr_{27}Fe_8$ and $Al_{67.2}Cr_{10.4}Cu_{12}Fe_{10.4}$ in 1M H_2SO_4 and 1M NaOH [49].

resistance in normal operation conditions ($Al_{67.2}Cr_{10.4}Cu_{12}Fe_{10.4}$) to be clearly distinguished.

6.4.2
Oxide Electronic Properties

As mentioned in the previous section, the formation of a stable surface oxide is responsible for the good corrosion resistance of $Al_{65}Cr_{27}Fe_8$, and $Al_{67.2}Cr_{10.4}Cu_{12}Fe_{10.4}$. The oxide properties play an important role in the cathodic oxygen and hydrogen reduction rate, but also on other surface properties such as surface energy, adhesion or friction. From Mott–Schottky experiments, where the capacitance of the solid/liquid interface is measured as a function of applied potential (inducing oxide formation), it can be further concluded that, in pH 8.4 borate buffer (Figure 6.11), the surface oxide formed on $Al_{65}Cr_{27}Fe_8$ presents a p-type semiconducting (decreasing capacitance with increasing potential) behavior. Pure chromium oxides also present a p-type semiconductor behavior in slightly alkaline media and this similarity supports the dominating role of chromium in the passive film formed on the $Al_{65}Cr_{27}Fe_8$ CMA. Surface oxidation seems to be quite different from that for stainless steel, although Cr and Fe are present in both compounds. Stainless steel usually shows an n-type semiconducting behavior for its passive surface oxide (increasing capacitance with increasing potential). This difference could again be crucial in the localized corrosion mechanisms. When a localized attack occurs, the surrounding oxide is the

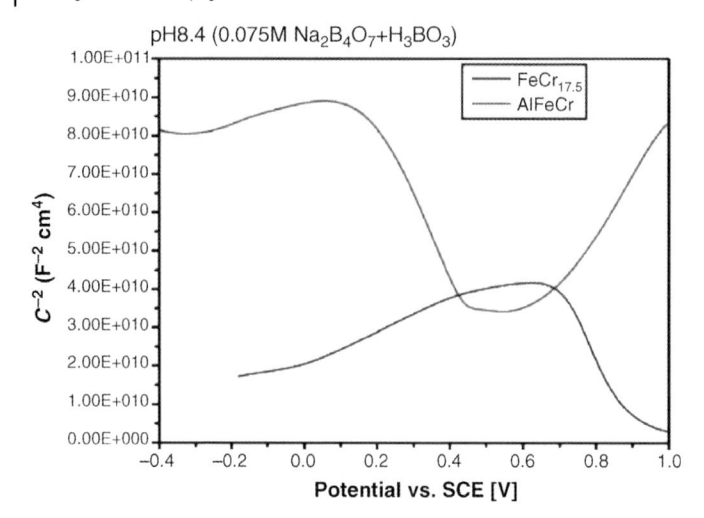

Figure 6.11 Mott–Schottky analysis of the metal–oxide–solution capacitance evolution as a function of applied potential for $Al_{65}Cr_{27}Fe_8$ and a ferritic Fe17.5Cr stainless steel [50].

site for the cathodic reduction reaction of O_2 or H^+. The 3–5-nm thick oxide can act like a diode; the n-type oxide will allow cathodic reduction at relatively high rate, speeding up the local dissolution. On the other side, a p-type oxide blocks the electron transfer for the cathodic reaction. Combined with the fact that Al-Cr-Fe alloys are difficult to activate at low pH, this can explain why the localized corrosion attack susceptibility seems low (see schematic model at the end of Section 6.4).

One criticism of the Mott–Schottky analysis is the fact that the sample surface needs to be polarized far away from the corrosion potential in order to get information about the passive oxide properties. An elegant alternative is the photo-electrochemical investigations where the current induced by light of different energy is recorded (see Figure 6.12). For semiconductors, as soon as the bandgap energy is reached, the incoming light on the surface induces electron transitions through the bandgap. This reaction results in an important current measured in addition to the normal passive current. For the case of $Al_{65}Cr_{27}Fe_8$ the semiconducting nature of the passive film around the corrosion potential could be demonstrated. Additionally, a bandgap of 4.2 eV is determined and, based on the phase shift of the photocurrent signal (described later), a p-type semiconducting behavior is again evidenced. The interesting fact is that, even if the passive layer is mostly Al (hydr-)oxide, the oxide properties are different from the insulating Al (hydr-)oxide. The influence of this semiconducting behavior on the other surface properties needs further investigations, because surface energy and friction behavior of these compounds are still unexplained up to now. Similar oxide properties have been found for the Cu-containing phase.

To characterize the difference between the two considered compounds $Al_{65}Cr_{27}Fe_8$ (designated as γ phase) and $Al_{67.2}Cr_{10.4}Cu_{12}Fe_{10.4}$ (designated as O phase), higher pH has also been investigated. In a pH 12 NaOH solution, the Cu-containing O phase

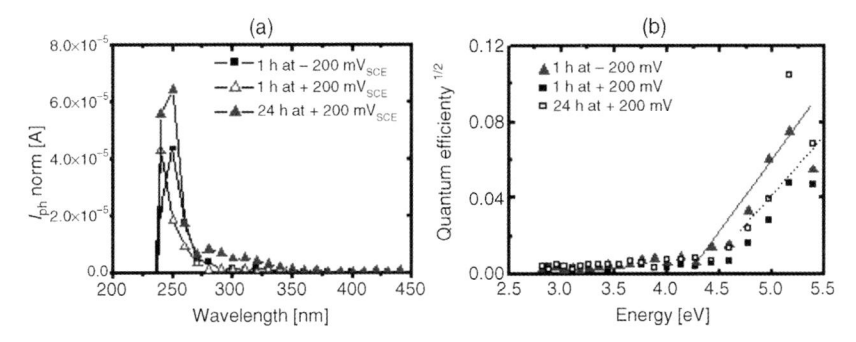

Figure 6.12 Photoelectrochemical behavior of the passive oxide formed on $Al_{65}Cr_{27}Fe_8$ in pH 8.4 Borate buffer at different applied potential: (a) measured photocurrent as function of light wavelength; (b) Quantum efficiency for the photoinduced transition and measured bandgap [50].

completely changed its oxide properties (see Figure 6.13a). The sample surfaces were electrochemically polarized (-350 mV vs. SCE) slightly above the corrosion potential to be able to measure a base passive current and, with a lock-in technique, to determine the additional current component induced by the light excitation. The $Al_{65}Cr_{27}Fe_8$ oxide maintained its 4.2 eV bandgap and its p-type semiconducting nature, whereas the bandgap dropped to 3.2 eV for the $Al_{67.2}Cr_{10.4}Cu_{12}Fe_{10.4}$ compound (Figure 6.14a). Here, the surface instability and dissolution of Al already play a role on the surface oxide properties even if the surface is still in its passive state. To discuss the question of the transition from p-type to n-type oxide, the O phase has been further polarized to $+100$ mV in this pH 12 solution (Figure 6.13b). The photocurrent has a similar evolution but with a lower intensity. The interesting fact is to be found in the phase shift, Figure 6.14b. The polarization clearly induced a switching from the p-type semiconducting behavior found at lower potential close to

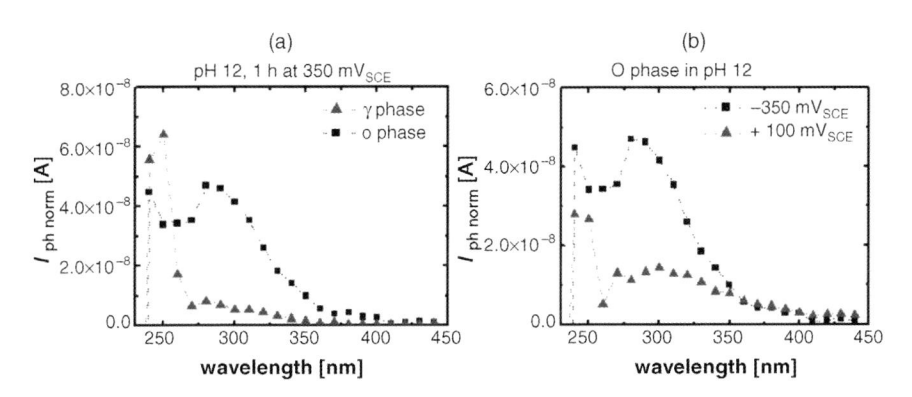

Figure 6.13 Photoelectrochemical behavior of the passive oxide formed in pH 12 NaOH solutions: (a) comparison of $Al_{65}Cr_{27}Fe_8$ (γ phase) and $Al_{67.2}Cr_{10.4}Cu_{12}Fe_{10.4}$ (O phase) polarized at -350 mV; (b) Influence of the electrochemical polarization potential on the measured photocurrent for the O phase [50].

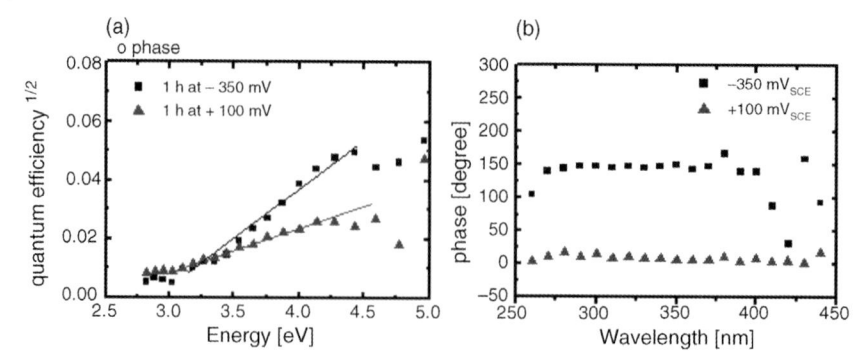

Figure 6.14 Oxide properties for the $Al_{67.2}Cr_{10.4}Cu_{12}Fe_{10.4}$ (O phase) CMA in pH 12 NaOH solution: (a) Quantum efficiency for the photoinduced transition and measured bandgap; (b) Photocurrent phase signal shift indicating the type of semiconductor on the surface [50].

corrosion potential (open-circuit potential in the passive state) to an n-type semi-conducting behavior. The bandgap stayed in the same range (3.2 eV), but the surface properties can drastically change between these two conditions. Engineering of surface properties could be performed in this way and this is certainly not very surprising that an alkaline treatment is the base of the high catalytic performance found for Al-Cu-Fe CMAs.

6.4.3
Localized Degradation Reactions

Uniform surface processes have been discussed in Sections 6.4.1 and 6.4.2. The use of powder metallurgically produced samples was possible. For the investigation of the localized breakdown of the surface oxides, impurity- and defect-free samples need to be used. For this purpose, $Al_4(Cr,Fe)$ polycrystalline and single-crystalline compounds have been used. The samples had a chromium content range from 12.5% Cr for the low-alloyed samples up to a maximum of 18.5% Cr for the highest alloyed $Al_4(Cr,Fe)$. The composition can be tuned by varying the ratio Cr to Fe. The localized corrosion investigations demonstrated a very good resistance against pitting for the $Al_4(Cr,Fe)$ polycrystalline compounds. In neutral NaCl, no localized corrosion attack (pitting) could be found when local microcapillary electrochemical potentiodynamic polarization measurements (capillary diameter 100 μm) are performed in the middle of the grains (see Figure 6.15a). Due to the small size of the sample, but also to the presence of cracks in the polycrystalline sample, all the localized corrosion investigations have been performed with a local electrochemical characterization methods. To perform a complete electrochemical characterization on selected submicrometer scales, the microcell technique is the ideal method. In this approach, the access of bulk solution to the surface is restricted to submicrometer areas by using a glass microcapillary filled with electrolyte. This microcell technique previously described by Suter et al. [51, 52] has proven to be very powerful for the identification of reaction kinetics of corrosion susceptible areas (e.g., inclusions in stainless steel) [51]. This

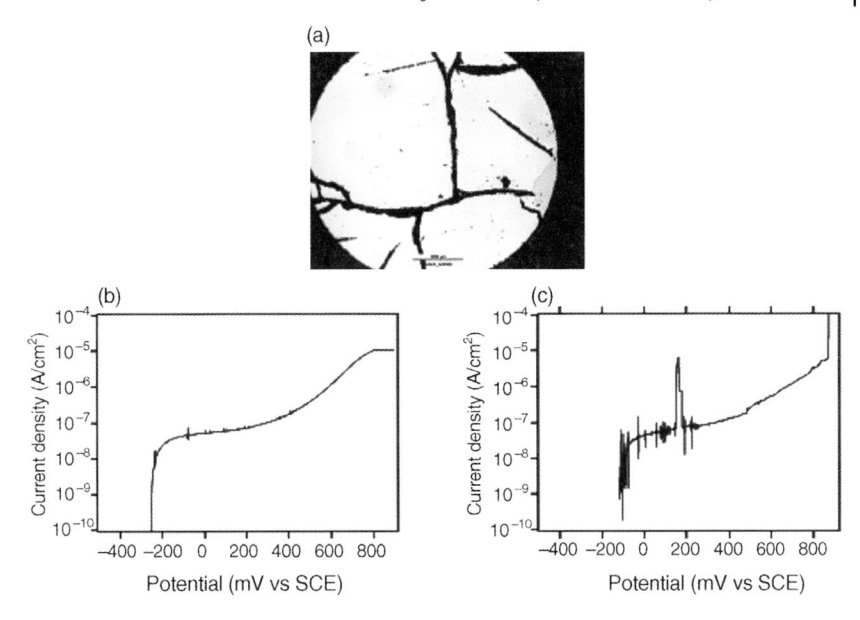

Figure 6.15 Electrochemical microcapillary (100 μm diameter) investigation of localized corrosion susceptibility in 0.1 M NaCl: (a) optical image of a etched polycrystalline $Al_4(Cr,Fe)$ sample; (b) potentiodynamic polarization measurement performed in a grain; (c) measurement on a defect (black boundary in a) [54].

method can be coupled to electrochemical sensors and provides essential information for the modeling of localized degradation processes [53]. Basically, the technique consists of a pulled glass microcapillary filled with an electrolyte acting as a microcell. It is fixed in an optical microscope stand at the revolving nosepiece replacing an objective. The specimen is mounted on the microscope stage. This arrangement enables a search for a site with different magnification before switching to the microcapillary measurements and ensures accurate positioning of the microcell. Reference and counter electrodes are connected to the capillary, allowing electrochemical control of the investigated surface. The counter electrode is a 0.5-mm platinum wire. An electrolyte bridge connects the SCE reference electrode.

Figure 6.15a shows a polycrystalline sample and it is obvious that some defects (black areas) are present on the sample corresponding to cracks induced by the cooling procedure. The exposed area is defined by the capillary diameter. Only the transpassive dissolution of chromium starting at 400 mV SCE is visible for this neutral pH after the long passive domain, see Figure 6.15b. This indicates clearly the influence of chromium on the surface protection, which is also confirmed by the presence of significant amount chromium in the oxide formed in aqueous media (XPS surface analytical investigation). Polarization measurements performed on the defects (small cracks) indicate the presence of localized dissolution (Figure 6.15c) and a stable localized attack (pitting) is observed at a potential of around 800 mV SCE after a first local transient dissolution peak at 200 mV SCE. The noise observed on the

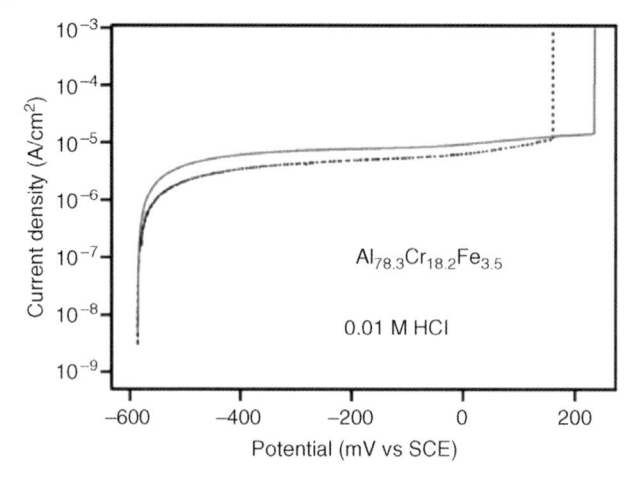

Figure 6.16 Electrochemical microcapillary (100 μm diameter) investigation of localized corrosion susceptibility in 0.01 M HCl.

curves is also related to the presence of defects, although it must be mentioned that the measured passive current is in the femtoampere (10^{-15} A) range and that current stabilization by the potentiostat and electromagnetic shielding of the measurement setup starts to be critical. An important statement concerning the Al-Cr-Fe localized corrosion susceptibility characterization is that the macroscopic electrochemical measurements on most of the samples (polycrystalline $Al_4(Cr,Fe)$ and powder metallurgical $Al_{65}Cr_{27}Fe_8$) do not make much sense in chloride-containing electrolytes. The reason for this is that only the influence of the largest defects are then analyzed and not the intrinsic alloy properties. This is also why the localized corrosion susceptibility of powder sintered $Al_{65}Cr_{27}Fe_8$ QC approximant compounds is much higher that of the polycrystalline $Al_4(Cr,Fe)$ even in the presence of higher chromium content. The alternative to the use of single crystals which are of very good quality but not suitable for large-scale testing, is certainly to consider CMA PVD coatings for the future electrochemical investigations.

If the cracks can be avoided with the microcapillary in the polycrystalline samples, then the inside of the grain shows a good resistance to localized corrosion even in aggressive low pH conditions (see Figure 6.16). Two measurements performed on different locations are displayed; they indicate that the breakdown potential is typically at around 200 mV for 0.01 M HCl. In this case, an additional parameter that is also important to mention is that the passive domain (from −600 mV up to 200 mV) is quite large and the risk of spontaneous localized attack is accordingly small.

On single-crystalline $Al_4(Cr,Fe)$ no evidence of localized corrosion even in 0.01 M HCl solutions could be found. Figure 6.17 shows examples of microcapillary electrochemical measurements in two pH 2 HCl solutions with different chloride contents. The passive current is obviously higher (10^{-5} A/cm^2 instead of 10^{-7} A/cm^2 in neutral solution) but this still corresponds to good passivation conditions. No

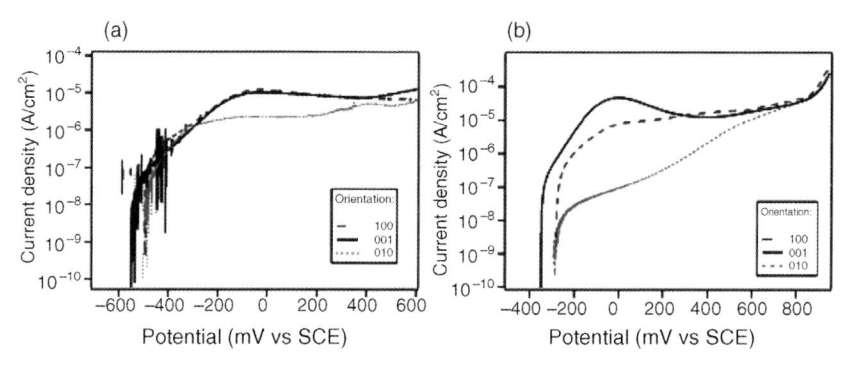

Figure 6.17 Electrochemical microcapillary (100 μm diameter) potentiodynamic polarization measurements performed on $Al_4(Cr,Fe)$ single crystals for different orientations of the surface: (a) in 0.01 M HCl; (b) in more aggressive 0.01 M HCl + 1 M NaCl [54].

current increase related to localized breakdown of the film could be seen. The single-crystalline samples have slightly different Cr content depending on the orientation:

- 100: $Al_{78.8}$ $Cr_{15.5}$ $Fe_{5.7}$
- 001: $Al_{78.3}$ $Cr_{18.2}$ $Fe_{3.5}$
- 010: $Al_{78.5}$ $Cr_{16.8}$ $Fe_{4.7}$.

The exact role of the Cr content in the susceptibility to localized corrosion was not addressed in detail but, from the measurement of Figure 6.17, it does not seem significant. The importance of grain boundaries and defects is higher, but for single crystals it is then possible to assess intrinsic materials properties. More interesting is the fact that there is a tendency observed in a whole range of solution for the 010 orientation to show a better passivation. The difference with other orientations becomes more and more significant when solution aggressivity increases (Figure 6.17b), the exact mechanisms for this is not yet clearly understood.

6.4.4
Summary: Localized Corrosion Model for the $Al_4(Fe,Cr)$ Compound

Summarizing all the previously described results, the following model for the corrosion mechanisms for the localized corrosion processes can be formulated (Figure 6.18).

Breakdown of the nm-thick passive film can be induced electrochemically but also in real applications as a result of mechanical damage of the surface. The absence of fast active dissolution in *acidic media* hinders the depth propagation of the localized attack. This is a *first* major advantage of this compound. Acidic media develop in localized corrosion attack as a result of the water hydrolysis reaction induced by the dissolved metallic ions.

The nature of the remaining oxide film also plays an important role in the control of a corrosion process. This nm-thick oxide acts as a diode and if a p-type oxide is present like for $Al_4(Fe,Cr)$ (*second* major advantage), it will block the cathodic reaction on the

- Resistant against localized corrosion initiation (except grain boundaries)

But... even if it should happen:

① slow active dissolution at low pH (difficulty to maintain high dissolution rates in the pits)

② p-type passive oxide in the neutral to alkaline pH domain?

Cathodic reaction blocked ②

O_2 or H^+ cathodic reduction possible

Alkaline pH

p- type oxide

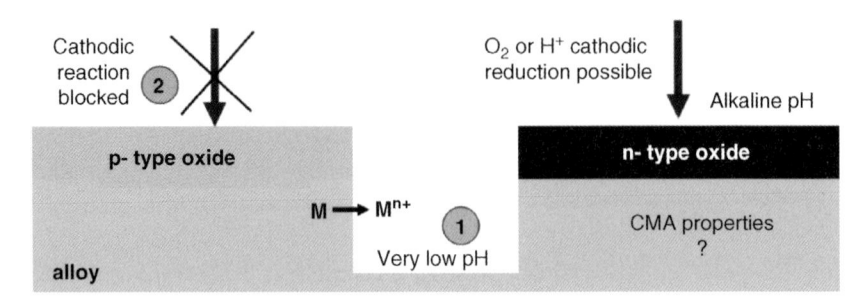

n- type oxide

alloy

$M \rightarrow M^{n+}$ ①

Very low pH

CMA properties ?

Figure 6.18 Schematic model of the localized corrosion processes on Al–Cr–Fe CMAs and benefit of its specific electrochemical behavior.

surface. In the presence of a cathodic reaction on the large oxide covered surface, the localized attack can be massively accelerated. This is for example the case of stainless steel, which shows n-type behavior of the surface oxide film.

6.5
High-Temperature Corrosion

6.5.1
Bulk Samples

Like aluminides, aluminum-based complex metallic alloys are expected to exhibit good resistance to high-temperature oxidation owing to the formation of a passivating alumina layer. The high-temperature corrosion characteristics of Al-rich alloys are often compared to those of pure aluminum for which the growth mechanisms of the alumina layer can be divided in two temperature ranges. Oxidation at low temperature (until 400 °C) leads to the growth of an amorphous alumina layer controlled by the diffusion of aluminum in the amorphous oxide. As mentioned previously (see Section 6.2.1), for temperatures above 400 °C, nucleation and growth of γ-Al_2O_3 crystallites in epitaxy with the subjacent aluminum structure is observed at the interface between the amorphous oxide and the metal substrate. Their growth is related to diffusion towards the metal/oxide interface of atomic oxygen through the amorphous layer and through the cracks in the amorphous layer generated by the crystallite growth. θ-Al_2O_3 and α-Al_2O_3 can also be observed during oxidation of aluminum alloys (Al-Fe, Al-Ni) at very high temperatures (higher than 800 °C).

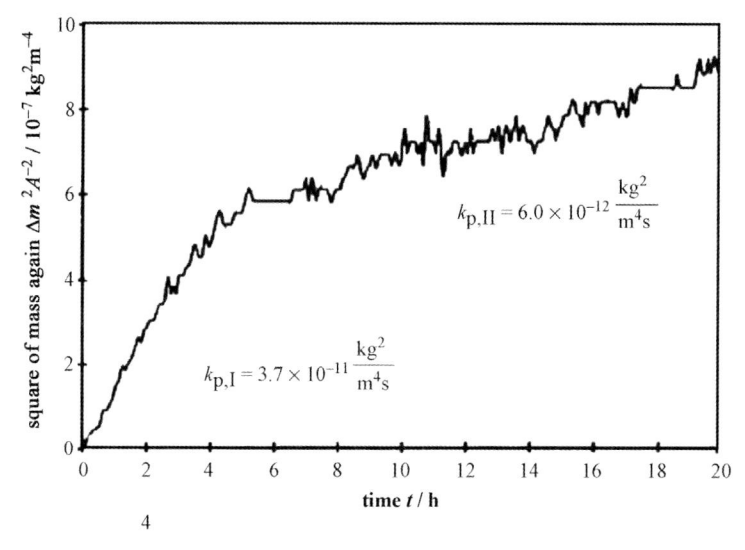

Figure 6.19 Thermogravimetric analysis of the oxidation of a $Al_{63}Cu_{25}Fe_{12}$ quasicrystal at 800 °C in synthetic air (21% O_2 + 79% N_2). From reference [55].

Similar formation of transient metastable crystalline aluminas was observed by Wehner et al. [55, 56] during oxidation of Al-Cu-Fe complex metallic alloys around 800 °C in environmental and synthetic air. In particular, a detailed investigation of the oxide layers formed on the bulk icosahedral $Al_{63}Cu_{25}Fe_{12}$ alloy showed that the early oxidation stage leads to the formation of strongly textured γ-Al_2O_3 presenting an orientational relationship with the underlying quasicrystal surface. As γ-Al_2O_3 cannot epitaxially grow on the aperiodic AlCuFe quasicrystal surface, only small and highly defected crystallites are observed. When the oxide layer is thicker than several hundred nanometers, θ-Al_2O_3 needles appear at the oxide/gas interface, whereas γ-Al_2O_3 is transformed into large hexagonal-shaped α-Al_2O_3 grains at the oxide/metal interface. As the α-Al_2O_3 grains are large and almost free of defects, they provide fewer short-circuit diffusion paths than the small γ-Al_2O_3 crystallites and the oxidation rate decreases as reflected in the change in the slope of the thermogravimetric plot (Figure 6.19).

Similar formation of α-Al_2O_3 nodules has been observed on crystalline Al-Cu-Fe alloys. At first sight, this oxidation behavior is comparable to the situation encountered in aluminides. However, the temperature at which γ-Al_2O_3 is observed to transform into α-Al_2O_3 is too low for direct transformation to occur. Therefore, an indirect mechanism due to the presence of Cu and involving the formation of a spinel and its further decomposition according to the equation $CuAl_2O_4 \leftrightarrow CuO + Al_2O_3$ has been proposed, based on the detection of Cu(II) species at the surface of the α-Al_2O_3 nodules.

By contrast with the thick crystalline layers obtained on AlCuFe alloys, much thinner oxides are formed during oxidation of Cr-containing alloys. For example, when carried out below 500 °C in dry or ambient air conditions, oxidation of the orthorhombic $Al_{70}Cu_9Fe_{10.5}Cr_{10.5}$ approximant phase leads to the formation of islands of an amorphous oxide scale, with the kinetics depending on temperature rather than on partial pressure of water [57]. After 70 h at 500 °C, the thickness of the oxide does not exceed 20–25 nm. Moreover, oxidation in air at very high temperatures (1040–1080 °C) of Cr-containing alloys of the Al-Cr-Fe family leads to oxide scales only a few nanometers thick, even for oxidation times longer than 100 h, due to the formation of a chromium oxide layer below the aluminum oxide scale that provides efficient protection against oxidation [16].

Due to their good resistance to high-temperature oxidation, much effort to understand the oxidation mechanisms of complex metallic alloys has been devoted so far to alumina-forming alloys. However, oxidation of the complex β-phase of Al_3Mg_2 in dry air at 420 °C has also been studied [58]. As for conventional aluminum–magnesium alloys, formation of a porous nonadherent nanocrystalline MgO scale having a cauliflower-like morphology is initially observed. It is followed by the growth of $MgAl_2O_4$ spinel crystallites at the interface between the substrate alloy and the MgO top layer until a continuous $MgAl_2O_4$ thin layer with a constant thickness (~10 nm) and a columnar microstructure is obtained. For long oxidation time (>24 h), fragmentation and cracking of the scale layer due to accumulation of Mg vapor under the spinel layer leads to a strong acceleration in the oxidation process.

Finally, it must be pointed out that oxidation mechanisms can be strongly influenced by evaporation of some of the alloy constituents that can either influence the oxidation kinetics, as illustrated above for β-Al_3Mg_2, or even lead to the formation of a new compound, as has been mentioned in the case of AlPdMn [56].

6.5.2
Powders, Thin Films and Oxidation-Induced Phase Transformations

Due to the intrinsic brittleness of complex metallic alloys at low temperatures, many applications for these materials are foreseen in the form of surface coatings and thin films. Similarly to what happens in bulk alloys, high-temperature oxidation of Al-based thin films leads to the formation of a protective aluminum oxide and segregation of aluminum to the surface of the film [59–62]. However, in the case of thin films, the Al reservoir is not infinite and segregation to the surface can thus lead to Al depletion in the bulk of the film. Although this is a general phenomenon that can be observed in any metallic alloy, the region of thermodynamic stability is often narrow for complex metallic alloys and even small Al consumption can then easily foster the formation of a new phase. Such oxidation-induced phase transformations have been evidenced during oxidation of Al-Cu-Fe thin films in air at 400 °C [59]. It must be noted that the presence of oxygen in the film, which can be due to either oxygen diffusion during the oxidation process or oxygen trapping during the deposition of the film, can also lead to similar phase transformations [63, 64]. Therefore, extreme caution must be taken during thin film deposition and/or subsequent annealing treatments to avoid the harmful influence of oxygen incorporation.

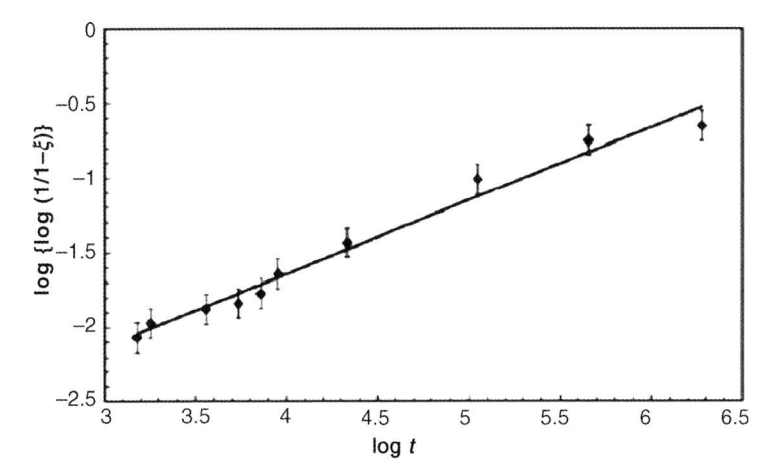

Figure 6.20 Avrami plot for oxidation of icosahedral $Al_{62.5}Cu_{25.3}Fe_{12.2}$ powders (particle size < 25 μm) at 500 °C in air. ξ the transformed volume fraction and t the oxidation time. Reprinted from reference 65, copyright © (2005), with permission of Elsevier.

The situation is very similar in the case of powders. For example, the behavior of icosahedral Al-Cu-Fe powders has been investigated by Dubois and coworkers during oxidation in air at 500 °C [65]. The formation of an amorphous aluminum oxide layer at the surface of the powder grains has been observed, due to long-range diffusion of aluminum. When the loss of aluminum within the grain is too high, the icosahedral phase is transformed into a cubic β-phase (CsCl type). An Avrami-law analysis of the process is shown in Figure 6.20, with ξ the transformed volume fraction and t the oxidation time. It can be seen that the slope of the Avrami plot is equal to 0.5, thus indicating that the transformation is diffusion controlled.

Several oxidation mechanisms have been proposed based on complete transformation of all crystallites of a particle or partial transformation localized near the grain surface or at the grain boundaries. It must be noted that the "reservoir effect" is strongly related to the size of the particle. When the grain size is higher than a critical value, depletion of aluminum is not observed and phase transformation does not occur. Here again, as atomized powders are often used in coating processing, care must be taken to avoid undesirable oxygen-induced phase transformation. However, it must be noted that surface oxidation of small particles can be turned as an advantage to avoid phase transformation when Al-based particles are used as reinforcement in Al-based composites [66]. Here, the oxide layer acts as a barrier for aluminum diffusion into the matrix, thus avoiding phase transformation of the particle.

Finally, it must be mentioned that small amounts of additional elements can strongly influence the oxidation behavior of CMA powders. For example, addition of Zn (Ce) in quasicrystalline $Al_{63}Cu_{25}Fe_{12}$ powders inhibits (promotes) air oxidation at 773 K [67]. In both cases, $FeAl_2O_4$ is formed, but the surface area of the oxidized powders decrease (increases) in the presence of Zn (Ce).

6.6
Conclusion

In spite of its huge importance for a number of applications, surface chemistry of CMAs is still poorly understood and most studies have been devoted so far to Al-based alloys.

The adsorption behavior of molecules other than oxygen or water suggests that a CMA surface can be more reactive than their pure metal constituents or conventional intermetallics, thus opening up promising perspectives in terms of catalysis.

Interaction with oxygen or water, either in the gas or liquid phases, leads to the formation of an aluminum (hydr-)oxide film whose chemical composition looks very similar to layers formed on conventional Al-based intermetallics in similar conditions. However, some major differences have been pointed out. In particular, specific electronic properties of this (hydr-)oxide film have been observed, which can prove very interesting in terms of corrosion protection. However, much more work is further required to understand the relationship between the observed properties of the (hydr-)oxide film and the complexity of the alloy substrate. Differences between CMAs and standard alloys of similar composition are certainly not to be found for the active dissolution processes. Specific oxides properties or surface structures resulting from corrosion processes after, for example, leaching out of the aluminum are much more likely to generate new CMA-specific industrial applications.

References

1 Thiel, P.A. (2004) *Prog. Surf. Sci.*, **75**, 69.
2 Dubois, J.-M., Kang, S.S., and Massiani, Y. (1993) *J. Non-Cryst. Solids*, **153–154**, 443.
3 Dubois, J.-M., Proner, A., Bucaille, B., Cathonnet, P., Dong, C., Richard, V., Pianelli, A., Massiani, Y., Ait-Yaazza, S., and Belin-Ferré, E. (1994) *Ann. Chim. France*, **19**, 3.
4 Chang, S.-L., Anderegg, J.W., and Thiel, P.A. (1996) *J. Non-Cryst. Solids*, **195**, 95.
5 Chang, S.-L., Chin, W.B., Zhang, C.-M., Jenks, C.J., and Thiel, P.A. (1995) *Surf. Sci.*, **337**, 135.
6 Chang, S.-L., Zhang, C.-M., Jenks, C.J., Anderegg, J.W., and Thiel, P.A. (1995) *Quasicrystals* (eds C. Janot and R. Mosseri) World Scientific, Singapore, p. 786.
7 Pinhero, P.J., Chang, S.-L., Anderegg, J.W., and Thiel, P.A. (1997) *Philos. Mag. B*, **75**, 271.
8 Popovic, D., Naumovic, D., Bovet, M., Koitzsch, C., Schlapbach, L., and Aebi, P. (2001) *Surf. Sci.*, **492**, 294.
9 Longchamp, J.-N., Burkardt, S., Erbudak, M., and Weisskopf, Y. (2007) *Phys. Rev. B*, **76**, 094203.
10 Dubot, P., Cenedese, P., and Gratias, D. (2003) *Phys. Rev. B*, **68**, 033403.
11 Pinhero, P.J., Anderegg, J.W., Sordelet, D.J., Besser, M.F., and Thiel, P.A. (1999) *Philos. Mag. B*, **79**, 91.
12 Pinhero, P.J., Anderegg, J.W., Sordelet, D.J., Lograsso, T.A., Delaney, D.W., and Thiel, P.A. (1999) *J. Mater. Res.*, **14**, 3185.
13 Rouxel, D. and Pigeat, P. (2006) *Prog. Surf. Sci.*, **81**, 488.
14 Gil-Gavatz, M., Rouxel, D., Pigeat, P., Weber, B., and Dubois, J.-M. (2000) *Philos. Mag. A*, **80**, 2083.
15 Rouxel, D., Gil-Gavatz, M., Pigeat, P., and Weber, B. (2005) *J. Non-Cryst. Solids*, **351**, 802.
16 Demange, V., Anderegg, J.W., Ghanbaja, J., Machizaud, F., Sordelet, D.J., Besser, M., Thiel, P.A., and Dubois, J.-M. (2001) *Appl. Surf. Sci.*, **173**, 327.

17 Demange, V., Machizaud, F., Dubois, J.-M., Anderegg, J.W., Thiel, P.A., and Sordelet, D.J. (2002) *J. Alloys Compd.*, **342**, 24.
18 Ko, J.S., Gellman, A.J., Lograsso, T.A., Jenks, C.J., and Thiel, P.A. (1999) *Surf. Sci.*, **423**, 243.
19 Smerdon, J.A., Wearing, L.H., Parle, J.K., Leung, L., Sharma, H.R., Ledieu, J., and McGrath, R. (2008) *Philos. Mag.*, **88**, 2073.
20 McGrath, R., Ledieu, J., and Diehl, R.D. (2004) *Prog. Surf. Sci.*, **75**, 131.
21 Ledieu, J., Dhanak, V.R., Diehl, R.D., Lograsso, T.A., Delaney, D.W., and McGrath, R. (2002) *Surf. Sci.*, **512**, 77.
22 Jenks, C.J., Lograsso, T.A., and Thiel, P.A. (1998) *J. Am. Chem. Soc.*, **120**, 12668.
23 Jenks, C.J. and Thiel, P.A. (1998) *J. Mol. Catal. A-Chem.*, **131**, 301.
24 McGrath, R., Ledieu, J., Cox, E.J., Haq, S., Diehl, R.D., Jenks, C.J., Fisher, I., Ross, A.R., and Lograsso, T.A. (2002) *J. Alloys Compd.*, **342**, 432.
25 Hoeft, J.T., Ledieu, J., Haq, S., Lograsso, T.A., Ross, A.R., and McGrath, R. (2006) *Philos. Mag.*, **86**, 869.
26 McGrath, R., Ledieu, J., Cox, E.J., Ferralis, N., and Diehl, R.D. (2004) *J. Non-Cryst. Solids*, **334–335**, 500.
27 Park, J.Y., Ogletree, D.F., Salmeron, M., Ribeiro, R.A., Canfield, P.C., Jenks, C.J., and Thiel, P.A. (2005) *Phys. Rev. B*, **71**, 144203.
28 Curtarolo, S., Setyawan, W., Ferralis, N., Diehl, R.D., and Cole, M.W. (2005) *Phys. Rev. Lett.*, **95**, 136104.
29 Ledieu, J., Muryn, C.A., Thornton, G., Diehl, R.D., Lograsso, T.A., Delaney, D.W., and McGrath, R. (2001) *Surf. Sci.*, **472**, 89.
30 Ledieu, J. and McGrath, R. (2003) *J. Phys.: Condens. Matter*, **15**, S3113.
31 Cox, E.J., Ledieu, J., Dhanak, V.R., Barrett, S.D., Jenks, C.J., Fisher, I., and McGrath, R. (2004) *Surf. Sci.*, **566–568**, 1200.
32 Krajci, M., and Hafner, J. (2008) *Surf. Sci.*, **602**, 182.
33 Fournée, V., Barthés-Labrousse, M.-G., and Dubois, J.-M. (2008) *Solid State Phenomena*, Trans Tech Publications, Switzerland, pp. 407–450.
34 Parle, J.K., Dhanak, V.R., Smerdon, J.A., and McGrath, R.(unpublished results).
35 Strohmeier, B.R. (1990) *Surf. Interface Anal.*, **15**, 51.
36 Pinhero, P.J., Sordelet, D.J., Anderegg, J.W., Brunet, P., Dubois, J.-M., and Thiel, P.A. (1999) *Mater. Res. Soc. Symp.*, **583**, 263.
37 Veys, D., Weisbecker, P., Domenichini, B., Weber, B., Fournée, V., and Dubois, J.-M. (2007) *J. Phys.: Condens. Matter*, **19**, 376207.
38 Simmons, G.W. and Beard, B.C. (1987) *J. Phys. Chem.*, **91**, 1143.
39 Veys, D. (2004) Evolution des proprietes physico-chimiques de surface des materiaux quasicristallins lors de sollicitations electrochimiques, PhD dissertation, Nancy, Institut National Polytechnique de Lorraine.
40 Dubois, J.-M., Fournée, V., and Belin-Ferré, E. (2004) Quasicrystals 2003-Preparation, Properties and Applications. in: MRS Symposium Proceedings, Warrendale, 805 (eds E. Belin-Ferré, M. Feuerbacher, Y. Ishii, and D.J. Sordelet), p. 287.
41 Rüdiger, A. and Köster, U. (1999) *J. Non-Cryst. Solids*, **250–252**, 898.
42 Rüdiger, A. and Köster, U. (2000) *Mater. Sci. Eng.*, **294–296**, 890.
43 Huttunen-Saarivirta, E. and Tiainen, T. (2004) *Mater. Chem. Phys.*, **85**, 383.
44 Massiani, Y., Ait Yaazza, S., Crousier, J.P., and Dubois, J.-M. (1993) *J. Non-Cryst. Solids*, **159**, 92.
45 Veys, D., Rapin, C., Li, X., Aranda, L., Fournée, V., and Dubois, J.-M. (2004) *J. Non-Cryst. Solids*, **347**, 1.
46 Balbyshev, V.N., King, D.J., Khramov, A.N., Kasten, L.S., and Donley, M.S. (2003) *Prog. Org. Coat.*, **47**, 357.
47 Balbyshev, V.N., King, D.J., Khramov, A.N., Kasten, L.S., and Donley, M.S. (2004) *Thin Solid Films*, **447**, 55.
48 Ura-Binczyk, E., Hauert, R., Lewandowska, M., Kurzydlowski, K.J., and Schmutz, P. (2010) *Electrochim. Acta*.
49 Ura-Binczyk, E., Homazeva, N., Ulrich, A., Lewandowska, M., Kurzydlowski, K.J., and Schmutz, P. (2010) *Corros. Sci.*
50 Ura-Binczyk, E., Quach-Vu, N.G., Lewandowska, M., Kurzydlowski, K.J., and Schmutz, P. (2010) *J. Electrochem. Soc.*
51 Suter, T. and Bohni, H. (1997) *Electrochim. Acta*, **42**, 3275.
52 Suter, T. and Bohni, H. (2005) (ed. F.M. Marcus) Los Angeles.

53 Webb, E.G., Suter, T., and Alkire, R.C. (2001) *J. Electrochem. Soc.*, **148**, B186.
54 Beni, A., DeRose, J., Bauer, B., Gille, P., and Schmutz, P. (2010) *Electrochim. Acta*.
55 Wehner, B.I. and Köster, U. (2000) *Oxid. Met.*, **54**, 445.
56 Wehner, B.I., Köster, U., Rüdiger, A., Pieper, C., and Sordelet, D.J. (2000) *Mater. Sci. Eng.*, **294–296**, 830.
57 Bonhomme, G., LeMieux, M., Weisbecker, P., Tsukruk, V.V., and Dubois, J.-M. (2004) *J. Non-Cryst. Solids*, **334–335**, 532.
58 de Noirfontaine, M.-N., Baldinozzi, G., Barthés-Labrousse, M.-G., Kusinski, J., Boëmare, G., Herinx, M., and Feuerbacher, M.(in preparation).
59 Haugeneder, A., Eisenhammer, T., Mahr, A., Schneider, J., and Wendel, M. (1997) *Thin Solid Films*, **307**, 120.
60 Kong, J., Zhou, C., Gong, S., and Xu, H. (2003) *Surf. Coat. Tech.*, **165**, 281.
61 Zhou, C., Cai, F., Xu, H., and Gong, S. (2004) *Mater. Sci. Eng. A*, **386**, 362.
62 Zhou, C., Cai, R., Gong, S., and Xu, H. (2006) *Surf. Coat. Tech.*, **201**, 1718–1723.
63 Kang, S.S. and Dubois, J.-M. (1995) *J. Mater. Res.*, **10**, 1071–1074.
64 Bonasso, N., Pigeat, P., Rouxel, D., and Weber, B. (2002) *Thin Solid Films*, **409**, 165.
65 Weisbecker, P., Bonhomme, G., Bott, G., and Dubois, J.-M. (2005) *J. Non-Cryst. Solids*, **351**, 1630.
66 Kenzari, S., Weisbecker, P., Geandier, G., Fournée, V., and Dubois, J.-M. (2006) *Philos. Mag.*, **86**, 287.
67 Yamasaki, M. and Tsai, A.P. (2002) *J. Alloys Compd.*, **342**, 473.

7
Mechanical Engineering Properties of CMAs

Jürgen Eckert, Sergio Scudino, Mihai Stoica, Samuel Kenzari, and Muriel Sales

7.1
Introduction

Within recent years, complex metallic alloys (CMAs), intermetallic compounds with giant unit cells, comprising up to more than a thousand atoms per unit cell [1], have attracted much attention ranging from scientific curiosity about their complex structure, physical and mechanical properties to technological aspects of preparation and potential applications [1–10]. In particular, CMAs exhibit several attractive properties for engineering applications, such as high strength-to-weight ratio, good oxidation resistance and high-temperature strength [7–10]. However, as with most intermetallic phases, one major drawback for their use in engineering applications is their limited plastic deformability at room temperature [9, 10]. In fact, although the principal loads should be borne at intermediate to elevated temperatures, brittleness or the lack of toughness of CMAs would impede manufacturing, handling and shipping. In this chapter, selected examples of the mechanical deformation behavior for single- as well as for multiphase intermetallic compounds with different structural complexity ranging from Laves phases, to complex metallic alloys and quasicrystals (QCs) are presented, revealing that several approaches, such as proper variation of the chemical composition, grain refinement to the nanometer regime and the development of a heterogeneous microstructure combining intermetallic particles with a ductile matrix phase, can be employed to improve the room-temperature ductility of CMAs and to produce materials with promising properties in terms of strength as well as of room-temperature ductility.

This chapter deals with several aspects of the mechanical properties of CMAs in view of their use in mechanical engineering applications. We start in the next section with the basics of structural and mechanical properties of intermetallics, either in the form of single-phase materials, or as multiphase materials. We then turn to their use as reinforcing components of engineering parts. We later address more specifically aluminum-based metal–matrix composites and complete the chapter with surface mechanical properties of CMAs, with a specific look at composite materials with lowered adhesion and friction properties.

Complex Metallic Alloys: Fundamentals and Applications
Edited by Jean-Marie Dubois and Esther Belin-Ferré
Copyright © 2011 WILEY-VCH Verlag GmbH & Co. KGaA, Weinheim
ISBN: 978-3-527-32523-8

7.2
Structure and Mechanical Properties of CMAs

7.2.1
Single-Phase Intermetallics

A typical example of intermetallics with promising high-temperature properties and brittleness at room temperature are Laves phases, the largest class of intermetallic compounds (for a review see references [11–13]). Laves phases are topologically close-packed (TCP) phases with ideal composition AB_2 (B is the small atom and A is the large atom) that crystallize in the cubic C15 structure or the hexagonal polymorphs C14 and C36 [14, 15]. Laves phases display several interesting properties, such as good corrosion resistance and high-temperature (HT) strength [11–13]. However, they are brittle at room temperature in single-phase form (RT) and have high ductile–brittle transition temperatures [11, 12], which limits their use in engineering applications. The RT brittleness of Laves phases is most likely due to their complex TCP structure and the resistance to dislocation motion [12]. Plastic deformation in Laves phases occurs only at temperatures above about two-thirds of the melting point [11] by the "synchroshear" mechanism, which involves the simultaneous motion of atoms on adjacent atomic planes [16, 17]. This mechanism is difficult at low temperatures [18], explaining the brittle behavior observed at room temperature.

Different approaches have been proposed for improving the low-temperature ductility of Laves phases. For example, Thoma et al. [19] observed for the C15 $HfCo_2$ Laves phase that the hardness values display a maximum at the stoichiometric composition (66–67 at.% Co), as shown in Figure 7.1(a), and they decrease moving away from stoichiometry. Similarly, the brittleness of the material (defined as the ratio of hardness/fracture toughness) decreases with increasing Co content from the stoichiometric composition (Figure 7.1(b)). Most likely, the substitution of Hf by the smaller Co atoms enhances the ability to accommodate deformation assisting the

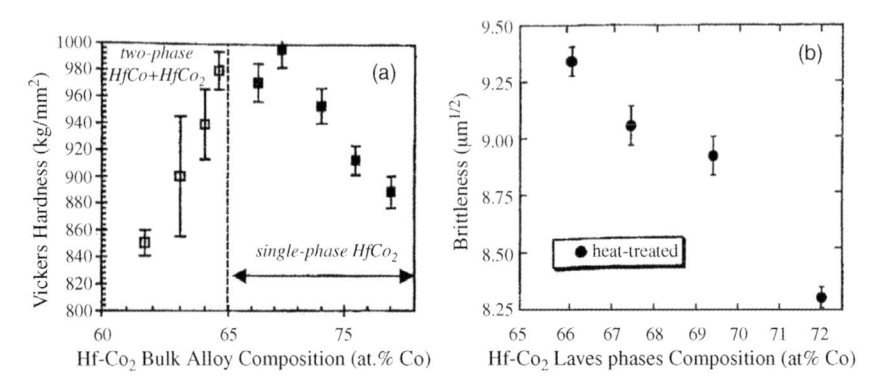

Figure 7.1 (a) Hardness values as a function of composition for the $HfCo_2$ alloys and (b) brittleness (defined as the ratio of hardness/fracture toughness) as determined from microhardness indentation testing (from reference [19]).

synchroshear mechanism [19]. Similar results have been observed for pseudobinary ZrCr$_2$-NbCr$_2$ Laves phase [20], which shows that alloying and off-stoichiometry introduce large amounts of atomic free volume, thus enhancing the RT fracture toughness of the material.

A different deformation mechanism has been reported for the C15 HfV$_2$-based Laves phase [21], which plastically deforms at room temperature through extensive twinning. The addition of Nb (intermediate in atomic size between Hf and V) to the HfV$_2$ Laves phase contributes to the ease of twinning in these alloys through the substitution of the ternary element on the Hf sublattice, which increases the free volume and leads to easier mechanical twinning via the synchroshear mechanism [1, 2, 21, 22]. Another low-temperature deformation mechanism has been observed for C36 Fe$_2$Zr [23, 24], which shows room-temperature ductility. During compression at room temperature the Fe$_2$Zr Laves phase undergoes a partial stress-induced phase transformation from the dihexagonal C36 to the cubic C15 phase, as revealed by X-ray diffraction and transmission electron microscopy [23, 24]. Although the latter examples are very promising for improving the RT ductility of intermetallic compounds, additional investigations on this aspect are needed to clarify whether such mechanisms for improving the low-temperature toughness of Laves phases can be used for different intermetallics or whether they are limited to only a few alloy systems.

Similarly to the Laves phases, CMAs display promising high-temperature mechanical properties but they lack room-temperature plastic deformation. A typical example is the β-Al$_3$Mg$_2$ phase (space group $Fd\bar{3}m$), an intermetallic compound with a giant unit cell ($a_0 = 2.824$ nm) containing about 1168 atoms [2], which has been extensively investigated with particular attention to its structure as well as to its physical and mechanical properties [2–4, 9]. Single-phase β-Al$_3$Mg$_2$ displays very interesting properties [9]. The material exhibits ductile behavior down to temperatures of about 500 K (Figure 7.2) [9]. At this temperature an upper yield stress of 780 MPa was observed, which is a very high value compared to commercial Al-Mg alloys [9]. In addition, the material displays low density (about 2.25 g/cm^3) [2] and high-temperature strength (∼300 MPa at 573 K) [9], thus making it an attractive candidate for structural applications. However, single-phase β-Al$_3$Mg$_2$ does not show RT plastic deformation.

Another interesting example of mechanical properties of CMAs is represented by the Al$_{13}$Co$_4$ orthorhombic phase (space group Pmn21) with lattice parameters $a = 8.2$ Å, $b = 12.3$ Å and $c = 14.5$ Å and 102 atoms in the unit cell [10]. The material displays a maximum yield stress of 790 MPa (Figure 7.3) [10] combined with a fracture strain of about 3% at 873 K, which can be considered the lower-temperature limit of the ductile regime. The stress then monotonically decreases with increasing temperature reaching a value of about 50 MPa at 1173 K.

Similarly to simple intermetallics, such as Laves phases, and complex metallic alloys, quasicrystals and related crystalline phases are very brittle at room and intermediate temperatures. As an example, the mechanical properties of crystalline and quasicrystalline phases in the ternary Al-Cu-Fe system are summarized in Table 7.1 [25]. Similar to other intermetallic phases, QCs display high brittle-to-ductile transition temperatures (BDTT), generally occurring at temperatures above

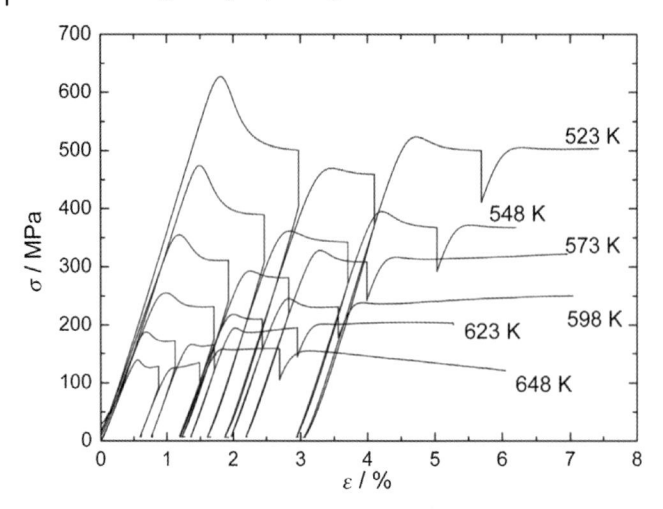

Figure 7.2 Stress–strain curves for β-AlMg at temperatures between 523 and 648 K and at a strain rate of $10^{-4}\,\mathrm{s}^{-1}$ (after reference [9]).

0.8 T_m (T_m = melting point) [26]. The Vickers microhardness of quasicrystalline Al–Cu–Fe has been correlated with the BDTT [27]. The values of the microhardness as a function of temperature (Figure 7.4) [27] are remarkably high and almost constant at ambient and intermediate temperatures corresponding to the brittle regime. Above 600 K a clear decrease of hardness with increasing temperature can be observed. This decrease is monotonous and extends over more than 200 K. The BDTT of the icosahedral Al-Cu-Fe lies in the interval between 723 and 760 K (indicated in Figure 7.4 as a gray area) and it coincides with the middle part of the temperature range associated with the sharp decrease of hardness.

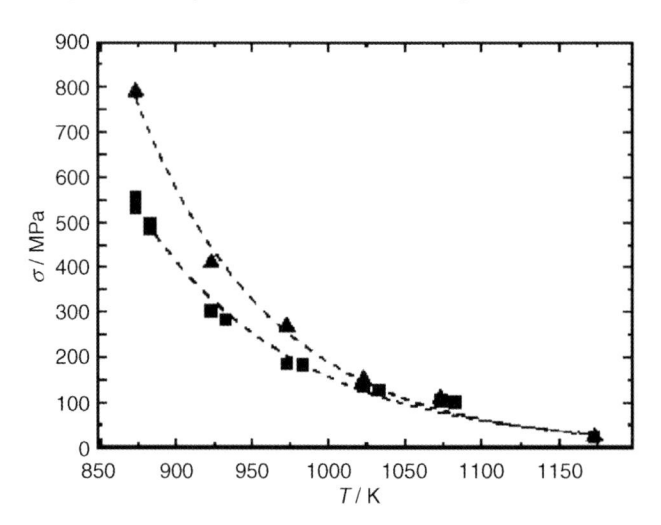

Figure 7.3 Upper yield stresses (triangles) and steady-state flow stresses (squares) at different temperatures and a strain rate of $10^{-5}\,\mathrm{s}^{-1}$ for the $Al_{13}Co_4$ orthorhombic phase (from reference [10]).

7.2 Structure and Mechanical Properties of CMAs

Table 7.1 Compositions and mechanical properties of crystalline and quasicrystalline phases in the ternary Al–Cu–Fe alloy system [25].

Composition	Phases	Structure	Hardness (H_V 0.25 N)	Toughness K_{IC} MPa $(m)^{1/2}$	Brittle-to-ductile temperature, °C
$Al_{72}Fe_{28}$	Al_5Fe_2	Orthorhombic	1100	1.05	~550
$Al_{75.5}Fe_{24.5}$	$Al_{13}Fe_4$	Monoclinic	1070	1.03	~750
$Al_{67.6}Cu_{32.4}$	Al_2Cu	Tetragonal	595	1.14	~400
$Al_{48}Fe_{52}$	AlFe(Cu)	Cubic (CsCl type)	775	>20	~430
$Al_{63}Cu_{25}Fe_{12}$	ψ-AlCuFe	Icosahedral	1000	1.64	~650

An alternative way to overcome the intrinsic RT brittleness of single-phase simple and complex intermetallics may be grain refinement of the structure down to the nanometer regime. There is an increasing amount of experimental evidence that indicates that ultrafine-grained (UFG; average grain size in the range 100–1000 nm and nanocrystalline materials (average grain size <100 nm) can lead to properties that are dramatically improved with respect to conventional grain-size (grain size >1 μm) polycrystalline or single-crystal materials of the same chemical composition [28, 29]. This behavior is schematically illustrated in Figure 7.5, which shows the variation of flow stress as a function of grain size from the μm to the nm regime [28]. The strength of conventional μm-sized and UFG materials is a function of the grain size. For these materials, the strengthening with grain refinement has been rationalized on the basis of the empirical Hall–Petch equation [30] ($\sigma_0 = \sigma_i + kd^{-1/2}$, where σ_0 is the yield stress, σ_i is the friction stress opposing dislocation motion, k is a constant, and d is the grain size). This empirical observation has been explained by the pile-up of dislocations at grain boundaries [28, 29]. As the microstructure is refined from μm and UFG regime into the nanocrystalline regime, this process breaks down and the flow stress versus grain size relationship departs markedly from

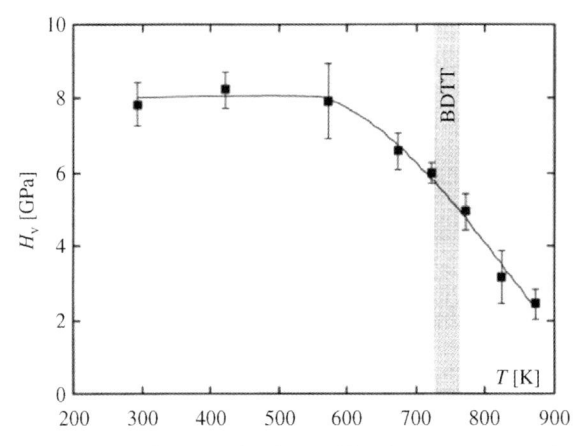

Figure 7.4 Vickers microhardness of the $Al_{63.5}Cu_{24}Fe_{12.5}$ icosahedral quasicrystal as a function of temperature (from reference [27]).

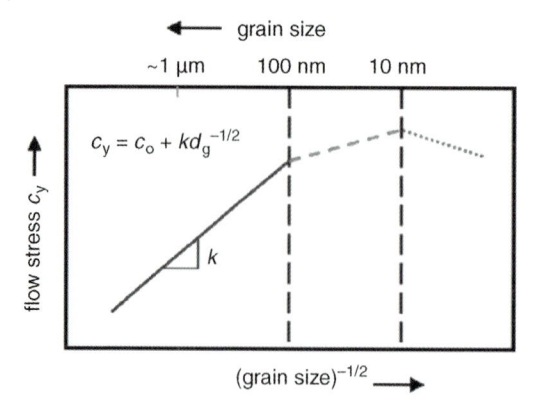

Figure 7.5 Schematic representation of the variation of yield stress as a function of grain size in μm-sized, UFG and nanocrystaline metals and alloys (from reference [28]).

that seen at higher grain sizes (Figure 7.5). With further grain refinement, the yield stress peaks in many cases at an average grain size value on the order of about 10 nm and a negative Hall–Petch slope, that is decreasing strength with decreasing grain size in the nanoscale grain size regime, is observed [28, 29]. A similar behavior has also been observed for the hardness [29].

Nanostructured single- or multiphase intermetallics can be prepared directly by solid-state techniques, such as ball milling of CMA precursors, or by controlled heat treatment of the milled powders. As a typical example, Figure 7.6 displays *in situ* X-ray diffraction experiments as a function of temperature for a mechanically milled single-phase β-Al_3Mg_2 CMA, revealing the formation of different nanocrystalline phases during heating [31]. Similar results have been achieved by mechanical alloying of elemental powder mixtures with the same composition [32, 33]. As already mentioned, lightweight nanostructured materials with composition $Al_{60}Mg_{40}$ (corresponding to the equilibrium β-Al_3Mg_2 phase) are of extreme interest for possible engineering applications due to their attractive combination

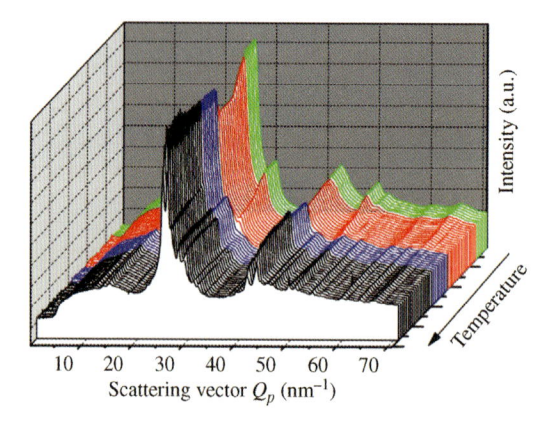

Figure 7.6 X-ray diffraction patterns for mechanically milled β-Al_3Mg_2 as a function of temperature (from reference [31]).

of low density and high strength [2, 9]. Such materials can be produced by consolidation of milled powders to achieve dense nanostructured specimens. For example, nanostructured $Al_{60}Mg_{40}$ material produced by mechanical alloying has been consolidated at different temperatures by spark plasma sintering (SPS) [34]. The choice of SPS as a consolidation technique was done because by this method sintering can be carried out at relatively low temperatures for a shorter time than in conventional sintering processes [35]. Therefore, the SPS process shows a large potential for achieving fast and full densification of nanostructured materials with limited grain growth [35].

Figures 7.7(a)–(e) show the typical microstructure of the powder samples consolidated by SPS and investigated by optical microscopy. The porosity characterizing the samples remarkably decreases with increasing sintering temperature from 473 to

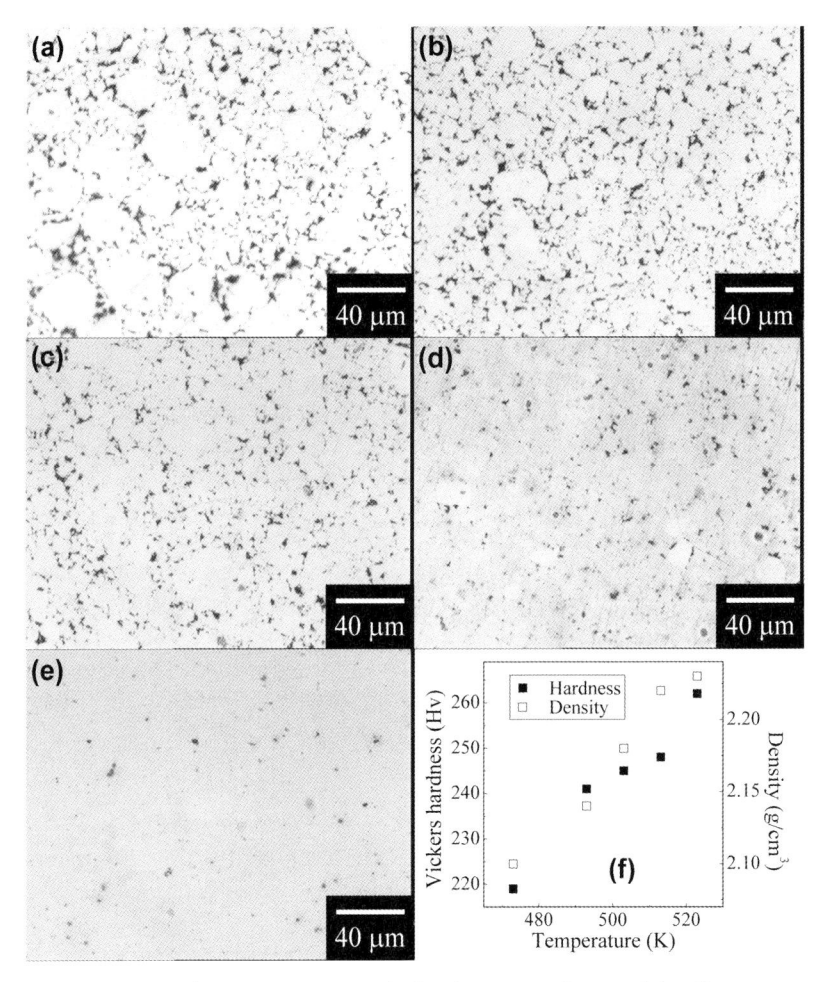

Figure 7.7 Optical microscopy micrographs for $Al_{60}Mg_{40}$ powder consolidated by SPS at (a) 463, (b) 473, (c) 493, (d) 513 and (e) 523 K. (f) hardness and density of the samples consolidated by SPS as a function of the sintering temperature (from reference [34]).

523 K. This is corroborated by an increase of the relative density with increasing sintering temperature from 94% for the sample sintered at 473 K to 98% for the sample sintered at 523 K (Figure 7.3(f)). XRD investigations [32, 33] reveal that the sample sintered at 473 and 493 K consists of a nanocrystalline Al(Mg) solid solution together with a small amount of γ-$Al_{12}Mg_{17}$ phase. On increasing the sintering temperature to 513 and 523 K, the samples display the formation and growth of the nanocrystalline hexagonal β'-phase [32, 33]. Hardness measurements reveal encouraging mechanical properties. The Vickers hardness (H_v) increases with increasing sintering temperature (Figure 7.3(f)) from 220 for the sample sintered at 473 K to 260 for the sample sintered at 523 K. Using the well-known relation $H_v = 3\sigma_y$ [36], this gives a yield strength σ_y ranging between 750 and 830 MPa, which is in good agreement with the values reported for single- and polycrystalline β-Al_3Mg_2 [9].

7.2.2
Multiphase Intermetallics

Although the approaches mentioned in the previous section are quite promising, the method that shows the largest potential for improving the ductility of intermetallics at low temperatures is the production of two- or multiphase microstructures consisting of a soft metallic matrix with second-phase intermetallic particles. The intermetallic phase acts as a strength-bearing component, while the metallic matrix supplies ductility. This type of microstructure can be achieved directly by solidification of the melt for near-eutectic compositions [37, 38]. Along this line, promising results have been achieved for composites containing Laves phases. For example, interesting properties have been reported for Cr-Cr_2Nb two-phase alloys [37]. The composites exhibit high strength and remarkable plastic deformation at all tested temperatures (Figure 7.8), which are comparable to those of conventional nickel-based superalloys [37]. In particular, the materials show high RT strength in the range of 800–1200 MPa combined with a ductility between 5 and 11%, which is almost unchanged up to 773 K. Extended room-temperature ductility has also been observed for Ta-HfV_2 two-phase alloys consisting of Laves phase and bcc solid solution [38]. The RT mechanical properties of the material can be tuned within a wide range of strength and ductility as a function of the volume fraction of the bcc phase (Figure 7.9) [38].

More recently, nanostructured or ultrafine eutectic–dendrite composites containing Laves phases, exhibiting a good combination of strength and ductility at room temperature, have been produced in the binary Fe-Zr [39], Ni-Zr [40] and Fe-Nb systems [41], which further indicates the validity of this approach for the improvement of the RT ductility of the Laves phases. Nanocrystalline and UFG eutectic structures are generally produced by copper mold casting, which is characterized by cooling rates on the order of 10–100 K/s [42]. As typical examples, Figure 7.10 shows the effect of composition on the nanostructure or ultrafine eutectic structure of $Fe_{100-x}Nb_x$ alloys with $x = 8, 10, 12$ [41]. For $Fe_{90}Nb_{10}$, with near-eutectic composition, the microstructure consists of alternating fine lamellae within micrometer-scale eutectic colonies (average size 30–40 mm) without any primary dendrites [Figure 7.10(b)]. The average eutectic lamellar spacing is

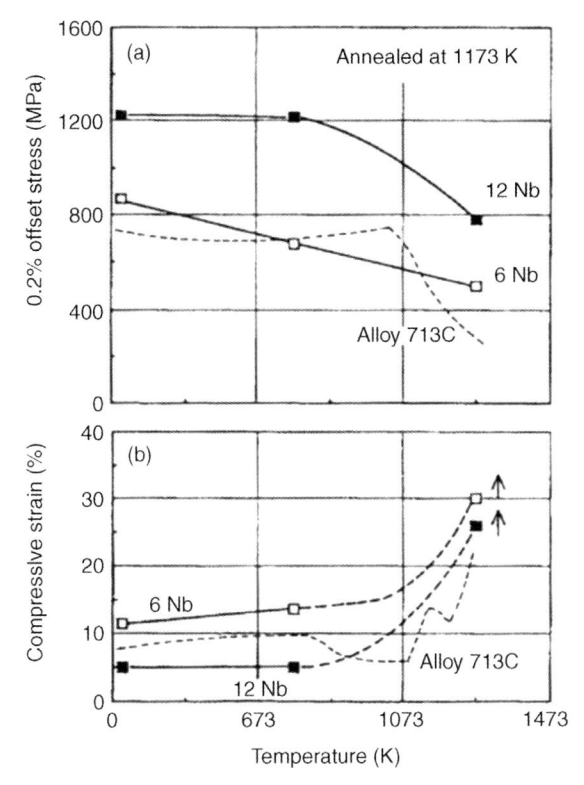

Figure 7.8 Yield strength and compressive strain of annealed Cr-6%Nb and Cr-12%Nb alloys as a function of temperature. For comparison, typical tensile properties for as-cast IN 713C are also shown (after reference [37]).

150–200 nm. On the other hand, the off-eutectic alloys ($Fe_{92}Nb_8$, $Fe_{88}Nb_{12}$) exhibit a composite microstructure consisting of primary dendrites that are homogeneously dispersed in a fine eutectic matrix, as shown in Figures 7.10(a) and (c). The XRD patterns (Figure 7.10(d)) confirm that the eutectic structure consists of the Fe_2Nb hexagonal Laves phase and an α-Fe solid solution. The microstructure has a remarkable impact on the RT mechanical properties, as shown in Figure 7.11(a). The $Fe_{90}Nb_{10}$ alloy with fully eutectic structure exhibits the highest yield strength ($\sigma_y \sim 1.1$ GPa) and ultimate fracture strength ($\sigma_{max} \sim 1.8$ GPa) but only a limited plastic strain of ~4%. The eutectic composite containing primary Fe_2Nb Laves phase particles ($Fe_{88}Nb_{12}$) also displays an ultimate fracture strength of $\sigma_{max} \sim 1.8$ GPa and a slightly improved plastic deformation ($\varepsilon_p \sim 6.5\%$). On the other hand, the $Fe_{92}Nb_8$ alloy containing ductile α-Fe solid solution exhibits lower yield strength ($\sigma_y \sim 0.75$ GPa) as well as ultimate fracture strength ($\sigma_{max} \sim 1.4$ GPa), but significantly larger plasticity ($\varepsilon_p \sim 13\%$) with respect to the $Fe_{90}Nb_{10}$ and $Fe_{88}Nb_{12}$ alloys [41]. Similar results have been reported for nanoeutectic $Fe_{100-x}Zr_x$ alloys with $x = 6, 8, 10$ and 12, as shown in Figure 7.11(b) [39].

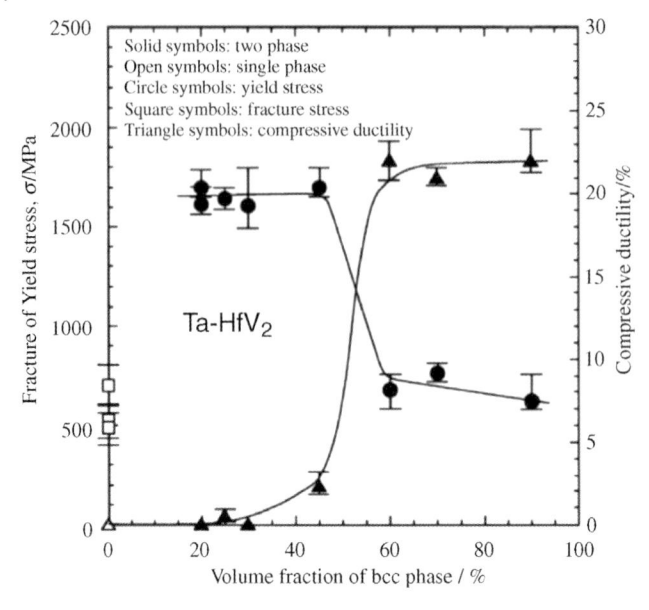

Figure 7.9 Room-temperature yield stress and compressive ductility of Ta-HfV$_2$ two-phase alloys consisting of Laves phase and bcc solid solution plotted against the total volume fraction of the bcc phase in each alloy (from reference [38]).

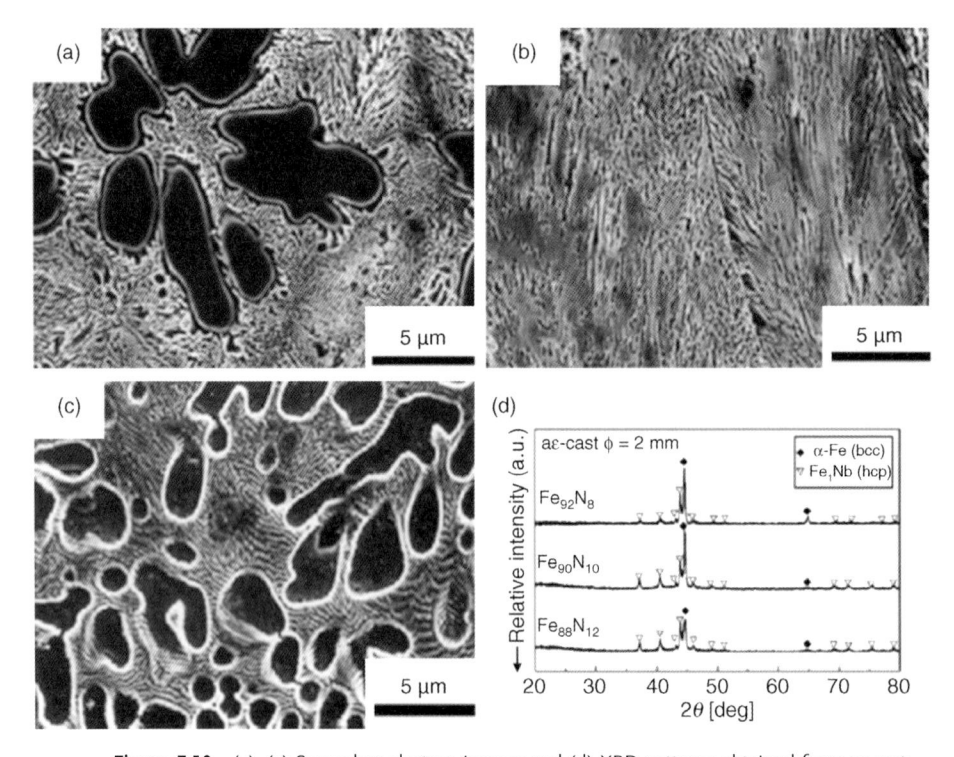

Figure 7.10 (a)–(c) Secondary electron images and (d) XRD patterns obtained from as-cast Fe$_{100-x}$Nb$_x$ ($x = 8$, 10, 12) alloys: (a) $x = 8$, (b) $x = 10$, (c) $x = 12$ (from reference [41]).

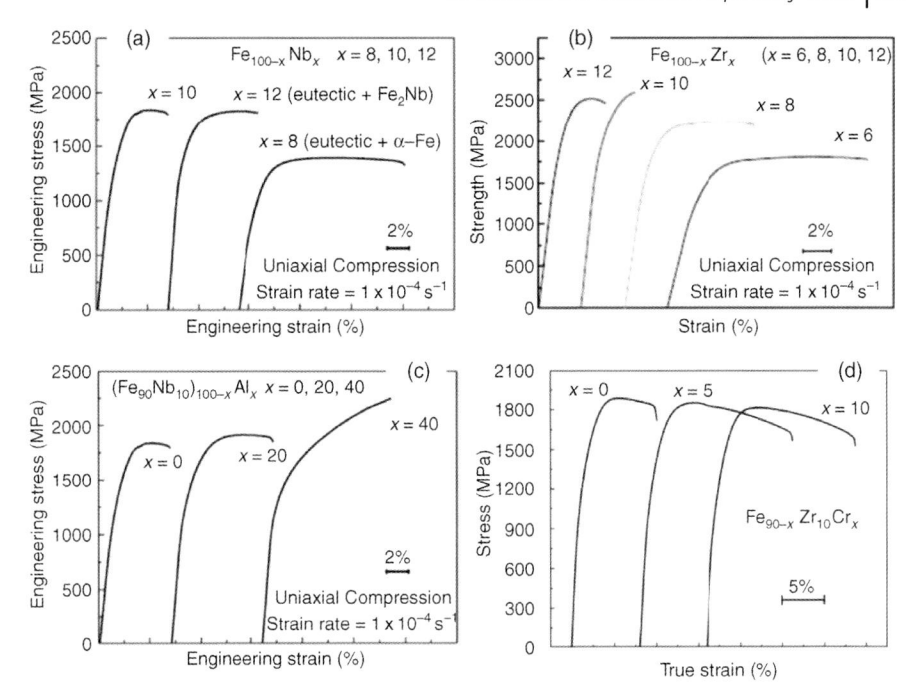

Figure 7.11 Compressive stress–strain curves for as-cast (a) $Fe_{100-x}Nb_x$ alloys with $x = 8$, 10 and 12 (from reference [41]), (b) $Fe_{100-x}Zr_x$ alloys with $x = 6$, 8, 10 and 12 (from reference [39]), (c) $(Fe_{90}Nb_{10})_{100-x}Al_x$ ($x = 0$, 20 and 40) (from reference [41]) and (d) $Fe_{90-x}Zr_{10}Cr_x$ samples with $x = 0$, 5 and 10 (from reference [43]).

In these materials, the application of a relatively high cooling rate (10–100 K/s) limits grain growth, consequently promoting the formation of nanocrystalline or UFG microstructures through the reduction of the eutectic lamellar spacing, which results in a marked improvement of the strength [39–41]. Further improvements can be achieved by the introduction of additional elements to the binary alloys. Figure 7.11 (c) shows the room-temperature stress–strain curves of $(Fe_{90}Nb_{10})_{100-x}Al_x$ ($x = 0$, 20 and 40) ternary alloys [41]. The addition of Al to binary $Fe_{90}Nb_{10}$ improves the compressive yield strength, the fracture strength and the plasticity. Also, the alloy with $x = 40$ displays marked work hardening, which is virtually absent in the other samples. With addition of Al, the lamellar eutectic structure of binary $Fe_{90}Nb_{10}$ [Figure 7.10(b)] drastically changes into a bimodal composite-type structure that consists of primary micrometer-scale dendrites and an ultrafine-scale lamellar eutectic [41]. The addition of Al promotes the formation of the soft dendritic α-Fe (Al) phase from 0 vol.% for binary $Fe_{90}Nb_{10}$ ($x = 0$) to 45 vol.% ($x = 20$) and 73 vol.% ($x = 40$), explaining the improved plastic deformation. The lamellar spacing is reduced with increasing Al content from 150–200 nm ($x = 0$) to 115–155 nm ($x = 20$) and 50–100 nm ($x = 40$). This suggests that the additional element inhibits the growth of the eutectic, favoring the refinement of the lamellar spacing and, therefore, increasing the strength of the material [41].

Promising results in terms of RT plastic deformation have also been reported for $Fe_{90-x}Zr_{10}Cr_x$ alloys ($x = 0$, 5 and 10) with a microstructure consisting of Laves phase particles with dimensions of about 2–4 μm embedded in a eutectic matrix made of lamellae of α-Fe and Laves phases with thickness ranging between 50 and 200 nm [43]. Two polymorphs of the Laves phase have been observed: the cubic C15 and the hexagonal C14/C36 phases. Figure 7.11(d) shows the effect of Cr on the RT mechanical properties of the $Fe_{90-x}Zr_{10}Cr_x$ alloys. The addition of Cr increases the plastic strain from 9% for the binary alloy up to 17% for the sample with 10 at.% Cr. The increase of plasticity does not occur to the expenses of the strength of the Cr-containing samples, which is reduced only by about 70 MPa with respect to the binary alloy (∼1900 MPa). The increased plastic deformation is linked to the specific structural features of the Laves phases [43]. The addition of Cr stabilizes the hexagonal Laves phase at the expense of the cubic polymorph, which decreases from 17 vol.% for the alloy with $x = 0$ to 2 vol.% for the sample with $x = 10$. During deformation of the samples containing Cr, the hexagonal C14/C36 phases do not show any dislocation activity and, most likely, is able to retard the propagation of cracks formed within the ferrite channels [43]. On the other hand, in the deformed binary alloy, the C15 Laves phase displays dislocation bands, which lead to the fracture within the Laves phase structure. This might represent an easy path for crack propagation through the sample, leading to early fracture and explaining the reduced RT plastic deformation of the alloy containing a large volume fraction of C15 phase [43].

All the multiphase microstructures presented above have been produced directly by solidification from the melt. However, multiphase microstructures can also be achieved by additional heat treatment of the solidified alloys to create or optimize the desired structure, such as in the case of (partial) devitrification of metallic glasses [29].

Among the different metallic glasses, Fe-based amorphous alloys have attracted much attention because of their excellent soft magnetic properties, high mechanical strength and good corrosion resistance [44]. Much research has endeavored to develop Fe-based bulk metallic glasses (BMGs) with improved glass-forming ability (GFA) [44]. Since 1995, several Fe-based multicomponent BMG systems have been reported. More recently, studies on ternary Fe-based BMGs indicated that a broad glass-forming range (GFR) can be found in relatively simple systems, especially in Fe-M-B (M = III-V group transition metal or rare-earth elements) alloys [44]. These BMGs possess excellent soft magnetic properties and good GFA. Interestingly, this type of BMGs develops CMA structures upon devitrification of the glass, which may show improved magnetic properties over the corresponding amorphous precursors [44].

Like liquids, glassy or amorphous materials possess a disordered structure lacking long-range order. That is, in a glass, there is no regular arrangement resulting from the distribution over long distances of a repeating atomic arrangement, as is characteristic for a crystal. There is only evidence of short-range order (SRO), which corresponds to the mutual arrangement of the nearest neighbors to a given atom and varies according to the atomic site considered [45]. In most cases, the SRO enhances the stability of a glass, as was observed for Pd-based metallic glasses [44] and for

Fe-based BMGs containing Nb [46]. The addition of Nb to the $Fe_{80}B_{20}$ glass enhances the GFA through the stabilization of the supercooled liquid [46]. The stable crystalline phases, which are expected to form during the devitrification of the binary $Fe_{80}B_{20}$ glass, are α-Fe and Fe_2B. The addition of Nb changes the crystallization behavior inducing the formation of the *fcc* $Fe_{23}B_6$ CMA with a large lattice parameter of about 1.2 nm and 96 atoms in the unit cell [46]. Its formation requires high thermal activation, explaining why the amorphous precursor shows such a good thermal stability and relatively high GFA. The $Fe_{23}B_6$ phase is metastable and transforms into the equilibrium phases at high temperature.

The Fe-B-Nb glass displays a kind of short-range order consisting of a network-like structure in which trigonal prisms consisting of Fe and B are connected to each other in edge- and plane-shared configuration modes through Nb glue atoms [46, 47]. The local triangular unit is quite similar to the Fe_3B crystal, as demonstrated by Matsubara *et al.* [48]. In the Fe_3B crystal, the triangular prisms are connected in two different ways, as schematically shown in Figure 7.12. One third of the Fe atoms is connected by sharing Fe at the vertex of the prism and the others by sharing the edge, whereas the Nb atoms occupy the vertices in a random manner. Beside Nb, other large early transition metal (ETM) elements, such as Zr or Cr, can be used to build this structure.

Poon *et al.* [49] considered the percentage and the radius of the atoms that are present in such Fe-based BMGs (atom size–composition relationship). The LTM-ETM-metalloid glasses (LTM = late transmission metal) contain midsize atoms as the majority component (Fe or Co, 60–70 at.%), small atoms as the next-majority component (the metalloids) and large-size atoms as the minority component (e.g., Nb < 10 at.%), leading to the "majority atom–small atom–large atom" (MSL) class. In these alloys the heat of mixing is negative. The percentage of small (S) atoms is

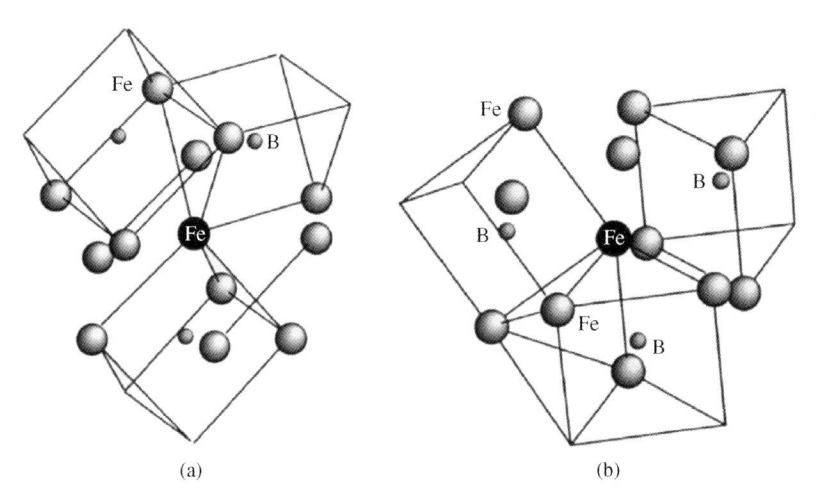

Figure 7.12 Schematic illustration of the atomic arrangements for (a) the vertex-sharing and (b) the edge-sharing triangular prisms in the Fe_3B crystal. Only atoms in the near-neighbor region around Fe at the center (solid circle) are shown (from reference [49]).

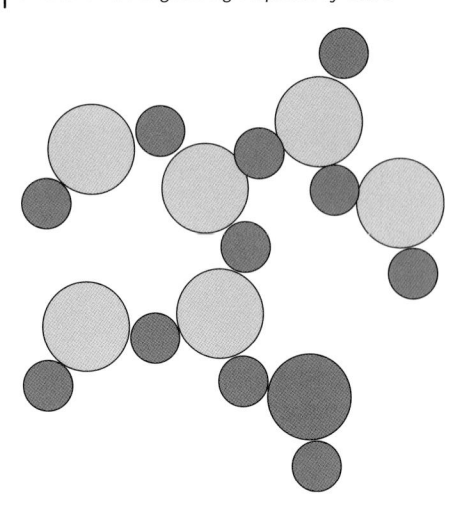

Figure 7.13 Schematic illustration of the atomistic network/backbone formed by the large atoms and small atoms in the MSL class of metallic glasses (from reference [49]).

around 20 at.%, while the large (L) atoms comprise less than 10 at.%. Within the structure, the L and S atoms may form a strong L–S percolating network or reinforced "backbone" structure, as illustrated in Figure 7.13 [49]. Presumably, the backbone structure can enhance the stability of the undercooled melt, which further suppresses crystallization [49]. However, if the concentration of the L atoms is significantly higher than 10 at.%, there will be an increasing tendency for the L atoms to cluster, which will effectively reduce the interaction between the L atoms and the M and S atoms. Thus, the optimum content of large atoms for the formation of BMGs of the MSL class is near 10 at.%.

It has been recently shown that the magnetic properties of the $Fe_{66}Nb_4B_{30}$ BMG can be improved by controlled glass devitrification [50]. The $Fe_{66}Nb_4B_{30}$ glass crystallizes through two steps: the first at 876 K and the second at 1067 K. The first crystallization event is related to the formation of the $(Fe,Nb)_{23}B_6$ CMA with *fcc* structure and with more than 110 atoms per unit cell (Figure 7.14) [50]. In this structure, the cuboctahedra and the cubes formed by metal atoms are connected through metalloid atoms. Thus, the B atoms are surrounded by 8 metal atoms to form an Archimedean square antiprism. These antiprisms ought to be symmetrically arranged in the $Fe_{23}B_6$–type structure. The $(Fe,Nb)_{23}B_6$ CMA structure can be retained by cooling to room temperature. However, it is metastable at high temperatures and transforms into the equilibrium Fe_2B, α-Fe and FeNbB phases during heating (Figure 7.14).

In the Fe-B glassy alloys containing more than 20 at.% B, the local atomic structure is characterized by the nonperiodic network of trigonal prisms with a coordination number around the B atoms of about 6. This type of locally ordered structure is expected to form in the amorphous $Fe_{66}Nb_4B_{30}$ alloy due to the high B content (>20 at.%). Therefore, during devitrification of the $Fe_{66}Nb_4B_{30}$ BMG a change in the short-range order from the trigonal prism of the glassy phase to the Archimedean square

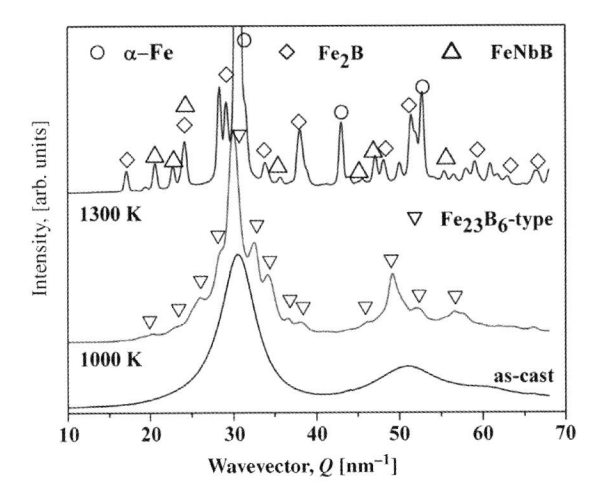

Figure 7.14 X-ray diffraction patterns in transmission configuration using a high-intensity monochromatic synchrotron beam (wavelength $\lambda = 0.0155$ nm) recorded in-situ during heating at room temperature, 1000 K (above the first crystallization event) and 1300 K (above the second crystallization event) for the Fe$_{66}$Nb$_4$B$_{30}$ BMG (from reference [50]).

antiprism of the (Fe,Nb)$_{23}$B$_6$ CMA as well as the simultaneous arrangement of these polyhedra to form the Fe$_{23}$B$_6$-type symmetry should occur.

This structure displays improved soft magnetic properties compared to its amorphous precursor. For example, the Curie temperature increases form 550 K for the parent glass to 720 K for the devitrified material containing the (Fe,Nb)$_{23}$B$_6$ CMA. In addition, the saturation magnetization is about 1.2 T, which is 15% higher than for the glass precursor, while the coercivity remains very low, at about 1 Oe [50]. The combination of such magnetic properties with the high strength typical for Fe-based metallic glasses (>3 GPa) are of great interest for industrial application, such as for transformer cores, magnetic clutches, magnetic shielding cases for sensors, sensors and actuators, and so on.

Besides Fe-based BMGs, multicomponent Zr-based alloys can be used for the production of bulk nanostructured materials [51]. However, the phase selection upon crystallization is strongly affected by the chemical composition of the glassy phase. To obtain nanostructured materials, the glassy specimens are typically annealed at temperatures within the supercooled liquid (SCL) region $\Delta T_x = T_x - T_g$, defined as the difference between the crystallization temperature (T_x) and the glass transition temperature (T_g), or close to the onset of crystallization. This procedure is based on the results first obtained for rapidly quenched thin ribbons, where sequential crystallization was observed for a variety of Zr-based bulk glass-forming alloys, such as Zr-Al-Ni-Cu-M (M = Ag, Au, Pt, Pd) [52], Zr-Cu-Al-Ni-Ti [53]. The stepwise crystallization behavior leads to primary precipitation of intermetallic or quasicrystalline phases from the supercooled liquid, which are embedded in a residual glassy phase with changed composition.

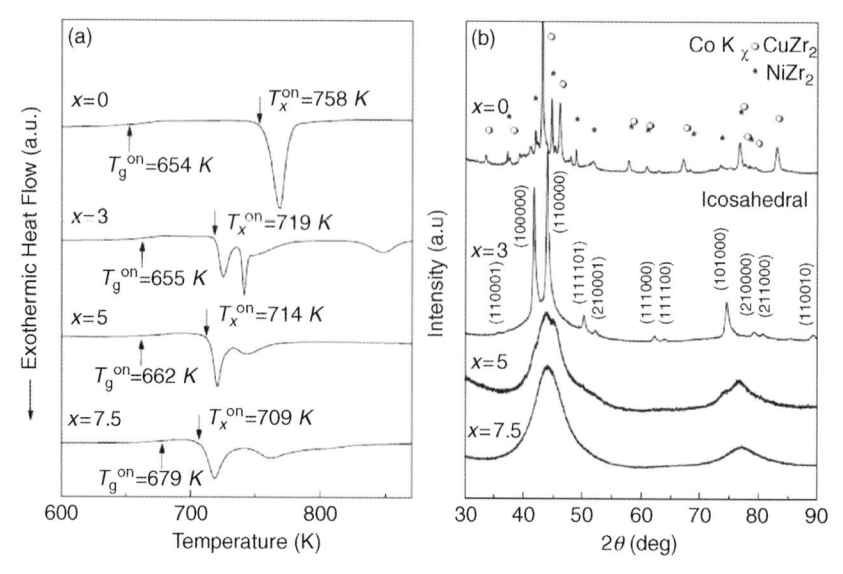

Figure 7.15 (a) DSC scans for $Zr_{62-x}Ti_xCu_{20}Al_{10}Ni_8$ glassy alloys. (b) corresponding XRD patterns after isothermal annealing: $x=0$, annealed at 723 K for 30 min; $x=3$, annealed at 703 for 5 min; $x=5$, annealed at 683 K for 30 min and $x=7.5$ annealed at 688 K for 40 min (from reference [53]).

Figure 7.15(a) displays differential scanning calorimetry (DSC) scans for as-cast glassy $Zr_{62-x}Ti_xCu_{20}Al_{10}Ni_8$ bulk samples ($x=0$, 3, 5 and 7.5) as typical examples for bulk glass-forming Zr-based alloys [53]. The sample with composition $Zr_{62}Cu_{20}Al_{10}Ni_8$ crystallizes through a single sharp exothermic peak pointing to the simultaneous formation of intermetallic compounds. Upon Ti addition, the crystallization mode changes towards a double-step process indicating a successive stepwise transformation into the equilibrium compounds while maintaining an extended supercooled liquid region between the glass transition temperature T_g^{on} and the crystallization temperature T_x^{on} (T_g^{on} and T_x^{on} are defined as the onset temperatures of the endothermic glass transition and the exothermic crystallization events, respectively). With increasing Ti content, the first DSC peak shifts to lower temperatures and the enthalpy related to the second exothermic peak decreases.

The nature of the crystallization products and the resulting microstructure after annealing were investigated by X-ray diffraction (XRD) (Figure 7.15(b)) [53]. $Zr_{62}Cu_{20}Al_{10}Ni_8$ transforms into cubic $NiZr_2$- and tetragonal $CuZr_2$-type compounds. Annealing the alloy with $x=3$ leads to primary precipitation of an icosahedral quasicrystalline phase with spherical morphology and a size of about 50–100 nm. For $x=5$, the diffraction peaks are weaker in intensity and broader because the precipitates are as small as about 5 nm. For $x=7.5$, the precipitates are about 3 nm in size. At first glance, the XRD pattern after annealing displays no obvious reflections but only broad amorphous-like maxima. However, careful examination of the annealed state clearly shows differences in scattering intensity compared to the as-cast state, pointing to the precipitation of a metastable cubic

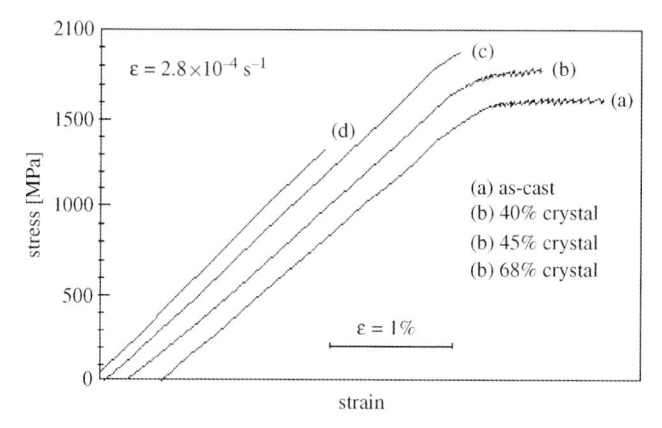

Figure 7.16 Compressive stress–strain curves of (a) as-prepared amorphous and partially crystallized $Zr_{57}Ti_5Cu_{20}Al_{10}Ni_8$ with (b) 40%, (c) 45%, and (d) 68 vol.% nanocrystals (from reference [54]).

complex phase with a grain size of about 2 nm coexisting with a residual glassy phase. Similar results were reported for other Zr-Cu-Al-Ni multicomponent alloys containing Ti, Ag, Pd or Fe [52]. This indicates that Zr-based multicomponent alloys are promising candidates for the production of bulk nanostructured quasicrystal-based two-phase materials.

The formation of nanostructured phases during partial devitrification changes the mechanical properties of metallic glasses. As an example, typical compressive stress–strain curves at room temperature for as-cast fully amorphous and partially crystallized $Zr_{57}Ti_5Cu_{20}Ni_8Al_{10}$ samples prepared by controlled annealing of the glass are presented in Figure 7.16 [54]. Even though the plastic deformation decreases with increasing volume fraction of precipitates and finally disappears when a critical volume fraction is reached, the dispersion of nanocrystals in the amorphous matrix can lead to a distinct strength increase [54]. For samples up to about 50 vol.% of nanocrystals, the fracture surface exhibits a well-defined vein pattern, indicating that the deformation mechanism is governed by the glassy phase and not by the nanocrystals [54]. When the volume fraction of the nanocrystalline precipitates increases to more than about 50 vol.%, the nature of the brittle intermetallic phases is likely to dominate the mechanical behavior, leading to a marked decrease in ductility [54].

Nanostructured or partially amorphous materials can also be produced by processing routes based on powder metallurgy (e.g., mechanical alloying, MA) [29, 51]. This synthesis route may directly lead to a two-phase nanostructure or, similarly to solidified metallic glasses, an additional heat treatment is needed to create the desired nanostructure. Depending on the milling intensity, MA can lead to glass formation or to the formation of nanoscale phases directly during milling. For example, $Zr_{57}Ti_8Nb_{2.5}Cu_{13.9}Ni_{11.1}Al_{7.5}$ glassy powders can be produced at a low milling intensity [55], whereas at higher intensity a composite consisting of *fcc* Ti_2Ni-type CMA particles embedded in an amorphous matrix can be achieved [56]. The *fcc*

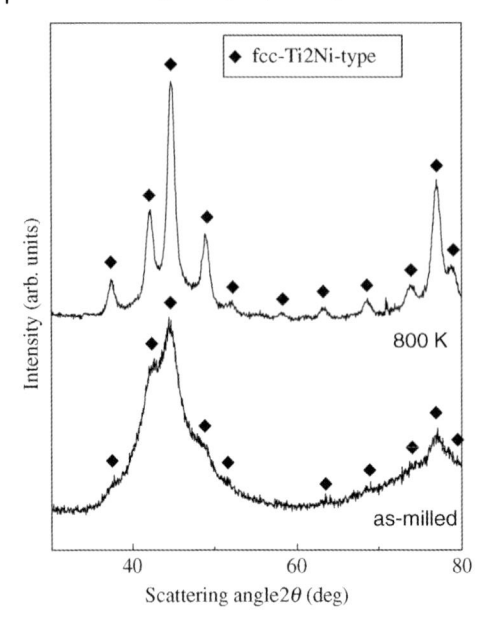

Figure 7.17 XRD patterns (Co K_α radiation) of as-milled $Zr_{57}Ti_8Nb_{2.5}Cu_{13.9}Ni_{11.1}Al_{7.5}$ powder and after heating up to the completion of the first crystallization event (800 K), revealing grain growth of the fcc-Ti_2Ni-type phase upon heating (from reference [56]).

particles in the as-milled powder are in the nanoscale regime (less than 10 nm), as indicated by the extremely broad diffraction peaks shown in Figure 7.17. The material is partially amorphous and the corresponding DSC curve [56] exhibits a distinct glass transition (T_g), followed by a supercooled liquid region before two exothermic events due to crystallization occur at higher temperatures. When the powder is heated to the completion of the first crystallization event (800 K), the diffraction peaks belonging to the *fcc* phase increase in intensity. This indicates grain growth of the *fcc* particles, which, nevertheless, are still in the nanometer regime (below 100 nm). The formation of a glassy-matrix composite containing nanoscale particles directly upon MA combined with the presence of a clear glass transition opens up the possibility to produce large amounts of composite material by a relatively simple route and to consolidate and shape the composite into bulk parts in the SCL regime. In addition, the particle size can be varied by proper heat treatment, giving the opportunity to tune the microstructure of the composite material.

7.3
Metal Matrix Composites Reinforced with CMAs

The main disadvantage of CMAs for technological applications in the form of single-phase bulk samples is their high hardness associated with low fracture toughness at room temperature. One possibility to circumvent their intrinsic brittleness and still to benefit from their atypical properties is to use CMAs in the form of coatings [57] or as reinforcement in composite materials. Examples of such composites include CMA precipitates in a matrix like maraging steel [58–60], Al-based [61, 62] or Mg-based [63, 64] alloys, or CMA particles [65] to reinforce a metal or a polymer matrix [66]. Other composites have been synthesized where a ductile phase is added to a CMA matrix, such as FeAl, $CuAl_{10}$ or Sn in CMA-based coatings [67–69] or in bulk CMA [70]. The general conclusion from these studies is that CMAs have a large potential as effective reinforcement particles in a ductile matrix.

The strong demand for lightweight engineering materials in the areas of automotive and aerospace industries has driven a considerable effort in the development of metal matrix composites (MMCs) [71, 72]. Typical reinforcements in MMCs are ceramics, such as Al_2O_3 and SiC [71, 72]. However, ceramic-reinforced composites suffer from low fracture toughness compared to conventional Al-based alloys. Other disadvantages in using ceramic reinforcements are the difficulty of recycling and the material cost penalty. These problems limit the range of application of these materials and motivate the development of alternative composites. Al-based CMAs are attractive substitutes to ceramics due to the low costs of the raw materials, no toxicity and compatibility with the Al matrix. However, interface bonding between CMA particles and the Al matrix is a critical point in the material processing as well as during the composite application. Among the different methods that have been employed to produce composites reinforced with CMA particles, the most frequently used approach is powder metallurgy, for example, mechanical alloying followed by consolidation. Among the variety of CMA compounds, the one that is most widely used as reinforcement is the stable icosahedral (i-)Al-Cu-Fe quasicrystalline (QC) phase discovered by Tsai et al. [73]. This phase displays very high hardness [74], which is a desirable feature of strengthening particles. Accordingly, an overview of processing, structural and mechanical properties of Al-based MMCs reinforced with quasicrystals is given in the next section.

7.3.1
Processing of Aluminum Matrix Composites Reinforced with CMAs

7.3.1.1 Thermal Stability of CMAs in Al-Based Matrix Composites
Aluminum matrix composites reinforced with i-$Al_{64}Cu_{24}Fe_{12}$ quasicrystalline particles with a particle size of about 5 μm (Al/(i-AlCuFe)p) were first prepared by sintering by Tsai et al. [65]. They found that the hardness of the composites increases by particles addition but they noticed the appearance of the ω-$Al_{70}Cu_{20}Fe_{10}$ tetragonal phase during the sintering process, which was done at 873 K for 3 h under low pressure (60 MPa). The i-AlCuFe phase does not coexist with fcc Al in the Al-Cu-Fe equilibrium phase diagram [75, 76]. Thermodynamic equilibrium is obtained by Al diffusion from the matrix to the quasicrystalline particles. As a result, the Al content in the particles increases, inducing a phase transformation from the i-phase to the ω-phase, in agreement with the phase diagram.

In the conventional casting method used to produce metal matrix composites (MMCs), Al-Cu-Fe particles are mixed to the molten aluminum [77, 78]. This

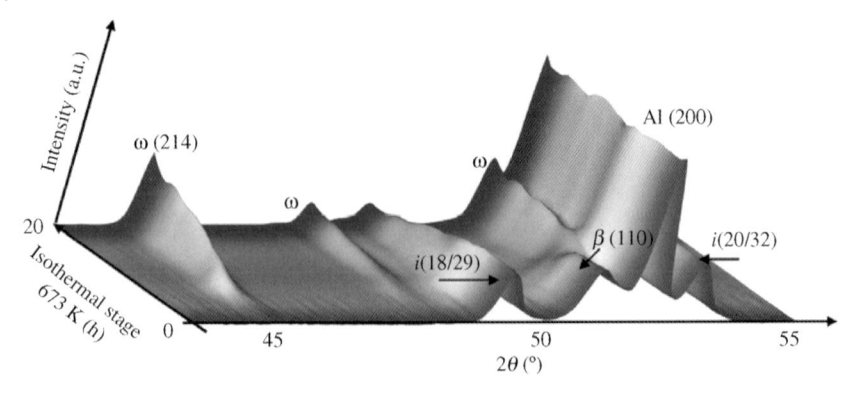

Figure 7.18 3D representation of the diffraction patterns (Co K$_{\alpha 1}$ radiation) acquired during an isothermal stage at 673 K for Al/30(AlCuFeB)p (from reference [80]). For clarity, only the main diffraction peaks are shown in the figure.

results in a mixture of phases, including the tetragonal ω-phase, and in the dissolution of the small Al-Cu-Fe particles into the matrix. The thermal stability of the i-phase in the Al matrix is limited due to the rapid formation of the ω-phase. The $i \rightarrow \omega$ transformation usually starts at about 673 K [79]. As an example, Figure 7.18 shows the diffraction patterns recorded during an isothermal stage at 673 K for an Al-based composite containing 30 vol.% of Al-Cu-Fe-B atomized particles (noted Al/30(AlCuFeB)p hereafter) prepared by cold isostatic pressing (1.6 GPa, 300 K) [80].

Initially, the sample mainly consists of fcc Al and quasicrystalline phases. In addition, the particles contain a small amount of β-Al$_{50-x}$(Cu, Fe)$_{50+x}$ (cubic, CsCl type $a \approx 2.9$ Å). The β-phase coexists with the quasicrystalline phase in the gas-atomized powders as a result of the peritectic reaction by which the icosahedral phase is formed. The addition of boron induces extra crystalline phases embedded in the i-AlCuFe matrix. These boron precipitates are not detected by X-ray diffraction due to a volume ratio lower than 1% [81–83]. The formation of the precipitates results in a higher fracture strength and in an increased hardness of the i-Al$_{59}$B$_3$Cu$_{25.5}$Fe$_{12.5}$ phase compared to the canonical i-Al$_{62}$Cu$_{25.5}$Fe$_{12.5}$ alloy [84–87]. The phase transformation (Al$_{matrix}$ + (i + β)$_{particles}$ → ω) is clearly seen in Figure 7.18, with the progressive disappearance of the i(18/29), i(20/32) and β(110) diffraction peaks and the simultaneous appearance of the ω-phase. The formation of the ω phase starts after an incubation time of about 1 h. These results highlight the extreme difficulty in preserving an Al-based CMA in an Al-based matrix, particularly for composites sintered at low pressure (30 to 100 MPa). The ω-phase systematically appears in several fabrication methods, such as hot consolidation [88], hot isostatic pressing [89], quasi-isostatic forging or vacuum hot pressing [90, 91].

7.3.1.2 Preserving Complex Phases in Al-Based Matrix Composites

To preserve the quasicrystalline structure of the reinforcement, it is necessary to either slow down the kinetics of the phase transformation or to adjust the temper-

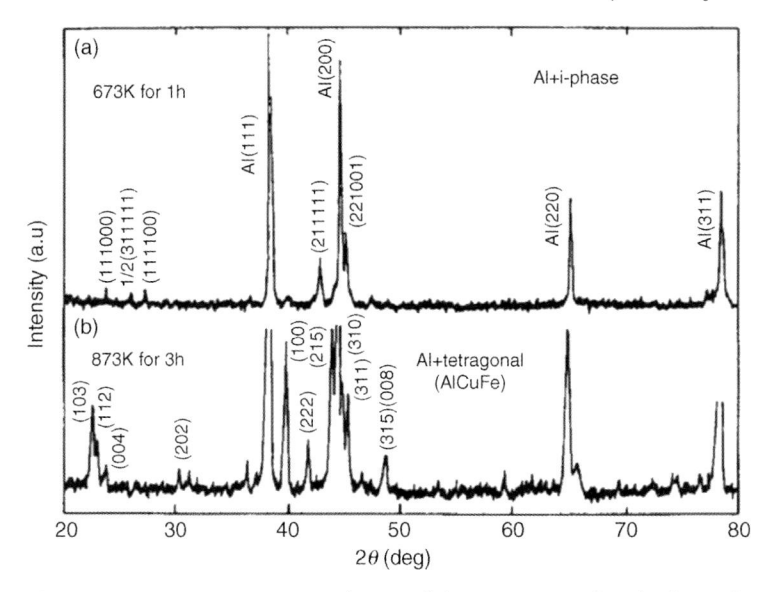

Figure 7.19 XRD patterns (Cu K$_\alpha$ radiation) of Al composites reinforced with 25 vol.% (Al-Cu-Fe) p. (a) Hot pressed at 673 K for 1 h and 260 MPa and (b) at 873 K for 3 h and 60 MPa (from reference [65]).

ature and/or the applied pressure during the sintering process. For example, a composite containing the i-phase dispersed in an Al matrix can be obtained by hot pressing using a temperature of 673 K for 1 h under 260 MPa [65], as shown in Figure 7.19, as well as by hot extrusion [78, 92]. In the latter case, a consolidation temperature of 623 K is required to preserve the i-phase and to obtain a dense extruded composite.

Therefore, the reduced stability of Al-based QCs in the Al matrix requires the use of a low temperature (<673 K) and/or a high pressure (>200 MPa) during the consolidation process. An alternative route is the use of a diffusion barrier at the matrix/particle interface to reduce the Al diffusion and to preserve the original i-AlCuFe structure in the composite. For example, Fleury et al. [78] have prepared MMCs with 15 vol.% of Al-Cu-Fe particles coated with a Ni layer of about 5 μm thickness by conventional casting. This approach effectively reduces the dissolution of the particles in the matrix and the particles preserve their original microstructure (Figure 7.20).

A second method is the formation of an oxide layer on the surface of the Al-Cu-Fe-B particles [93]. The oxide layer acts as an efficient barrier against Al diffusion. Under specific conditions, the formation of the ω-Al$_7$Cu$_2$Fe phase can be completely inhibited. By adjusting the oxidation conditions (and thus the oxide thickness), it is possible to control the content of ω-phase in the particles. A typical pretreatment is oxidation in air at 873 K and up to 100 h. The diffraction patterns obtained from preoxidized powders do not show additional diffraction peaks, suggesting that the oxide layer is either amorphous or poorly crystallized and very thin [94, 95]. Moreover,

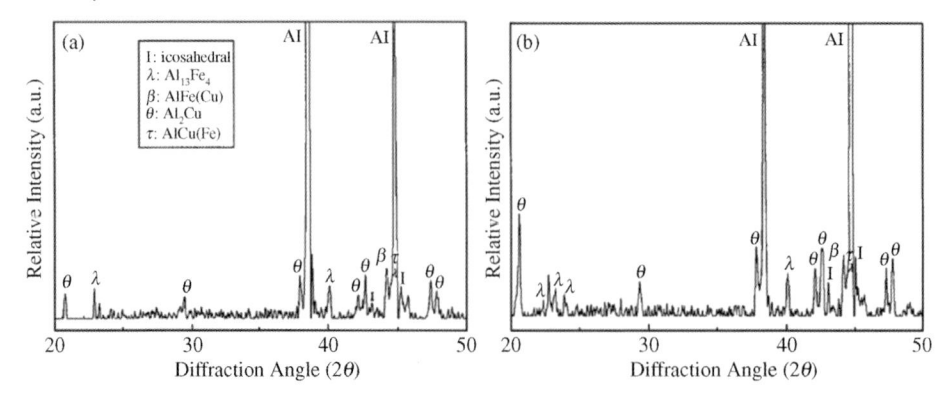

Figure 7.20 Al matrix composites prepared by conventional casting. (a) XRD pattern of as-cast Al/(Al-Cu-Fe)p and (b) XRD pattern of as-cast Al composites with Ni coated 15 vol.% (Al-Cu-Fe)p (reprinted from reference [78]). Note that the reinforcement content is higher in (b).

the kinetics of Al diffusion from the matrix to the quasicrystalline particles is significantly reduced for sintering temperatures up to 823 K [96]. As an example, Figure 7.21 compares the evolution of the volume fraction of the ω-phase during the transformation ($i \rightarrow \omega$) for two sets of composites prepared by cold isostatic pressing [96]. The formation of the ω-phase is much faster for composites containing nonoxidized Al-Cu-Fe-B particles (series 1) compared to preoxidized particles (series 2). Consequently, composites made of i-particles dispersed in a pure aluminum matrix with different volume fractions could be prepared by solid-state sintering with low uniaxial pressure (LUP). This positive result is, however, weakened due to the negative influence of the oxide on the interface strength [96], as shown in the following section.

7.3.2
Mechanical Properties of Al-Based Composites Reinforced with CMAs

In general, the mechanical properties of composite materials depend on the properties of the matrix, the nature of the reinforcement, the particle size, their dispersion in the matrix and on the heat treatment and fabrication process [71]. The improvement of the mechanical properties relies essentially on three main factors: (i) the effect of decreasing size of the matrix due to the presence of the reinforcement, (ii) the internal stresses generated by the difference in expansion rates between the matrix and the particles and (iii) the predominant direct strengthening factor, such as the volume fraction of the reinforcement [71].

A detailed study of Al composites reinforced by i-particles obtained by extrusion at 623 K has been performed by Schurack et al. [92]. The authors concluded that the strength increases proportionally to the particle volume fraction according to the rule of mixtures. Several strengthening mechanisms have been proposed: solid solution hardening resulting from the dissolution of small particles in the Al matrix,

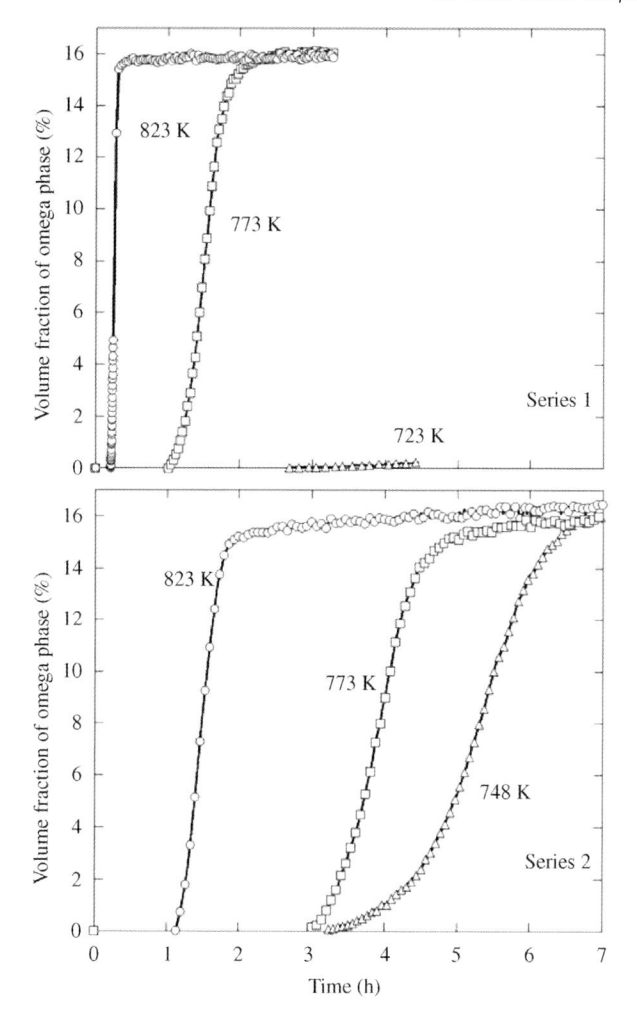

Figure 7.21 Volume fraction of ω-phase as a function of time and isothermal temperatures for the composites containing 15 vol.% of *i*-particles: (series 1) sample annealed in vacuum for 60 min; (series 2) sample oxidized in air for 60 min. Initially, both series 1 and 2 contain only fcc Al and *i*-phase (from reference [96]).

precipitation hardening due to the formation of the ω-phase and dispersion hardening induced by the presence of the icosahedral phase [77, 78]. The influence of the matrix on the mechanical behavior of QC-reinforced composites is shown in Figure 7.22(a) [78]. When an $Al_{96}Cu_4$ alloy is used as matrix in place of pure Al, the yield stress increases by about 65% [78]. As-cast composites reinforced with Ni-coated QC particles have better compressive properties compared to composites prepared by using noncoated particles [78]. An average increase of 6% in yield stress is achieved by preserving the *i*-phase (Figure 7.22(b)). Note that in the latter case, the strengthening effect is smaller than that obtained by using the $Al_{96}Cu_4$ alloy matrix.

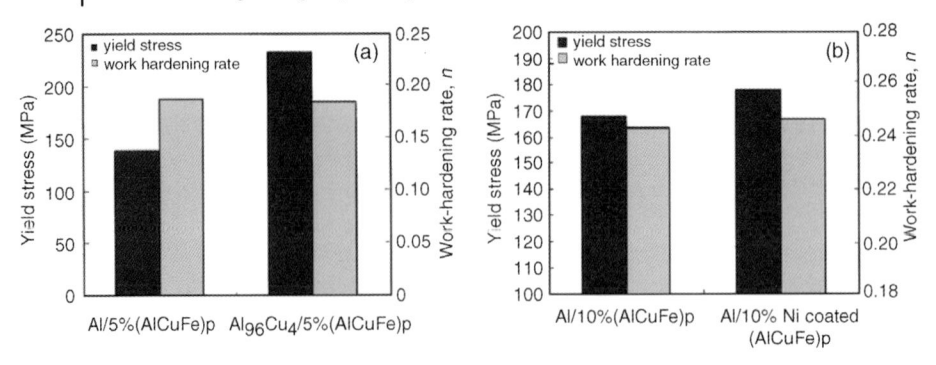

Figure 7.22 Al matrix composites prepared by conventional casting. Influence of the Al matrix on (a) yield stress and work hardening and (b) influence of the Ni coating layer on the yield stress and work hardening (reprinted from reference [78]). Note that the reinforcement content is higher in (b).

Composites reinforced with volume fractions of i-phase higher than 15 vol.% (produced by isostatic pressing at 673 K for 1 h and 260 MPa) exhibit higher hardness with respect to composites reinforced with the ω-phase [65]. For Al/25(i-AlCuFe)p, the microhardness reaches 120 kg mm^{-2}, which is about five times higher than that of pure Al and 33% higher than that of Al/25(ω-Al-Cu-Fe)p. Tsai et al. [65] proposed that the plasticity of these composites could be controlled by the Orowan mechanism due to a good correlation between the strengthening effect and the distribution of the particles. Similar hardness properties have been obtained using different QC systems, such as in Al/(Al$_{65}$Cu$_{20}$Cr$_{15}$)p composites [97].

In the case of sintered composites containing 15 vol.% of preoxidized i-Al-Cu-Fe-B particles prepared by LUP, the mechanical properties are poorer (Figure 7.23), even though the i-phase is preserved [97]. The degradation of the mechanical properties is attributed to the negative influence of the oxide layer, which weakens the Al/(i-AlCuFeB)p interface [96]. These results are consistent with several recent studies of Tang et al. [91, 98, 99] demonstrating the importance of the oxide-layer thickness on the mechanical properties of Al/(AlCuFe)p composites.

The best mechanical properties are obtained when the reinforcement particles are totally transformed into ω-phase, due to enhanced bonding strength between the matrix and the particles resulting from interparticle diffusion during the sintering process [77, 91, 98–100].

For example, a high yield strength of 405 MPa is obtained for the sintered composite containing 45 vol.% of i-particles completely transformed into the ω-phase. This corresponds to an increase of the yield stress ($\Delta\sigma_y$) and of the Brinell hardness of 300 MPa and 143 HB, respectively, with respect to the unreinforced matrix [96]. However, the fracture strain of this composite is rather poor ($\varepsilon_f \approx 1\%$). A linear correlation can be made between Brinell hardness and yield stress as a function of volume fraction of (AlCuFeB)p, as shown in Figure 7.24 [96]. For comparison, Al composites reinforced with 40 vol.% of SiC particles (10–21 μm) have a $\Delta\sigma_y$ of about 150 MPa [101]. The increase of yield stress can be explained by the model proposed by Tang et al. [91]. This model corresponds to a simple summation of three strength-

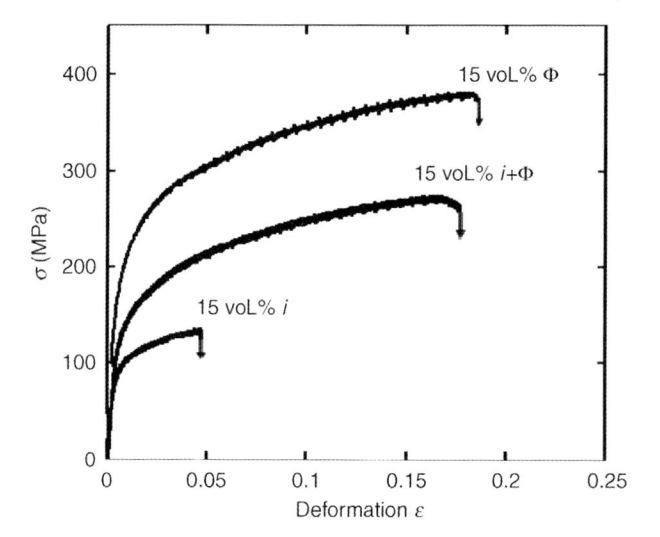

Figure 7.23 Comparison of the stress–strain curves of the Al/(Al-Cu-Fe-B)p composites with particles containing 15 vol.% of preoxidized Al-Cu-Fe-B particles. The curves were obtained from room temperature compression tests at a constant strain rate $\dot{\varepsilon} = 10^{-4}\,\text{s}^{-1}$. The composites were prepared by LUP with 32 MPa at 773 K under a helium atmosphere during 3 h 30 min (from reference [96]).

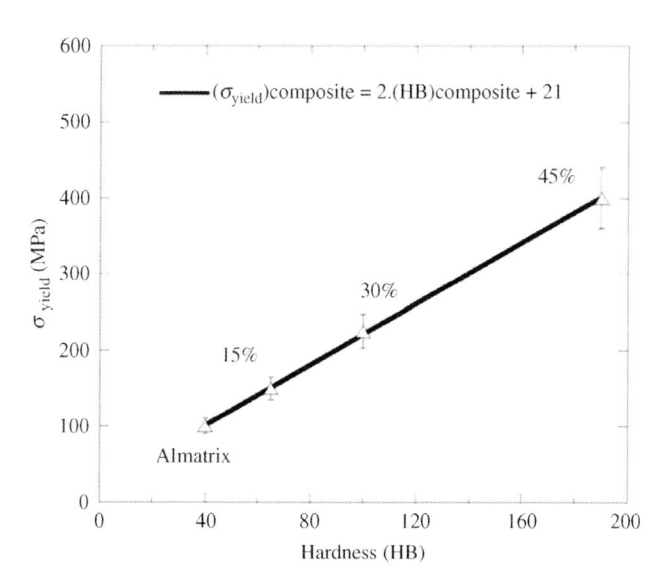

Figure 7.24 Dependence of σ_y on the Brinell hardness and the volume fraction of particles for Al/(ω-Al-Cu-Fe)p composites sintered by LUP. The hardness was measured by Brinell indentation with a 2.5 mm ball and a load of 62.5 daN applied for 5 s and σ_y values were evaluated by compression tests performed on composites samples ($3 \times 3 \times 10\,\text{mm}^3$) under a constant strain rate ($\dot{\varepsilon} = 10^{-4}\,\text{s}^{-1}$).

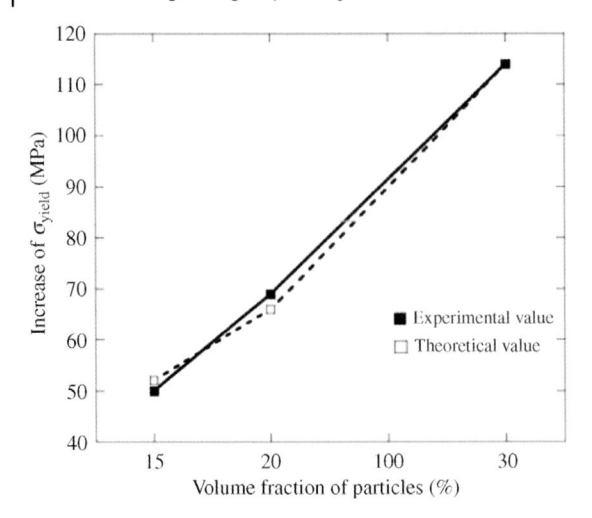

Figure 7.25 Comparison of theoretical and experimental $\Delta\sigma_y$ values for Al/(ω-AlCuFe)p as a function of reinforcement content. Composite samples were consolidated by a vacuum hot technique at 823 K for 6 h, using a pressure of 175 MPa after the first hour at 823 K. Redrawn from reference [91].

ening mechanisms that are the load transfer mechanism, thermal expansion mismatch and geometrically necessary dislocation strengthening [91]. A very good agreement between experimental and predicted $\Delta\sigma_y$ is observed, as shown in Figure 7.25.

The ω-phase acts as an effective reinforcement compared to commercial Al composites. Contrary to Al/(SiC)p composites, the formation of the ω-phase induces compressive residual stresses in the Al matrix, which is associated with the unusual strengthening [99, 100]. The compressive stress may be promoted by the volume expansion of the reinforcement particles caused by transformation of the i-phase into the lower density ω-phase and the corresponding stiffness mismatch of the matrix and the reinforcement phases [99].

The first investigations of the mechanical properties as a function of temperature for similar composites produced by hot isostatic pressing (HIP) have been performed by Kabir et al. [89]. The initial composition Al/30(i-AlCuFe)p is transformed into Al/40(ω-AlCuFe)p during the HIP process. The authors found a sharp decrease of σ_y with increasing temperature. Figure 7.26 shows the evolution of the compressive properties of the composites in the temperature range 293–773 K. They concluded that the contribution to the total flow stress arising from the reinforcement particles is mainly thermally activated, which is not compatible with the Orowan bypass mechanism. This suggests that the $i \rightarrow \omega$ phase transformation contributes to the enhancement of the material strength. It is proposed that plastic deformation is controlled by thermally activated motion of matrix dislocations by cross-slip and/or climb mechanisms [89].

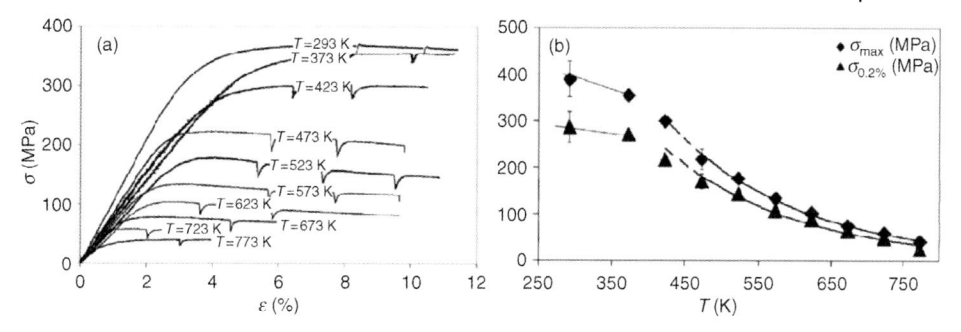

Figure 7.26 (a) Comparison of the stress–strain curves of the Al/40(ω-AlCuFe)p composites prepared by HIP as a function of temperature and (b) evolution of the yield stress and maximum stress as a function of temperature. The compression tests were performed at a nominal strain rate $\dot{\varepsilon} = 1.4 \times 10^{-4}\,\text{s}^{-1}$ (reprinted from reference [89]).

Compared to commercial Al/(SiC or Al_2O_3)p composites, the use of CMAs as strengthening particles into an Al-based matrix provides an unusual increase in strength. For example, the increase of the yield stress of Al/(10–45 vol.% CMA)p composites varies from about 40 to 300 MPa without any heat treatment [77, 96, 100]. These properties can be further improved when the original structure of the CMA phase is preserved during the consolidation process [78] but this implies the use of a diffusion barrier and thus an additional step in the material production. Compared to similar Al/(SiC or Al_2O_3)p composites, the increase of the yield stress varies from about 40 to 150 MPa [71, 101, 102], depending on the particle size, the post-thermal treatment and the Al-matrix used. Therefore, the use of CMAs as reinforcement particles appears as a promising alternative for technological applications, especially for automotive and aerospace industries.

7.4
Surface Mechanical Testing and Potential Applications

7.4.1
Fretting Tests (Cold Welding) of CMAs

On spacecrafts, a variety of engineering mechanisms exhibit ball-to-flat surface contacts, which are periodically closed for several thousands of times. Vibrations occurring during launch or during movement of for example, antennas in space, can lead to small oscillating movements in the contact, which is referred to as "fretting" [103]. This fretting movement can eventually degrade the materials surface whether they are natural oxides, chemical conversion films or even metallic coatings. This can dramatically increase the tendency of these contacting surfaces to "cold weld". This lateral motion causes severe destruction and adhesion forces [103].

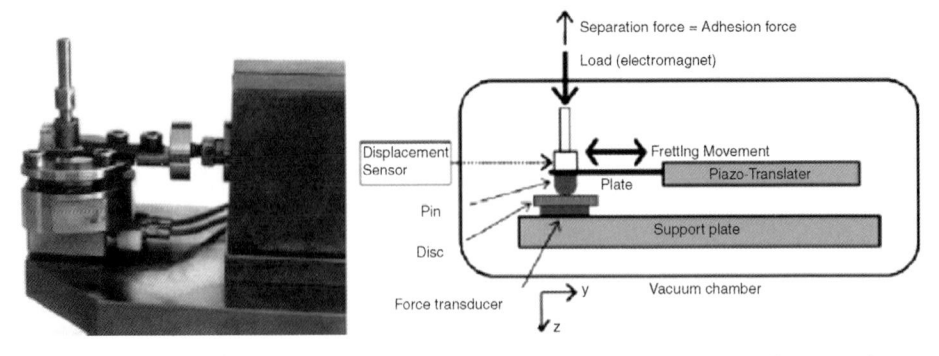

Figure 7.27 Image of the fretting device (left) and corresponding schematic drawing (right) showing how pin (upper rod) and disc (mounted directly on a force transducer) are fixed. The piezo-translator generates the fretting movement (from reference [103]).

In order to gain experience in this effect, a special device – called "fretting facility" – has been developed at the Austrian Research Centre (ARC) and was used to investigate several combinations of bulk materials for their tendency towards cold welding. The test philosophy (described in detail in an in-house specification of ARC [103]) is based on repeated closing and opening of a pin-to-disc contact. In general, a pin is brought into contact with a disc for several thousand times (Figure 7.27). The static load and impact energy are fixed for each pair with respect to the elastic limit (EL) of the contact materials. Hertz's theory is used to calculate the contact pressure in the ball-to-flat contact. Using the yield strength of the softer material, the von Mises criterion defines an elastic limit (EL): if the load (contact pressure) exceeds the EL, plastic yielding would occur. Similarly, for the contact energy, a limit can be deduced above which yielding occurs [103, 104]. Based on parameter studies [105, 106], the ARC-standard was defined: one static load is selected to achieve a contact pressure of 60%EL for a fretting test and is applied for 5000 cycles.

The European Co-operation for Space Standardization (ECSS) has released specifications on contact surfaces [107], suggesting the following main requirements:

a) the peak Hertzian contact pressure shall be below 93% of the yield limit of the weakest material (this refers to a contact pressure of 58% of the elastic limit);
b) the actuator shall be demonstrated to overcome two times the worst possible adhesion force.

Therefore, results obtained from cold-welding tests according to the ARC in-house specification [103], can be used to address the necessary opening forces for actuators in mechanisms (fretting tests are done at 60% EL).

During a fretting test, the contact is closed (without impact) followed by a static load that is held for 10 s [103]. During this time, fretting is applied: the pin is moved with a sinusoidal frequency of 210 Hertz and a stroke of 50 μm. After stopping fretting, the pin is separated from the disc and the force necessary for separation is measured. A resting time of about 10 s is achieved at a vacuum pressure of less than 5×10^{-7} mbar.

This kind of adhesion is not visible in air, because in air only fretting corrosion is visible: the oxide layer is always present at the top of the material and protects it against cold welding [103]. All these fretting tests are done in vacuum, because on the one hand no adhesion is detected for a fretting test in air (whatever the substrate), and on the other hand, these materials are aimed to be used later on in space (coating elaborated for space applications). For tests where no adhesion was found, that is where the adhesion force was below the detection threshold, the noise of the measurement is given as a worst-case approximation. The indication of low and high adhesion is related to comparison of adhesion force values themselves (data from general materials [108]). A fretting test is "successful" if there is no adhesion during at least 5000 cycles.

To check if an adhesion value is true or just noise, "buffer files" are exploited: a buffer file represents the end of each cycle, when the pin is separating from the disc. This is the force during the unloading as a function of time. In the case of true adhesion, a negative force between the pin and the disc is observed, which refers to tension between pin and disc [103]. The breaking force is referred to as an "adhesion force" (i.e. the jump of the force). If the jump is high, the adhesion is high. If the jump is not very high, the adhesion is low. Finally, if there is no jump, there is no adhesion, only noise.

During the fretting test, there is a microplastic deformation. After each cycle the shape of the contact area is different, because of this microplastic deformation. Therefore, wear of the materials (wear of disc and/or of pin) is observed. There is a time-dependent change of "adhesion forces". For each cycle, one value of the adhesion force is measured. From all values of "adhesion force", only the maximum is taken (for bulk materials or resisting coatings). This approach reflects engineering objectives, that is to provide values for design of mechanisms [108]. Details can be found in reference [103].

Among the materials with complex structure, quasicrystals and approximant phases have been extensively investigated under fretting (Table 7.2) [109, 110]. Representative compositions of such materials are $Al_{62.5}Cu_{25.3}Fe_{12.5}$, $Al_{59.5}Cu_{25.3}Fe_{12}B_3$ and $Al_{70}Pd_{20}Mn_{10}$ (in at.%). The progressive loss of the metallic character is imposed by the formation of icosahedral atomic clusters and their connection (and interpenetration) according to a point group symmetry that is no longer compatible with translational order pertaining to normal crystals. As a result, the density of conduction states is depressed at the Fermi edge and quasicrystals resemble more semiconductors than their metallic constituents. In these materials, propagation of phonons is hindered by the absence of periodicity, except for long wavelengths. Specific atom movements, inherent to this kind of crystallographic structures and called phason jumps, take place furthermore in a number of atomic sites located in double pit potentials. Both types of dynamic excitations dominate the transport of heat (which is very low as compared to normal metals), whereas phasons may also contribute to plastic strain.

7 Mechanical Engineering Properties of CMAs

Table 7.2 Material mechanical data for the discs (three icosahedral phases and five approximants) and the pins (different steels, aluminum and titanium alloys) investigated under fretting [109].

Sample Short name	Compositions (at%)	$H_V \pm 50$ (daN/mm^2)	$Y \pm 0.5$ (GPa)	Poisson ratio	$E \pm 10$ (GPa)
AlCrFeBMo	$Al_{74.5}Cr_{15}Fe_6B_3Mo_{1.5}$	520	19.5	0.3	140
ξ'-AlPdMn	$ξ'-Al_{73.9}Pd_{22.2}Mn_{3.9}$	1000	38.5	0.4	100
i-AlPdMn	$i-Al_{70}Pd_{22.1}Mn_{7.9}$	1000	38.5	0.4	100
i-AlCuFeB	$i-Al_{59.5}Cu_{25.3}Fe_{12.2}B_3$	580	22.2	0.3	140
AlCuB	$Al_{41.8}Cu_{57.2}B_1$	710	27.3	0.3	140
i-AlCuFe	$i-Al_{62.5}Cu_{25.3}Fe_{12.2}$	530	20.4	0.3	140
γ-AlCrFe	$γ-Al_{67.6}Cr_{23.3}Fe_{9.1}$	840	32.4	0.3	140
β-AlCuFe	$β-Al_{50}Cu_{40}Fe_{10}$ LZ	540	20.8	0.3	140
AISI 316L	FeMnCrSiMoNi	175	6.7	0.28	190
Al 7075	AlMnZn	170	6.5	0.33	70
A286	FeCrNiTi	300	11.4	0.28	200
Ti-IMI 318	Ti6Al4V	320	12.2	0.32	115
AISI 52100	FeCr	700	26.9	0.28	200

Bulk quasicrystals tend to be rather brittle, at temperatures below a few hundred degrees Celsius. They behave more like a covalently bonded material than like metals. Quasicrystals are extremely poor electrical and thermal conductors. They are also hard, and their surfaces have very low coefficients of friction, low surface energy, good wear resistance, and good oxidation and corrosion resistance [110, 111]. They can also prevent cold welding.

Single-crystal icosahedral and approximant phases were tested under fretting tests in vacuum. Figure 7.28 [109, 110] shows that the adhesion force of Al-Pd-Mn

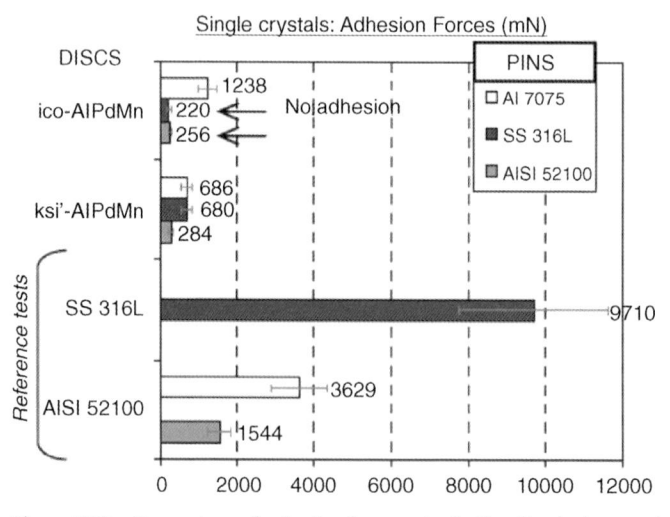

Figure 7.28 Comparison of adhesion force under fretting for single crystals [109].

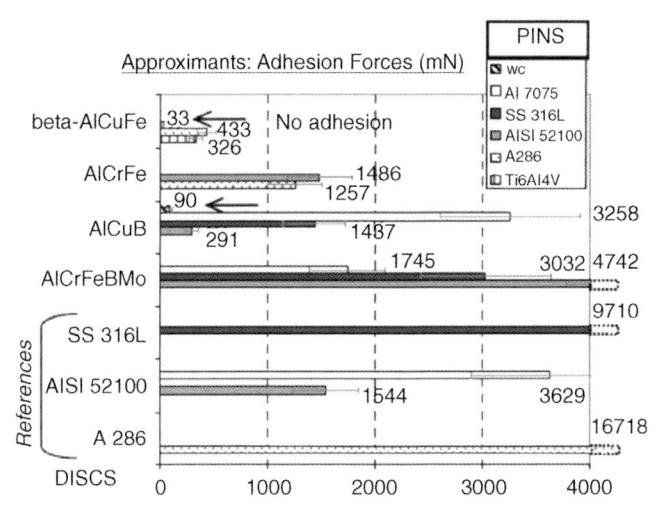

Figure 7.29 Comparison of adhesion forces under fretting for approximants [109].

quasicrystal and approximant versus metallic alloys is lower than for these alloys versus themselves. The icosahedral phase i-$Al_{70}Pd_{22.1}Mn_{7.9}$ shows no adhesion whatsoever versus the bearing steel AISI 52100 and the stainless steel SS 316L. For tests with metals, low or high adhesion is detected.

Figure 7.29 [109, 110] shows that for approximants, adhesion is against all kinds of counterparts, except for β-$Al_{50}Cu_{40}Fe_{10}$ versus steel AISI52100 and $Al_{41.8}Cu_{57.2}B_1$ versus WC. For comparison, very high adhesion for steel A286 versus itself was found because of the high Ni content.

Figure 7.30 [109, 110] shows the tests on two sintered quasicrystals (i-$Al_{62.5}$-$Cu_{25.3}Fe_{12.2}$ and i-$Al_{59.5}Cu_{25.3}Fe_{12.2}B_3$) and one single crystal (i-$Al_{70}Pd_{22.1}Mn_{7.9}$). For

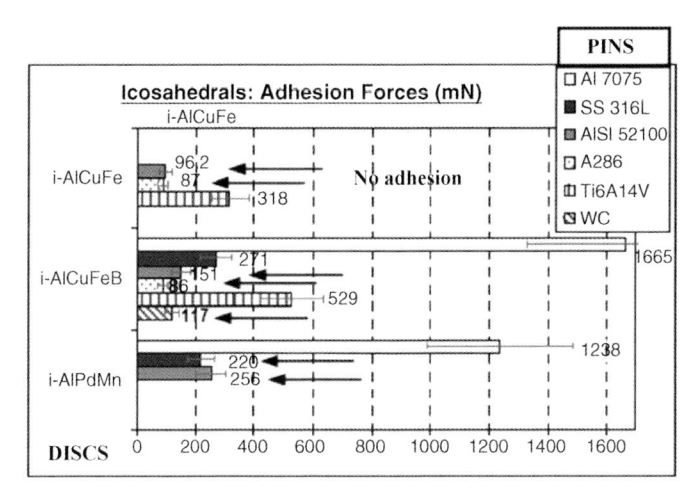

Figure 7.30 Comparison of adhesion forces under fretting for icosahedral quasicrystals [109].

these icoshaedral phases, no adhesion versus steels AISI 52100, A286 and SS 316L was detected (except for i-Al$_{59.5}$Cu$_{25.3}$Fe$_{12.2}$B$_3$ versus SS316L, where a very low adhesion is observed). However, with Ti6Al4V and Al7075 counterparts, high adhesion was found.

It is remarkable that the adhesion between stainless steels and icosahedral crystals is much lower than for stainless steels in contact to themselves. A similar conclusion is valid for the aluminum alloy (Al7075 versus itself: 7330 mN). Therefore, the adhesion forces between quasicrystals and steels or titanium-based alloys (even Al7075) were found to be negligible in comparison to typical metals used for space applications.

To complete fretting-test knowledge, wear volumes of both pin and disc are measured with an optical profiler with a 3D topography [110]. This system enables calculation of volumes, that is a volume below or above the "zero plane". After a fretting test, the disc wear mark is often a hole (Figure 7.31, top row), or sometimes a hole combined with a deposition of pin material, whereas the pin wear mark is often a "hill" (due to the disc material deposition, see Figure 7.31, bottom row). Then, the worn volume refers to a hole or a hole with a deposition. Nevertheless, the volume is easily calculated by the system (Figure 7.31). Afterwards, wear data is related to adhesion (Table 7.3) [109, 110].

The general tendency, shown in Table 7.3, is that the absence of adhesion is related to a low wear volume. Conversely, high adhesion reveals high wear volume (with debris too). As a result, from the viewpoint of fretting tests, all approximant systems are much more brittle than icosahedral quasicrystals. Moreover, especially for samples with high adhesion, a large amount of loose debris was formed, which would be detrimental for space applications. Under nongravity, these debris particles would fly around and would contaminate the whole spacecraft.

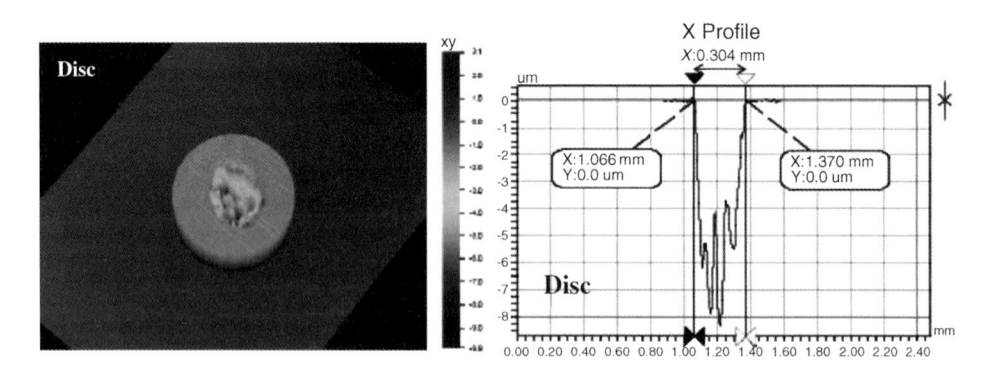

Figure 7.31 Contact area of the disc (top) and pin (bottom) after fretting test on i-Al$_{62.5}$Cu$_{25.3}$Fe$_{12.2}$ versus A286 (Maximum of noise: 87 mN – No adhesion). The disc contact area is a hole with a depth of $-8.3\,\mu m$ and a diameter of 0.304 mm, whereas the pin contact area is a hill (some material is deposited) with a height of $8\,\mu m$ and a diameter of 0.344 mm [109].

Table 7.3 Wear volumes (10^{-5} mm^3) of discs. Legend: adhesion, low adhesion, no adhesion (noise).

	Wear volumes (10^{-5} mm^3 ±20%) [of worn discs]					
Discs/Pins	AISI 52100	AISI316L	Al7075	WC	Ti$_6$Al$_4$V	A286
AlCrFeBMo	>55 000	>55 000	51 106	/	/	/
ξ′-AlPdMn	2369	22 523	16	/	/	/
AlCuB	18	9	4	8	/	/
AlCrFe	>55 000	/	/	/	/	>55 000
β-AlCuFe	8	/	/	/	60	32
i-AlPdMn	2042	140	73	/	/	/
i-AlCuFeB	13	39	3	29	75 367	11
i-AlCuFe	15	/	/	/	11 123	37
	Adhesion	Low Adhesion	No Adhesion			

Gray colours refer to the graphs shown in Figures 7.28–7.30, indicating the level of adhesion, to see the relation between the wear volume (volume of disc which is lost), and results of fretting test [109].

In fretting tests, icosahedral phases show no adhesion versus bearing steel AISI52100 (Fe, Cr) and even not for stainless steel A286 (Fe, Cr, Ti, Ni). The latter is made of an austenitic phase and is therefore prone to strong adhesion, if it would be in contact to itself. However, low adhesion is noticed versus stainless steel SS316, and high adhesion is found versus aluminum (Al7075) and titanium (Ti6Al4V) alloys. Thus, for i-Al$_{62.5}$Cu$_{25}$Fe$_{12.5}$ and i-Al$_{59.5}$Cu$_{25.3}$Fe$_{12.2}$B$_3$, no adhesion versus steels AISI52100 and A286 is noticed, combined with a low wear volume. For the approximant β-Al$_{50}$Cu$_{40}$Fe$_{10}$, there is no adhesion versus steel AISI52100 and low adhesion versus space steel A286, with low wear volumes too.

After the fretting tests, there is a hole on the disc and a deposition of material on the pin, consisting of Al, Cu and Fe, as shown in the SEM images of Figure 7.32. Similarly, Figure 7.33, EDX analysis proves that the worn and lost disc material sticks on the pin [109, 110].

Thus, the surface is strongly changed due to fretting. For comparison, a fretting test of an A286 steel disc versus itself shows strong surface destruction due to adhesive wear (the contact area of the disc after the fretting test exhibits a hole with a depth of −80 μm and a diameter of 0.655 mm), combined with high adhesion force (maximum of adhesion 16 718 mN). Material is worn out of the surface and pressed back or adheres to the contact partner [110].

The duration of all these fretting tests is similar (~5000 cycles). Not only is there no adhesion between quasicrystals and steels AISI52100 and A286, but also the wear of a quasicrystal disc versus steel is much lower than steel versus itself (see Table 7.3 for wear volumes). Several quasicrystalline compounds, especially icosahedral Al-Cu-Fe-(B), were shown to have the best performance in terms of fretting tests. Regarding aerospace applications, or vacuum technology, the two bulk sintered icosahedral phases i-Al$_{62.5}$Cu$_{25}$Fe$_{12.5}$ and i-Al$_{59.5}$Cu$_{25.3}$Fe$_{12.2}$B$_3$ present a potential for antifretting applications in vacuum against a steel counterpart, because they avoid adhesion

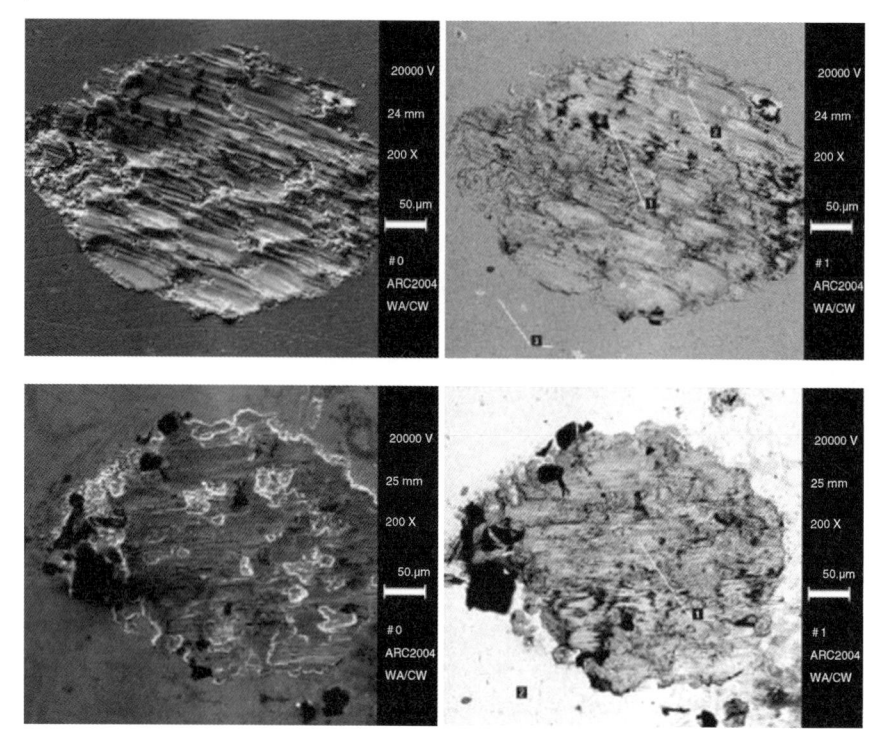

Figure 7.32 Top row: SEM picture (left) and backscattered electron picture (right) of the contact area of the disc after the fretting test of sample i-$Al_{62.5}Cu_{25.3}Fe_{12.2}$ versus A286 space alloy (maximum of noise: 87 mN – no adhesion) [109]. Bottom row: same data, respectively, but for the pin after the same test.

versus steels AISI52100 and A286, combined with a low wear volume (thus no emission of debris). Furthermore, low adhesion is detected versus the stainless steel SS316L, combined with a low wear volume too (again with no debris). But both icosahedral phases cannot prevent cold welding against titanium Ti6Al4V and aluminum Al7075 alloys, because high adhesion is noticed (combined with a high wear volume for Ti6Al4V and formation of debris).

7.4.2
Friction Properties of Composites

7.4.2.1 CMA Matrix Composites
The first investigations of the tribological behavior of CMA composites were performed on quasicrystalline composite coatings by Sordelet *et al.* [67]. The main objective of this work was to examine the effect of the addition of a ductile phase on the abrasion resistance of the Al-Cu-Fe quasicrystalline coatings. Composite coatings were deposited by the plasma-arc spraying technique. The volume fraction of the ductile Fe-Al phase was varied from 1 to 20 vol.%. The powders used for the

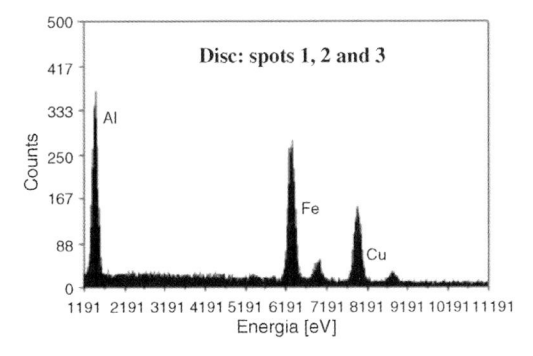

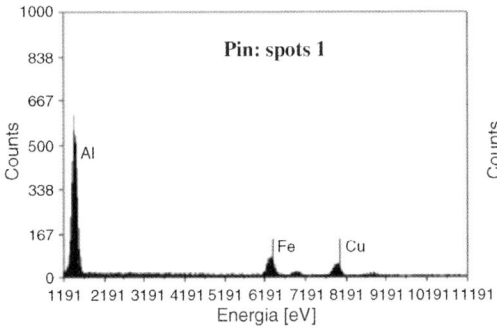

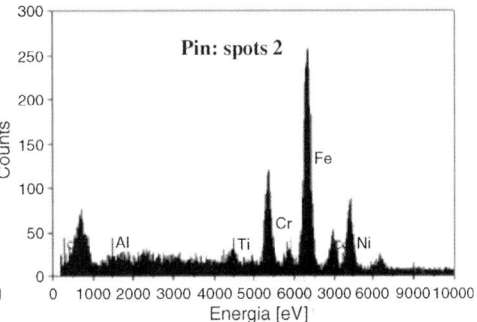

Figure 7.33 EDX analysis of the contact area of the disc (top, analysis spots 1, 2 and 3 marked in Figure 7.32, top row, right) and of the steel pin (bottom, spot 1 and spot 2 marked in Figure 7.32, bottom row, right) after the fretting test i-Al$_{62.5}$Cu$_{25.3}$Fe$_{12.2}$ versus A286 (maximum of noise: 87 mN – no adhesion). Only Al, Cu and Fe from the quasicrystal are present (no pin material) on the disc contact area, whereas only Al, Cu and Fe from the quasicrystal disc are present on the pin contact area (no steel pin material, made of Fe, Cr, Ni and Ti, is detected, therefore the quasicrystal is sticking on the pin) [109].

deposition consisted of a mixture of phases (i, β and Fe-Al phases), which are preserved in the final coatings. Earlier studies have shown that plasma-arc sprayed Al-Cu-Fe coatings are very brittle and exhibit poor abrasive wear resistance. The addition of the ductile phase significantly improves the abrasion resistance by increasing the effective fracture toughness. Adding just 1 vol.% of Fe-Al phase into the Al-Cu-Fe matrix induces a plastic flow wear mode. This particular coating possesses the best abrasion resistance associated with the highest average hardness (Figure 7.34).

Tin is another ductile metal that has been added to quasicrystalline-based composite coatings. The effect of Sn addition on the tribological properties of plasma-sprayed AlCuFe-based composite coatings was studied by Shao et al. [69]. Friction, hardness and wear behaviors of the composites were investigated as a function of Sn content up to 30 vol.%. Figure 7.35 shows a comparison of the volume loss after 30 m of sliding distance for Al-Cu-Fe and (AlCuFe + Sn) composite coatings and their corresponding counterpart balls. The wear of the balls was found to be almost independent of the composition of the coatings. The best wear performance was achieved for the Al-Cu-Fe + 20 vol.% Sn composite coating. The results of this study

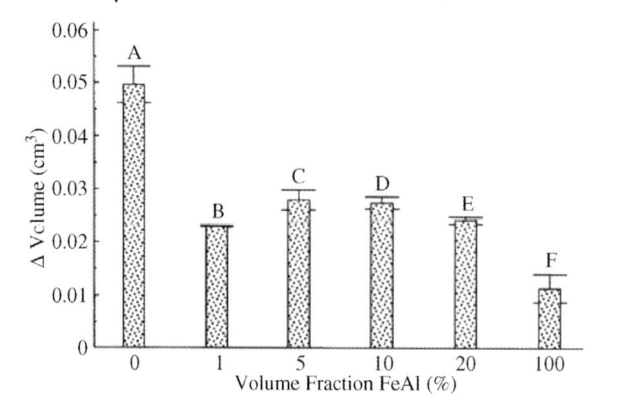

Figure 7.34 (a) Volume loss as a function of FeAl content during abrasion testing of air plasma-sprayed (Al-Cu-Fe + Al-Fe) composite coatings and (b) corresponding average hardness values. Abrasive wear tests were investigated with equipment described in ASTM-G65, rubber wheel abrasion test using a load of 87.2 N with a total of 300 revolutions at a wheel speed of 200 rpm. Silica particles (+200 μm, −300 μm) were used as abrasive powder. Reprinted from reference [67].

Coating Label	$H_{v0.025}$ GPa
A	6.9±1.1
B	8.1±0.9
C	6.1±0.8
D	6.0±0.8
E	5.8±0.7
F	4.1±0.7

show that the addition of Sn modifies the wear mechanism from brittle cracking to plastic deformation. This corresponds to an enhancement of the fracture toughness of (AlCuFe/Sn) composite coatings while their microhardness decreases linearly with the increase of the Sn volume fraction. Similar results were obtained for as-cast quasicrystalline composites; the addition of about 10 vol.% Sn enables a three-fold increase of the fracture toughness and a two-fold increase of the compressive strength for Al-Cu-Fe-B and Al-Cu-Fe composites, respectively [70].

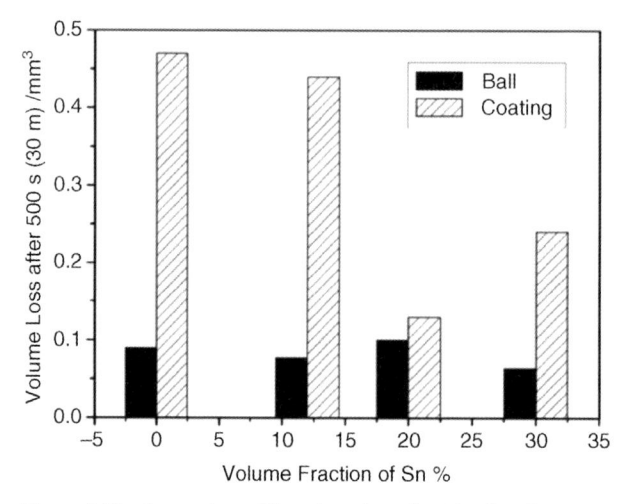

Figure 7.35 Comparison of the volume loss after 30 m for (AlCuFe + Sn) composite coatings as a function of Sn content and their corresponding counterpart balls. (GCr15 bearing steel ball, ø 12.7 mm, normal load = 10 N, reciprocating stroke of 0.6 mm). Redrawn from reference [69].

Similar AlCuFe/Sn quasicrystalline coatings have been prepared by the laser cladding technique [112]. The effect of Sn on the microhardness and the frictional behavior of these coatings are similar to those mentioned above for plasma-sprayed samples. Again, the best properties are obtained for the coating with 20 vol.% of Sn [112].

The best-performing quasicrystalline composite coating known so far has been developed by Lynntech Inc. using the electrocodeposition process on an Al-3004 alloy substrate [113]. Atomized $Al_{65}Cu_{23}Fe_{12}$ quasicrystalline powder (7 vol.%; $\emptyset \approx 10\,\mu m$) is poured into a nickel-plating electrolyte (Ni-434). The resulting composite coating is approximately 25 μm thick and consists of Al-Cu-Fe particles surrounded by a thin layer of Ni-P. This coating possesses poor wetting against polar liquids (average contact angle of water droplets is 117°) and excellent friction and wear resistance. The friction coefficient (μ) is exceptionally low (average value of μ is 0.05), with no visible wear track on the composite samples. The thermal stability of these composite coatings and their properties might be of great interest for various practical applications such as cookware, bearings, landing gear and engine parts.

7.4.2.2 Al-Based Composites Reinforced with CMAs

Friction properties of MMCs reinforced by quasicrystalline powders prepared by hot pressing were first investigated by Qi et al. [97]. As already mentioned, the hardness increases with increasing volume fraction of quasicrystalline particles in the range of $Al/10-30(Al_{65}Cu_{20}Cr_{15})p$. Reduced friction coefficients and improved wear resistance are observed for Al/(AlCuCr)p compared to the Al matrix. The best properties are obtained for the highest volume fraction investigated. Compared to commercial $Al/(SiC)_p$, Al/(AlCuCr)p composites have higher hardness and exhibit lower friction coefficient [114].

Composites reinforced with i-AlCuFeB particles also display improved friction properties. The quasicrystalline AlCuFeB alloy has a lower friction coefficient than the traditional AlCuFe alloy and is thus a better candidate for tribological applications [115]. Figure 7.36 shows the variation of the friction coefficient for composites containing 15, 30 and 60 vol.% of AlCuFeB particles transformed into the $\omega\text{-}Al_7Cu_2Fe$ sintered by LUP. In all cases and for limited sliding distances, composites have a friction coefficient much lower than that of sintered fcc aluminum $\mu_{Al} > 1$ and a significant decrease of the friction coefficient with increasing volume fraction of Al-Cu-Fe-B particles. However, after five meters of sliding, all coefficients tend towards the same value ($\mu \cong 0.6$). Chemical analyses of the worn surface of the composites and hard steel balls show that the composites and counterpart balls are covered by a transferred layer that is made up primarily of aluminum and oxygen (see inset in the Figure 7.36).

This can be explained by the high contact pressures induced by the experimental conditions (500 MPa), producing severe plastic deformation resulting in the coating of the indenter by Al transferred from the matrix. Therefore, the sliding body is no longer hard steel but essentially Al. Only at the beginning of the friction experiment, before material transfer occurs, does the nature of the sliding bodies have an

7 Mechanical Engineering Properties of CMAs

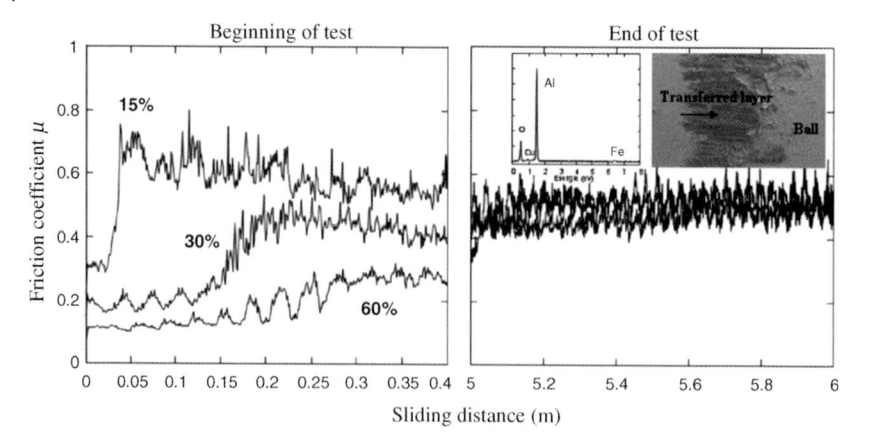

Figure 7.36 Variation of the friction coefficient for Al/(ω-Al-Cu-Fe-B)p composites with sliding distance. At the end of test, all friction curves merge to the same average value of μ. Friction tests were carried out at room temperature in air (humidity 50%) and under nonlubricated conditions using a normal load of 2 N and 6 mm 100C6 hard steel balls. The sliding velocity was 0.5 cm s^{-1}.

influence on friction [93, 96]. Note that similar frictional properties were observed for Al/AlCuFeB plasma-sprayed composite coatings [116]. Figure 7.37 shows the variation of the work of the friction force, calculated by integrating each friction curve over the first five meters of sliding. The presence of the i-phase in the Al/(AlCuFeB)p composites improves the friction and hardness properties less significantly than when the ω-phase is present. This is due to the embrittlement of the Al/(i-AlCuFeB)p interface caused by the oxide layer (see Section 7.3.2 for more details).

For all composites, material transfer is observed resulting in a sharp increase of the friction coefficient. The sliding distance before the occurrence of this transition defines a distance α [93]. This increases with the increasing volume fraction and with

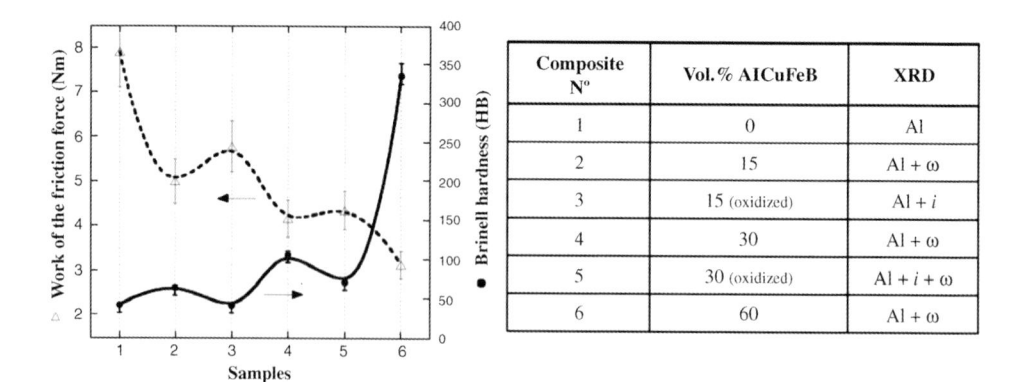

Composite N°	Vol.% AlCuFeB	XRD
1	0	Al
2	15	Al + ω
3	15 (oxidized)	Al + i
4	30	Al + ω
5	30 (oxidized)	Al + i + ω
6	60	Al + ω

Figure 7.37 Brinell hardness and work of the friction force of the various samples listed in the table. The lines are only guides for the eye (from reference [96]).

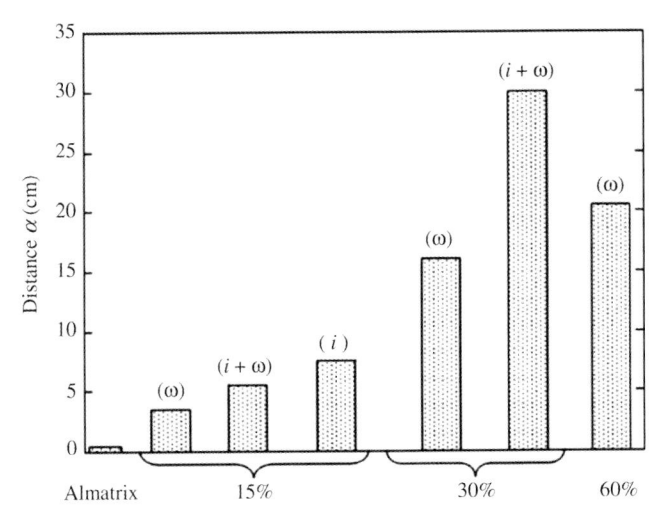

Figure 7.38 Dependence of the distance α with the volume fraction of (AlCuFeB)p and the icosahedral phase content (see text for details).

the presence of the i-phase (Figure 7.38). The dependence of α on the presence of the i-phase may be attributed to the low surface adhesion reported for quasicrystals.

7.5
Conclusions

In this overview, phenomenological results concerning the formation and the structure of single- or multiphase intermetallics and the resulting mechanical properties were presented. Such materials may contain intermetallic compounds with different structural complexity ranging from simple Laves phases to complex metallic alloys and quasicrystals. Regarding the processing techniques, both solidification from the melt or solid-state processing can be utilized. These synthesis routes may directly lead to single- or multiphase nanostructures. In other cases, additional heat treatment has to be employed in order to create or optimize the desired microstructure, such as in the case of devitrification of metallic glasses. Whereas solidification techniques can directly yield bulk samples with the desired microstructure, rapidly quenched ribbons, gas-atomized powders or mechanically alloyed powders and composites have to be subsequently consolidated to achieve dense bulk specimens.

Single-phase intermetallics display several attractive properties for engineering applications, such as high strength-to-weight ratio, good oxidation resistance and high-temperature strength. However, one major drawback for their use in engineering applications is the limited plastic deformability at room temperature. Several approaches, such as proper variation of the chemical composition and grain refinement to the nanometer regime, can be used to improve the room-temperature ductility of intermetallics. However, the method that shows the largest potential for improving the ductility of intermetallics at low temperatures is the production of two- or multiphase microstructures consisting of a soft metallic matrix with second-phase intermetallics. The mechanical properties of such two- or multiphase alloys are very encouraging regarding the combination of high strength and good ductility at room temperature. For example, Al- or Fe-based alloys containing intermetallic phases (e.g., Laves phases or quasicrystals) can exhibit very high room-temperature strength together with good ductility. In addition, interesting surface properties have been observed for several Al-based quasicrystals and composites, which further increase the importance and potential applications of intermetallics as engineering materials.

Finally, this class of materials provides many interesting topics for the study of microstructure–property relations, offering both the possibility to discover and develop new materials and properties, and a way to test models and understanding of mechanical deformation in single-phase materials as well as in composites consisting of different phases with different physical and mechanical properties, such as high strength in one phase and good ductility in the other.

Acknowledgments

Thanks are due to the European Commission for supporting this work in part under contract no. G5RD-CT-2001-0584.

References

1 Urban, K. and Feuerbacher, M. (2004) *J. Non-Cryst. Solids*, **334–335**, 143.
2 Feuerbacher, M. et al. (2007) *Z. Kristallogr.*, **222**, 259.
3 Dolinšek, J., Apih, T., Jeglič, P., Smiljanić, I., Bilušić, A., Bihar, Ž., Smontara, A., Jagličić, Z., Heggen, M., and Feuerbacher, M. (2007) *Intermetallics*, **15**, 1367.
4 Bauer, E. et al. (2007) *Phys. Rev. B*, **76**, 014528.
5 Smontara, A., Smiljanić, I., Bilušić, A., Jagličić, Z., Klanjšek, M., Roitsch, S., Dolinšek, J., and Feuerbacher, M. (2007) *J. Alloys Compd.*, **430**, 29.
6 Maciá, E. and Dolinšek, J. (2007) *J. Phys.: Condens. Matter.*, **19**, 176212.
7 Feuerbacher, M., Thomas, C., and Roitsch, S. (2008) *Intermetallics*, **16**, 943.
8 Demange, V., Machizaud, F., Dubois, J.M., Anderegg, J.W., Thiel, P.A., and Sordelet, D.J. (2002) *J. Alloys Compd.*, **342**, 24.
9 Roitsch, S., Heggen, M., Lipińska-Chwaøek, M., and Feuerbacher, M. (2007) *Intermetallics*, **15**, 833.
10 Heggen, M., Deng, D., and Feuerbacher, M. (2007) *Intermetallics*, **15**, 1425.
11 Livingston, J.D. (1994) *Mater. Res. Soc. Symp. Proc.*, **322**, 395.
12 Livingston, J.D. (1992) *Phys. Status Solidi A*, **131**, 415.
13 Von Keitz, A. and Sauthoff, G. (2002) *Intermetallics*, **10**, 497.
14 Kumar, K.S. (1997) *Mater. Res. Soc. Symp. Proc.*, **460**, 677.
15 Liu, C.T., Zhu, J.H., Brady, M.P., McKamey, C.G., and Pike, L.M. (2000) *Intermetallics*, **8**, 1119.
16 Hazzledine, P.M., Kumar, K.S., Miracle, D.B., and Jackson, A.G. (1993) *Mater. Res. Soc. Symp. Proc.*, **288**, 591.

17. Kumar, K.S. and Hazzledine, P.M. (2004) *Intermetallics*, **12**, 763.
18. Hazzledine, P.M. (1994) in *Twinning in Advanced Materials*, (ed M.H. Yoo and M. Wuttig), The Minerals, Metals and Materials Society, Warrendale/PA, p. 403.
19. Thoma, D.J., Chen, K.C., Baskes, M.I., and Petersen, E.J. (2001) in The Fourth Pacific Rim International Conference on Advanced Materials and Processing (PRICM4), The Japan Institute of Metals (eds S. Hanada, Z. Zhong, S.W. Nam, and R.N. Wright), p. 995.
20. Nakagawa, Y., Ohta, T., Kaneno, Y., Inoue, H., and Takasugi, T. (2004) *Metal. Mater. Trans. A*, **35**, 3469.
21. Livingston, J.D. and Hall, E.L. (1990) *J. Mater. Res.*, **5**, 5.
22. Chu, F. and Pope, D.P. (1993) *Mater. Sci. Eng. A*, **170**, 39.
23. Liu, Y., Livingston, J.D., and Allen, S.M. (1992) *Metall. Mater. Trans. A*, **23**, 3303.
24. Liu, Y., Allen, S.M., and Livingston, J.D. (1993) *Mater. Res. Soc. Symp. Proc.*, **288**, 203.
25. Huttunen-Saarivirta, E. (2004) *J. Alloys Compd.*, **363**, 150.
26. Takeuchi, S., Edagawa, K., and Tamura, R. (2001) *Mater. Sci. Eng. A*, **319–321**, 93.
27. Giacometti, E., Baluc, N., Bonneville, J., and Rabier, J. (1999) *Scr. Mater.*, **41**, 989.
28. Kumar, K.S., Van Swygenhoven, H., and Suresh, S. (2003) *Acta Mater.*, **51**, 5743.
29. Koch, C.C. (ed.) (2002) *Nanostructured Materials: Processing, Properties and Potential Applications*, Noyes Publications/William Andrew Publishing, Norwich, NY.
30. Courtney, T.H. (1990) *Mechanical Behavior of Materials*, McGraw-Hill, New York, NY.
31. Scudino, S., Sakaliyska, M., Stoica, M., Surreddi, K.B., Ali, F., Vaughan, .G., Yavari, A.R., and Eckert, J. (2008) *Phys. Status Solidi (RRL)*, **2**, 272.
32. Scudino, S., Sakaliyska, M., Surreddi, K.B., and Eckert, J. (2009) *J. Alloys Compd.*, **483**, 2.
33. Scudino, S., Sakaliyska, M., Surreddi, K.B., and Eckert, J. (2009) *J. Phys.: Conf. Series*, **144**, 012019.
34. Sakaliyska, M., Scudino, S., Nguyen, H.V., Surreddi, K.B., Bartusch, K.B., Ali, F., Kim, J.-S., and Eckert, J. (2009) *Mater. Res. Soc. Symp. Proc.*, **1128**, U05.
35. Mamedov, V. (2002) *Powder Metall.*, **45**, 322.
36. Davies, L.A. (1986) *Mechanical Behavior of Rapidly Solidified Materials*, The Metallurgical Society, Warrendale, PA.
37. Takeyama, M. and Liu, C.T. (1991) *Mater. Sci. Eng. A*, **132**, 61.
38. Kim, W.Y., Luzzi, D.E., and Pope, D.P. (2003) *Intermetallics*, **11**, 257.
39. Park, J.M., Sohn, S.W., Kim, T.E., Kim, D.H., Kim, K.B., and Kim, W.T. (2007) *Scr. Mater.*, **57**, 1153.
40. Park, J.M., Kim, T.E., Sohn, S.W., Kim, D.H., Kim, K.B., Kim, W.T., and Eckert, J. (2008) *Appl. Phys. Lett.*, **93**, 031913.
41. Park, J.M., Kim, K.B., Kim, W.T., Lee, M.H., Eckert, J., and Kim, D.H. (2008) *Intermetallics*, **16**, 642–650.
42. Srivastava, R., Eckert, J., Löser, W., Dhindaw, B.K., and Schultz, L. (2002) *Mater. Trans. JIM*, **43**, 1670.
43. Scudino, S., Donnadieu, P., Surreddi, K.B., Nikolowski, K., Stoica, M., and Eckert, J. (2009) *Intermetallics*, **17**, 532.
44. Inoue, A. (2000) *Acta Mater.*, **48**, 279.
45. Elliott, S.R. (1990) *Physics of Amorphous Materials*, Longman Scientific and Technical, Essex.
46. Inoue, A., Shen, B.L., and Chang, C.T. (2004) *Acta Mater.*, **52**, 4093.
47. Imafuku, M., Sato, S., Koshiba, H., Matsubara, E., and Inoue, A. (2001) *Scr. Mater.*, **44**, 2369.
48. Matsubara, E., Sato, S., Imafuku, M., Nakamura, T., Koshiba, H., Inoue, A., and Waseda, Y. (2001) *Mater. Sci. Eng. A*, **312**, 136–144.
49. Poon, S.J., Shiflet, G.J., Guo, F.Q., and Ponnambalam, V. (2003) *J. Non-Cryst. Solids*, **317**, 1.
50. Stoica, M., Hajlaoui, K., LeMoulec, A., and Yavari, A.R. (2006) *Philos. Mag. Lett.*, **86**, 267.
51. Eckert, J. and Scudino, S. (2007) in Chapter 6/1–27, *Materials Processing Handbook* (eds J.R. Groza, J.F. Schackelford, E.J. Lavernia, and M.T. Powers), Taylor & Francis CRC Press.
52. Inoue F A., Zhang, T., Saida, J., Matsushita, M., Chen, M.W., and

Sakurai, T. (1999) *Mater. Trans. JIM*, **40**, 1181.

53 Eckert, J., Kuhn, U., Mattern, N., Reger-Leonhard, A., and Heilmaier, M. (2001) *Scr. Mater.*, **44**, 1587.

54 Leonhard, A., Xing, L.Q., Heilmaier, M., Gebert, A., Eckert, J., and Schultz, L. (1998) *Nanostruct. Mater.*, **10**, 805.

55 Scudino, S., Mickel, C., Schultz, L., Eckert, J., Yang, X.Y., and Sordelet, D.J. (2004) *Appl. Phys. Lett.*, **85**, 4349.

56 Eckert, J., Scudino, S., Yu, P., and Duhamel, C. (2007) *Mater. Sci. Forum*, **534–536**, 1405.

57 Dubois, J.M., Kang, S.S., and Massiani, Y. (1993) *J. Non-Cryst. Solids*, **153–154**, 443.

58 Liu, P., Stigenberg, A.H., and Nilsson, J.-O. (1994) *Scr. Metall. Mater.*, **31**, 249.

59 Nilsson, J.-O., Stigenberg, A.H., and Liu, P. (1994) *Metall. Mater. Trans. A*, **25**, 2225.

60 Liu, P., Stigenberg, A.H., and Nilsson, J.-O. (1995) *Acta Metall. Mater.*, **43**, 2881.

61 Inoue, A. and Kimura, H. (2000) *Mater. Sci. Eng. A*, **286**, 1.

62 Inoue, A., Kimura, H.M., and Zhang, T. (2000) *Mater. Sci. Eng. A*, **294–296**, 727.

63 Bae, D.H., Kim, S.H., Kim, D.H., and Kim, W.T. (2002) *Acta Mater.*, **50**, 2343.

64 Zhang, Y., Yu, S., Song, Y., and Zhu, X. (2008) *J. Alloys Compd.*, **464-1-2**, 575.

65 Tsai, A.P., Aoki, K., Inoue, A., and Masumoto, T. (1993) *J. Mater. Res.*, **8**, 5.

66 Bloom, P.D., Baikerikar, K.G., Otaigbe, J.U., and Sheares, V.V. (2000) *Mater. Sci. Eng. A*, **294–296**, 156.

67 Sordelet, D.J., Besser, M.F., and Logsdon, J.L. (1998) *Mater. Sci. Eng. A*, **255**, 54.

68 Reimann, A. and Lugscheider, E. (2001) in Proceedings of the 2001 International Thermal Spray Conference, ASM International (eds C.C. Berndt, K.A. Khor, and E. Lugscheider), p. 33.

69 Shao, T., Cao, X., Fleury, E., Kim, D.H., Hua, M., and Se, D. (2004) *J. Non-Cryst. Solids*, **334–335**, 466.

70 Fleury, E., Kim, Y.C., Kim, D.H., and Kim, W.T. (2004) *J. Non-Cryst. Solids*, **334–335**, 449.

71 Clyne, T.W. and Withers, P.J. (1993) *An Introduction to Metal Matrix Composites*, Cambridge University Press, Cambridge.

72 Kainer, K.U. (2006) *Metal Matrix Composites. Custom-made Materials for Automotive and Aerospace Engineering*, Wiley-VCH, Weinheim.

73 Tsai, A.P., Inoue, A., and Masumoto, T. (1987) *J. Mater. Sci. Lett.*, **6**, 1403.

74 Köster, U., Liu, W., Liebertz, H., and Michel, M. (1993) *J. Non-Cryst. Solids*, **153–154**, 446.

75 Bradley, A.J. and Goldschmidt, H.J. (1939) *J. Inst. Metals*, **65**, 403.

76 Quiquandon, M., Quivy, A., Faudot, F., Sâadi, N., and Calvayrac, Y. (1995) in Proceedings of the 5th International Conference on Quasicrystals, World Scientific, Singapore (eds C. Janot and R. Mosseri), p. 152.

77 Lee, S.M., Jung, J.H., Fleury, E., Kim, W.T., and Kim, D.H. (2000) *Mater. Sci. Eng. A*, **294–296**, 99.

78 Fleury, E., Lee, S.M., Choi, G., Kim, W.T., and Kim, D.H. (2001) *J. Mater. Sci.*, **36–4**, 963.

79 Cherdyntsev, V.V., Kaloshkin, S.D., Tomilin, I.A., Shelekhov, E.V., Baldokhin, Yu.V., and Afonina, E.A. (2004) *Phys. Met. Metallogr.*, **97–5**, 479.

80 Kenzari, S. (2006) Institut National Polytechnique de Lorraine. PhD thesis, unpublished.

81 Sordelet, D.J., Bloomer, T.A., Kramer, M.J., and Unal, O. (1996) *J. Mater. Sci. Lett.*, **15**, 935.

82 Huang, S.Y. and Shield, J.E. (1997) *Philos. Mag. B*, **75**, 157.

83 Brien, V., Khare, V., Herbst, F., Weisbecker, P., Ledeuil, J.B., de Weerd, M.C., Machizaud, F., and Dubois, J.M. (2004) *J. Mater. Res.*, **19**, 2974.

84 Kang, S.S. and Dubois, J.M. (1992) *Philos. Mag. A*, **66**, 151.

85 Belin-Ferré, E., Dubois, J.M., Fournée, V., Brunet, P., Sordelet, D.J., and Zhang, L.M. (2000) *Mater. Sci. Eng. A*, **294–296**, 818.

86 Brunet, P., Zhang, L.M., Sordelet, D.J., Besser, M., and Dubois, J.M. (2000) *Mater. Sci. Eng. A*, **294–296**, 74.

87 Dubois, J.M. (2000) *Mater. Sci. Eng. A*, **294–296**, 4.

88 Kaloshkin, S.D., Tcherdyntsev, V.V., Laptev, A.I., Stepashkin, A.A., Afonina, E.A., Pomadchik, A.L., and Bugakov, V.I. (2004) *J. Mater. Sci.*, **39-16-17**, 5399.

89. El Kabir, T., Joulain, A., Gauthier, V., Dubois, S., Bonneville, J., and Bertheau, D. (2008) *J. Mater. Res.*, **23–4**, 904.
90. Tang, F., Meeks, H., Spowart, J.E., Gnaeupel-Herold, T., Prask, H., and Anderson, I.E. (2004) *Mater. Sci. Eng. A*, **386**, 194.
91. Tang, F., Anderson, I.E., Gnaupel-Herold, T., and Prask, H. (2004) *Mater. Sci. Eng. A*, **383**, 362.
92. Schurack, F., Eckert, J., and Schultz, L. (2003) *Philos. Mag.*, **83–11**, 1287.
93. Kenzari, S., Weisbecker, P., Geandier, G., Fournée, V., and Dubois, J.M. (2006) *Philos. Mag.*, **86**, 287.
94. Bonhomme, G., LeMieux, M., Weisbecker, P., Tsukruk, V.V., and Dubois, J.M. (2004) *J. Non-Cryst. Solids*, **334–335**, 532.
95. Weisbecker, P., Bonhomme, G., Bott, G., and Dubois, J.M. (2005) *J. Non-Cryst. Solids*, **351**, 1630.
96. Kenzari, S., Weisbecker, P., Curulla, M., Geandier, G., Fournée, V., and Dubois, J.M. (2008) *Philos. Mag.*, **88**, 755.
97. Qi, Y.H., Zhang, Z.P., Hei, Z.K., Yan, L., and Dong, C. (1998) *Mocaxue Xuebao/Tribol.*, **18–2**, 129.
98. Tang, F., Anderson, I.E., and Biner, S.B. (2003) *Mater. Sci. Eng. A*, **363**, 20.
99. Tang, F., Gnäupel-Herold, T., Prask, H., and Anderson, I.E. (2005) *Mater. Sci. Eng. A*, **399**, 99.
100. Tang, F., Anderson, I.E., and Biner, S.B. (2002) *J. Light Met.*, **2**, 201.
101. Mcdanels, D.L. (1985) *Metall. Trans. A*, **16**, 1105.
102. Arsenault, R.J., Wang, L., and Feng, C.R. (1991) *Acta Metal. Mater.*, **39**, 47.
103. Merstallinger, A. and Semerad, E.,In-house-standard by Austrian Research Centre Seibersdorf, Issue 1(1995), Issue 2 (1998).
104. Johnson, K.H. (1985) *Contact Mechanics*, Cambridge University Press, Cambridge.
105. Merstallinger, A., Semerad, E., Scholze, P., and Schmidt, C. (2000) ESTEC Contract No. 11760/95/NL/NB, CO 21.
106. Merstallinger, A., Semerad, E., Dunn, B.D., and Störi, H. (1995) 6th Europ. Space Mechanisms & Tribology Symposium, Proceedings, ESA SP-374 Zürich.
107. European Co-operation for Space Standardisation (ECSS) (2000) ECSS-E-30Mechanical, Part 3A Mechanisms, section 4.7.4.4.5 Separable contact surfaces, p. 32.
108. Merstallinger, A., Semerad, E., and Dunn, B.D. (2003) Proc. 9th European Space Symposium on Material in a Space Environment, ESTEC Noordwijk (NL).
109. Sales, M., Merstallinger, A., Brunet, P., De Weerd, M.C., Khare, V., Traxler, G., and Dubois, J.M. (2006) *Philos. Mag.*, **86**, 965.
110. Dubois, J.M. (2005) *Useful Quasicrystals*, World Scientific, Singapore.
111. Dubois, J.M., Brunet, P., Costin, W., and Merstallinger, A. (2004) *J. Non-Cryst. Solids*, **334–335**, 475.
112. Feng, L.P., Shao, T.M., Jin, Y.J., Fleury, E., Kim, D.H., Chen, D.R., and Wei, Y.G. (2005) *T. Nonferr. Metal. Soc.*, **15–2**, 432.
113. Minevski, Z., Tennakoon, C.L., Anderson, K.C., Nelson, C.J., Burns, F.C., Sordelet, D.J., Hearing, C.W., and Pickard, D.W. (2003) in *Quasicrystals 2003-Preparation, Properties and Applications, MRS Proc. Series*, vol. 805 (eds E. Belin-Ferré, M. Feuerbacher, Y. Ishii, and D.J. Sordelet), Warrendale, p. 345.
114. Qi F Y.H., Zhang, Z.P., Hei, Z.K., and Dong, C. (2000) *T. Nonferr. Metal. Soc.*, **10–3**, 358.
115. Dubois, J.M., Archambault, P., Bresson, L., and Cathonnet, P. (1996) French Patent No. 9602224.
116. Stihle, A., Liao, H., Bertrand, P., Allain, N., Coddet, C., and Kenzari, S. (2006) in Proceedings of the 2006 International Thermal Spray Conference, ASM International (eds B.R. Marple, M.H. Hyland, Y.C. Lau, R.S. Lima and J. Voyer), p. 1089.

8
CMA's as Magnetocaloric Materials
Spomenka Kobe, Benjamin Podmiljšak, Paul John McGuiness, and Matej Komelj

8.1
Introduction

Magnetocalorics (MC) are materials that show a magnetocaloric effect (MCE), or adiabatic temperature change (ΔT_{ad}), which is defined as the heating or cooling of a magnetic material in response to the application of a magnetic field.

The MCE was first discovered in iron samples in 1881 by Warburg [1] and was later explained independently by Debye [2] and Giauque [3]. They also suggested the first practical use of the MCE: adiabatic demagnetization, used to reach temperatures lower than that of liquid helium, which had been the lowest achievable experimental temperature.

Most magnetic materials exhibit a large MCE only at low temperatures, making them unsuitable for practical use in everyday life. But with the discovery of the giant MCE in $Gd_5Si_2Ge_2$ in 1997 by Pecharsky and Gschneidner [4], magnetic refrigeration (MR) became a viable and competitive technology with vapor cycle refrigeration. $Gd_5Si_2Ge_2$ is a ferromagnetic material with a spontaneous ordering temperature of 276 K.

In recent years, much research has been carried out to find new materials with higher MCEs around room temperature, with the goal of making the first magnetic refrigerator. Because such an apparatus must be small, energetically favorable and economically justified, we need MC materials that are not too expensive, are easy to produce and have a high MCE so that they can be operated using permanent magnets in the room-temperature region.

The physical origin of the MCE is the coupling of the magnetic sublattice to the applied magnetic field, H, which changes the magnetic contribution to the entropy of the solid. This process is thermodynamically equivalent to the process that occurs in a gas in response to changing pressure. If we isothermally compress gas, the entropy decreases. This is analogous to the isothermal magnetization of a paramagnet or a soft ferromagnet. While adiabatically expanding the gas, the temperature (T) decreases. This is equivalent to adiabatic demagnetization, where we remove the magnetic field H. The total entropy remains constant, but

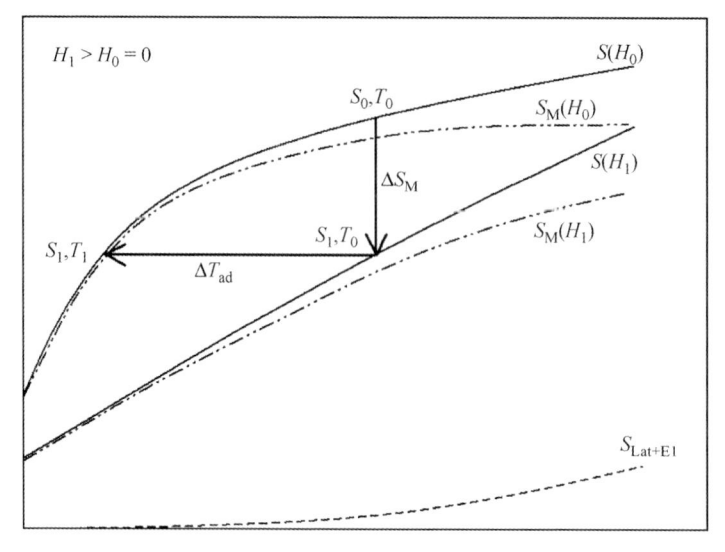

Figure 8.1 Temperature dependence of the total entropy for a ferromagnet at two different fields showing the MCE. The total entropy ΔS is the contribution of the lattice entropy, S_{Lat}, the electronic entropy, S_{El}, and the magnetic entropy, S_{M}.

the temperature decreases because the magnetic entropy increases. These two processes are shown in Figure 8.1. The isothermal magnetization is represented by the vertical arrow. This represents the magnetic entropy change, ΔS_M, which is defined as:

$$\Delta S = S(T_0, H_0) - S(T_0, H_1) \tag{8.1}$$

The adiabatic demagnetization is represented by the horizontal arrow. This represents the adiabatic temperature change, ΔT_{ad}, which is defined as:

$$\Delta T_{ad} = T_0 - T_1 \tag{8.2}$$

This reduction in the temperature gives the cooling effect. Both parameters, the ΔT_{ad} and the ΔS_M, describe the MCE.

The relation between H, the magnetization, M, and T, and the MCE values, ΔT_{ad} $(T, \Delta H)$ and $\Delta S_M(T, \Delta H)$, is given by one of Maxwell's equations,

$$\left(\frac{\partial S(T, H)}{\partial H}\right)_T = \left(\frac{\partial M(T, H)}{\partial T}\right)_H \tag{8.3}$$

which for an isothermal-isobaric process after integration yields

$$\Delta S_M(T, \Delta H) = \int_{H_1}^{H_2} \left(\frac{\partial M(T, H)}{\partial T}\right) dH \tag{8.4}$$

Equation 8.4 shows that a high ΔS_M can be expected for large magnetic-field changes and when $|(\partial M(T, H)/\partial T)_H|$ has the highest value, that is, around the temperature of the magnetic transformation of a conventional ferromagnet, T_C, or near the absolute zero temperature of a paramagnet.

If we express the total differential as a function of T, H and p:

$$\Delta S = \left(\frac{\partial S}{\partial T}\right)_{H,p} dT + \left(\frac{\partial S}{\partial H}\right)_{T,p} dH + \left(\frac{\partial S}{\partial p}\right)_{H,T} dp \qquad (8.5)$$

combining Equations 8.3 and 8.6

$$C_{H,p} = T\left(\frac{\partial S}{\partial T}\right)_{H,p} \qquad (8.6)$$

where $C_{H,p}$ is the heat capacity at constant magnetic field and pressure, we can write for an adiabatic-isobaric process, the temperature change due to the MCE:

$$dT = -\frac{T}{C_{H,p}}\left(\frac{\partial M}{\partial T}\right)_{H,p} dH \qquad (8.7)$$

After integrating we obtain:

$$\Delta T_{ad}(T, \Delta H) = -\int_{H_0}^{H_1} \left(\frac{T}{C(T,H)}\right)_H \left(\frac{\partial M(T,H)}{\partial T}\right)_H dH \qquad (8.8)$$

The same conclusions can be made for ΔT_{ad} as for ΔS_M. Additionally, it is clear that ΔT_{ad} is large at high temperatures or low heat capacities, providing that all the other parameters remain the same. The largest MCE is expected when the heat capacity of a material is strongly influenced by the magnetic field.

These assumptions are only valid for a continuous second-order magnetic phase transformation. In the vicinity of a discontinuous first-order magnetic phase transformation these equations fail to describe the MCE, because the bulk magnetization is expected to undergo a discontinuous change at constant temperature, making the heat capacity in the function $(T/C(T,H))_H$ and the function $(\partial M(T,H)/\partial T)_H$ become infinite for this transformation.

In reality, these changes occur over a few degrees, and both functions can be measured experimentally. The largest values of ΔT_{ad} are predicted to occur when the Curie temperature is strongly affected by the magnetic field. For a more thorough explanation of the MCE, the reader is referred to Tishin and Spichkin [5].

The MCE can be described in terms of the magnetic entropy change (ΔS_M) using the units of J kg^{-1} K^{-1}, but for engineers the unit J cm^{-3} K^{-1} may be preferable, as it describes the cooling of a MC material per unit volume. Because many reports are still in J kg^{-1} K^{-1}, with no data on the density, the comparison is not straightforward. In Table 8.1 a comparison of the units for two materials, Gd and Gd$_5$Si$_2$Ge$_2$, is presented. Gd is a representative of materials with a second-order magnetic transformation (SOMT), while Gd$_5$Si$_2$Ge$_2$ has a first-order magnetic transformation

Table 8.1 The two references described with different units for a better comparison with other MC materials [4, 6].

Field change MCE	2 T ΔS_M	ΔT_{ad}	5 T ΔS_M		ΔT_{ad}
unit	mJ cm^{-3} K^{-1} J kg^{-1} K^{-1}	K	mJ cm^{-3} K^{-1}	J kg^{-1} K^{-1}	K
Gd ($\varrho = 7.901$ g cm^{-3})	−39.5 −5	5.7	−77.4	−9.8	11.5
Gd$_5$Si$_2$Ge$_2$ ($\varrho = 7.52$ g cm^{-3})	−105.3 −14	7.4	−139.1	−18.5	15.2

(FOMT). Both MCE parameters, ΔS_M and ΔT_{ad}, are reported together with the field to which the material was exposed.

When comparing the ΔS_M values the problem is that ΔT_{ad} is not taken into consideration. A better parameter for comparing magnetocaloric materials is the refrigerant capacity (RC), which is defined as

$$q = \int_{T_1}^{T_2} \Delta S_M(T) dT \qquad (8.9)$$

where T_1 and T_2 are the temperatures of the hot and cold sinks, respectively, and $\Delta S_M(T)$ is the refrigerant's magnetic entropy change as a function of temperature. The refrigerant capacity, therefore, is a measure of how much heat can be transferred between the cold and hot sinks in one ideal refrigeration cycle [7]. To get the total refrigerant capacity we need to take into account the hysteresis losses that occur during the FOMT. These losses have to be subtracted from the RC to compare the values with materials that have a SOMT, where those losses do not occur.

8.2
Materials

8.2.1
Theoretical Investigation of the Magnetocaloric Effect

In order to develop a magnetocaloric (MC) material the application of which might yield the desired performance, it is important to understand the intrinsic properties. A possible approach is to model the MC behavior of the investigated material theoretically. Nowadays, the state-of-the-art simulations of real materials are based on the application of the density-functional theory (DFT) [8], which makes it possible to calculate the electronic ground state in solids without any additional parameters to the chemical composition and the crystal structure. However, the MC effect is not an electronic ground-state phenomenon, since it depends on the temperature and it involves the presence of an external magnetic field. The question is whether the ground-state properties, for example, the magnetic moments, magnetic-exchange-coupling constants, or simply the electronic density of states, calculated *ab initio* by using one of the many DFT-based methods, can be related to the complex MC effect?

On the other hand, the characteristic isothermal entropy change (ΔS_M) and the adiabatic temperature change (ΔT) with magnetic-field variation can be investigated on a model Hamiltonian in combination with the mean-field approximation or the Monte-Carlo simulation. Such an approach, of course, requires a set of input parameters, which are usually determined so that the theoretical results fit to the experimental data best. Nevertheless, at least some of the required parameters might be obtained from a DFT calculation.

8.2.1.1 $Gd_5Si_2Ge_2$

The most investigated MC materials, besides elemental gadolinium, are the alloys of the type $Gd_5Si_xGe_{4-x}$. Therefore, it is not surprising that there have been also a lot of activities aimed at a theoretical description of the experimentally observed behavior. A standard DFT calculation with the local-spin-density approximation (LSDA) does not describe the Gd 4f-states properly [9], therefore the resulting mean-field phase-transition temperature of 61 K is significantly lower than the experimental value 276 K [4]. The theoretical position of the 4f bands and the ferromagnetic ground state can be corrected by adding the on-site Coulomb repulsion, for example, in terms of the LSDA + U method, which indeed significantly improves the agreement with the experiment, yielding a phase-transition temperature between 180 and 230 K, depending on the considered crystal phase [10]. It should be pointed out though that the application of the LSDA + U method requires the two additional parameters, namely Hubbard's U and J. In principle, these two parameters could be calculated *ab initio* too, whereas a mean-field model Hamiltonian investigation [11] with the parameters chosen to reproduce the experimental results as well as possible could include the influence of magnetic field, pressure and magnetoelastic deformation to interpret their influence on the MC effect. The mean-field approximation neglects the short-range interactions; therefore, it does not reproduce the magnetic part of the heat capacity around the magnetic phase transition correctly. In order to avoid the limitations of the mean-field approximation, Monte-Carlo simulations on various first-order $Gd_5Si_xGe_{4-x}$-type alloys were performed [12], in general, with a tendency to improve the agreement with experiment. A state-of-the-art, systematic, theoretical investigation of the magnetocaloric effect in $Gd_5Si_2Ge_2$ was performed as a combination of an *ab initio* calculation and the application of a magnetothermodynamic model. In addition to a good agreement with the measured data it was shown that the orthorhombic phase was more stable than the monoclinic one at low temperatures, due to a lower total energy. The calculated Gd magnetic moments and magnetic-exchange-coupling constants were found to be higher for the orthorhombic phase than for the monoclinic phase. Different values of the Gd magnetic moments in different, or even within the same, phases were ascribed to a variation of the 5d-electrons' contribution at different sites.

8.2.1.2 $LaFe_{13-x}Si_x$

Although a first-order-phase-transition material, it exhibits mainly the properties associated with a second-order-phase-transition material, which makes it unique,

Figure 8.2 Fermi surface calculated within the framework of the DFT and LSDA.

and hence attractive from the experimental as well as from the theoretical points of view. However, theoretical investigations have so far been quite rare, in part probably because the parent compound, $LaFe_{13}$, is not stable, which requires the presence of additional elements, for example, silicon, yielding a not-so-well-defined crystal structure. Nevertheless, based on a standard DFT calculation it was revealed that the observed first-order magnetic transition was in fact a series of three consecutive first-order transitions [13]. The reason for that is the shape of the energy–magnetization curve, with an extensive flat-bottom part and several shallow minima and maxima. Furthermore, low energy barriers between the various spin states make possible a faster magnetization and/or demagnetization with less hysteresis. Although the presence of silicon is necessary to stabilize the material, it is not believed that it is essential for the MC effect, due to its influence on the electronic structure, which is anyhow complex, as demonstrated with the Fermi surface calculated within the framework of the DFT and LSDA, as shown in Figure 8.2 [14].

8.2.2
Elemental Magnetocalorics

Gadolinium often serves as a standard to compare the magnetocaloric effect in new materials. With a Curie temperature T_C of 294 K and a ΔS_M of -39.5 mJ cm^{-3} K^{-1} and ΔT_{ad} of 5.7 K (both for a 2-kOe field change), Gd is a good MC material [15]. These results, however, are only possible with Gd of high purity. Commercial Gd generally has about a two times lower MCE than high-purity samples [7]. The high price of Gd is its major drawback. For example, the price for 300 g of high-purity Gd lumps with 99.9% purity (approximately the mass needed for a magnetic refrigerator) is around €1000 [16].

Other rare-earth elements, which have a smaller MCE, also require lower temperatures and are not usable for room-temperature refrigeration [17]. Studies on Gd and solid solutions of Gd, have given promising results. Doping Gd with Tb and Nd lowered the T_C of Gd [18]. Additions of B to Gd expanded the unit-cell volume, thus increasing the T_C as well as the refrigeration capacity [19].

8.2.3
Intermetallic Compounds

8.2.3.1 Laves Phases

Much research has been done on RCo_2-based systems (where R = Dy, Ho or Er) because these systems exhibit a first-order paramagnetic–ferromagnetic transition. The highest ΔT_{ad} is found for the $DyCo_2$ compound (4.5 K for a field change from 0 to 20 kOe), but this compound has a T_C value of only 142 K [20]. $HoCo_2$ exhibits a 8.8 K temperature change for a 50-kOe field change at 83 K. $ErCo_2$ has the highest ΔS_M (-331 mJ cm^{-3} K^{-1} for 0 to 50 kOe), but a low ΔT_{ad}. Most recent studies involve the substitution of a rare-earth metal for one of the magnetic lanthanides [21] or the substitution of a nonrare-earth metal for Co [22]. Substituting Al for Co in $DyCo_2$, for example, an increase of T_C up to 206 K was achieved for the compound $Dy(Co_{1-x}Al_x)_2$ where $x = 0.1$ [23]. ΔS_M drops off for $x > 0.02$ because the FOMT is destroyed. It can be said that the maximum magnetic entropy change, $\Delta S(max)_M$, is the highest in the compound with the lowest T_C. Figure 8.3 illustrates this dependence. Studies of MCE

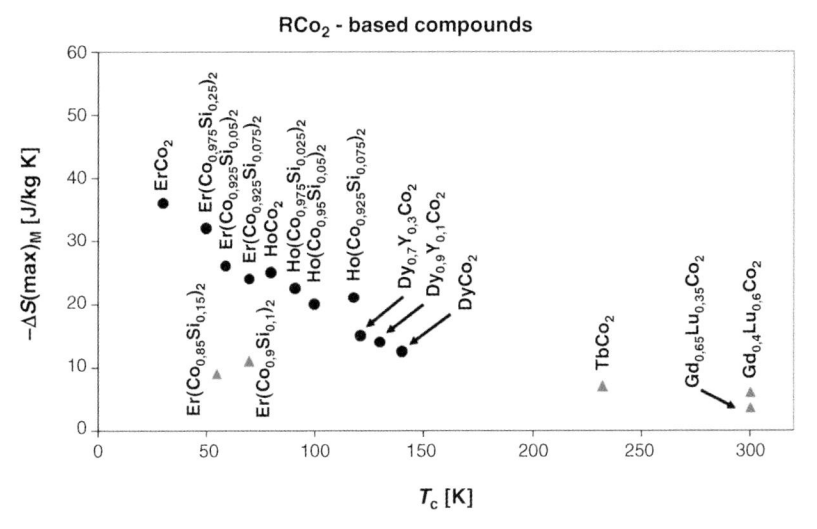

Figure 8.3 $\Delta S(max)_M$ vs. T_C of RCo_2-based compounds. The circles indicate materials that show a FOT whereas the circles indicate materials that show a SOT [29].

under applied pressure always show a drop in the T_C and an increase in the ΔS with increasing pressure [24]. A good overview of RCo_2-based systems is given by Singh et al. [25].

For RAl_2-based systems, only a few papers have been published. In all the available reports, the RAl_2 compounds show a MCE only in the low-temperature region, and MCE values for heavy lanthanides are higher than for light ones [26, 27]. RNi_2 compounds are only suitable for cooling in the 7 K to 22 K temperature range [28].

As shown, all these materials are suitable only for low-temperature magnetic refrigeration.

8.2.3.2 CMAs [$Gd_5(Si_{1-x}Ge_x)_4$ Alloys and Related 5:4 Materials]

$Gd_5(Si_{1-x}Ge_x)_4$ alloys are among the most extensively researched MC materials. After the discovery of the useful properties of $Gd_5Si_2Ge_2$, the whole range of stoichiometries was investigated. When the temperature or the applied magnetic field is changed, these alloys undergo a simultaneous change of crystallographic symmetry and magnetic order. This type of transformation, termed magnetic-martensitic, is extremely rare. For alloys with $0.24 \leq x \leq 0.5$, a transformation from a paramagnetic monoclinic $Gd_5(Si_2Ge_2)$-type structure to a ferromagnetic orthorhombic Gd_5Si_4-type structure occurs. For alloys with $0 \leq x \leq 0.2$, a transformation from an antiferromagnetic Sm_5Ge_4-type structure to a ferromagnetic orthorhombic Gd_5Si_4-type structure occurs, but only at low temperatures.

These two groups of alloys exhibit a GMCE. The first group, alloys with $0.24 \leq x \leq 0.5$, is more interesting due to its higher transition temperatures. The transition temperature ranges from 140 K for $x = 0.24$ to 276 K for $x = 0.5$ and ΔT_{ad} up to 19 K ($x = 0.43$) at $\Delta H = 50$ kOe. The alloy $Gd_5(Si_{1-x}Ge_x)_4$ with $x \geq 0.5$ exhibits a SOMT. However, these results depend strongly on the proper heat treatment and purity of the starting elements [30]. The available rare-earth metals have a significant content of interstitial impurities, primarily H, C, N and O. This content varies between 2 and 5 at.%. H. Fu et al. studied the effect of low-purity Gd. XRD studies showed that alloys with low-purity Gd consist of multiple phases, including $Gd_5Si_2Ge_2$-type, Gd_5Si_3-type, and GdGe-type, making it difficult to prepare a single-phase material [31]. Wu et al. studied the influence of oxygen in $Gd_5(Si_xGe_{1-x})_4$ with $x = 0.475$ and 0.43. This group found that oxygen favored the decomposition of the monoclinic phase into the $GdSi_yGe_{1-y}$ (1:1) and the $Gd_5Si_zGe_{3-z}$ (5:3) phases, in the process, destroying the GMCE by suppressing the structural transition [32].

The adiabatic temperature rise for $Gd_5Si_2Ge_2$ was measured by Gschneidner et al. [33] to be 16.5 K when the magnetic field was ramped up to a rate of 20 kOe min^{-1} up to 50 kOe. The latter value agreed quite well with the ΔT_{ad} value of 16.8 K calculated from heat-capacity measurements made on the same sample. The drawbacks of using these alloys are the high prices and the high hysteresis losses.

Provencano [34] reported a reduction of hysteresis losses (over 90%) in $Gd_5Ge_2Si_2$ by adding a small amount of iron to the alloy, forming $Gd_5Ge_{1.9}Si_2Fe_{0.1}$. Also, the peak magnetic entropy change shifted from 275 to 305 K, broadening its width but reducing its value. The T_C shift for the iron-added alloy was reported in 1997 by

Pecharsky et al. [35], but they did not measure the hysteresis losses. Due to the reduction in hysteresis losses, a greater total net capacity of 355 J kg^{-1} was obtained for the iron-containing alloy. The iron suppressed the formation of the orthorhombic phase, making the transition second order, which reduced the hysteresis losses.

After the 1997 Pecharsky report, many researchers published studies substituting with different elements. Shull et al. [36], for example, reported similar reductions in hysteresis losses with other elements (Co, Cu, Ga, Mn and Al), although no change was observed when substituting Bi or Sn. Zhuang et al. added Pb and found that the first-order structural/magnetic transition was preserved, which increased the T_C and the ΔS [37]. Zhang et al. [38] investigated the thermal hysteresis and phase composition of $Gd_5Ge_2Si_2$ and $Gd_5Ge_{1.9}Si_2T_{0.1}$ (where T stands for Mn, Fe, Co or Ni) series of alloys. With the addition of Fe and Co, the alloys have a monoclinic $Gd_5Ge_2Si_2$-type and an orthorhombic Gd_5Si_4-type diphase structure. With Mn and Ni addition, only the orthorhombic Gd_5Si_4-type structure is observed. In all these cases the thermal hysteresis of $Gd_5Ge_2Si_2$ was significantly reduced. Also, Ga was observed to be homogenously distributed throughout the matrix and suppressed the structural transformation [39]. This was not the case in further studies with Fe by Podmiljsak et al. [40]. Iron was observed only in the newly formed grain-boundary phase. Also, when substituting Si for Fe, no suppression of the transformation was observed.

Zhang et al. melt spun $Gd_5Ge_{1.8}Si_{1.8}Sn_{0.4}$ and observed a reduction in the magnetic and thermal hysteresis [41]. The morphology and composition of the phases in a $Gd_5Ge_2Si_2$ alloy depend strongly on the solidification rate.

Carvalho et al. [42] measured the magnetic properties and the MCE of $Gd_5Ge_2Si_2$ under hydrostatic pressures up to 9.2 kbar. Contrary to the observed effect with MnAs, pressure increased the Curie temperature of $Gd_5Ge_2Si_2$ up to 305 K, did not affect the saturation magnetization and markedly decreased its magnetocaloric effect.

Yue et al. [43] studied the MCE in $Gd_x(Gd_5Si_2Ge_2)_{1-x}$ ($x = 0.3$, 0.5, and 0.7) composite materials produced with a spark-plasma sintering technique. Increasing x from 0.3 to 0.7 was accompanied by a shift in the ΔT_{ad} peak temperature from 286 to 293 K, and the peak value of ΔT_{ad} slowly increased from 1.6 to 2.0 K at a magnetic field change of 1.5 T.

Gd was substituted with other rare-earth metals. For $Nd_5(Si_{1-x}Ge_x)_4$ alloys Thuy et al. [44] reported a $T_C = 110$ K and $\Delta S(max)_M = -39$ mJ cm^{-3} K^{-1}, which is substantially lower than for Gd-based alloys. $Tb_5(Si_{1-x}Ge_x)_4$ alloys have been the second most widely studied R_5T_4 system. Here, the structure exhibits a SMOT and the MCE values are comparable with other materials that have SMOT ($Gd_5(Si_{1-x}Ge_x)_4 x \geq 0.5$). A range of $\Delta S(max)_M$ values have been reported for $Tb_5Si_2Ge_2$ (for example, -97 mJ cm^{-3} K^{-1} [45], and -171 mJ cm^{-3} K^{-1} [46], for $\Delta H = 50$ kOe). Tb_5Si_4 showed the highest T_C among the studied alloys of 225 K and a ΔT_{ad} of 6.8 K [47], while Tb_5Ge_4 was ordered in an antiferromagnetic fashion with a Néel temperature of 91 K. Dy substitution reported by Ivtchenko [48] greatly lowered the T_C and only $Dy_5(Si_3Ge)$ was found to undergo a FOMT with a $\Delta S(max)_M$ close to RCo_2. All the values are for a field change of 50 kOe. Other rare-earth substitutions have not yet been reported. Deng et al. [49] substituted Gd with Tb, and the first-order transition was retained. Although the values of the transition temperature decreased, the as-cast $(Gd_{0.74}Tb_{0.26})_5(Si_{0.43}Ge_{0.57})_4$ still displayed a large magnetic entropy change up to 18.89 J kg^{-1} K^{-1}.

Morellon et al. [50] studied the effects of pressure on $Tb_5Ge_2Si_2$. With a pressure of 8.6 kbar, the high-temperature second-order ferromagnetic transition was found to be coupled with the low-temperature first-order structural phase change into a single first-order magnetic-crystallographic transformation, transforming the material from an ordinary material into a GMCE material.

Ryan et al. [51] reported that Gd_5Sn_4, which has the Sm_5Ge_4 orthorhombic-type structure, exhibits a GMCE ($\Delta S(max)_M = -336$ mJ cm^{-3} K^{-1} at $\Delta H = 5$ kOe) at $T_C = 82$ K, which is still significantly smaller than the $Gd_5(Si_{1-x}Ge_x)_4$ values at the same temperature.

8.2.4
Mn-Based Compounds

A number of different metallic manganese compounds exhibit interesting MCE behavior. Several of these compounds have quite large MCE values; others exhibit fairly strong negative MCEs.

8.2.4.1 Mn(As$_{1-x}$Sb$_x$) Alloys

The base material MnAs undergoes a coupled structural/magnetic FOMT at 318 K. The ferromagnetic hexagonal NiAs-type structure transforms to the paramagnetic orthorhombic MnP-type structure upon heating or demagnetizing. The MCE values are large enough to consider these compounds in the GMCE class of magnetic refrigerants: $\Delta S(max)_M = -218$ mJ cm^{-3} K^{-1} and $\Delta T_{ad} = 13$ K for a 0 to 50 kOe field change at T_C [52].

Wada et al. [53, 54] have studied the effect of substituting Sb for As in MnAs. They note that Sb stabilizes the NiAs-type structure when $x \geq 0.1$, and the FOMT changes to a SOMT, resulting in a reduction of ΔS_M and ΔT_{ad}. Adding Sb also lowers the T_C. The MnAs$_{1-x}$Sb$_x$ system behaves differently from most families of magnetic refrigerant materials in that the ΔS_M decreases with decreasing T_C.

The MCE properties of MnAs$_{1-x}$Sb$_x$ for $0.1 \leq x \leq 0.2$ are outstanding; these alloys are among the leading candidates for near-room-temperature magnetic refrigerant materials. However, the high vapor pressure of As, whose boiling point is 876 K, makes it difficult to economically prepare large quantities (tons) of MnAs. A second problem is the fact that As is a governmentally regulated poison, which means special handling facilities would be required for preparing MnAs$_{1-x}$Sb$_x$ materials, and special environmental regulations would have to be met to place such cooling devices into commercial use.

Under hydrostatic pressure, up to 2.23 kbar, Gama et al. [55] enhanced the isothermal entropy changes in MnAs, reaching values up to 267 J kg^{-1} K^{-1}. These values are far greater than predicted by the magnetic limit, which is calculated by assuming magnetic-field independence of the lattice and electronic entropy contributions. The origin of this so-called colossal magnetocaloric effect (CMCE) is the contribution to the entropy variation coming from the lattice through the magnetoelastic coupling. De Campos et al. reported even higher values for $Mn_{1-x}Fe_xAs$ compounds, which exhibited the colossal effect at ambient pressure [56]. The MCE peak varied from 285 to 310 K, depending on the Fe concentration. De Campos et al. also observed a large thermal hysteresis. They also discovered a colossal effect in $Mn_{1-x}Cu_xAs$ compounds, revealing a peak of $-175\,J\,kg^{-1}\,K^{-1}$ or a 5-T field variation at 318 K and ambient pressure. The Cu atoms act as a small hydrostatic pressure, inducing the colossal magnetocaloric effect at ambient pressure [57].

Sun et al. [58] substituted Mn with Cr. $Mn_{1-x}Cr_xAs$ compounds with $x=0.006$ and 0.01 exhibited a giant room-temperature MCE with reduced (or even without) hysteretic behavior. A maximum $\Delta S(max)_M$ value of $20.2\,J\,kg^{-1}\,K^{-1}$ is observed at 267 K for a 5-T field change for $Mn_{0.99}Cr_{0.01}As$. However, a study from Liu et al. [59] regarding the reliability of the Maxwell equation near the T_C has called into question the correctness of this result.

8.2.4.2 MnFe($P_{1-x}As_x$) Alloys

$MnFeP_{0.45}As_{0.55}$ showed interesting MCE properties, namely $\Delta S(max)_M = -132$ $mJ\,cm^{-3}\,K^{-1}$ for a 0 to 50 kOe field change, and an ordering temperature of 307 K. The ΔS_M and T_C values indicate that $MnFeP_{0.45}As_{0.55}$ is a competitive magnetic refrigerant for near-room-temperature applications. As the As content (x) decreases, T_C decreases and $\Delta S(max)_M$ generally increases and seems to peak at $x=0.35$, but it is still smaller than that of $Gd_5(Si_xGe_{1-x})_4$ alloys (-230 versus $-310\,mJ\,cm^{-3}\,K^{-1}$). Substituting 10 at.% of Fe for Mn does not change the T_C, but increases $\Delta S(max)_M$ by ~40% for a field change of 0 to 50 kOe [60]. On the other hand, the substitution of Ge for As, that is, $MnFe(P_{0.5}As_{0.5-x}Ge_x)$, has just the opposite effects – a large increase of T_C from 282 K for $x=0$ to 570 K for $x=0.5$, and a reduction of the MCE [61]. But further studies of the $MnFe(P_{1-x}Ge_x)$ ($0.1 \le x \le 0.5$) showed comparable results with the As system. This has eliminated the problem of As toxicity [62]. The same group continued to research with the addition of Si [MnFe(P,Si,Ge) compounds]. They found a nonlinear dependence of the Curie temperature on the Si concentration. This dependence is associated with the change in the lattice parameters a and c, and their ratio c/a. Compounds with larger a parameter and smaller c/a ratio have a higher T_C [63, 64]. Tegus et al. [65] studied the effect of Cr and Co substitutions for Fe in $MnFe(P_{1-x}As_x)$ and found that both Cr and Co lowered the T_C and ΔS_M. Cr even changed the FOMT to a SOMT. Melt-spun $Mn_{1.1}Fe_{0.9}P_{1-x}Ge_x$ ($x=0.2, 0.24$) ribbons were studied by Yan et al. [66]. The maximum magnetic entropy change $-\Delta S_M$ was found for $Mn_{1.1}Fe_{0.9}P_{0.76}Ge_{0.24}$ synthesized by mechanical alloying ($30\,J\,kg^{-1}\,K^{-1}$ at 306 K) with a value of $35.4\,J\,kg^{-1}\,K^{-1}$ in a field change from 0 to 5 T at around 317 K.

Pressure experiments were made by Yabuta et al. [67] and Brueck et al. [68]. Applying pressure slightly increased the T_C for MnFe(P,As), but reduced it for the

MnFe(P,Ge) system. Pressure also increased the magnetic hysteresis. The amount of As played an important role in the pressure dependence. Small amounts of As with increasing pressure reduced the ΔS and broadened the curve, while higher concentrations did not effect the ΔS when pressure was applied.

The preparation of the P-containing alloys is similar to that for the $MnAs_{1-x}Sb_x$ alloys and the problems are the same. Phosphorus presents some special handling problems.

8.2.4.3 Ni$_2$MnX (X = Ga, In, Sn, Sb) Heusler Alloys

These compounds are first-order ferromagnetically at a rather high Curie temperature, with a usually negative ΔS_M (SOMT). At lower temperatures they undergo a first-order magnetic phase transition corresponding to a reversible structural transition from the high-temperature cubic austenite phase to the low-temperature tetragonal martensite phase. The transition causes a distortion of the crystal lattice structure, leading to hardening of the magnetic saturation process. Consequently, the magnetization of the martensitic phase is essentially lower than that of the austenitic one, which leads to an inverse MCE and a positive ΔS_M. Such a magnetization change is very sharp, which provides high ΔS_M values.

In 2001 Hu et al. [69] reported a MCE in Ni-Mn-Ga Heusler alloys (ratio 2:1:1). Their results suggested that a negative MCE is associated with a first-order martensitic transition, which is quite large at low magnetic fields and decreases as the magnetic change becomes larger. Zhou et al. [70] reported a large ΔS_M for an alloy of the composition $Ni_{55.2}Mn_{18.6}Ga_{26.2}$. They obtained a value of $-168\,mJ\,cm^{-3}\,K^{-1}$ for $\Delta H = 50\,kOe$ at $T_C = 317\,K$. These data suggest that Ni-Mn-Ga Heusler alloys might be good magnetic regenerator alloys operating between $\sim$300 and $\sim$350 K. A magnetic regenerator serves to expand a conventional refrigerator temperature span by transferring the heat between the parts of the refrigeration cycle in opposite directions [71]. These alloys are also very easy to prepare. The drawback is their narrow ΔS_M maximum peak (three times narrower than $Gd_5Si_2Ge_2$). Yu et al. substituted Ni with Co in this type of alloy, resulting in a higher T_C and a lower ΔS [72]. Stadler et al. researched the substituted Heusler alloy $Ni_2Mn_{1-x}Cu_xGa$, which shows a very high maximum magnetic entropy change of $\Delta S_M = -64\,J\,kg^{-1}\,K^{-1}$ (per unit volume $\Delta S_M = -532\,mJ\,cm^{-3}\,K^{-1}$) at 308 K for a magnetic field change $\Delta H = 50\,kOe$ [73].

A large inverse MCE was found in the $Ni_{0.50}Mn_{0.50-x}Sn_x$ (0.13 $\le x \le$ 0.15) alloys in 2005, when Krenke et al. [74] reported a large positive ΔS_M of 19 J kg^{-1} K^{-1} near 300 K for a magnetic field change of 5 T in $Ni_{50}Mn_{37}Sn_{13}$. Han et al. investigated magnetic entropy changes in $Ni_{50-x}Mn_{39+x}Sn_{11}$ alloys (x = 5, 6, and 7). Under an applied magnetic field of 10 kOe they found high magnetic entropy changes of 6.8, 10.1, and 10.4 J kg^{-1} K^{-1} for x = 5, 6, and 7, respectively [75].

Oikawa et al. were the first to report on the magnetic and martensitic transition behaviors of $Ni_2Mn_{1+x}In_{1-x}$ Heusler alloys. This group detected a martensitic transition from the ferromagnetic austenite phase to the antiferromagnetic-like martensite phase, and a large positive magnetic entropy change, which reached 13 J kg^{-1} K^{-1} at 9 T [76]. Sharma et al. found in $Ni_{50}Mn_{34}In_{16}$ a ΔS_M value of

19 J kg^{-1} K^{-1} around 240 K for the martensitic transition, while near room temperature this alloy showed a conventional MCE. The ΔS_M at 300 K was -7.5 J kg^{-1} K^{-1}. Sharma et al. also noted that the Curie temperature of the ferromagnetic martensite phase of NiMnIn lies in the same temperature region as the martensitic transition, thus increasing the overall MCE and explaining the high ΔS_M [77]. The transition temperature can be easily tuned by changing the composition [78] or substituting Co for Ni [79]. Moya et al. directly measured the adiabatic temperature change in the Ni$_{50}$Mn$_{34}$In$_{16}$ alloy and showed the possibility of both cooling and heating in a giant inverse magnetocaloric compound. It has been shown that the irreversibility of the first-order structural transition gives rise to measured temperature changes that are lower than those calculated using equilibrium thermodynamics [80].

A value of $\Delta S_M = 8$ J kg^{-1} K^{-1} was found for Ni$_{45.4}$Mn$_{41.5}$In$_{13.1}$ for a field change of 10 kOe [81]. Pathak et al. reported high ΔS_M values for the FOMT of this alloy, but with high hysteresis losses in Ni$_{50}$Mn$_{50-x}$In$_x$. At the SOMT temperature, they found a decrease of the $-\Delta S_M$, but with a higher net-refrigeration-capacity value of 280 J kg^{-1} K^{-1} for samples with $x = 16$ around room temperature for a magnetic field change of 0–5 T [82].

Du et al. investigated the magnetocaloric effect of ferromagnetic Heusler alloys Ni$_{50}$Mn$_{50-x}$Sb$_x$ ($x = 12$, 13 and 14). A large positive ΔS_M was observed in the vicinity of the martensitic transition. The maximum value of ΔS_M was 9.1 J kg^{-1} K^{-1} in Ni$_{50}$Mn$_{37}$Sb$_{13}$ at 287 K for a magnetic field change of 5 T [83]. After doping with Co (Ni$_{50-x}$Co$_x$Mn$_{39}$Sb$_{11}$), the martensitic transition temperature decreased rapidly, while the T_C of the austenitic phase increased linearly [84]. Krenke et al. introduced small amounts of Fe and Co in place of Ni. Adding 3 at.% of Fe increased the ΔS_M, while increasing the thermal hysteresis. Co had the opposite effect [85]. Xuan et al. studied the effect of annealing Ni$_{44.1}$Mn$_{44.2}$Sn$_{11.7}$ ribbons. By changing the annealing temperature of the ribbons, they could control the transition temperature and the ΔS_M [86]. Yasuda et al. showed that applying pressure increased the martensitic transition temperature linearly, while the T_C did not change with increasing pressure [87].

Among the ferromagnetic Heusler alloys showing a martensitic transition, those systems in which the T_C of the martensite phase lies in the same temperature region or below that of the martensitic transition temperature are likely to show a large inverse magnetocaloric effect.

8.2.4.4 Miscellaneous Compounds

Songlin et al. [88] measured the MCE in Mn$_5$Ge$_3$, which ferromagnetically orders at 298 K. The ΔS_M for a 0 to 50 kOe field change was -67 mJ cm^{-3} K^{-1}. They also found that when substituting Sb for Ge, the T_C increased to 312 K and the ΔS_M decreased to -40 mJ cm^{-3} K^{-1} for $x = 0.1$.

Tohei et al. [89] studied the MCE of Mn$_3$GaC. This material exhibits a FOMT from an antiferromagnetic (AF) to a ferromagnetic (FM) state with increasing temperature, which is the opposite of what occurs in most magnetic materials. As a result, Mn$_3$GaC exhibits a fairly large negative MCE ($\Delta S_M = 103$ mJ cm^{-3} K^{-1} for a

0 to 20 kOe field change at 160 K). Substituting Co for Mn lowers the T_C and the ΔS_M.

Zhao et al. studied the magnetic properties and magnetocaloric effects of $Mn_5Ge_{3-x}Si_x$ alloys with $x = 0.1$, 0.3, 0.5, 1.0, 1.5 and 2.0. T_C decreased with increasing x. The highest ΔS_M was found for $x = 0.5$ (7.8 J kg^{-1} K^{-1} at $T_C = 299$ K for magnetic field changes from 0 to 1 T and from 0 to 5 T) [90].

8.2.5
La(Fe$_{13-x}$M$_x$)-Based Compounds

The LaFe$_{13}$ phase does not exist; in fact, no intermetallic compounds form in the La-Fe binary system, and La and Fe form immiscible liquids at the Fe-rich side between 8 and 20 at% La above 1460 °C [91]. Consequently, other elements must be added to La–Fe alloys in order to form the intermetallic La(Fe$_{13-x}$M$_x$) phases.

Palstra et al. reported some unusual magnetic properties for the La(Fe$_{13-x}$Si$_x$) [92]. The T_C increased from 198 K at $x = 1.5$ to 262 K at $x = 2.5$, while the saturation magnetic moment decreased from 2.08 to 1.85 μB/Fe as x increased from 1.5 to 2.5. In 1999, Fujita et al. [93] observed a large volume change of ~1.5% in La(Fe$_{11.44}$Si$_{1.56}$), just above the Curie temperature (195 K). They claimed this behavior was due to an itinerant-electron metamagnetic (IEM) transition. This result suggested that this alloy might have interesting magnetocaloric properties.

In 2001, Hu et al. [94] were the first to find the GMCE in La(Fe$_{13-x}$Si$_x$) alloys. They reported that La(Fe$_{11.4}$Si$_{1.6}$), which orders at 208 K, had a ΔS_M value of -140 mJ cm^{-3} K^{-1} for a magnetic field change of 0 to 50 kOe. They also found that when x was increased (i.e. more of the Fe was replaced by Si), the magnetic ordering temperature increased and the MCE was substantially reduced. The Curie temperature increased monotonically from 180 K at $x = 1.3$ to 250 K at $x = 2.6$. For La(Fe$_{10.2}$Si$_{2.8}$) [95] the T_C fell abruptly to 195 K. This may be due to a change in the magnetic properties of the material. ΔS_M dropped rather rapidly on increasing x, from -215 mJ cm^{-3} K^{-1} at $x = 1.3$ to -100 mJ cm^{-3} K^{-1} at $x = 1.8$. For $x > 1.8$, ΔS_M was small and appeared to level off at -40 mJ cm^{-3} K^{-1} (all at $\Delta H = 50$ kOe).

Although the low x value La(Fe$_{13-x}$Si$_x$) compounds exhibit large ΔS_M values, it does not follow that these materials will necessarily have large values of ΔT_{ad}. ΔT_{ad} varies from 8.6 to 12.1 K for a magnetic field change of 0 to 50 kOe [96] which is 20–30% smaller than for the Gd$_5$(Si$_x$Ge$_{4-x}$) alloys.

It should be pointed out that all the La(Fe$_{13-x}$Si$_x$) samples prepared to date are two-phase alloys containing up to 5% α-Fe. This is not surprising, considering that La and Fe form immiscible liquids. Even long-term anneals, up to 30 days, do not eliminate the second phase α-Fe.

This can, however, be overcome to some extent by melt spinning the alloy. Gutfleisch et al. reduced the annealing time to 2 h by melt spinning the material, presumably because the elements could be more homogeneously distributed in this way. A very large magnetic entropy change of $\Delta S_M = -31$ J/kg K was obtained at 201 K under 5 T in LaFe$_{11.8}$Si$_{1.2}$ melt-spun ribbons, which is much higher than for the bulk sample [97]. Also, thermal- and/or field-hysteresis can be practically overcome by melt spinning [98].

Sun et al. studied the effects of pressure on the MCE in a polycrystalline $LaFe_{11.6}Si_{1.4}$ sample [99]. The Curie temperature of the sample rapidly decreased from 191 K at ambient pressure to 80 K under 8.3 kbar. The giant magnetocaloric effect in $LaFe_{11.6}Si_{1.4}$ was greatly enhanced by pressure, especially at low magnetic fields. For a field variation of just 1 T, the maximum value of the entropy change was as high as 34 J kg^{-1} K^{-1}.

Substituting Fe with Co led to an increase in the T_C up to 274 K for the composition $LaFe_{11.2}Co_{0.7}Si_{1.1}$, which still showed a large ΔS_M [100]. By melt spinning the slightly different composition $LaFe_{11}Co_{0.8}Si_{1.2}$, an even higher T_C value of 290 K was found; the ΔS_M was also higher than in the bulk sample [101]. Substituting Si with Ga in $LaFe_{11.2}Co_{0.7}Si_{1.1-x}Ga_x$ reduced the ΔS_M (-11.9 J kg^{-1} K^{-1}), but increased the refrigerant capacity significantly (254.8 J kg^{-1}) [102].

Many researchers have investigated substituting La with other elements. Fujieda et al. showed that partially substituting La for Pr in $La(Fe_{0.88}Si_{0.12})_{13}$ enhances the MCE. The large MCE was attributed to the ferromagnetic coupling between the magnetic moments of Pr and Fe and the increase of magnetization change at the Curie temperature T_C due to a volume contraction [103]. Pr also reduced the hysteresis losses [104]. Liu et al. experimentally studied the entropy changes in the compounds $La_{1-x}Pr_xFe_{11.5}Si_{1.5}$ ($x = 0.3$ and 0.4) [105]. They reported different values for ΔS_M with different measurement or computation techniques. A tower-shaped entropy change with a height of -27 J kg^{-1} K^{-1} was obtained by analyzing the heat capacity, while the Maxwell relation predicted an extra entropy peak with a height of -99 J kg^{-1} K^{-1}, slightly varying with Pr content. They concluded that the Maxwell relation cannot be used in the vicinity of the Curie temperature due to the coexistence of paramagnetic and ferromagnetic phases, and the huge entropy peak is a spurious result. Similar conclusions are applicable to the experimental results of MnAs and $Mn_{1-x}Fe_xAs$ calculated with the Maxwell equation, which were reported before.

Anh et al. substituted Nd for La and reported a gradual increase in T_C and a lower MCE [106]. Fujieda et al. investigated a single-phase cubic $NaZn_{13}$-type $La_{1-z}Nd_z(Fe_{0.88}Si_{0.12})_{13}$ with $z = 0.2$. With a magnetic field change from 0 to 5 T, the isothermal magnetic entropy change and the refrigerant capacity increased to -27 J kg^{-1} K^{-1} and 518 J kg^{-1}, respectively [107].

Passamani et al. substituted 5 at.% of La with Y. The $\Delta S(max)_M$ of the Y compound was found to be roughly the same as the La compound (-18 J kg^{-1} K^{-1}), but with a significantly broader magnetic entropy-change peak [108].

By partially substituting Ce for La, the lattice constant was reduced and the Curie temperature T_C decreased. In addition, the isothermal ΔS_M and the ΔT_{ad} due to the IEM transition were enhanced because of the increase in the entropy change caused by the latent heat. Values of -28 J/kg K could be achieved using $La_{0.7}Ce_{0.3}(Fe_{0.88}Si_{0.12})_{13}$, which is 50% higher than for the compound without Ce [109].

Balli et al. substituted Er for La. They observed that the Curie temperature increased slightly with increasing Er content up to $x = 0.3$. For low Er contents, a large ΔS_M was observed. However, a decrease of ΔS_M with increasing Er concentration was observed because the metamagnetic transition was eliminated [110].

The addition of interstitial hydrogen or carbon has a large effect on the magnetic properties of these alloys. Hydrogen-addition studies were carried out on various La$(Fe_{13-x}Si_x)H_y$ samples, which are represented below [111, 112].

Figure 8.4 illustrates the dependence of the La$(Fe_{13-x}Si_x)H_y$ properties on H content. An excellent correlation can be seen between T_C and the hydrogen content, y, regardless of the Fe : Si ratio (a). For a fixed Fe : Si ratio, ΔT_{ad} and ΔS_M show a dependence on y (b,c). ΔS_M slightly decreases with increasing y, but ΔT_{ad} increases by about 50% when $y = 1.5$ compared with $y = 0$. The Fe : Si ratio at a fixed y also has a strong effect on the magnetocaloric properties: the higher is the Fe content, the larger are ΔS_M and ΔT_{ad}.

Fukamichi et al. reported that the T_C of La$(Fe_xSi_{1-x})_{13}H_y$ was scarcely influenced by thermal cycles. They also found that this alloy's thermal conductivity, which is important for the heat flow in the refrigerating cycle, was much higher than the conductivity of other candidate materials, but was still lower than Gd [113]. Mandal et al. were able to prepare hydrides of a LaFe$_{11.57}$Si$_{1.43}$ intermetallic compound using reactive milling and could tune the Curie temperature from 199 to 346 K [114].

Fujita et al. tested hydrogen-absorbed spherical particles of La$(Fe_{0.86}Si_{0.14})_{13}$ in an active magnetic regenerator (AMR) and achieved a relatively wide temperature span of 16 K [115].

Saito et al. prepared spherical particles of La(Fe,Co,Si)$_{13}$ with diameters ranging widely from 0.1 to 1.2 mm by using the rotating-electrode process. The high MCE was still preserved. This shape of particles is particularly useful for the AMR cycle refrigeration [116].

Balli et al. combined the LaFe$_{11.3}$Co$_{0.4}$Si$_{1.3}$, LaFe$_{11.2}$Co$_{0.5}$Si$_{1.3}$, LaFe$_{11.1}$Co$_{0.6}$Si$_{1.3}$, LaFe$_{11.1}$Co$_{0.8}$Si$_{1.1}$ and LaFe$_{11}$Co$_{0.9}$Si$_{1.1}$ compounds in order to form a composite refrigerant working in the 240–300 K temperature range. The calculated optimal mass ratios depended on the external magnetic field, and the isothermal entropy change of the composite material remained approximately constant within the temperature range 240–300 K [117].

The addition of carbon was studied in La(Fe$_{11.6}$Si$_{1.4}$)C$_y$ [118]. Adding C was found to increase T_C from 195 K for $y = 0$ to 250 K for $y = 0.6$, but lowered ΔS_M with a rapid drop for $y \geq 0.4$ because the first-order IEM transition changed to a second-order transition.

8.2.6
Manganites

The rare-earth manganites (R$_{1-x}$M$_x$MnO$_3$, where R = La, Nd, or Pr and M = Ca, Sr, Ba, etc.) have been extensively studied with a variety of different elements. To date, the highest reported ΔS_M values are reasonable, but not outstanding. Several of the manganites have ΔS_M values comparable to Gd, but most are smaller.

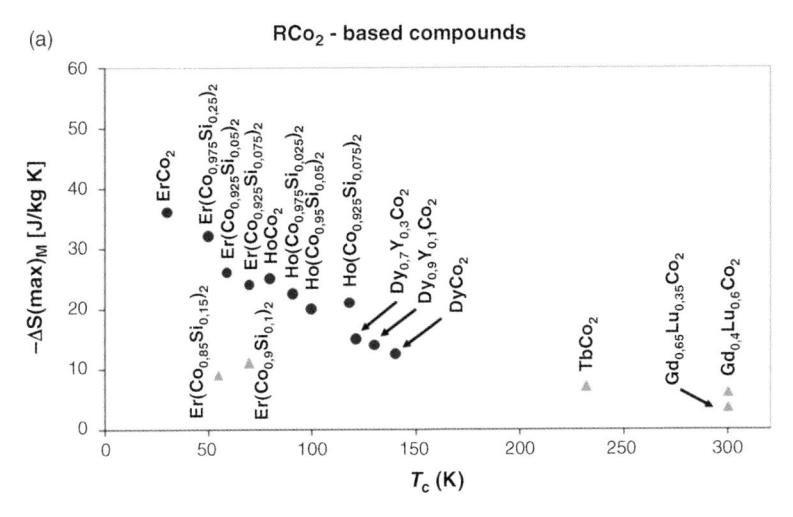

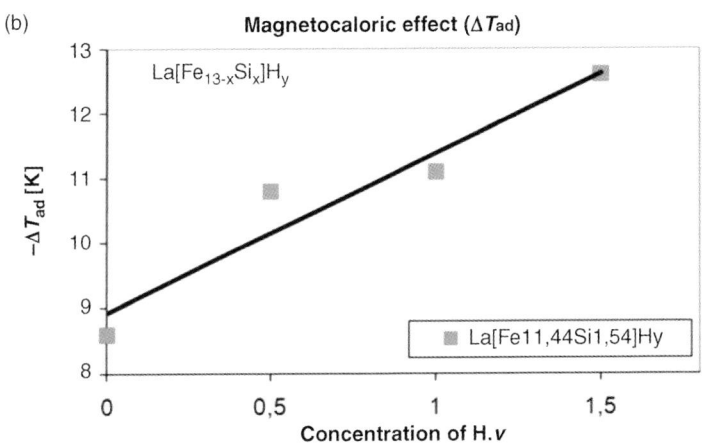

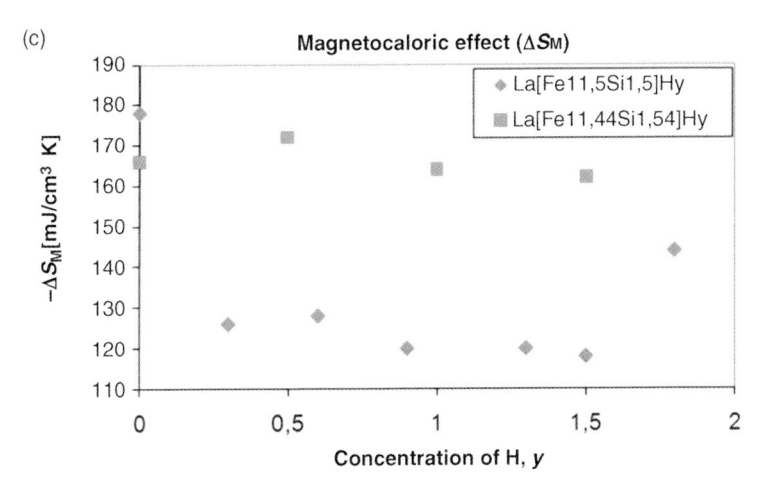

Figure 8.4 Curie temperature versus hydrogen concentration, y (a) and the adiabatic temperature rise (b) and the magnetic entropy change (c) versus the hydrogen concentration, y for the $La(Fe_{13-x}Si_x)H_y$ alloy system.

Among the existing rare-earth manganites, the $La_{1-x}Ca_xMnO_3$ phases exhibited the largest MCEs, but their Curie temperatures were below room temperature. For example, the maximum $T_C = 267$ K for $La_{0.67}Ca_{0.33}MnO_3$ [119]. It was shown that, by substituting Ca with a small amount of Sr, the MCE peak temperature could be tuned in the temperature range 150–300 K, while retaining relatively large MCE values [120]. A large ΔS_M of -2.26 J kg^{-1} K^{-1} at 354 K for $\Delta H = 1$ T was also observed for $La_{0.6}Sr_{0.2}Ba_{0.2}MnO_3$ [121]. In general, any substitution of M (M = Sr, Ba and Pb) for Ca in (La–Ca–M)MnO$_3$ manganites usually leads to an increase in T_C but to a reduction in the MCE. Certain combinations can result in useful magnetocalorics. The substitution of Bi (5 at%) for La in $La_{0.67}Ca_{0.33}MnO_3$ was found to significantly improve the MCE. For $\Delta H = 1$ T, the sample exhibited a large ΔS_M of -3.5 J kg^{-1} K^{-1}, but the T_C was reduced to 248 K [122].

When Mn was substituted with Si [$La_{2/3}Ca_{1/3}Mn_{1-x}Si_xO_3$ ($x = 0.05$–0.20)], the T_C was lowered without significantly changing the ΔS_M [123]. Substitution of Al and Ti for Mn in $La_{0.7}Sr_{0.3}MnO_3$ effectively lowered the ferromagnetic ordering temperature T_C from 364.5 K to well below 300 K, making this compound appropriate for RT applications. In this case, the ΔS_M remained unchanged only when Ti was used [124, 125].

Potassium-doped manganese perovskites of the type $La_{1-x}K_xMnO_3$ ($x = 0.05$, 0.10, 0.15) with a nanometer crystallite size have also been investigated [126]. These alloys showed T_C tunability from 260.4 K (for $x = 0.05$) to 309.7 K (for $x = 0.15$), but with ΔS_M lower than pure Gd.

Chen et al. [127] measured the change in the MCE of rhombohedral $(La_{0.8}Na_{0.2})$ $MnO_{3-\delta}$ as a function of the oxygen deficiency, δ. They found that ΔS_M increased from -22.8 mJ cm^{-3} K^{-1} for $\delta = 0$ to -23.2 mJ cm^{-3} K^{-1} for $\delta = 0.06$ and T_C increased from 278 K ($\delta = 0$) to 364 K ($\delta = 0.06$). Both values are for $\Delta H = 10$ kOe.

Chen et al. [128] also studied the effect of La deficiencies on T_C and the MCE of $(La_{0.8-y}Ca_{0.2})MnO_3$. The Curie temperature increased from 182 K for no La deficiency to 260 K for $y = 0.05$. As y increased, the nature of the magnetic transformation changed from a SOMT (for $y = 0$ and 0.01) to a FOMT (for $y = 0.03$ to 0.10), and the maximum value of the MCE increased from $\Delta S_M = -7.7$ at $y = 0$ to $\Delta S_M = -22.3$ mJ cm^{-3} K^{-1} at $y = 0.03$ for $\Delta H = 10$ kOe.

The MCE properties of $(La_{0.67}Ba_{0.33})MnO_3$ were studied by Zhong et al. They reported that at $T_C = 337$ K, $\Delta S_M = -18.8$ mJ cm^{-3} K^{-1} for $\Delta H = 10$ kOe.

8.2.7
Miscellaneous Intermetallic Compounds

The highest MCE value to date was observed in a quenched $Fe_{0.49}Rh_{0.51}$ alloy at the AF-to-FM first-order phase transition at 313 K [129, 130]. Its maximum $\Delta T/\Delta H$ reached 7.08 K/T and maximum $\Delta S_M/\Delta H = -24.62$ J kg^{-1} K^{-1} T^{-1} for $\Delta H = 0.65$ T. The problem with these alloys is that their MCE is irreversible.

Another candidate for MR is the alloy Nd_2Fe_{17}. With a T_C of 325 K and ΔS_M slightly lower than those of the Si-rich $Gd_5(Si_{1-x}Ge_x)_4$ alloys, it has potential. The drawback of this material is its low ΔT_{ad} [131].

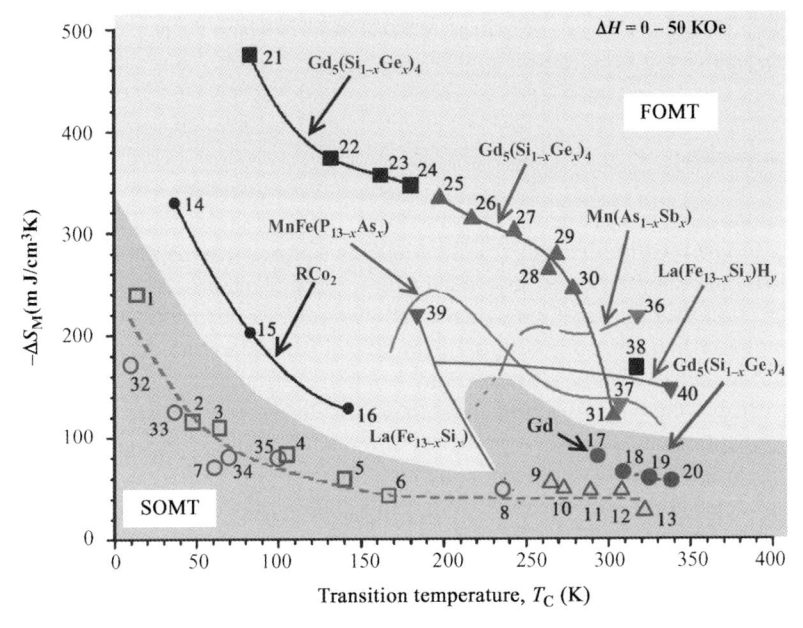

Figure 8.5 The magnetic entropy change for $\Delta H = 50\,\text{kOe}$ for most known MC materials.

Gd_7Pd_3, which orders at 318 K, has MCE values of $\Delta S_M = 2.5$ and 6.5 J/K kg and $\Delta T_{ad} = 3.0$ and 8.5 K at 320 K, respectively, at 2 and 5 T. These are comparable with those of Gd_5Si_3Ge (Figure 8.5), which has a similar Curie temperature [132].

Niu et al. [133] and Niu [134] examined a series of alloys of the $Gd_4(Bi_xSb_{1-x})_3$ system, which have a cubic anti-Th_3P_4 type structure. All of the alloys are ferromagnetic below their respective Curie temperatures, which vary from 266 K for $x=0$ to 332 K for $x=1.0$. The MCE values for Gd_4Bi_3 are significantly smaller than those of the $Gd_5(Si_3Ge)$ phase, which orders at about the same temperature. The MCE values for $Gd_4(Bi_xSb_{1-x})_3$ for $0 \leq x \leq 0.75$, however, are significantly larger than for the pure Gd_4Bi_3 phase.

Chau et al. [135] reported a GMCE in amorphous soft magnetic ribbons of $Fe_{73.5-x}Cr_xSi_{13.5}B_9Nb_3Cu_1$ ($x=1-5$), which were fabricated by rapid quenching on a single copper wheel. For $x=1$, the ΔS_M was 9.8 J kg^{-1} K^{-1}, which is comparable to that of pure Gd ($T_C = 327$ K).

Many other intermetallic compounds have been studied and found to have poor MCEs with a T_C or T_N well below 200 K. These compounds may be interesting for low-temperature cooling, but not for room-temperature use. A good overview of these binary, ternary and quaternary intermetallic compounds can be found in the review paper by Gschneidner [136].

In the last two years, Gd bulk metallic glasses have been extensively studied [137–140]. Most of these materials show proper ΔS_M and high RC but in a temperature range below 200 K. Certain reports have claimed that these materials may be usable up to 300 K [141], but further research is needed to see if they have the potential for RT magnetocaloric materials.

8.2.8
Nanocomposites

Nanocomposites must be dispersed in a matrix in order to prevent agglomeration associated with the minimization of surface energy. The matrix usually shows little, if any, magnetocaloric activity. In a typical nanocomposite material, the concentration of nanoparticles is usually less than 50% by volume. Thus, the measured MCE will be reduced by a factor proportional to the ratio of the volumes (or the masses) of the inactive matrix and the active particulate. Matrix effects aside, nanoparticles, which usually show superparamagnetic behavior, have been known to exhibit enhancement of the MCE when compared with conventional paramagnets. It is, therefore, feasible that basic research on the MCE of nanocomposites will lead to a better understanding of the relationships among the structure, magnetism and thermodynamics of solids. Poddar *et al.* [142], for example, reported on the MCE of two magnetic nanoparticles systems – cobalt ferrite and manganese zinc ferrite – with mean sizes of around 5 and 15 nm, respectively. The MCE values are still lower than conventional MC materials, but quite a bit higher than previously reported values for nanomaterials.

8.2.9
Comparison of MCE Materials

Gschneidner *et al.* compared the ΔS_M, ΔT_{ad} and the cooling capacity of MC materials studied up to 2005 [143].

As noted earlier, a better parameter for comparing magnetic materials is the refrigerant capacity q. Figure 8.7 plots q versus T_C for many different materials. As might be expected, q generally increases with decreasing temperature.

A common mistake in comparing MC materials is to contrast two materials with T_C values that are more than 25 K apart. Such a comparison is not valid since ΔS_M has strong temperature dependence. The same is true for q.

In GMCE materials, the first-order magnetostructural transition is responsible for the GMCE. But this transition also results in major problems. First, there is the large volume change [144]. Because intermetallic compounds are very brittle, the large volume change could destroy the form of the MC. Secondly, there is the problem with the large hysteresis losses that occur in materials with a FOMT. These losses can be reduced by substitutions in the alloy, but at the expense of ΔS_M [145]. Of even greater concern is the time dependence of ΔT_{ad}. In the cases of $Gd_5(Si_2Ge_2)$ and $La(Fe_{11.44}Si_{1.56})$, the directly measured ΔT_{ad} values are significantly smaller than those obtained indirectly from heat-capacity measurements because of the kinetics of the transformation – the more rapidly ΔT_{ad} is measured, the smaller the apparent value of ΔT_{ad}, by 30–50%. This could be a serious problem because MRs will operate between 1 and 10 Hz and much of the MCE will be lost (i.e. not utilized) during the

Table 8.2 Advantages and disadvantages of various magnetic refrigerants [147]. Relative merits of various near-room-temperature magnetocaloric materials.

Property	Gd	Gd$_5$(Ge, Si)$_4$	LaFeSi	MnAs
Materials costs	✓	x	✓✓	✓✓
Production costs	✓	x	x	x x
Large-scale production	✓	✓	✓	?
MCE, ΔS	✓	✓✓	✓	✓
Hysteresis	✓	x x	x	x
Environment	✓	✓	✓	x x

magnetic field increase and the field decrease [146]. Gd metal, on the other hand, shows no time dependence of ΔT_{ad}.

Another problem with the intermetallic Mn refrigerants containing As and/or P is the fact that both have high vapor pressures (the boiling point of As is 876 K and that of P is 550 K). This presents an additional challenge in the handling of these elements in the production of the appropriate compound; As, P and Sb are also toxic.

The advantages and disadvantages of a number of known candidate magnetic regenerator materials are summarized in Table 8.2. The comparison is made with Gd metal, the prototype magnetic refrigerant. A tick (✓) indicates that the factor is essentially the same as for Gd; a cross (×) means worse behavior than Gd. The table reveals that there is no clear favorite GMCE material as a replacement for Gd and Gd-based solid solution alloys, Gd–R. As a matter of fact, Gd and its solid-solution alloys continue to hold their own as the near-room-temperature magnetic refrigerants of choice.

8.2.10
Conclusions

A number of new materials with GMCE properties have been developed. Among these candidate materials, there is no clear winner as a replacement for Gd metal, the prototype 298 K magnetic refrigerant material. Nevertheless, the current research on MR compounds leaves much to be explored and improved. Two critical areas that need to be addressed are (1) engineering materials to overcome the limitations of the currently available magnetic refrigerant regenerator materials and (2) increasing the magnetic field strength of permanent magnets while reducing the size, mass and cost.

8.3
Magnetocaloric Effect and Hysteresis Losses of CMAs

The size of the field-induced entropy change, ΔS_m, is not the only important parameter when it comes to the application of a magnetocaloric material. Although

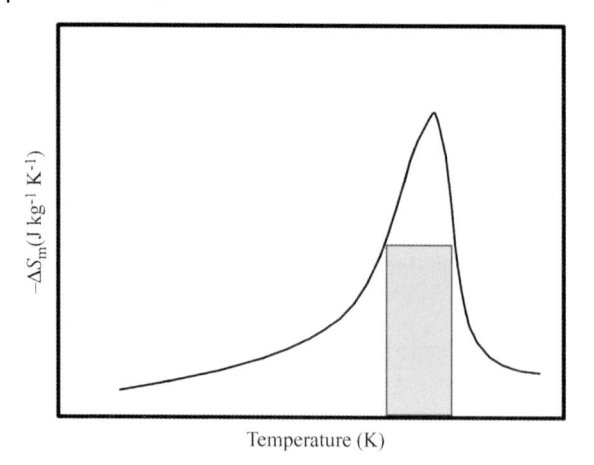

Figure 8.6 Computed refrigerant capacity (RC) value for a magnetocaloric material.

it is true that a larger ΔS_m material will offer more cooling power than a material with a lower value of ΔS_m, it is a mistake to overlook the question of the losses associated with the magnetic hysteresis. Perhaps a more useful measure of a magnetocaloric material is the refrigerant capacity (RC) value. There is no single accepted way to determine this RC value for a particular material; however, the basic idea is to numerically integrate the area under the ΔS_m versus T curves, obtained from hysteresis-loop measurements on the material. In the method described by Kittel [148], the limits of the temperature integration are set by the half-maximum of the ΔS_m peak. Figure 8.6 illustrates an example of such an area integration.

In another method, Wood and Potter [149] defined the RC for a reversible system between T_h, the upper temperature, and T_c, the lower temperature, as: $RC = \Delta S_m \Delta T$, where ΔS_m is the magnetic entropy change at the hot and cold ends of the cycle, and $\Delta T = T_h - T_c$. For most ΔS_m versus T curves these two methods give similar results.

In Section 8.2.3.2 a detailed literature survey is presented. Provenzano et al. [34] reported on a very dramatic reduction in the losses associated with $Gd_5Ge_2Si_2$ by substituting very small amounts of Fe for the Ge. Using the methods described above, they found that the losses associated with the ternary $Gd_5Ge_2Si_2$ material were about 65 J kg^{-1}, but for the $Gd_5Ge_{1.9}Si_2Fe_{0.1}$ they were less than 4 J kg^{-1}. As these are the energy costs to make one cycle of the hysteresis loop, they must be carefully considered when looking at the effectiveness of a magnetocaloric material under real operating conditions. In order to determine the usefulness of their $Gd_5Ge_2Si_2$ and $Gd_5Ge_{1.9}Si_2Fe_{0.1}$ materials, the authors simply subtracted the losses from the corresponding RC values. The results were 355 J kg^{-1} and 235 J kg^{-1} for the Fe-containing alloy, and 240 J kg^{-1} and 200 J kg^{-1} for the alloy without any Fe [36]. Therefore, on this basis the $Gd_5Ge_{1.9}Si_2Fe_{0.1}$ material is a much more effective magnetocaloric material than the $Gd_5Ge_2Si_2$ material, even though it has a significantly lower ΔS_m peak. In their microstructural studies they found the $Gd_5Ge_2Si_2$ material to be single phase, while the $Gd_5Ge_{1.9}Si_2Fe_{0.1}$ material was composed of three phases: the matrix phase (with no Fe), a light grain-boundary phase and a dark grain-boundary phase (both rich in Fe). These observations were supported by X-ray diffraction experiments. The authors of the paper concluded that the iron suppresses the formation of the orthorhombic phase, making it a second-order phase transition, which as a result, reduced the hysteresis losses.

8.3.1
Substituting Ge and Si with Various Elements to Reduce the Hysteresis Losses

After the report of Provenzano et al., many papers were published, reporting substitutions with different elements. Shull et al. [36] reported similar findings with other elements, that is, Co, Cu, Ga, Mn and Al; only when substituting with Bi or Sn were no changes observed. Zhuang et al. [37] added Pb, and they managed to retain the first-order structural/magnetic transition, which increased the T_C and the ΔS_m. Zhang et al. [38] investigated the thermal hysteresis and phase composition of the $Gd_5Ge_2Si_2$ and $Gd_5Ge_{1.9}Si_2T_{0.1}$ (T: Mn, Fe, Co, Ni) series of alloys. With the addition of Fe and Co the alloys had monoclinic $Gd_5Ge_2Si_2$-type and orthorhombic Gd_5Si_4-type diphase structures, whereas with Mn and Ni only the orthorhombic Gd_5Si_4-type structure was observed. Also, the thermal hysteresis of $Gd_5Ge_2Si_2$ was significantly reduced. Ga was also found to suppress the structural transformation, and it was observed that it homogenously distributes throughout the matrix [39].

8.3.2
Phase Formation and Magnetic Properties of $Gd_5Si_2Ge_2$ with Fe Substitutions

In response to the work described above, our own investigations have focused on a better understanding of the effects of substituting Fe on the microstructure of the $Gd_5Si_2Ge_2$-type materials over a much wider range than was looked at by previous investigators. The compositions investigated are collected in Table 8.3, together with

Table 8.3 Composition of arc-melted alloys.

	at.% Gd	at.% Si	at.% Fe	at.% Ge
S0	55.6	22.2	/	22.2
S1.4	55.6	20.8	1.4	22.2
S2.8	55.6	19.4	2.8	22.2
S5.6	55.6	16.6	5.6	22.2
S8.4	55.6	13.8	8.4	22.2
S11.1	55.6	11.1	11.1	22.2
G0	55.6	22.2	/	22.2
G0.7	55.6	22.2	0.7	21.5
G1.4	55.6	22.2	1.4	20.8
G2.8	55.6	22.2	2.8	19.4
G5.6	55.6	22.2	5.6	16.6
G8.4	55.6	22.2	8.4	13.8

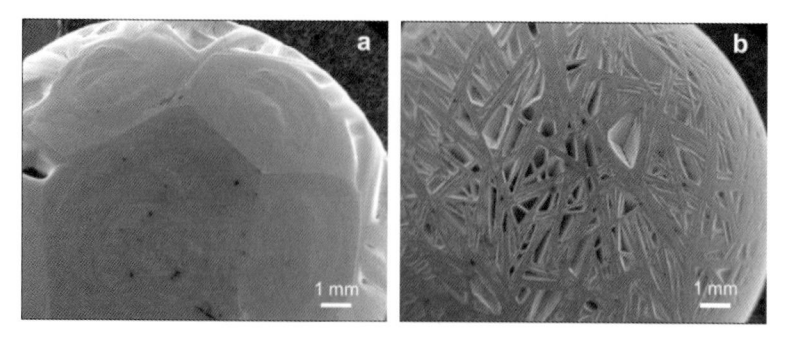

Figure 8.7 SEI of arc-melted buttons with the S0 (a), S2.8 (b).

their respective sample codes, S0, S1.4, S2.8, S5.6, S8.4, and S11.1 for the Si substitutions and G0, G0.7, G1.4, G2.8, G5.6 and G8.4 for the Ge substitutions. All the samples were prepared from high-purity starting elements of gadolinium (99.99 wt.%), silicon (99.995 wt.%), germanium (99.999 wt.%) and iron (99.99 wt.%). The samples were prepared as 5-g buttons by arc melting a mixture of pure elements on a water-cooled copper hearth in an argon atmosphere at a pressure of 0.5 bar. Each sample was re-melted three times and after each re-melting the samples were turned over to ensure their homogeneity. The buttons were then homogenized for 1 h at 1300 °C in argon.

The electron microscopy secondary-electron images in Figure 8.7 show the macrostructures of two of the arc-melted buttons. The differences in the upper surfaces of the button samples are very obvious. The S0 sample exhibits regular pentagons and hexagons, reminiscent of a "buckyball". The S2.8 sample's surface shows a sinew-like effect.

In the case of alloy S0, where no iron is present, the surface of the button is typical for the solidification of so-called semimetals, such as silicon and germanium and intermetallic phases with low crystallographic symmetry. Both silicon and germanium have melting entropies of the order of ~ 3.6 kB/atom, which gives approximately 30.0 J K^{-1} mol^{-1}, whilst most metals, including gadolinium and iron, have melting entropies of around 1 kB/atom, that is, ΔS_m Fe $= 8.48$ J K^{-1} mol^{-1} and ΔS_m Gd $= 5.51$ J K^{-1} mol^{-1}. This means that the high melting entropies for both silicon and germanium will increase the tendency for the established intermetallic phases to form facets during the process of solidification. Since the phases in the examined alloys have predominantly monoclinic (lower symmetry) and orthorhombic structures at room temperature, they will also solidify forming a faceted solidification front in the absence of iron. The reason for the sinew-like structure remains unclear; however, we believe that solidification shrinkage of one of the main phases plays a crucial role.

Optical micrographs of the samples are shown in Figure 8.8, and were first reported in 2007 [150]. It is clear from the six images that all the samples are composed of multiphase structures. The microstructure of the S0 sample consists of the Gd$_5$(Si,Ge)$_4$ matrix phase, A, and a grain-boundary phase, GB1 (Gd$_{51.5}$Si$_{31.7}$Ge$_{16.8}$). A new matrix phase, the B phase, appears with the smallest addition of iron, that is,

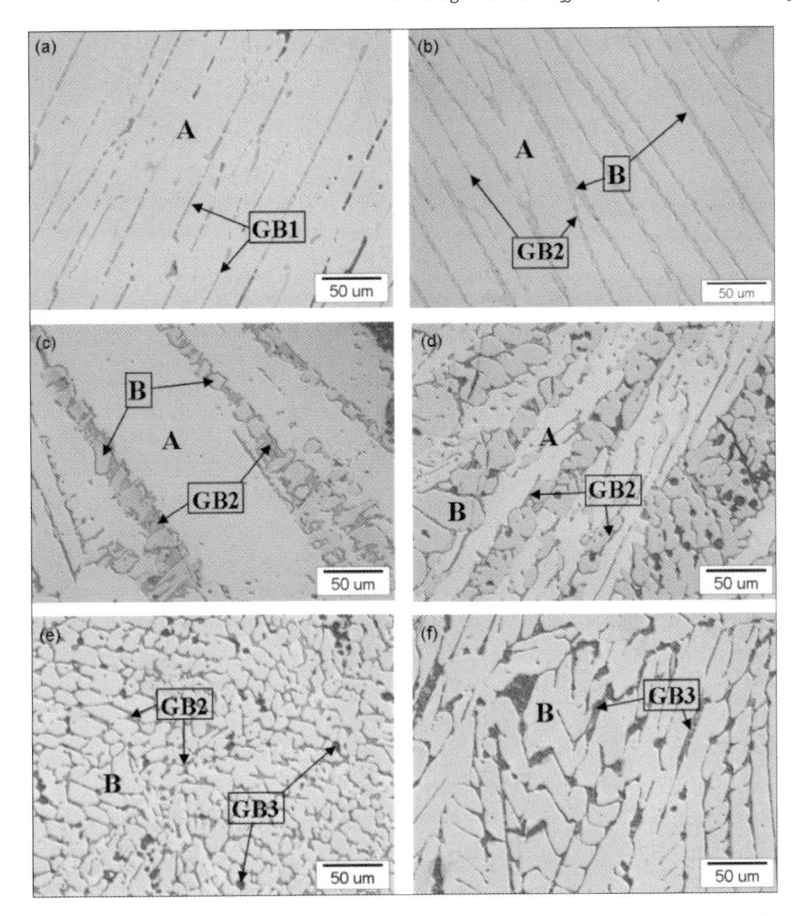

Figure 8.8 Optical images of the etched microstructures of arc-melted S0 (a), S1.4 (b), S2.8 (c), S5.6 (d), S8.6 (e) and S11.1 (f).

the S1.4 sample, together with the grain-boundary phase GB2. The GB1 phase was not observed here, not in any of the subsequent samples. The composition of the matrix phase B suggests that it is a $Gd_5(Si,Ge)_3$-type phase. With increasing amounts of added iron the amount of matrix phase A decreases, until it disappears completely in sample S8.6. It was found that approximately half of the matrix phase A is replaced by matrix phase B in the S5.6 sample.

The samples where the germanium is substituted with iron are shown in Figure 8.9.

The G0 sample is the same as the S0 sample, and is not shown in Figure 8.9. The main phase in the G0.7 sample, that is, Phase C, has the composition $Gd_{60.6}Si_{19.1}Ge_{20.3}$, the same as that observed by Provenzano et al. [27] As the amount of iron increases, the amount of Phase C is seen to decrease, although the composition remains almost constant. The amount of Phase D is seen to increase with the increasing amounts of iron, and its composition gets slightly richer in silicon as more and more iron is

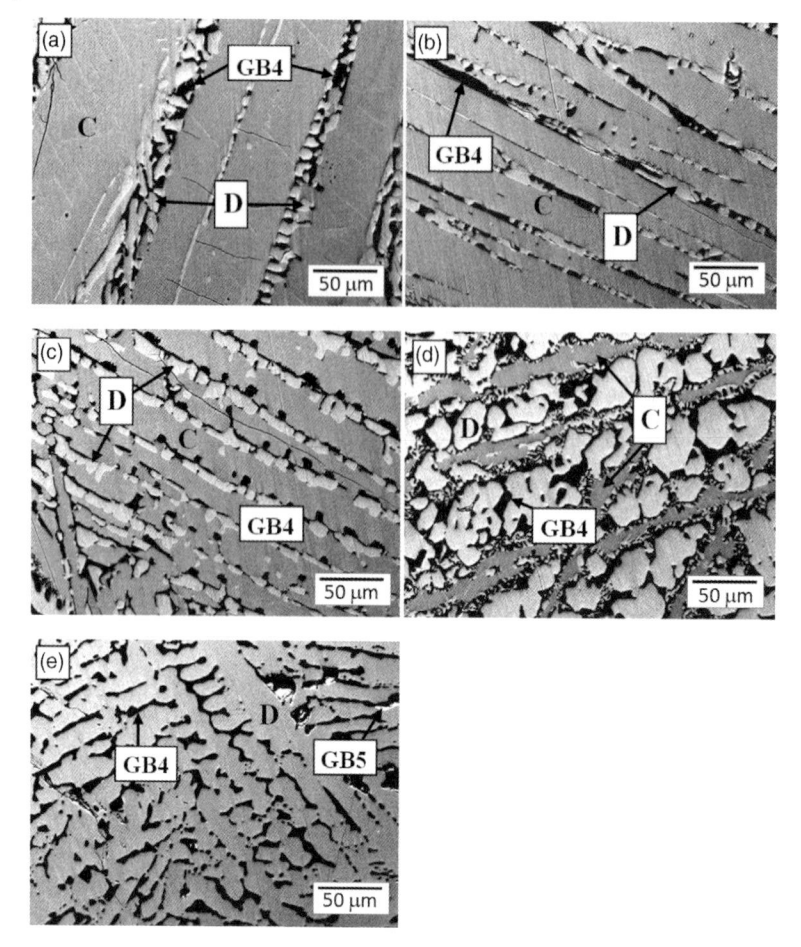

Figure 8.9 Optical images of the etched microstructures of arc-melted G0.7 (a), G1.4 (b), G2.81.4 (c), G5.6 (d) and G8.4 (e). All images have the same magnification.

added. No Fe could be detected in Phase D. A summary of the compositions of the major phases observed in the different samples is given in Table 8.4.

The grain-boundary phases in the G samples are referred to as GB4 and GB5. The GB4 phase appears as black in the microstructures and becomes increasingly rich in iron as more iron is substituted for the germanium. The GB5 phase is very light in color and was only observed in the G8.4 sample, which contained the largest amount of iron. A summary of the compositions of the observed grain-boundary phases is given in Table 8.5.

The optical micrographs in Figure 8.8 show the gradual changes with compositional variations. The second matrix phase, B, begins to form between the grains of the original matrix phase, A, and then gradually comes to dominate the microstructure as the amount of iron in the sample increases. The EDS measurements clearly

Table 8.4 Elemental compositions in at.% of the main phases present in the S and G samples evaluated using EDS. Analyzing errors to be considered are Gd ±0.2, Si ±0.1, Fe ±0.2 and Ge ±0.3.

	Phase A				Phase B			
	Gd	Si	Fe	Ge	Gd	Si	Fe	Ge
S0	55.8	23.7	/	20.5	–			
S1.4	55.2	20.7	0.5	23.6	61.5	16.1	0.6	21.8
S2.8	56.2	20.0	0.5	23.3	62.8	16.2	1.0	20.6
S5.6	56.2	17.6	0.8	25.4	63.2	13.8	0.6	22.4
S8.4	56.3	15.2	0.7	27.8	61.8	11.1	0.7	26.4
S11.1	–				61.8	11.7	1.3	25.2
	Phase C				Phase D			
G0.7	60.6	19.1	—	20.3	66.3	12.8	—	21.9
G1.4	60.0	19.4	—	20.6	66.0	12.0	—	22.0
G2.8	60.9	19.9	—	20.2	66.0	13.0	—	21.0
G5.6	59.1	21.2	—	19.7	65.2	14.8	—	21.0
G8.4	–				64.7	15.9	—	20.4

reveal the presence of iron in both matrixes, A and B. However, the amount of iron does not vary much, being between 0.5 wt.% and 1.3 wt.% for both phases in all the samples. Much more dramatic changes are seen in the amounts of Si and Ge in the matrix phases. In both cases the addition of iron at the expense of Si causes Ge to dominate over Si in the 5 : 4 phase. In the S1.4 sample, for example, the relative amounts of Si and Ge in the two phases, A and B, were 20.7 : 23.6 and 16.1 : 21.8

Table 8.5 Elemental compositions in at.% of the grain-boundary phases present in the samples S and G evaluated using EDS. Analyzing errors to be considered are Gd ±0.2, Si ±0.1, Fe ±0.2 and Ge ±0.3.

	GB2 phase				GB3 phase			
	Gd	Si	Fe	Ge	Gd	Si	Fe	Ge
S0	–				–			
S1.4	34.8	27.4	33.3	4.5	–			
S2.8	35.5	28.0	30.8	5.6	–			
S5.6	35.5	26.9	31.7	5.9	–			
S8.4	35.3	26.3	31.4	7.0	33.1	5.5	59.9	1.5
S11.1	33.2	28.3	34.5	4.0	31.3	13.2	53.1	2.5
	GB4 phase				GB5 phase			
G0.7	51.0	21.8	9.8	17.4	–			
G1.4	36.8	30.7	20.9	11.7	–			
G2.8	44.9	23.0	16.6	15.5	–			
G5.6	44.1	23.1	17.7	15.1	–			
G8.4	33.1	24.9	31.7	10.3	54.4	20.1	—	25.5

respectively. By the time we reach sample S8.4, the ratios have shifted to samples that are much richer in Ge, that is, 15.2 : 27.8 and 11.1 : 26.4. This change in the Ge : Si ratio is very important because the decrease, rather than increase, in the T_c, is a consequence of this change. This relatively large increase in the proportion of Ge in the phase causes the T_c to fall, with the fall being in good agreement with the magnetic phase diagram of the Gd_5Si_4–Gd_5Ge_4 pseudobinary system reported in 1997 [4].

8.3.3
X-Ray Diffraction Measurements

The XRD patterns of the Si-substituted samples can be seen in Figure 8.10. The peaks in the pattern that belong to the S0 sample confirm that the main phase seen in the microstructure (Figure 8.8a) is the $Gd_5Si_2Ge_2$ monoclinic phase. The vast majority of the peaks agree with the calculated pattern for the $Gd_5Si_2Ge_2$ compound. The positions of the calculated peaks are identified by the circles in Figure 8.10. New peaks can be observed in the other patterns as a result of the formation of new phases in all the samples with added iron. Some shifts in the peaks, while maintaining the same structure, can also be seen, and these shifts result from the iron entering both the A and B matrix phases and forming solid solutions [151]. The pattern of the S11.1 sample indicates that the matrix phase B is the only matrix phase still present in the sample.

8.3.4
Magnetic Measurements

Magnetization *vs.* temperature measurements were made on the samples S0–S8.4, see Figure 8.11. The results indicate that the Curie temperature reduces with

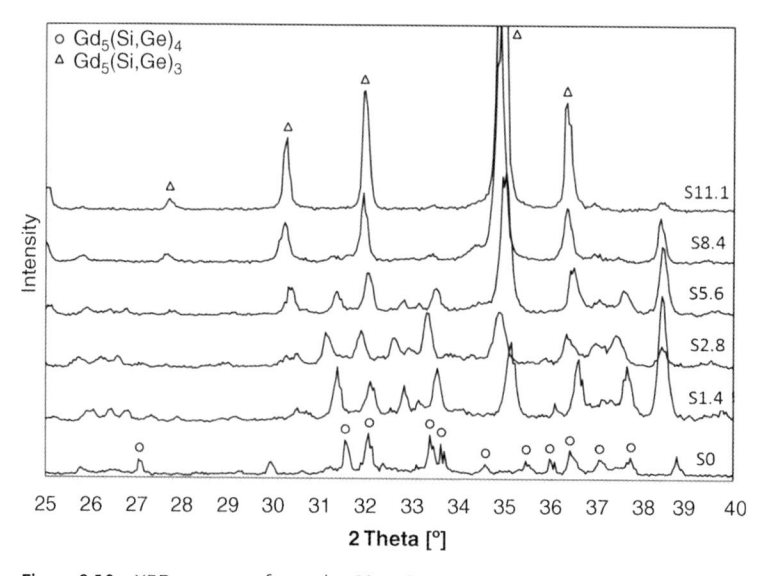

Figure 8.10 XRD patterns of samples S0 to S11.1.

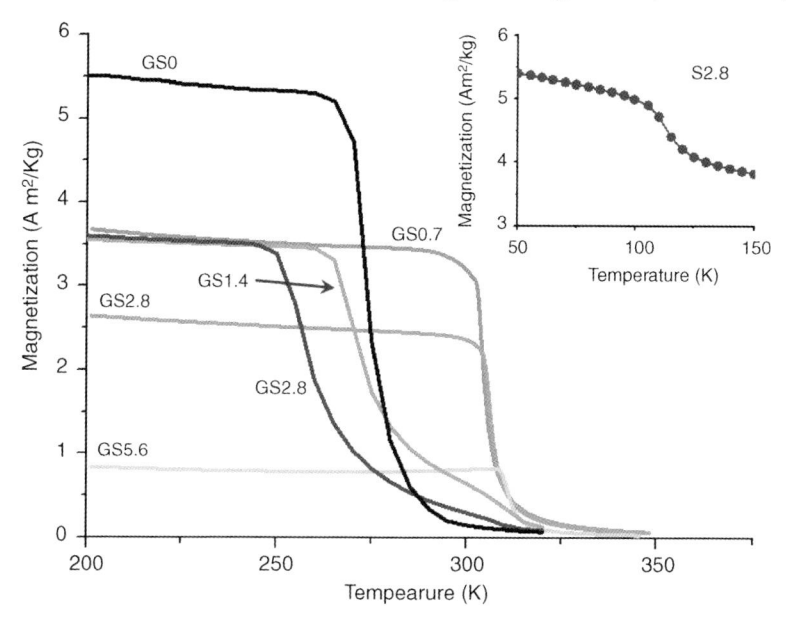

Figure 8.11 Curie-temperature dependence of the various phases in the Si- and Ge-substituted alloys.

increasing amounts of added iron when substituting for Si. This is consistent with an increasingly Ge-rich matrix phase, and is confirmed by our measurements in Table C. For all the samples except S0 the Néel temperature of the 5 : 3 phase can be observed at about 110 K. An example of the transition is shown in the inset in Figure 8.11.

We also conducted similar experiments for some of the G alloys, looking at the Curie temperatures for the G0, G0.7 and G2.8 alloys. It is clear from Figure 8.12 that substituting the Ge with iron leads to a significant increase in the Curie temperature of the samples [40].

The increase in the Curie temperature when the Ge is substituted by Fe is in excellent agreement with Provenzano [34]. We believe that one of the major effects of adding Fe is for it to combine primarily with the silicon (so leading to the formation of two or three grain-boundary phases), resulting in a matrix phase that is depleted in silicon compared to the alloy without any addition of iron.

We also looked at the maximum values of ΔS_m for the S series of alloys. From Figure 8.12 it is clear that the ΔS_m values tend to follow the Curie temperatures and decrease in line with the amount of iron. S0 has the maximum ΔS_m value of almost 10 J/kg K. Samples S2.8 and S5.6 have their maximum ΔS_m values in the same temperature range as S1.4, but their values are 50% lower. The samples S8.4 and S11.1 are not suitable for room-temperature magnetic refrigeration because of their low magnetic transition temperatures. Thus, no ΔS_m calculations were made for them. Pure Gd was measured as a standard and is shown in the figure as a dashed line. Measurements of the hysteresis losses showed that they tended to decrease with the amount of added iron. We did observe a very modest increase for the S1.4 sample, but the increase could not be described as significant.

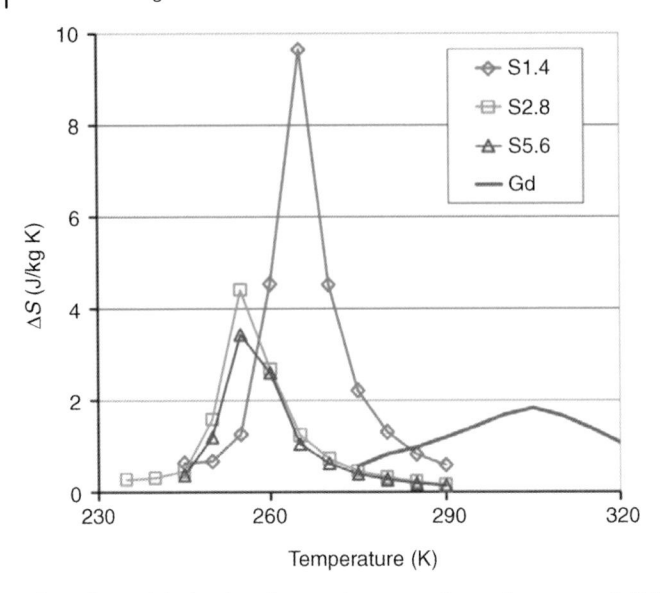

Figure 8.12 Calculated total magnetic entropy change for a magnetic field change of 1.5 Tesla.

8.3.5
Hysteresis Losses

The results of the entropy changes (ΔS_m) and the measurements of the hysteresis losses for substitutions of Ge or Si with Fe are shown in Figure 8.13. The peak value for ΔS_m occurs, not unsurprisingly, at the composition $Gd_5Si_2Ge_2$. For small substitutions of Si both the ΔS_m and the losses are seen to decrease; however, the fall off in the losses is less than that observed for the substitution of Ge by Fe. In this case we observed a decrease in the losses by 90% for G0.3, 99.4% for G1.4 and 99.9% for G2.8. The reduction in ΔS_m for the same samples was 69.3, 81.6, and 94.3%, respectively. These results are in good agreement with those of Provenzano et al., but provide a clearer representation of the extent and the type of the substitution required to bring about the necessary reduction in the hysteresis losses.

8.4
TEM Investigation of CMAs

8.4.1
TEM Research on - $Gd_5Si_2Ge_2$ Alloys

The crystal structure of the $Gd_5(Si_xGe_{1-x})_4$ phases consist of 36 atoms per unit cell distributed among six to nine independent crystallographic sites [152]. At one end, the Si-rich compounds ($0.5 < x \leq 1$) display the Gd_5Si_4-type orthorhombic (space group Pnma) structure, O(I). The intermediate region, $0.24 \leq x \leq 0.5$, displays a

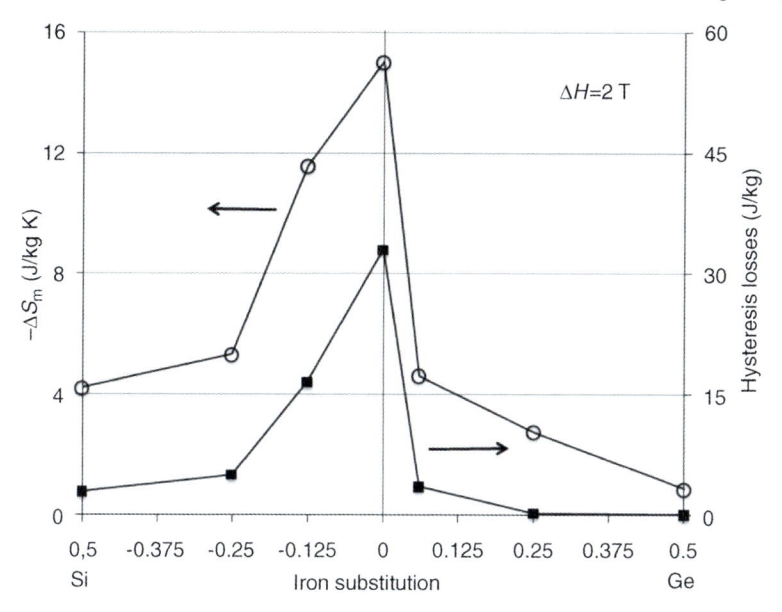

Figure 8.13 Magnetic entropy changes and hysteresis losses *versus* the amount of substituted Fe in both alloys. The arrows indicate the corresponding axes.

monoclinic structure, M, space group P112$_1$/a, which is a subgroup of Pnma, where the cline plane n and the mirror plane m are lost. Gd$_5$Si$_2$Ge$_2$ also has this structure. At the other end the Ge-rich compounds ($x \leq 0.2$) display a Sm$_5$Ge$_4$-type orthorhombic (space group Pnma) phase. In the range $0.2 < x < 0.24$, O(II) and M structures coexist [153].

The silicide and the germanide differ from each other a great deal in terms of bonding arrangements. However, they are built from essentially equivalent layers (slabs) that are infinite in two dimensions (*a* and *c*), as shown in Figure 8.14a.

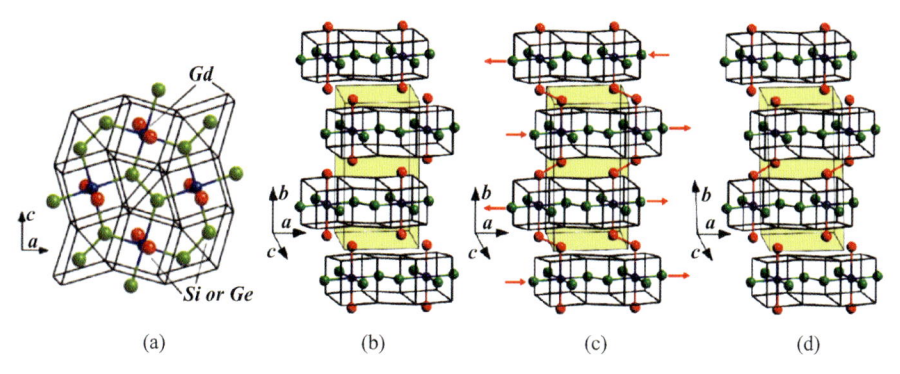

Figure 8.14 The basic building block (slab) of Gd$_5$(Si$_x$Ge$_{1-x}$)$_4$ phases (a) and relationships between their room temperature crystal structures: $0.5 < x \leq 1$ (b), $0.24 \leq x \leq 0.5$ (c), and $0 \leq x \leq 0.2$ (d). Figures taken from reference [154].

The Gd atoms are located in the corners of the cubes and trigonal prisms (the Gd atoms are not shown for clarity), and inside the cubes. Both Si and Ge inside the slab form partially covalent bonds. The other Si(Ge) atoms are located on the slab surface, and they play a crucial role in the interslab bonding, thus controlling both the crystal structure and properties of the alloys. Figure 8.14b shows that in the crystal structure between $0.5 < x \leq 1$ the slabs are all interconnected via partially covalent interslab Si(Ge)-Si(Ge) bonds. In the intermediate phase ($0.24 \leq x \leq 0.5$) one half of the interslab Si(Ge)–Si(Ge) bonds are broken, as shown in Figure 8.14c. The layers of the Gd_5Ge_4-based solid solution ($x \leq 0.2$) are no longer interconnected, that is, there are no interslab Si(Ge)-Si(Ge) bonds (Figure 8.14d).

$Gd_5Si_2Ge_2$ has a monoclinic crystal structure at room temperature known as the β: space group $P112_1/a$ with lattice parameters $a = 7.5808(5)$, $b = 14.802(1)$, $c = 7.7799(5)$ Å, and $\gamma = 93.190(4)°$ [155]. At lower temperatures (below 276 K) the material has an orthorhombic crystal structure (α-$Gd_5Si_2Ge_2$) with all the interslabs connected. Choe et al. [156] show that sites inside the slabs are preferred by Si: it has an about 60% occupancy and Ge occupies the remaining 40%, while the sites responsible for bonding between the slabs are occupied by 60% of Ge and only 40% of Si. Upon the α → β phase transformation, these slabs undergo a shear movement along the crystallographic a direction in the orthorhombic α phase, which breaks one-half of the (Si,Ge)-(Si,Ge) covalent bonds between the slabs, and becomes the monoclinic β phase. Interatomic distances change less than 3% during α → β transformation. The α → β transformation can be induced by changing the magnetic field, as was reported for $Gd_5Si_{1.8}Ge_{2.2}$ [157]. When the field is applied to the β phase at temperatures just exceeding 276 K, the paramagnetic β phase transforms to the ferromagnetic α one. Reducing the magnetic field to 0 T at constant temperature creates the paramagnetic β phase from the ferromagnetic α phase. Therefore, this structural transition is controlled by composition, temperature and magnetic field.

Another interesting structural feature in β-$Gd_5Si_2Ge_2$, found by Choe et al. [156] is nonmerohedral twinning, for which only a small portion of reflections from two twin components are exactly superimposed, while the remaining reflections are not. Twinning can be classified as merohedral or nonmerohedral. In the case of nonmerohedral twinning only some subsets of the diffraction reflections from different domains overlap, whereas for merohedrally twinned crystal diffraction patterns from different domains overlap exactly in three dimensions.

The nonmerohedral twinning in β-$Gd_5Si_2Ge_2$ disappears and reappears, respectively, as the material is cycled across the transition temperature. Figure 8.15 illustrates a model of the possible twin mechanism during the phase transition.

For no twinning in the β-phase adjacent pairs of slabs shift in the opposite directions along the a-axis (see the arrows in Figure 8.15(a)). On the other hand, if two neighboring interslab Si(Ge)–Si(Ge) bonds are broken, then two different orientations of equivalent monoclinic unit cells are generated, which are related to each other by a mirror operation in the a–c plane (perpendicular to the b-axis; see Figure 8.15(b)), and nonmerohedral twinning occurs.

Meyers et al. [158] investigated twinning in β-$Gd_5Si_2Ge_2$ using transmission electron microscopy and selected-area diffraction analyses. They confirmed the twin

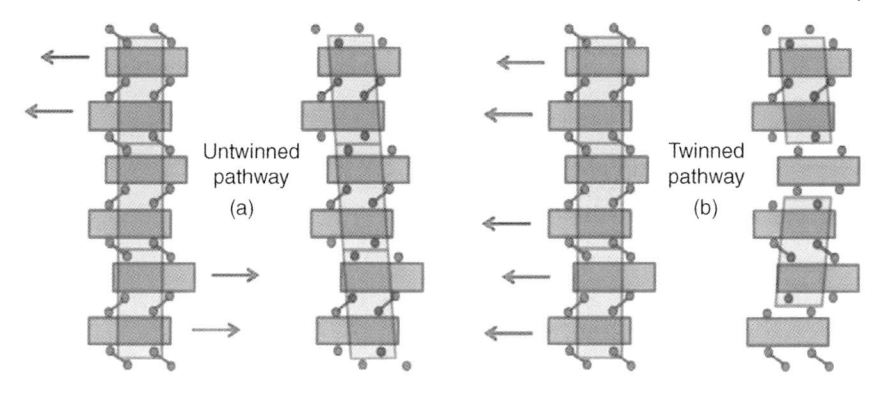

Figure 8.15 Two possible pathways for the $\alpha \rightarrow \beta$ structural transformation by the shear movement along the **a**-axis. The first and the third figure are the α structure and the second and the fourth figure are the β structure models.

patterns in β-$Gd_5Si_2Ge_2$, as suggested in the single-crystal structure study by Choe. All of the single-crystal specimens that were examined were nonmerohedral twins at room temperature, suggesting that twinning is an intrinsic feature of the monoclinic β phase. Direct observation of the transformation with a discussion on the mechanism of the reversible, low-temperature phase transformation was studied with transmission electron microscopy also by Meyers et al. [159].

Ugurlu et al. [160] explained, using SEM and TEM, the existence of long linear features that form during the solidification of $Gd_5Si_2Ge_2$. It has been shown that these linear features grow as thin plates oriented in specific directions. The results confirmed an earlier study that suggested the features might possibly be a Widmanstatten structure [161].

8.4.2
Imaging in the TEM

The fundamental principle of imaging with transmission electron microscopy (TEM) is first to view the diffraction pattern; this indicates how the observed specimen is scattering the incident beam. From the TEM image we can see different contrasts, which we obtain by selecting specific electrons or excluding them from the imaging system. We can form bright-field (BF) images by selecting the direct electrons or dark-field (DF) images by selecting scattered electrons. The direct or scattered electron beam is selected with the objective aperture. Without the aperture the contrast is poor. Also, the size of the aperture governs which electrons contribute to the image and thus controls the contrast. To create a DF image we usually tilt the incident beam in such a way that the scattered electrons remain on axis and create a centered dark-field (CDF) image.

In a scanning transmission electron microscope (STEM) we select the direct or scattered beams in an equivalent way, but we use detectors rather than apertures. Figure 8.16 shows different operational modes.

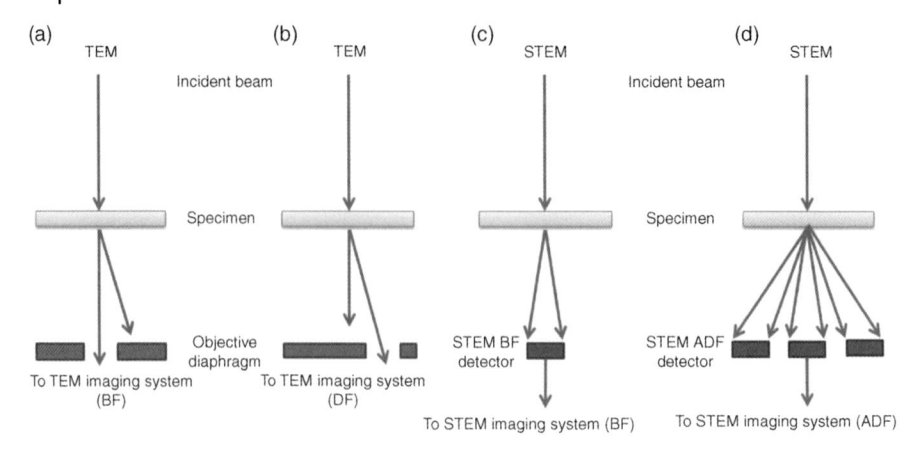

Figure 8.16 The use of an objective aperture in TEM to select the direct (a) or scattered (b) electrons and an on-axis detector (c) or an annular detector (d) in STEM to perform equivalent operations. A and B form a BF image, while B and D form a DF image [162].

8.4.2.1 Mass-Thickness Contrast

Mass-thickness contrast arises from incoherent (Rutherford) elastic scattering of electrons. Rutherford scatter is a strong function of the number Z, that is, the mass or the density, ϱ, as well as the thickness, t, of the specimen.

Usually, mass-thickness contrast images are used for qualitative analysis. Here, we can expect high-Z regions of a specimen to scatter more electrons than low-Z regions of the same thickness. Also, thicker regions will scatter more electrons than thinner regions of the same average Z. That is why thicker and/or higher mass areas will appear darker than thin and/or lower mass areas in a BF image (Figure 8.17). The opposite is true for DF images.

If we analyze noncrystalline materials, mass-thickness contrast is the most important feature, but because it is impossible to prepare a uniformly thin sample, mass-thickness contrast cannot be completely avoided.

8.4.2.2 TEM Images

A good technique for observing the shape of the particles on a film is shadowing. Shadowing introduces some mass contrast to a thickness-contrast image and with this method we can also make 3D images. If we shadow polymers with heavy metals, they can segregate on unsaturated bonds, thus making them darker in BF images.

The TEM variables that affect the mass-thickness contrast for a given specimen are the objective aperture and the voltage (kV). Large apertures collect more scattered electrons; the contrast is lowered and the overall image intensity increases. A lower kV will increase the scattering angle and the cross-section. Hence, the contrast increases, but lowers the intensity. DF images show complementary contrast to BF images. They are darker, but with excellent contrast.

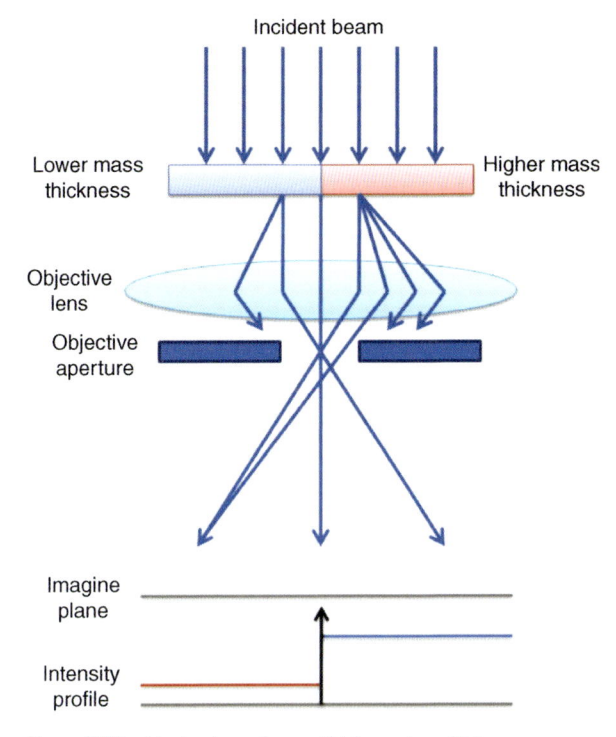

Figure 8.17 Mechanism of mass-thickness in a BF image.

8.4.2.3 STEM Images

Scanning-transmission electron microscopy (STEM) is more flexible then TEM, because we can control L, which changes the collection angle of the detector. In this way we can control which electrons contribute to the image. They can enhance the contrast, but show poorer resolution. To get a reasonable intensity in a scanning image we have to use a larger beam. The STEM is usable for:

- thick samples when the chromatic aberration limits the TEM resolution;
- a beam-sensitive specimen;
- a specimen that has inherently low contrast in the TEM and we cannot digitalize the TEM image.

8.4.2.4 TEM Diffraction Contrast

Bragg diffraction is controlled by the crystal structure and orientation of the specimen. This can be used for creating contrast in TEM images. Diffraction contrast is simply a special form of amplitude contrast, because the scattering occurs at special (Bragg) angles. Here, the coherent elastic scattering produces diffraction contrast. The incident electrons must be parallel in order to give sharp diffraction spots and strong diffraction contrast.

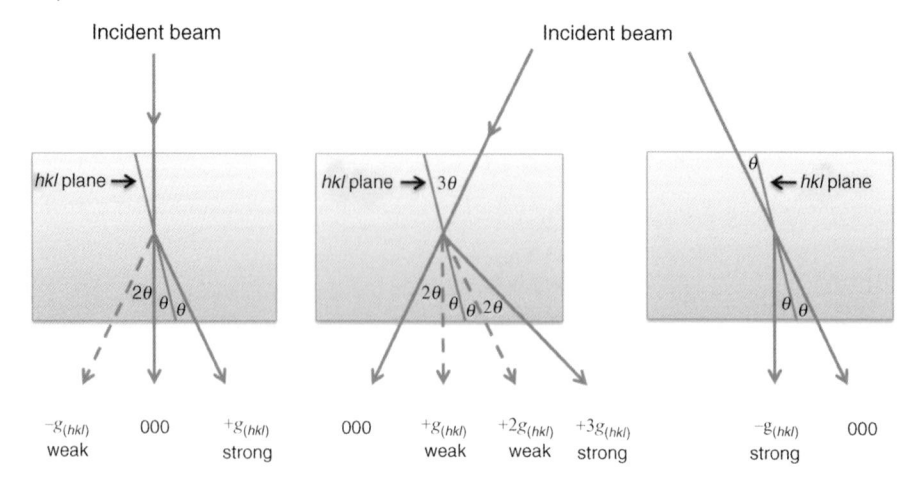

Figure 8.18 (a) Standard two-beam condition with a bright 000 and hkl spot. To get a strong two-beam CDF condition we have to move the $-g_{hkl}$ spot (c) into the on-axis position and not the g_{hkl} spot (b).

For the mass-thickness contrast we use any scattered electrons to form a DF image. For diffraction contrast we tilt the specimen to two-beam conditions, in which only one diffracted beam is strong. The area that appears bright in the DF image is the area where the hkl planes are at the Bragg condition. We can tilt the specimen to set up different two-beam conditions and get different DF images. For crystalline materials precise tilting is necessary. To get the best contrast from defects, the specimen should not be exactly at the Bragg's condition ($s = 0$). So the deviation parameter s should be small and positive; with large s the defect images become narrower, but the contrast is reduced. With negative s the defects will be difficult to see.

Two-beam CDF images are more difficult to get, because if the incident beam tilt is in a way that strong hkl reflection moves onto the optic axis, this reflection becomes weaker and the 3h3k3l reflection becomes strong (weak beam condition). We need to tilt in to the reflection to get a strong on-axis reflection (Figure 8.18). This technique is crucial to obtain and to interpret diffraction-contrast images.

8.4.2.5 STEM Diffraction Contrast

The principle is the same as for mass-thickness contrast. We use a BF detector to pick up the direct beam and the annular DF detector to pick up the diffracted beam. To preserve two-beam conditions, the ADF detector must only pick up one strong diffracted beam.

However, the diffraction contrast observed in the STEM image will generally be much poorer than the TEM contrast. To explain this, we have to remember the three conditions that have to be fulfilled for a strong contrast:

- the incident beam must be parallel, that is i.e. the convergence angle (α) must be very small;
- the specimen must be tilted to a two-beam condition;

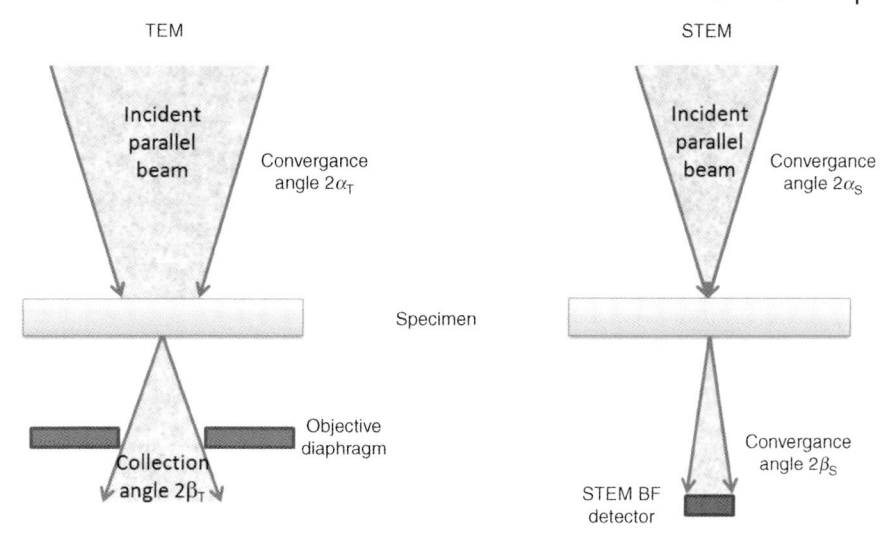

Figure 8.19 Comparison of the beam-convergence and divergence angles (a) in TEM and (b) in STEM.

only the direct beam or the one strong diffracted beam must be collected by the objective aperture (Figure 8.19).

To get the same conditions in TEM and STEM, α_{TEM} must equal α_{STEM} and β_{TEM} (the objective aperture collection semiangle) must equal β_{STEM}. Since this is not possible, because of the high convergence angle in STEM, we can overcome this if we fulfill the condition $\alpha_{STEM} = \beta_{TEM}$ and $\alpha_{TEM} = \beta_{STEM}$. To satisfy the second equation we have to make a very small β_{STEM}, which reduces the signal falling on the STEM and we get a noisy STEM image. This can be improved with a field emission gun (FEG), but generally diffraction-contrast images of crystal defects are the domain of a TEM investigation.

8.4.3
Experimental Results of TEM Analyses

The samples were prepared from elements with purity of Gd(99.99%), Si(99.9995%) and Ge(99.9999%) (total mass around 5 g). They were arc melted in an argon atmosphere into the shape of buttons and annealed at 1000 °C for 22 h in vacuum in a Carbolite furnace. The samples were subsequently cut, polished, dimpled and ion milled to obtain transparency, which is required for TEM analysis.

TEM images were performed on a JOEL TEM 2010 F. Figure 8.20 shows a BF image of the $Gd_5Si_2Ge_2$ sample. We can observe planar defects all over the sample. The lines are twin boundaries. From the morphology of the defects it is very clear that a substantial number of dislocations are present (the curved lines with the nonuniform contrast, indicating the strain fields). The isolated "spots" with black-and-white contrast can be interpreted as dislocations that are perpendicular to the normal of

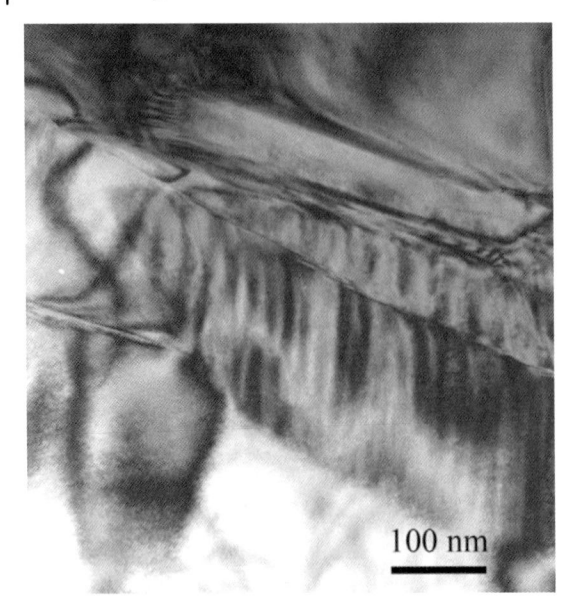

Figure 8.20 TEM image of $Gd_5Si_2Ge_2$ sample.

the prepared disc sample. As well as these dislocations there are some parallel lines with a more uniform contrast, which can be attributed to planar defects, such as stacking faults or twin boundaries. However, for an exact determination of these types of defects a much more detailed examination is required. Nevertheless, our TEM investigation strongly suggests that the structure of Fe-substituted materials is even more complex than we predicted. In addition to the suppression of the orthorhombic-to-monoclinic transition, the development of additional phases, and the shift in the T_C, we also have amorphous regions, dislocations, planar faults and crystallographically related grains.

Figure 8.21 shows a nice example of the difference between BF and DF images of the sample. As mentioned above, for BF images we select only the direct electrons and the white areas are where there is no scattering. The opposite is true for the DF image.

Figure 8.22 shows the diffraction pattern of $Gd_5Si_2Ge_2$ and we can confirm the existence of twins in this material. The angle between the twins is 25°. In Figure 8.23, we have indexed the DF and defined $Z = [2\ -4\ 7]$.

8.4.4
TEM of Low Loss Samples

Not many papers with TEM research on $Gd_5Si_2Ge_2$ have been published so far. TEM imaging is a useful tool for analyzing planar defects and twins that appear in every $Gd_5Si_2Ge_2$ sample. A proper indexation can be a very difficult and time-consuming procedure. In Provenzano's earliest report [145] about hysteresis losses it was

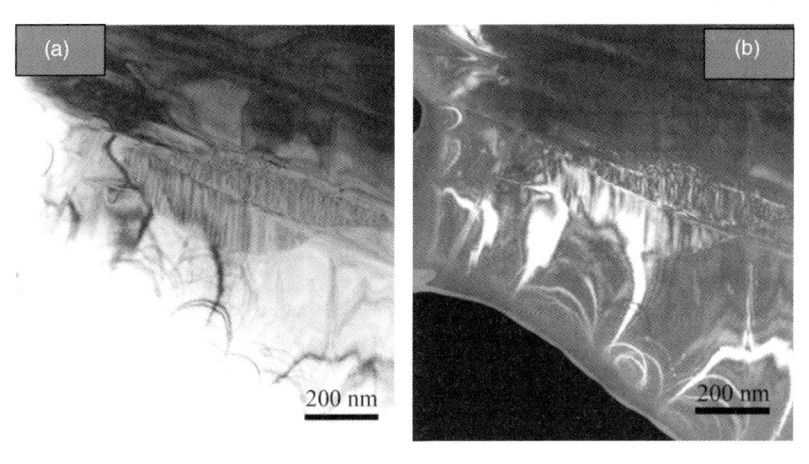

Figure 8.21 BF (a) and DF (b) image of the same area for the Gd$_5$Si$_2$Ge$_2$ sample.

mentioned that the addition of Fe creates a magnetic nanostructure, based on the superparamagnetic behavior of the material at higher temperature; the authors also speculated that such samples might contain an inhomogeneous distribution of vacancies or defects in the Gd–Ge–Si matrix phase.

For this reason we performed a detailed TEM investigation on the samples with the lowest hysteresis losses. Figure 8.24 shows a high-resolution TEM micrograph of a powdered sample of Gd$_5$Si$_2$Ge$_{2-y}$Fe$_y$ ($y=0.25$). The inset shows a fast Fourier transformation (FFT) of the area indicated by the white square. Surprisingly, the micrograph exhibits the characteristics of an amorphous structure; there are no lattice fringes and no indication of any long-range order. The FFT, with its diffuse rings, confirms the amorphous nature of this area. A chemical analysis of this

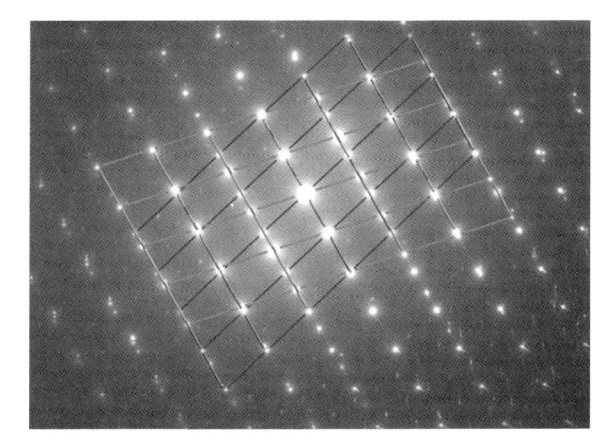

Figure 8.22 Diffraction pattern of a twin boundary.

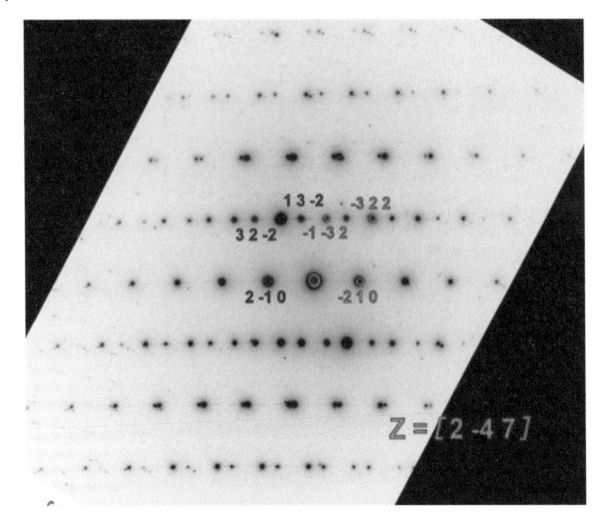

Figure 8.23 Indexation of the diffraction pattern.

amorphous phase showed the $Gd_{59}Si_{21}Ge_{20}$ composition with only trace amounts of Fe being detected.

The technique used to prepare this TEM specimen precludes any possibility of us knowing where exactly the amorphous regions are, relative to the other phases, or any assessment of the amount of amorphous material. Furthermore, an X-ray diffractogram did not indicate that there was any amorphous material in the sample. However, the X-ray penetration depth is limited and X-ray diffraction is not able to reliably

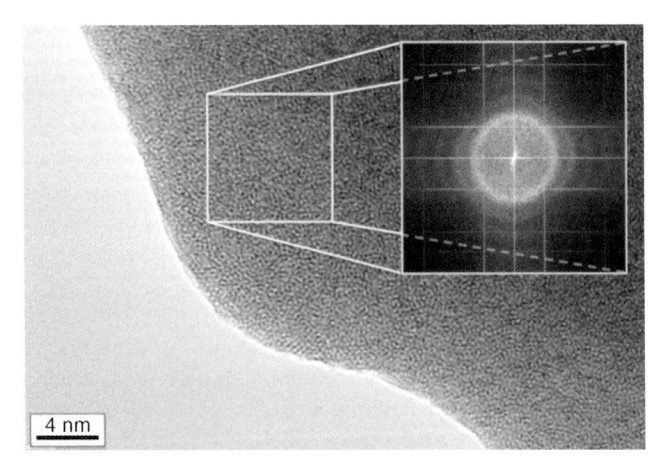

Figure 8.24 High-resolution TEM micrograph of a powder sample of $Gd_5Si_2Ge_{2-y}Fe_y$ ($y=0.25$) on a carbon-coated copper grid and the corresponding electron diffraction pattern.

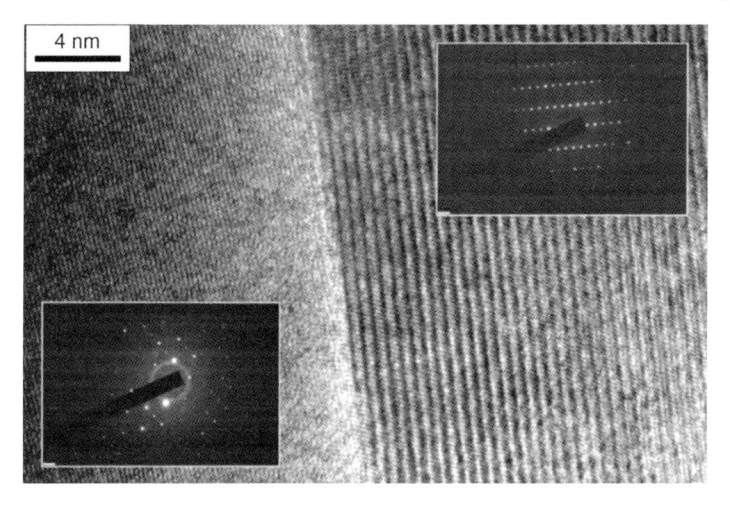

Figure 8.25 High-resolution TEM micrograph of a grain-boundary region from the sample $Gd_5Ge_2Si_{2-y}FeO_y$ ($y = 0.25$) and the corresponding electron-diffraction patterns.

confirm the presence of a few per cent of any particular phase or structure. From this we can conclude that the regions of amorphous material must represent a very small fraction of the material. Figure 8.25 shows a grain-boundary region from a conventionally prepared TEM sample $Gd_5Ge_2Si_{2-x}Fe_x$ ($x = 0.125$). The insets are the respective selected-area electron-diffraction (SAED) patterns. The right-hand grain has the composition $Gd_{55}Ge_{25}Si_{20}$, while the left-hand grain is the Fe-containing $Gd_{35}Si_{25}Fe_{30}Ge_{10}$. This figure shows clearly that there is no evidence of any nanoscale grain-boundary phases, any nanosized grains or unusual features. In the left-hand inset, where the SAED pattern of the iron-containing phase is displayed, narrow circles can be resolved. Such detail is characteristic of randomly oriented nanocrystals. Closer inspection of the diffraction pattern suggests that the rings could be attributed to the presence of α-iron. In addition to the circles with their center at the central spot, an echo (a diffuse circle) can also be observed around the strong diffracting spots, characteristic of double diffraction. From these observations we can conclude that α nanoparticles are present in the upper and lower sections of the sample, but the method of its preparation must always be considered, since ion milling can be responsible for such artefacts. There is, however, clear evidence of a crystallographic relationship between the two grains; most probably here we are dealing with a low-angle grain boundary.

From the results presented we are able to conclude that $Gd_5Si_2Ge_2$ alloys with substitutions of Fe show very significant differences in terms of both their macrostructures and microstructures. The matrix phase A with the $Gd_5(SiGe)_4$ composition for S0 and G0 (Table 8.2) becomes the B matrix phase with the $Gd_5(SiGe)_3$ composition for S11.1 for the Si-substitution samples. A similar change was seen when we substituted Ge with Fe. The substituted iron was found in all the matrix and grain-boundary phases, although the amounts in the matrix phases were very low.

The iron contributes mainly to the grain-boundary phases that are formed and to a change in the relative amounts of Si and Ge in the matrix phases.

The substitution of silicon with iron reduces the Curie temperature; this is due to the relatively large change in the Ge : Si ratio for the $Gd_5(SiGe)_4$ phase, leading to much more Ge than Si being present. Substituting Ge with Fe leads to an increase in the Curie temperature.

Our microstructural investigations have helped to clarify some of the magnetic changes observed when adding Fe at the expense of Si and Ge in $Gd_5Si_2Ge_2$-type alloys. However, we also see that very similar microstructures can lead to substantially different magnetic properties. It is clear, therefore, that we will only be able to improve our understanding of the behavior of these materials by applying a wide range of analytical techniques.

8.5 Conclusions

We have investigated a wide range of Fe substitutions of the $Gd_5Si_2Ge_2$ magnetocaloric alloy with the aim to quantitatively clarify the effect of iron with respect to the reduction in entropy and hysteresis losses. We found that Fe has contrasting effects on the T_C, depending on whether it substitutes for Ge or Si. We also found that it produces a $Gd_5(Ge,Si,Fe)_3$-type phase with a T_C at approximately 110 K. Substituting Ge with Fe was found to reduce the hysteresis losses almost to zero; substituting Si with Fe was somewhat less effective, because it tended to maintain the first-order transition. Our TEM studies revealed the presence of features not reported previously, such as amorphous regions, dislocations, planar faults and crystallographically related grains. The results of our studies suggest that there is much more to discover about magnetocaloric materials, and that some of the problems associated with first-order transitions and hysteresis losses in Gd-Ge-Si materials can be overcome with small compositional modifications involving the substitution of Si with Fe and, to a greater extent, for Ge.

Acknowledgments

The authors would like to acknowledge the financial support of European Network of Excellence "Complex Metallic Alloys" (CMA) in FP6 and the Slovenian Agency for Research and Development.

References

1 Warburg, E. (1881) *Ann. Phys. (Leipzig)*, **13**, 141.
2 Debye, P. (1926) *Ann. Phys.*, **81**, 1154.
3 Giauque, W.F. (1927) *J. Am. Chem. Soc.*, **49**, 1864.
4 Pecharsky, V.K. and Gschneidner Jr., K.A. (1997) *Phys. Rev. Lett.*, **78**, 4494.
5 Tishin, A.M. and Spichkin, Y.I. (2003) *The Magnetocaloric Effect and its*

Applications, Institute of Physics Publishing, Bristol.

6 Dan'kov, S.Y., Tishin, A.M., Pecharsky, V.K., and Gschneider Jr., K.A. (1998) *Phys. Rev. B*, **57**, 3478.

7 Gschneidner Jr., K.A., Pecharsky, V.K., and Tsokol, A.O. (2005) *Rep. Prog. Phys.*, **68**, 1479–1539.

8 Kohanoff, J. (2006) *Electronic Structure Calculations for Solids and Molecules*, Cambridge University Press, Cambridge.

9 Pecharsky, V.K., Samolyuk, G.D., Antropov, V.P., Pecharsky, A.O., and Gschneidner Jr., K.A. (2003) *J. Solid State Chem.*, **171**, 57.

10 Samolyuk, G.D. and Antropov, V.P. (2005) *J. Appl. Phys.*, **97**, 10A310.

11 von Ranke, P.J., de Oliveira, N.A., and Gama, S. (2004) *J. Magn. Magn. Mater.*, **277**, 78.

12 Nóbrega, E.P., de Oliveira, N.A., von Ranke, P.J., and Troper, A. (2005) *Phys. Rev. B*, **72**, 134426.

13 Kuz'min, M.D. and Richter, M. (2007) *Phys. Rev. B*, **76**, 092401.

14 Smontara, A., Smiljanić, I., Ivkov, J., Stanić, D., Komelj, M., and Dolinšek, J. (2008) *Phys. Rev. B*, **78**, 104204.

15 Dan'kov, S.Y., Tishin, A.M., Pecharsky, V.K., and Gschneider Jr., K.A. (1998) *Phys. Rev. B*, **57**, 3478.

16 http://www.goodfellow.com (16.03.2009).

17 Tishin, A.M. and Spichkin, Y.I. (2003) Chapter 8, *The Magnetocaloric Effect and its Applications*, Institute of Physics Publishing, Bristol.

18 Zhang, X., Yang, L., Zhou, S., Qi, L., and Liu, Z. (2001) *Mater. Trans.*, **42**, 2622.

19 Wang, D., Huang, S., Han, Z., Su, Z., Wang, Y., and Du, Y. (2004) *Solid State Commun.*, **131**, 97.

20 Pecharsky, A.O., Gschneidner Jr., K.A., and Pecharsky, V.K., MCE of RCo2 (R=Tb, Dy, Ho and Er) unpublished.

21 Wang, D., Liu, H., Tang, S., Yang, S., Huang, S., and Du, Y. (2002) *Phys. Lett. A*, **297**, 247.

22 Duc, N.H., Anh, D.T.K., and Brommer, P.E. (2002) *Physica B*, **319**, 1.

23 Liu, H., Wang, D., Tang, S., Cao, Q., Tang, T., Gu, B., and Du, Y. (2002) *J. Alloys Compd.*, **346**, 314.

24 Singh, N.K., Kumar, P., Suresh, K.G., Nigam, A.K., Coelho, A.A., and Gama, S. (2007) *J. Phys.: Condens. Matter*, **19**, 036213.

25 Singh, N.K., Suresh, K.G., Nigam, A.K., Malik, S.K., Coelho, A.A., and Gama, S. (2007) *J. Magn. Magn. Mater.*, **317**, 68.

26 von Ranke, P.J., de Oliveira, N.A., Costa, M.V.T., Nobrega, E.P., Caldas, A., and de Oliveira, I.G. (2001) *J. Magn. Magn. Mater.*, **226–230**, 970.

27 von Ranke, P.J., de Oliveira, I.G., Guimaraes, A.P., and da Silva, X.A. (2000) *Phys. Rev. B*, **61**, 447.

28 von Ranke, P.J., Grangeia, D.F., Caldas, A., and de Oliveira, N.A. (2003) *J. Appl. Phys.*, **93**, 4055.

29 Duc, N.H., Anh Kim, D.T., and Brommer, P.E. (2002) *Physica B*, **319**, 1.

30 Pecharsky, A.O., Gschneidner Jr., K.A., and Pecharsky, V.K. (2003) *J. Appl. Phys.*, **93**, 4722.

31 Fu, H., Chen, Y., Tu, M., and Zhang, T. (2005) *Acta Mater.*, **53**, 2377.

32 Wu, W., Tsokol, A.O., Gschneidner Jr., K.A., and Sampaio, J.A. (2005) *J. Alloys Compd.*, **403**, 118.

33 Gschneidner Jr., K.A., Pecharsky, V.K., Brück, E., Duijn, H.G.M., and Levin, E.M. (2000) *Phys. Rev. Lett.*, **85**, 4190.

34 Provenzano, V., Shapiro, A.J., and Shull, R.D. (2004) *Nature*, **429**, 853.

35 Pecharsky, V.K. and Gschneidner, K.A. (1997) *J. Magn. Magn. Mater.*, **167**, L179.

36 Shull, R.D., Provenzano, V., Shapiro, A.J., Fu, A., Lufaso, M.W., Karapetrova, J., Kletetschka, G., and Mikula, V. (2006) *J. Appl. Phys.*, **99**, 09K908.

37 Zhuang, Y.H., Li, J.Q., Huang, W.D., Sun, W.A., and Ao, W.Q. (2006) *J. Alloys Compd.*, **421**, 49.

38 Zhang, T., Chen, Y., Tang, Y., Du, H., Ren, T., and Tu, M. (2007) *J. Alloys. Comps.*, **433**, 18.

39 Kumar, D.M.R., Manivel Raja, M., Gopalan, R., Balamuralikrishnan, R., Singh, A.K., and Chandrasekaran, V. (2008) *J. Alloys Compd.*, **461**, 14–20.

40 Podmiljsak, B., Škulj, I., Markoli, B., Žužek Rožman, K., McGuiness, P.J., and Kobe, S. (2009) *J. Magn. Magn. Mater.*, **321**, 300.

41 Zhang, T., Chen, Y., and Tang, Y. (2007) *J. Phys. D: Appl. Phys.*, **40**, 5778.

42 Carvalho, A.M.G., Campoy, J.C.P., Coelho, A.A., Plaza, E.J.R., Gama, S., and von Ranke, P.J. (2005) *J. Appl. Phys.*, **97**, 10M320.

43 Yue, M., Zhang, J., Zeng, H., Chen, H., and Liu, X.B. (2006) *J. Appl. Phys.*, **99**, 08Q104.

44 Thuy, N.P., Tai, L.T., Hien, N.T., Nong, N.V., Vinh, T.Q., Thang, P.D., Nguyen, T.P., and Molinié, P. (2001) Proc. 8th Asia-Pacific Physics Conf. (Taipei, Taiwan) (World Scientific, Singapore), p. 354.

45 Zhuo, Yi., Chahine, R., and Bose, T.K. (2003) *IEEE Trans. Magn.*, **39**, 3358.

46 Morellon, L., Magen, C., Algarabel, P.A., Ibarra, M.R., and Ritter, C. (2001) *Appl. Phys. Lett.*, **79**, 1318.

47 Huang, H., Pecharsky, A.O., Pecharsky, V.K., and Gschneidner Jr., K.A. (2002) *Adv. Cryog. Eng.*, **48**, 11.

48 Ivtchenko, V.V., Pecharsky, V.K., and Gschneidner Jr., K.A. (2000) *Adv. Cryog. Eng.*, **46**, 405.

49 Deng, J.Q., Zhuang, Y.H., Li, J.Q., and Zhu, Q.M. (2007) *Mater. Lett.*, **61**, 2359.

50 Morellon, L., Arnold, Z., Magen, C., Ritter, C., Prokhnenko, O., Skorokhod, Y., Algarabel, P.A., and Ibarra, M.R. (2004) *J. Phys. Rev. Lett.*, **93**, 137201.

51 Ryan, D.H., Elouneg-Jamr'oz, M., van Lierop, J., Altounian, Z., and Wang, H.B. (2003) *Phys. Rev. Lett.*, **90**, 117202.

52 Wada, H. and Tanabe, Y. (2001) *Appl. Phys. Lett.*, **79**, 3302.

53 Wada, H., Taniguchi, K., and Tanabe, Y. (2002) *Mater. Trans.*, **43**, 73.

54 Wada, H., Morikawa, T., Taniguchi, K., Shibata, T., Yamada, Y., and Akishige, Y. (2003) *Physica B*, **328**, 114.

55 Gama, S., Coelho, A.A., de Campos, A., Carvalho, A.M.G., Gandra, F.C.G., von Ranke, P.J., and de Oliveira, N.A. (2004) *Phys. Rev. Lett.*, **93**, 237202.

56 de Campos, A., Rocco, D.L., Carvalho, A.M.G., Caron, L., Coelho, A.A., Gama, S., da Silva, L.M., Gandra, F.C.G., dos Santos, A.O., Cardoso, L.P., von Ranke, P.J., and de Oliveira, N.A. (2006) *Nature Mater.*, **5**, 802.

57 Rocco, D.L., de Campos, A., Carvalho, A.M.G., Caron, L., Coelho, A.A., Gama, S., Gandra, F.C.G., Dos Santos, A.O., Cardoso, L.P., Von Ranke, P.J., and de Oliveira, N.A. (2007) *Appl. Phys. Lett.*, **90**, 242507.

58 Sun, N.K., Cui, W.B., Li, D., Geng, D.Y., Yang, F., and Zhang, Z.D. (2008) *Appl. Phys. Lett.*, **92**, 072504.

59 Liu, G.J., Sun, J.R., Shen, J., Gao, B., Zhang, H.W., Hu, F.X., and Shen, B.G. (2007) *Appl. Phys. Lett.*, **90**, 032507.

60 Brück, E., Tegus, O., Li, X.W., de Boer, F.R., and Buschow, K.H.J. (2003) *Physica B*, **327**, 431.

61 Li, X.W., Tegus, O., Zhang, L., Dagula, W., Brück, E., Buschow, K.H.J., and de Boer, F.R. (2003) *IEEE Trans. Magn.*, **39**, 3148.

62 Dagula, W., Tegus, O., Fuquan, B., Zhang, L., Si, P.Z., Zhang, M., Zhang, W.S., Brück, E., de Boer, F.R., and Buschow, K.H.J. (2005) *IEEE Trans. Magn.*, **41**, 2778.

63 Cam Thanh, D.T., Brück, E., Tegus, O., Klaasse, J.C.P., Gortenmulder, T.J., and Buschow, K.H.J. (2006) *J. Appl. Phys.*, **99**, 08Q107.

64 Cam Thanh, D.T., Brück, E., Tegus, O., Klaasse, J.C.P., and Buschow, K.H.J. (2007) *J. Magn. Magn. Mater.*, **310**, e1012.

65 Tegus, O., Brück, E., Li, W.X., Zhang, L., Dagula, W., de Boer, F.R., and Buschow, K.H.J. (2004) *J. Magn. Magn. Mater.*, **272**, 2389.

66 Yan, A., Müller, K.-H., Schultz, L., and Gutfleisch, O. (2006) *J. Appl. Phys.*, **99**, 08K903.

67 Yabuta, H., Umeo, K., Takabatake, T., Chen, L., and Uwatoko, Y. (2007) *J. Magn. Magn. Mater.*, **310**, 1826.

68 Brück, E., Kamarad, J., Sechovsky, V., Arnold, Z., Tegus, O., and de Boer, F.R. (2007) *J. Magn. Magn. Mater.*, **310**, 1008.

69 Hu, F.-X., Shen, B.-G., Sun, J.-R., and Wu, G.-H. (2001) *Phys. Rev. B*, **64**, 132412.

70 Zhou, X., Li, W., Kunkel, H.P., and Williams, G. (2004) *J. Phys.: Condens. Matter*, **16**, L39.

71 Tishin, A.M. and Spichkin, Y.I. (2003) Chapter 11, *The Magnetocaloric Effect and its Applications*, Institute of Physics Publishing, Bristol.

72 Yu, S.Y., Cao, Z.X., Ma, L., Liu, G.D., Chen, J.L., and Wu, G.H. (2007) *Appl. Phys. Lett.*, **91**, 102507.

73 Stadler, S., Khan, M., Mitchell, J., Ali, N., Gomes, A.M., Dubenko, I., Takeuchi,

A.Y., and Guimarães, A.P. (2006) *Appl. Phys. Lett.*, **88**, 192511.

74 Krenke, T., Duman, E., Acet, M., Wassermann, E.F., Moya, X., Manosa, L., and Planes, A. (2005) *Nature Mater.*, **4**, 450.

75 Han, Z.D., Wang, D.H., Zhang, C.L., Xuan, H.C., Gu, B.X., and Du, Y.W. (2007) *Appl. Phys. Lett.*, **90**, 042507.

76 Oikawa, K., Ito, W., Imano, Y., Sutou, Y., Kainuma, R., Ishida, K., Okamoto, S., Kitakami, O., and Kanomata, T. (2006) *Appl. Phys. Lett.*, **88**, 122507.

77 Sharma, V.K., Chattopadhyay, M.K., Kumar, R., Ganguli, T., Tiwari, P., and Roy, S.B. (2007) *J. Phys.: Condens. Matter*, **19**, 496207.

78 Oikawa, K., Ito, W., Imano, Y., Sutou, Y., Kainuma, R., Ishida, K., Okamoto, S., Kitakami, O., and Kanomata, T. (2006) *Appl. Phys. Lett.*, **88**, 122507.

79 Kainuma, R., Imano, Y., Ito, W., Sutou, Y., Morito, H., Okamoto, S., Kitakami, O., Oikawa, K., Fujita, A., Kanomota, T., and Ishida, K. (2006) *Nature*, **439**, 957.

80 Moya, X., Mañosa, L., Planes, A., Aksoy, S., Acet, M., Wassermann, E.F., and Krenke, T. (2007) *Phys. Rev. B*, **75**, 184412.

81 Han, Z.D., Wang, D.H., Zhang, C.L., Tang, S.L., Gu, B.X., and Du, Y.W. (2006) *Appl. Phys. Lett.*, **89**, 182507.

82 Pathak, A.K., Khan, M., Dubenko, I., Stadler, S., and Ali, N. (2007) *Appl. Phys. Lett.*, **90**, 262504.

83 Du, J., Zheng, Q., Ren, W.J., Feng, W.J., Liu, X.G., and Zhang, Z.D. (2007) *J. Phys. D: Appl. Phys.*, **40**, 5523.

84 Han, Z.D., Wang, D.H., Zhang, C.L., Xuan, H.C., Zhang, J.R., Gu, B.X., and Du, Y.W. (2008) *J. Appl. Phys.*, **104**, 053906.

85 Krenke, T., Duman, E., Acet, M., Moya, X., Mañosa, L., and Planes, A. (2007) *J. Appl. Phys.*, **102**, 033903.

86 Xuan, H.C., Xie, K.X., Wang, D.H., Han, Z.D., Zhang, C.L., Gu, B.X., and Du, Y.W. (2008) *Appl. Phys. Lett.*, **92**, 242506.

87 Yasuda, T., Kanomata, T., Saito, T., Yosida, H., Nishihara, H., Kainuma, R., Oikawa, K., Ishida, K., Neumann, K.-U., and Ziebeck, K.R.A. (2007) *J. Magn. Magn. Mater.*, **310**, 2770.

88 Songlin, D., Tegus, O., Brück, E., de Boer, F.R., and Buschow, K.H.J. (2002) *J. Alloys Compd.*, **337**, 269.

89 Tohei, T., Wada, H., and Kanomata, T. (2003) *J. Appl. Phys.*, **94**, 1800.

90 Zhao, F.Q., Dagula, W., Tegus, O., and Buschow, K.H.J. (2006) *J. Alloys Compd.*, **416**, 43.

91 Massalski, T.B. (1990) *Binary Alloy Phase Diagrams*, 2nd edn, vol. **2**, Materials Park Ohio, USA, ASM International, p. 1718.

92 Palstra, T.T.M., Mydosh, J.A., Nieuwenhuys, G.J., van der Kraan, A.M., and Buschow, K.H.J. (1983) *J. Magn. Magn. Mater.*, **36**, 290.

93 Fujita, A., Akamatsu, K., and Fukamichi, K. (1999) *J. Appl. Phys.*, **85**, 4756.

94 Hu, F.-X., Shen, B.-G., Sun, J.-R., Cheng, Z.-H., Rao, G.-H., and Zhang, X.-X. (2001) *Appl. Phys. Lett.*, **78**, 3675.

95 Wen, G.H., Zheng, R.K., Zhang, X.X., Wang, W.H., Chen, J.L., and Wu, G.H. (2002) *J. Appl. Phys.*, **91**, 8537.

96 Fujita, A., Fujieda, S., Hasegawa, Y., and Fukamichi, K. (2003) *Phys. Rev. B*, **67**, 104416.

97 Gutfleisch, O., Yan, A., and Müller, K.-H. (2005) *J. Appl. Phys.*, **97**, 10M305.

98 Lyubina, J., Gutfleisch, O., Kuz'min, M.D., and Richter, M. (2008) *J. Magn. Mater.*, **320**, 2252.

99 Sun, Y., Arnold, Z., Kamarad, J., Wang, G.-J., Shen, B.-G., and Cheng, Z.-H. (2006) *Appl. Phys. Lett.*, **89**, 172513.

100 Hu, F., Shen, B., Sun, J., Wang, G., and Cheng, Z. (2002) *Appl. Phys. Lett.*, **80**, 826.

101 Yan, A., Müller, K.-H., and Gutfleisch, O. (2008) *J. Alloys Compd.*, **450**, 18.

102 Deng, J., Chen, X., and Zhuang, Y. (2008) *Rare Metals*, **27**, 345.

103 Fujieda, S., Fujita, A., and Fukamichi, K. (2007) *J. Appl. Phys.*, **102**, 023907.

104 Fujieda, S., Fujita, A., and Fukamichi, K. (2007) *J. Magn. Magn. Mater.*, **310**, 1004.

105 Liu, G.J., Sun, J.R., Shen, J., Gao, B., Zhang, H.W., Hu, F.X., and Shen, B.G. (2007) *Appl. Phys. Lett.*, **90**, 032507.

106 Anh, D.T.K., Thuy, N.P., Duc, N.H., Nhien, T.T., and Nong, N.V. (2003) *J. Magn. Magn. Mater.*, **262**, 427.

107 Fujieda, S., Fujita, A., and Fukamichi, K. (2007) PRICM 6: Sixth Pacific Rim

International Conference on Advanced Materials and Processing, 561, p. 1093.
108 Passamani, E.C., Takeuchi, A.Y., Alves, A.L., Demuner, A.S., Favre-Nicolin, E., Larica, C., and Proveti, J.R. (2007) *J. Appl. Phys.*, **102**, 093906.
109 Fujieda, S., Fujita, A., Fukamichi, K., Hirano, N., and Nagaya, S. (2006) *J. Alloys Compd.*, **408**, 1165.
110 Balli, M., Fruchart, D., Gignoux, D., Rosca, M., and Miraglia, S. (2007) *J. Magn. Magn. Mater.*, **313**, 43.
111 Fujita, A., Fujieda, S., Hasegawa, Y., and Fukamichi, K. (2003) *Phys. Rev. B*, **67**, 104416.
112 Chen, Y.-F., Wang, F., Shen, B.-G., Hu, F.-X., Sun, J.-R., Wang, G.-J., and Cheng, Z.-H. (2003) *J. Phys.: Condens. Matter*, **15**, L161.
113 Fukamichi, K., Fujita, A., and Fujieda, S. (2006) *J. Alloys Compd.*, **408–412**, 307.
114 Mandal, K., Pal, D., Gutfleisch, O., Kerschl, P., and Müller, K.-H. (2007) *J. Appl. Phys.*, **102**, 053906.
115 Fujita, A., Koiwai, S., Fujieda, S., Fukamichi, K., Kobayashi, T., Tsuji, H., Kaji, S., and Saito, A.T. (2007) *Jpn. J. Appl. Phys.*, **46**, L154.
116 Saito, A.T., Kobayashia, T., and Tsuji, H. (2007) *J. Magn. Magn. Mater.*, **310**, 2808.
117 Balli, M., Fruchart, D., and Gignoux, D. (2007) *J. Phys.: Condens. Matter*, **19**, 236230.
118 Chen, Y.-F., Wang, F., Shen, B.-G., Wang, G.-J., and Sun, J.-R. (2003) *J. Appl. Phys.*, **93**, 1323.
119 Guo, Z.B., Du, Y.W., Zhu, J.S., Huang, H., Ding, W.P., and Fen, D. (1997) *Phys. Rev. Lett.*, **78**, 1142.
120 Phan, M.H., Yu, S.C., and Hur, N.H. (2005) *Appl. Phys. Lett.*, **86**, 072504.
121 Phan, M.H., Tian, S.B., Hoang, D.Q., Yu, S.C., Nguyen, C., and Ulyanov, A.N. (2003) *J. Magn. Magn. Mater.*, **258**, 309.
122 Gencer, H., Atalay, S., Adiguzel, H.I., and Kolat, V.S. (2005) *Physica B*, **357**, 326.
123 Li, L., Nishimura, K., Hutchison, W.D., and Mori, K. (2008) *J. Phys. D: Appl. Phys.*, **41**, 175002.
124 Nam, D.N.H., Bau, L.V., Khiem, N.V., Dai, N.V., Hong, L.V., Phuc, N.X., Newrock, R.S., and Nordblad, P. (2006) *Phys. Rev. B*, **73**, 184430.
125 Nam, D.N.H., Dai, N.V., Hong, L.V., Phuc, N.X., Yu, S.C., Tachibana, M., and Takayama-Muromachi, E. (2008) *J. Appl. Phys.*, **103**, 043905.
126 Das, S., and Dey, T.K. (2007) *J. Alloys Compd.*, **440**, 30–35.
127 Chen, W., Zhong, W., Hou, D.L., Gao, R.W., Feng, W.C., Zhu, M.G., and Du, Y.W. (2002) *J. Phys.: Condens. Matter*, **14**, 11889.
128 Chen, W., Zhong, W., Pan, C.F., Chang, H., and Du, Y.W. (2001) *Acta Phys. Sin.*, **50**, 319.
129 Nikitin, S.A., Myalikguliev, G., Tishin, A.M., Annorazov, M.P., Astaryan, K.A., and Tyurin, A.L. (1990) *Phys. Lett. A*, **148**, 363.
130 Annaorazov, M.P., Asatryan, K.A., Myalikgulyev, G., Nikitin, S.A., Tishin, A.M., and Tyurin, A.L. (1992) *Cryogenics*, **32**, 867.
131 Dan'kov, S.Y., Ivtchenko, V.V., Tishin, A.M., GschneidnerJr., K.A., and Pecharsky, V.K. (2000) *Adv. Cryog. Eng.*, **46**, 397.
132 Canepa, F., Napoletano, M., and Cirafici, S. (2002) *Intermetallics*, **10**, 731.
133 Niu, X.J., Gschneidner Jr., K.A., Pecharsky, A.O., and Pecharsky, V.K. (2001) *J. Magn. Magn. Mater.*, **234**, 193.
134 Niu, X.J. (1999) MS Thesis Iowa State University, Ames Iowa, USA.
135 Chau, N., Thanh, P.Q., Hoa, N.Q., and The, N.D. (2006) *J. Magn. Magn. Mater.*, **304**, 36.
136 GschneidnerJr., K.A., Pecharsky, V.K., and Tsokol, A.O. (2005) *Rep. Prog. Phys.*, **68**, 1479.
137 Luo, Q., Zhao, D.Q., Pan, M.X., and Wang, W.H. (2006) *Appl. Phys. Lett.*, **89**, 081914.
138 Liang, L., Hui, X., Wu, Y., and Chen, G.L. (2008) *J. Alloys Compd.*, **457**, 541.
139 Liang, L., Hui, X., and Chen, G.L. (2008) *Mater. Sci. Eng. B*, **147**, 13.
140 Gorsse, S., Chevalier, B., and Orveillon, G. (2008) *Appl. Phys. Lett.*, **92**, 122501.
141 Wang, Y.T., Bai, H.Y., Pan, M.X., Zhao, D.Q., and Wang, W.H. (2008) *Sci. China Ser. G-Phys. Mech. Astron.*, **51**, 337.
142 Poddar, P., Gass, J., Rebar, D.J., Srinath, S., Srikanth, H., Morrison, S.A., and Carpenter, E.E. (2006) *J. Magn. Magn. Mater.*, **307**, 227.

143 GschneidnerJr., K.A., Pecharsky, V.K., and Tsokol, A.O. (2005) *Rep. Prog. Phys.*, **68**, 1479.
144 Morellon, L., Magen, C., Algarabel, P.A., Ibarra, M.R., and Ritter, C. (2001) *Appl. Phys. Lett.*, **79**, 1318.
145 Provenzano, V., Shapiro, A.J., and Shull Reduction, R.D. (2004) *Nature*, **429**, 853.
146 Gschneidner Jr., K.A., Pecharsky, V.K., Brück, E., Duijn, H.G.M., and Levin, E.M. (2000) *Phys. Rev. Lett.*, **85**, 4190.
147 GschneidnerJr., K.A. and Pecharsky, V.K. (2008) *Int. J. Refrig.*, **31**, 945.
148 Kittel, C. (1951) *Phys. Rev.*, **82**, 565.
149 Wood, M.E. and Potter, W.H. (1985) *Cryogenics*, **25**, 667.
150 Podmiljšak, B., McGuiness, P.J., Škulj, I., Markoli, B., Dražić, G., and Kobe, S. (2007) 2nd Inter. Conf. on Magnetic Refrigeration at Room Temperature: proceedings, Portorož, Slovenia, 11–13, 2007 (Science et technique du froid, 2007-1). Paris: Institut International du Froid, p. 145.
151 Podmiljšak, B., McGuiness, P.J., Škulj, I., Markoli, B., Žužek Rožman, K., and Kobe, S. (2008) *IEEE Trans. Magn.*, **44**, 4529.
152 Pecharsky, V.K. and Gschneidner Jr., K.A. (2001) *Adv. Mater.*, **13**, 683.
153 Casanova, F. (2001) Magnetocaloric effect in $Gd_5(Si_xGe_{1-x})4$ alloys, doctoral dissertation, University of Barcelona.
154 Pecharsky, V.K. and Gschneidner Jr., K.A. (2001) *Adv. Mater.*, **13**, 683.
155 Pecharsky, V.K. and Gschneidner Jr., K.A. (1997) *J. Alloys Compd.*, **260**, 98.
156 Choe, W., Pecharsky, V.K., Pecharsky, A.O., Gschneidner Jr., K.A., Young Jr., V.G., and Miller, G.J. (2000) *Phys. Rev. Lett.*, **84**, 4617.
157 Morellon, L., Algarabel, P.A., Ibarra, M.R., Blasco, J., García-Landa, B., Arnold, Z., and Albertini, F. (1998) *Phys. Rev. B*, **58**, R14 721.
158 Meyers, J.S., Chumbley, S., Choe, W., and Miller, G.J. (2002) *Phys. Rev. B*, **66**, 012106.
159 Meyers, J.S., Chumbley, S., Laabs, F., and Pecharsky, A.O. (2003) *Acta Mater.*, **51**, 61.
160 Ugurlu, O., Chumbley, L.S., Schlagel, D.L., and Lograsso, T.A. (2005) *Acta Mater.*, **53**, 3525.
161 Meyers, J.S., Chumbley, L.S., Laabs, F., and Pecharsky, A.O. (2002) *Scr. Mater.*, **47**, 509.
162 Williams, D.B. and Carter, C.B. (1996) *Transmission Electron Microscopy*, Plenum Press, New York.

9
Recent Progress in the Development of Thermoelectric Materials with Complex Crystal Structures
Silke Paschen, Claude Godart, and Yuri Grin

9.1
Introduction

To date, only a few materials have made their way to thermoelectric applications: Bi- and Te-based materials are used in Peltier coolers and microgenerators around room temperature, and Si- and Ge-based materials in thermoelectric generators for deep space missions. However, many more material classes with high potential for thermoelectric applications are known and are being intensively investigated in laboratories all around the world. The dimensionless thermoelectric figure of merit, which characterizes a material's thermoelectric performance, is shown for the most important materials as a function of temperature in Figure 9.1.

Any thermoelectric device employs couples of n- and p-type material. Thus, in each targeted temperature range, both kinds of material are needed. For practical reasons (e.g., to avoid problems due to different thermal expansion coefficients or corrosion effects of two different materials) it is desirable to combine very similar materials for the n- and p-branches in a device, ideally the same material which can be both n- and p-doped. The efficiencies are, however, in most such cases considerably lower for one type than for the other.

Before discussing the different material classes (Section 9.4), we shall introduce the relevant physical quantities (Section 9.2) and discuss common design principles (Section 9.3).

9.2
Thermoelectric Figure of Merit

The dimensionless thermoelectric figure of merit ZT is given by

$$ZT = S^2 \frac{\sigma}{\kappa} T = \frac{S^2}{\varrho} \frac{1}{\kappa} T \tag{9.1}$$

Complex Metallic Alloys: Fundamentals and Applications
Edited by Jean-Marie Dubois and Esther Belin-Ferré
Copyright © 2011 WILEY-VCH Verlag GmbH & Co. KGaA, Weinheim
ISBN: 978-3-527-32523-8

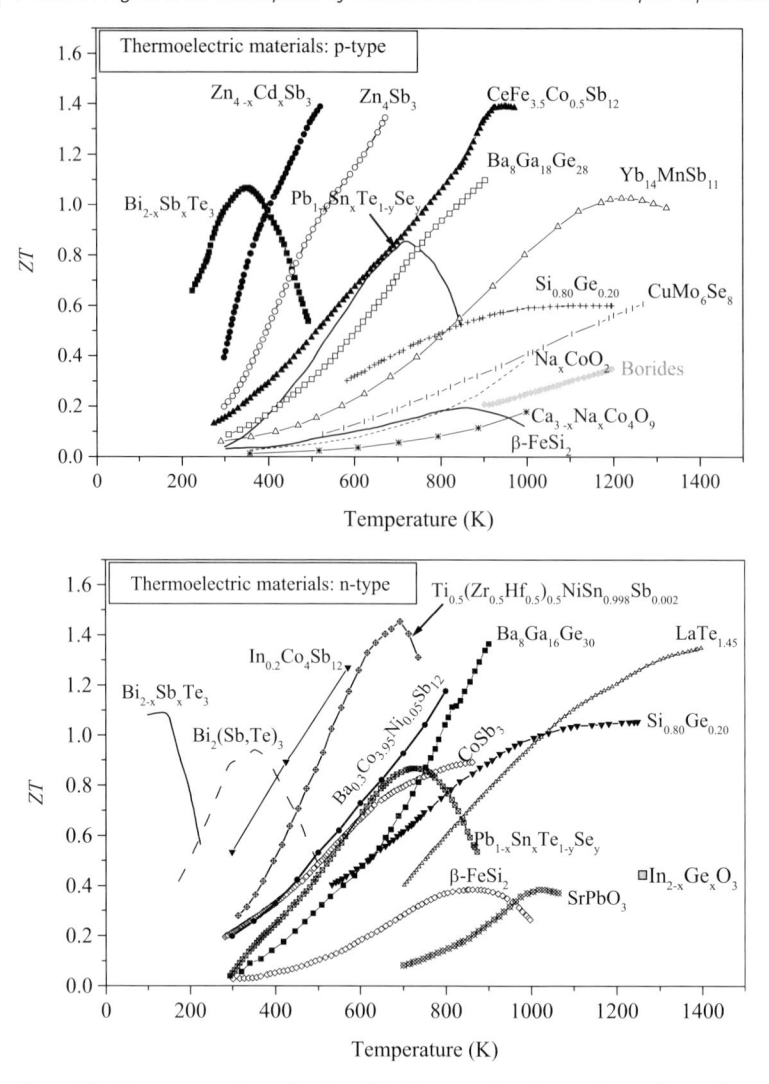

Figure 9.1 Temperature dependencies of the dimensionless thermoelectric figure of merit, $ZT(T)$, of p- (top) and n-type (bottom) materials [1].

where S is the Seebeck coefficient or thermopower, σ the electrical conductivity, $\varrho = 1/\sigma$ the electrical resistivity, κ the thermal conductivity, and T the absolute temperature. The first term on the right-hand side, S^2/ϱ, is referred to as the power factor.

Materials with $ZT > 1$ are typically considered as promising for thermoelectric applications. Of course, this is a simplistic view since each application has different thresholds above which the efficiency becomes competitive with respect to other coolers or energy generators and, even more importantly, since many factors can keep materials from being applied even if their ZT is high enough: insufficient thermal and/or mechanical stability, price, availability of the constituent elements, toxicity, etc.

In spite of the simplicity of Equation 9.1 it is, unfortunately, not straightforward to optimize ZT. This is due to the fact that S, σ, and κ are interrelated. Optimizing one may adversely affect the others. For simple semiconductors this is illustrated in Figure 9.2.

The quantities S, σ, and κ have different functional forms of the charge-carrier concentration n. High n leads to high σ, but also to small S and large κ. Low n, on the other hand, optimizes both S and κ but leads to too low σ values.

More specifically, in the simplest case of a single parabolic band with energy-independent scattering, the dependencies on n are

$$S \sim m^* \frac{1}{n^{2/3}} T \tag{9.2}$$

$$\sigma = \frac{\tau \cdot e}{m^*} n \tag{9.3}$$

$$\kappa = \kappa_e + \kappa_l \tag{9.4}$$

$$\kappa_e = L_0 \sigma T \tag{9.5}$$

$$\kappa_l = \frac{1}{3} v_l \tau_l c_l \tag{9.6}$$

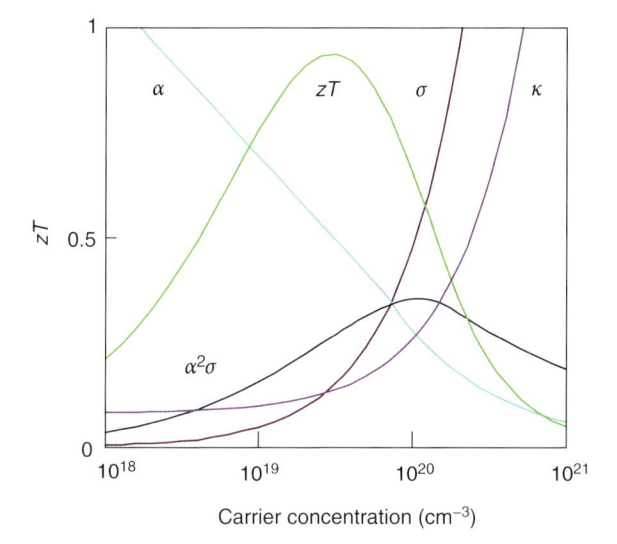

Figure 9.2 Thermopower, here denoted by α, electrical conductivity, σ, thermal conductivity, κ, and dimensionless thermoelectric figure of merit, ZT, for a simple semiconductor (see text, from Refs. [2] and [3]).

where m^* is the effective charge-carrier mass, τ the scattering time of the charge carriers, κ_e and κ_l the electronic and lattice contributions to the thermal conductivity, $L_0 = \pi^2 k_B^2/(3e^2)$ the Sommerfeld value of the Lorenz number, and v_l, τ_l, and c_l the effective velocity, scattering time, and specific heat of the phonons, respectively. Equation 9.5 is the Wiedemann–Franz law that relates κ_e to σ. Despite its strict validity only in the limits of low and high temperatures, it is frequently used to estimate the electronic contribution to the total thermal conductivity also at intermediate temperatures.

9.3
Design Principles

While each material requires specific approaches to optimize its thermoelectric performance there are also several more general design principles that are relevant for various material classes. These are discussed in the following.

9.3.1
Phonon Engineering

Much discussed is the design concept of a "phonon glass–electron crystal" (PGEC), a material that conducts heat poorly, as a glass, but charge very well, as a crystal [4]. To realize this concept, scattering centers that act much more strongly on phonons than on electrons have to be introduced into a material. This can be accomplished in different ways.

If in the original material the mean free path of the phonons l_l is longer than the mean free path of the charge carriers l_e and if κ_l contributes sizably to κ then the introduction of a reduced dimension l with $l_l > l > l_e$ is a useful approach. This has been nicely demonstrated with Si nanowires (Figure 9.3).

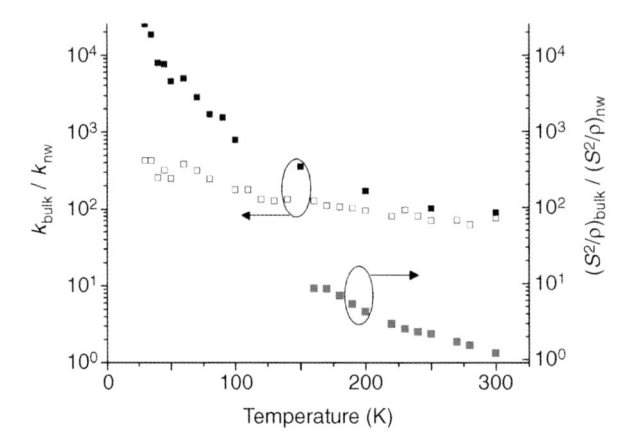

Figure 9.3 Thermal conductivity, κ, (left) and power factor, S^2/ϱ, (right) ratios of bulk (black symbols for intrinsic and open and grey symbols for doped) Si to highly doped Si nanowire (from Ref. [5]).

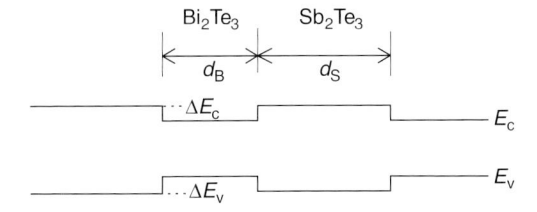

Figure 9.4 Heterojunction band diagram for a p-type Bi_2Te_3/Sb_2Te_3 superlattice showing the energy differences ΔE_c and ΔE_v of the conduction and valence band, respectively, in Sb_2Te_3 (of thickness d_s) with respect to those in Bi_2Te_3 (of thickness d_B; from Ref. [6]).

In a similar fashion, Bi_2Te_3/Sb_2Te_3 superlattices have been claimed to be efficient in suppressing the thermal conductivity via coherent backscattering of phonons [6]. The charge carriers were argued not to be significantly affected due to the heterojunction band diagram shown in Figure 9.4 with $\Delta E_c, \Delta E_v < k_B T$ at 300 K. This "phonon-blocking electron-transmitting" version of the PGEC approach is appealing but remains to be confirmed.

Various bulk implementations of these nanostructuring approaches have recently been realized for materials with simple crystal structure and have led to very attractive ZT values [7, 8]. However, to date not much is known on the stability of the nanostructures within bulk materials.

Yet another way to hinder phonon transport more than electron transport is the use of rattling atoms in oversized polyhedra, leading to excessively low lattice thermal conductivities (Figure 9.5) but to less dramatically reduced electrical conductivities. While the existence of rattling modes in a number of cage compounds has been demonstrated by various means [9], including X-ray scattering [10], Raman scattering [11], Mössbauer spectroscopy [12], and resonant ultrasound spectroscopy [13], the mechanism of the reduction of the lattice thermal conductivity is still debated [14, 15].

A neutron-scattering investigation of the low-energy phonon dispersion curves of the clathrate $Ba_8Ga_{16}Ge_{30}$ revealed an avoided crossing of flat rattler modes with acoustic phonon branches, thereby reducing the phonon velocity (Figure 9.6). To appreciate this effect the simple kinetic theory expression in Equation 9.6 should be replaced by the microscopic formulation [16]

$$\kappa_l = \frac{1}{V} \sum_{i,q} v_{i,x}^2(\mathbf{q}) \cdot \tau_i(\mathbf{q}, T) \cdot C_i(\mathbf{q}, T) \tag{9.7}$$

where V is the unit cell volume, $v_{i,x}(\mathbf{q}) = \partial \omega_i(\mathbf{q})/\partial q_x$ the group velocity of phonon mode i at wavevector $\mathbf{q}$ along the thermal gradient (here taken along x), $\tau_i(\mathbf{q}, T)$ the phonon lifetime at temperature T, and $C_i(\mathbf{q}, T)$ the specific heat of phonon mode i at $\mathbf{q}$ and T. From Equation 9.7 it is clear that not just the sound velocity but the group velocities at all $\mathbf{q}$ vectors contribute to κ_l. The effect of lifetime reduction associated with the avoided crossing was argued to be of minor importance [15]. Thus, this study suggests that apart from the more common approach to decrease the phonon lifetime

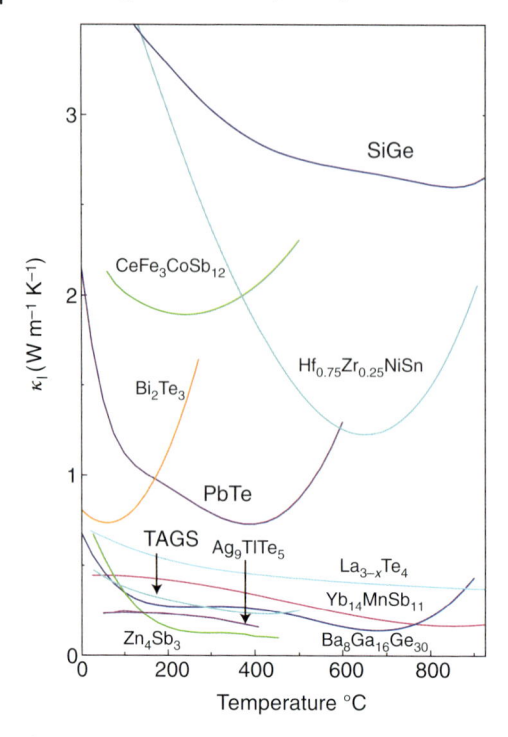

Figure 9.5 Temperature dependencies of lattice thermal conductivities, $\kappa_l(T)$, of various thermoelectric materials (from Ref. [3]).

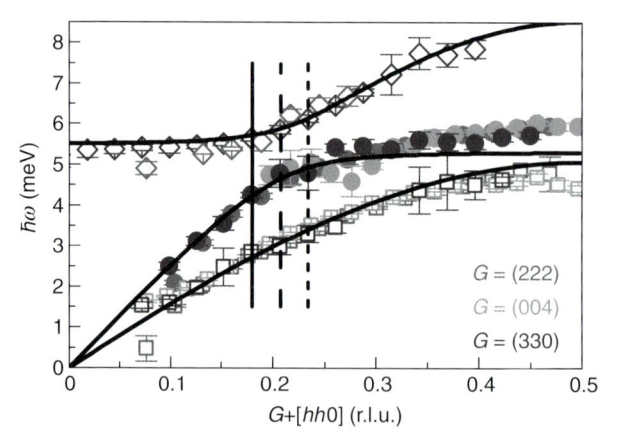

Figure 9.6 Phonon dispersion branches of $Ba_8Ga_{16}Ge_{30}$ showing the avoided crossing of the rattler mode (diamonds) with the longitudinal acoustic mode (full circles) at a reduced wavevector of about 0.2 (from Ref. [15]).

also the reduction of the phonon group velocity should be regarded as a feasible "phonon-engineering" route.

A simpler but related way to engineer the phonon density of states (DOS) might be to enhance the complexity of the crystal structure. This will *increase* the number of optical phonon modes, which do not carry heat, and thus *decrease* the weight of acoustic modes with large group velocity. However, in materials with complex crystal structures and a large number of atoms per unit cell there is typically also enhanced disorder [17, 18]. Thus, it is difficult to disentangle both effects and prove that the former process can efficiently lower the thermal conductivity. A simple reduction of the thermal conductivity via disorder is generally accompanied by a similar reduction of the electrical conductivity and is therefore not a valid approach to enhance ZT. Further investigations into this issue are clearly needed.

Most of these phonon-engineering approaches have received a lot of attention in recent years and some implementations of these approaches in bulk materials will be described in Section 9.4.

9.3.2
Electron Engineering

A much less appreciated route towards highly efficient thermoelectric materials is "electron engineering," that is, tailoring of the *electronic* band structure in the vicinity of the Fermi level, to maximize the thermopower and/or to enhance the charge-carrier mobility. The effect of the electronic DOS on the (diffusion) thermopower is, to a good approximation, described by the Mott formula [19]

$$S = \frac{\pi^2 \cdot k_B^2 \cdot T}{3 \cdot e} \left(\frac{\partial \ln \sigma}{\partial E}\right)_{E_F} = \frac{\pi^2 \cdot k_B^2 \cdot T}{3 \cdot e} \left(\frac{1}{\sigma}\frac{\partial \sigma}{\partial E}\right)_{E_F}. \tag{9.8}$$

Thus, in addition to the absolute value of the electrical conductivity σ also its energy derivative at the Fermi energy, E_F, comes into play. According to Equation 9.3, σ is proportional to the electronic DOS and to the scattering time. While in simple metals and semiconductors the electronic DOS evolves smoothly with energy there are a number of situations in which sharp features in the electronic DOS may arise.

The most drastic modification of the electronic DOS near the Fermi level occurs in so-called Kondo systems. In this family of materials a strong hybridization between conduction electrons and local magnetic moments, steming from elements with partially filled $4f$, $5f$, or sometimes also $3d$ shells, leads to the formation of a sharp resonance (the Kondo resonance) near E_F. If the local magnetic moments reside on a regular lattice (Kondo lattice systems), a gap opens within the Kondo resonance. In the case of "half-filling" where the Fermi level lies within the gap, a Kondo insulator results, otherwise heavy fermion metals are obtained. These phenomena occur below a characteristic temperature, the Kondo temperature T_K, which in most systems is below room temperature. Thus, this route of electronic band-structure engineering appears most promising for low-temperature applications (Peltier cooling).

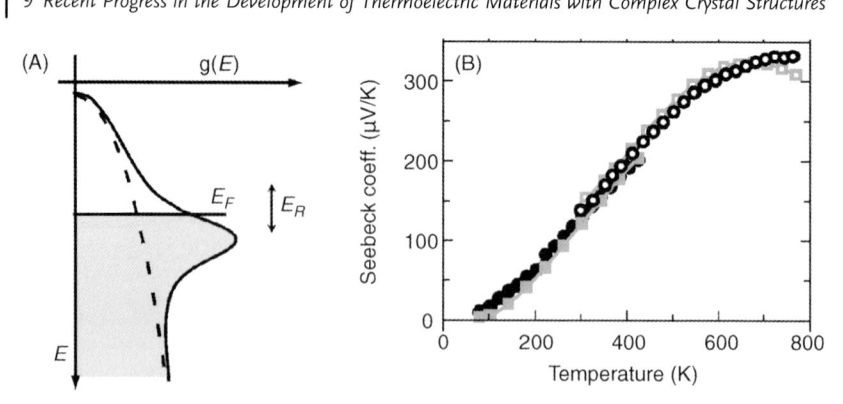

Figure 9.7 Electronic DOS $g(E)$ vs. energy E (A) suggested to explain the large Seebeck coefficient (B) of $Tl_{0.01}Pb_{0.99}Te$ (squares) and $Tl_{0.01}Pb_{0.99}Te$ (circles, from Ref. [26]).

There is experimental proof of "giant" thermopower values in such strongly correlated materials [20]. Particularly drastic are the cases of FeSi [21] and $FeSb_2$ [22], where the thermopower peaks sharply with maximum values of 700 µV/K and −45 mV/K at 40 K and 10 K, respectively.

Thermopower (absolute) values of the order of 100 µV/K below room temperature are also found for a number of valence-fluctuating compounds such as $CePd_3$ and $YbAl_3$ [23]. However, due to the high thermal conductivities of these compounds their ZT values can, in unstructured bulk form, not compete with the ones of Bi_2Te_3-based materials.

For Na_xCoO_2 it was suggested [24, 25] that the peculiar shape of one of the bands (a_{1g} band), referred to as "pudding mold" shape, leads simultaneously to high thermopower and low electrical resistivity.

Yet another mechanism – "resonant scattering" of mobile charge carriers from energy levels of localized impurity atoms – has recently been claimed to enhance the electronic DOS in Tl-substituted PbTe [26] (Figure 9.7A). Unlike in strongly correlated systems, here the Seebeck coefficient (Figure 9.7B) is a smooth function of temperature and continues to increase up to 800 K.

Finally, it has been agrued recently [27] that in systems with strong electron–phonon coupling both phonon and electron engineering may work simultaneously. In $In_4Se_{3-\delta}$ a charge-density wave (CDW) instability is held responsible for both a reduction of the thermal conductivity and an enhancement of the Seebeck coefficient. The (in-plane) lattice distortion due to a Peierls transition is held responsible for the former, the gap opening due to the CDW formation for the latter.

9.4
Thermoelectric Materials

As illustrated in Figure 9.1 various different material classes show promising thermoelectric properties. For the compounds Bi_2Te_3, PbTe, TAGS and related materials, as well as for Si–Ge solid solutions the optimization strategy typically follows the rules for simple semiconducting thermoelectrics (cf. Figure 9.2). In addition, nanostructuring techniques have been successfully applied to further enhance ZT of these materials [28–30]. The structural features of complex intermetallic phases, which are at focus here, allow in some cases to (partially) decouple the electronic and lattice degrees of freedom and, thus, to separately tune the thermoelectric properties entering ZT already in the bulk (non-nanostructured) material. In the following, five groups of materials will be discussed: Zintl phases, Sb-based filled skutterudites, intermetallic clathrates, the intermetallic phases around the composition Zn_4Sb_3, and representatives of the MgAgAs-type (half-Heusler) crystal structure.

9.4.1
Zintl Phases

According to the widely accepted picture of chemical bonding in Zintl phases, these materials should be electron balanced and thus show either semiconducting behavior or metallic behavior with very low charge-carrier concentrations [31]. In both cases, it is possible to change the charge-carrier type by doping.

$Yb_{14}MnSb_{11}$ (Figure 9.8, top) and compounds of the $CaAl_2Si_2$-type structure (Figure 9.8, bottom) have recently attracted considerable attention. The former compound has a complex tetragonal structure (space group $I4_1/acd$), built up from several different structural units. $ZT \approx 1$ is reached for p-type material [32] at temperatures sizably higher than in skutterudites (Figure 9.1). The flexibility of the structure towards the exchange of elements opens the possibility for further ZT enhancement.

Several ternary compounds with $CaAl_2Si_2$-type structure reach relatively high ZT values at elevated temperatures. In particular, $ZT = 0.92$ was reached for $EuZn_2Sb_2$ at 700 K [33] and $ZT = 0.66$ for $EuCd_2Sb_2$ at 616 K [34]. The formation of solid solutions between different ternary phases has been shown to further increase ZT. For instance, the thermoelectric performance of $EuZn_2Sb_2$ and $EuCd_2Sb_2$ was optimized by mixed occupation of the transition metal position in $Eu(Zn_{1-x}Cd_x)_2Sb_2$. The highest ZT value of 1.06 at 650 K is obtained for the solid solution with $x = 0.1$ [35].

9.4.2
Skutterudites

This series of materials has been extensively investigated during the past 15 years. The electropositive element A inserted in the binary cubic $CoAs_3$-type structure (M_4E_{12} space group $Im\bar{3}$) leads to the filled cubic variant $A_yM_4E_{12}$ (Figure 9.9) where A occupies the 2a site, the transition metal M of the groups 8, 9, or 10 of the periodic table is located at the 8c site, and E = As, P, Sb occupies similar vertice positions of the ME_6 polyhedra as in the binary structure. Atoms with various oxidation numbers can be employed as the component A: monovalent (K, Na), divalent (Ca, Sr, Ba), trivalent

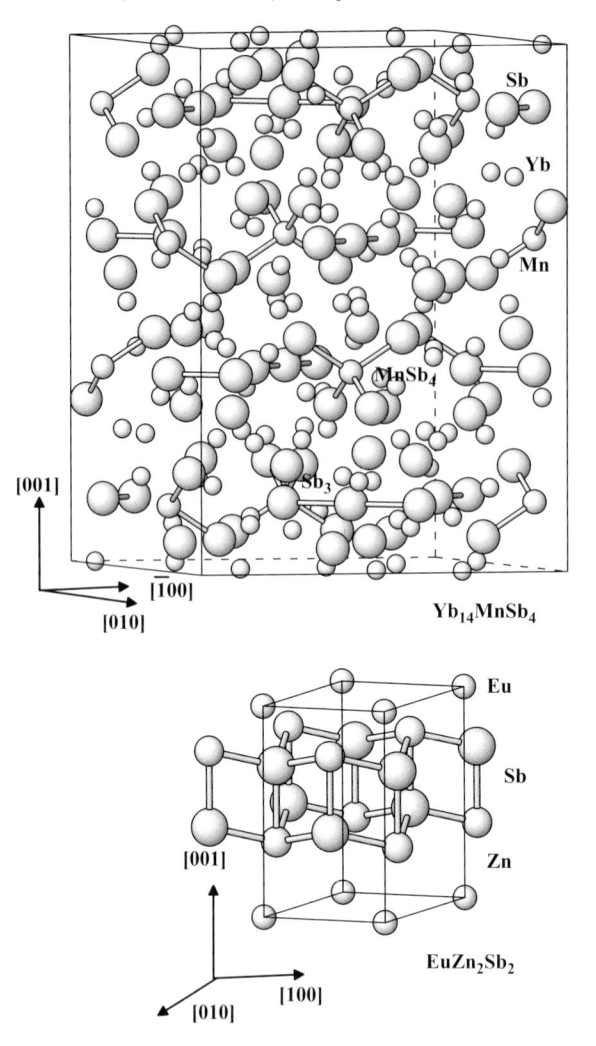

Figure 9.8 Crystal structures of Yb$_{14}$MnSb$_{11}$ (top) and EuZn$_2$Sb$_2$ (bottom).

(La, Ce, Pr, Nd, Th, U), and mixed-valent (Yb [36]) elements. The occupancy of the A site depends on the nature of A – it is close to 1 for K and Na [37, 38], but smaller than 1 for rare-earth elements: ≈ 0.2 for $M =$ Co (Table 9.1), ≈ 0.9 for $M =$ Fe, and close to 1 for $M =$ Pt [39].

The cage size in CeFe$_4$Sb$_{12}$ may be determined from the distances between the atoms in the structure. The distance between the Ce position (cage center) and the Sb position (cage border), $d(\text{Ce–Sb}) = 3.39$ Å, minus the covalent radius of Sb, $d(\text{Sb–Sb})/2 = 1.46$ Å, corresponds to the space available for Ce. With 1.93 Å this is much larger than the ionic radius of Ce^{3+}, which is 1.02 Å, allowing Ce to undergo large thermal oscillations ("rattling"). Indeed, experimental evidence for rattling was provided by neutron diffraction, where the atomic displacement parameters (ADPs)

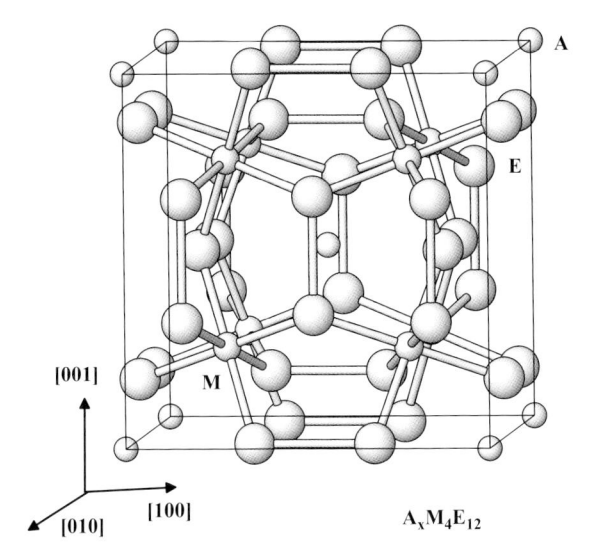

Figure 9.9 Crystal structure of filled skutterudites.

were shown to be three times larger for the guest atoms A than for the host atoms M and E [46, 47]. Phenomenological models [48, 49] have been put forward that describe the guest atoms as Einstein oscillators and the host structure as a Debye solid. More recently, inelastic neutron-scattering experiments revealed a larger than anticipated guest–host interaction [14] and thus the need for a refined microscopic picture.

The thermal conductivity of filled skutterudites is strongly suppressed compared to that of the unfilled binary structure. In particular, it takes very small values in Sb-based filled skutterudites, which have the largest cages. In addition to guest motion, vacancies and substitutions on the M position may lower the thermal conductivity. In $A_y(Fe,Co)_4Sb_{12}$ this was accounted for by a combination of several normal modes [50].

By appropriate substitutions on the M and E sites, accompaning the filling of the cages, it is possible to maintain the semiconducting/semimetallic ground state of $CoSb_3$. The ZT values are strongly increased from approx. 0.4 in $CoSb_3$ to above 1 for both n- and p-type compounds. Some of the highest ZT values of binary and ternary skutterudites are reported in Table 9.2. The temperature dependencies of ZT for the best skutterudite-based materials are shown in Figure 9.1.

Table 9.1 Occupation limits y in $A_yCo_4Sb_{12}$.

A = Ba	La	Ce	Eu	Yb	Tl
0.44 [40]	0.23 [41]	0.1 [42]	0.54 [43]	0.25 [44]	0.22 [45]

Table 9.2 Higest ZT in unfilled and filled skutterudites.

Materials	Type	ZT_{max}	$T(ZT_{max})$ (K)	Reference
$(CoSb_3)_{0.75} + (FeSb_2)_{0.25}$	n	0.37	773	[51]
$Co_{0.94}Ni_{0.04}Sb_3$	n	0.5	750	[52]
$CoSb_3$	p	0.21	600	[53]
$Co(Sb,Te)_3$ (0.75% Te)	n	0.5	600	[54]
$Co(Sb,Te)_3$ (4% Te)	n	0.8	750	[55]
$IrSb_3$	p	0.15	800	[56]
$Ba_{0.24}Co_4Sb_{11.87}$	n	1.1	850	[40]
$Ba_{0.3}Co_{3.95}Ni_{0.05}Sb_{12}$	n	1.2	800	[57]
$Ca_yCo_{4-x}Ni_xSb_{12}$	p	1	800	[58]
$Ce_{0.28}Co_{2.5}Fe_{1.5}Sb_{12}$	p	1.1	800	[59]
$Ce_xFe_{3.5}Co_{0.5}Sb1_{12}$	p	1.2	870	[60]
$Eu_{0.42}Co_4Sb_{11.37}Ge_{0.50}$	n	1.1	700	[61]
$In_{0.25}Co_4Sb_{12}$	n	1.2	570	[62]
$La_xFe_{4-y}Co_ySb_{12}$	p	1	800	[63]
$Nd_xCo_4Sb_{12}$	n	0.45	700	[64]
$Tl_xCo_4Sb_{12}$	n	0.2	300	[45]
$Yb_{0.8}Fe_{3.4}Ni_{0.6}Sb_{12}$	p	1	800	[65]
$Yb_{0.19}Co_4Sb_{12}$	n	>1	600	[66]
$(Ce,Yb)_{0.4}Fe_3CoSb_{12}$	p	1	800	[67]

9.4.3
Clathrates

The crystal structures of clathrates consist of three-dimensional frameworks made up of three- and four-bonded atoms (host). By their arrangement large cavities are formed [68]. These cavities are usually filled with guest atoms. Only recently, an "empty" Ge clathrate was synthesized [69]. Having complex crystal structures with large unit cells and different kinds of vacancies or substitutions (Pearson symbols $cI408$ for $Ba_8Ge_{43}\square_3$ [70, 71] and $Rb_8Ge_{44}\square_2$ [72] where $\square$ represents a vacancy, $cP124$ for Ba_6Ge_{25} [73], $Ba_{24}Si_{100}$ [74] and $K_{6-8}Sn_{25}$ [75–77], and $tP193$ for $Si_{130}P_{42}Te_{21}$ [78]), clathrates represent a class of complex intermetallic compounds.

According to the charge distribution between the host and the guest species, clathrates may be classified into three groups. In cationic clathrates, the host plays the role of a polycation and the guest is an anion, for example, $Ge_{38}P_8Te_8$ [79]. In neutral clathrates the host is formed by hydrogen-bonded water molecules and the guests are neutral atomic or molecular species (e.g., noble gases, methan) [80]. In anionic clathrates the host has a negative charge that is compensated by the cationic guest (see, e.g., [81]). While the properties of cationic clathrates are only scarely investigated (see, e.g., [78, 82–84]), most interesting with respect to thermoelectric applications are anionic clathrates. The first representative of this family – Na_xSi_{136} – was found in 1965 [85]. Ge- and Sn-based clathrates were characterized only much later (see, e.g., [86]). The crystal structures of anionic clathrates belong to five basic types, which are denoted as the clathrate-I, clathrate-II, clathrate-III, clathrate-VIII and clathrate-IX type. The common structural feature of the frameworks I, II, III and VIII is the presence of four-bonded species only and the formation of large cages with 20, 24, 26 and 28 vertices filled by the electropositive element (Figure 9.10).

The type-I crystal structure $A_8 E_{46}$ is formed by two E_{20} and six E_{24} polyhedra. A hypothetical empty framework of group 14 elements (e.g., Si, Ge, Sn) would be charge balanced in the Zintl–Klemm [87] sense: each group 14 element uses its four valence electrons for the four bonds, leading to the fulfillment of the octet rule for each atom. The presence of guest cations leads to excess electrons, filling the antibonding band and leading to a metallic character. To restore semiconducting behavior it is necessary to partially substitute E by an electron acceptor. If A is divalent, this can be realized by replacing 16 of the group 14 E atoms by group 13 elements (e.g., Ga, Al, In). The charge balance of such a compound is represented by the notation $[A^{2+}]_8[Ga^{1-}]_{16}[Ge^0]_{30}$ (cubic structure, space group $Pm\bar{3}n$).

There is ample experimental evidence for the usefulness of this concept. For example, a systematic investigation of $Eu_8 Ga_{16\pm x}Ge_{30\mp x}$ revealed a direct correspondence of the deviation x from charge balance and the charge-carrier concentration in the compound [88, 89].

Also substitutions of A by other than group 13 elements, including transition metals [90–96], have been much investigated. Particularly appealing in this approach is the ease to obtain both p- and n-type conductors [97].

In the larger E_{24} polyhedra, the guest atoms have large amplitudes of thermal motion, as evidenced by large ADP values [10, 98]. The large residual ADP values at zero temperature are considered to be due to static disorder induced by the split positions of the guest atoms in the E_{24} polyhedra. Tunneling between these has been clearly evidenced by Mössbauer spectroscopy [12].

The lattice thermal conductivity of clathrates is typically very low and "glass-like" [41, 99], albeit marked differences between p- and n-type material have been observed [97, 100]. Type-I clathrates filled with Eu or Ba have reached ZT values above 1 (Figure 9.1) [101–103].

9.4.4
Zn$_4$Sb$_3$

Four phases exist around the composition Zn_4Sb_3. Room-temperature stable, disordered β-Zn_4Sb_3 undergoes a phase transition at 254 K to α-Zn_4Sb_3 and at 765 K to γ-Zn_4Sb_3. Below 235 K, a second low-temperature phase (α'-Zn_4Sb_3) can be detected. While the α phase is ordered and has – *de facto* – an ideal composition $Zn_{13}Sb_{10}$, the α' phase is characterized by a slight Zn deficiency ($Zn_{13-\delta}Sb_{10}$, $\delta > 0$) with respect to $Zn_{13}Sb_{10}$ [104]. The sequence of phase transitions β–α–α' is reversible. The β phase is stable up to $+492\,°C$ (765 K), the γ phase up to the melting point at $566\,°C$.

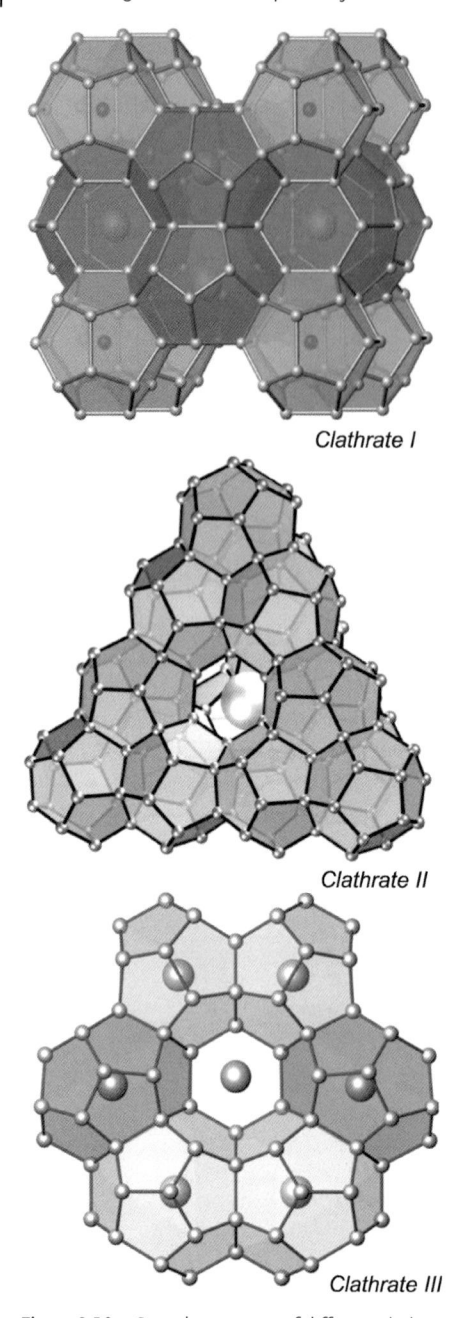

Figure 9.10 Crystal structures of different clathrate types with cages in polyhedral representation.

The high ZT values of p-type samples of the γ phase (1.25 at 650 K for Zn_4Sb_3 and 1.4 at 525 K for $Zn_{3.2}Cd_{0.8}Sb_3$ [105]) result from an extraordinarily low thermal conductivity, combined with the electronic structure of a heavily doped semiconductor. Both the electronic and the phononic properties of Zn_4Sb_3 can be understood in terms of unique structural features. The identification of six Sb^{3-} ions and two $(Sb^{2-})_2^{4-}$ dimers per formula unit reveals that "Zn_4Sb_3" is a valence balanced semiconductor with the ideal stoichiometry $Zn_{13}Sb_{10}$. The structure contains significant disorder, with Zn atoms distributed over multiple positions. In particular the partial occupation of interstitial sites with Zn and a low energy phonon mode associated with Sb_2 dimers [106] are held responsible for the extremely low lattice thermal conductivity.

So far, no competitive n-type material can be produced. Also, the mechanical properties need to be improved before use in a thermoelectric device can be considered [107].

9.4.5
MgAgAs-type (half-Heusler) Phases

The cubic structure of the Heusler phases A_2BC (space group $Fm\bar{3}m$) consists of the three sites $A = (1/4, 1/4, 1/4)$, $B = (0, 0, 0)$, and $C = (1/2, 1/2, 1/2)$. If the site A is only half-occupied the symmetry reduces to the space group $F\bar{4}3m$. The resulting cubic MgAgAs-type structure is frequently referred to as half-Heusler structure (Figure 9.11).

Among the MgAgAs-type compounds those with 18 valence electrons have a band structure of a semiconductor. The suppression of the (lattice) thermal conductivity is realized by complex chemical substitutions promoting the scattering of phonons by mass fluctuations.

For example, the n-type derivatives of ZrNiSn, $Hf_{0.5}Zr_{0.5}Ni_{0.8}Pd_{0.2}Sn_{0.99}Sb_{0.01}$ and $Hf_{0.6}Zr_{0.4}NiSn_{0.98}Sb_{0.02}$, reach maximum ZT values of ≈ 0.7 at 800 K and ≈ 1 at

ZrNi$_2$Sn
(AlCu$_2$Mn type)

ZrNiSn
(MgAgAs type)

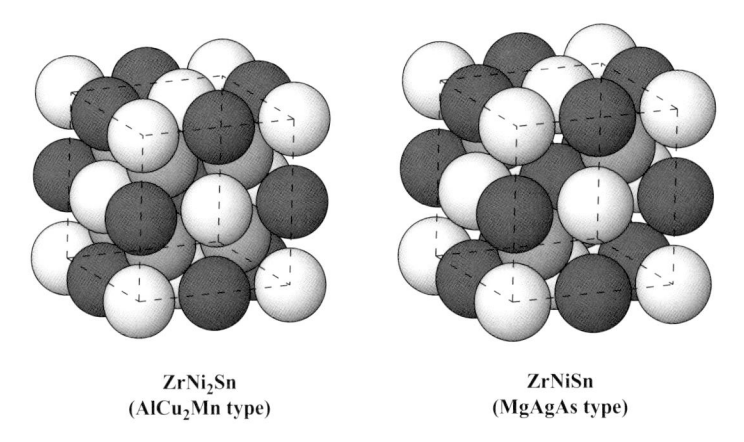

Figure 9.11 Crystal structures of the phases ZrNi$_2$Sn (MgCu$_2$Al type) and ZrNiSn (MgAgAs type): black spheres – Zr, dark gray – Ni and light gray – Sn.

1000 K, respectively [108]. n-type $Ti_{0.5}(Zr_{0.5}Hf_{0.5})NiSn_{0.98}Sb_{0.02}$ reach the highest ZT value of ≈ 1.4 at 700 K [109].

In LnPdSb (Ln = Sr, Y or rare earth, especially Er [110, 111]) high Seebeck coefficients have been observed at room temperature, leading to a ZT value of 0.15 for ErPdSb at 300 K [112] and of 0.29 for $ErPdSb_{0.99}Sn_{0.01}$ at 669 K [113]. A further enhancement by appropriate substitutions appears feasible.

9.5
Concluding Remarks

Today, we are able to control materials with complex crystal structures and hundreds or even thousands of atoms per unit cell. This enables us to combine their functionalities, in particular extremely low lattice thermal conductivity in compounds with nanocaged structures or interstitials with high electrical conductivity. It is now time to transfer further phonon and electron engineering concepts, proven for simpler materials in the past decade, to these complex materials, to boost efficiency and bring thermoelectric applications from niche to large-scale applications.

References

1 Godart, C., Gonçalves, A.P., Lopes, E.B., and Villeroy, B. (2009) Proc. Spring Meeting MRS 2009, San Francisco, Vol. 1166, N08-01, p. 183.
2 Ioffe, A.F. (1960) *Semiconductor Thermoelements*, AN USSR, Moscow.
3 Snyder, G. and Toberer, E. (2008) *Nature Mater.*, **7**, 105.
4 Slack, G.A. (1995) Chapter 34 (New materials and performance limits for thermoelectric cooling), in *CRC Handbook of Thermoelectrics* (ed. D.M. Rowe) CRC Press, Boca Raton, p. 407.
5 Hochbaum, A.I., Chen, R., Delgado, R.D., Liang, W., Garnett, E.C., Najarian, M., Majumdar, A., and Yang, P. (2008) *Nature*, **451**, 163.
6 Venkatasubramanian, R., Siivola, E., Colpitts, T., and O'Quinn, B. (2001) *Nature*, **413**, 597.
7 Dresselhaus, M.S., Chen, G., Tang, M.Y., Yang, R., Lee, H., Wang, D., Ren, Z., Fleurial, J.-P., and Gogna, P. (2007) *Adv. Mater.*, **19**, 1043.
8 Minnich, A.J., Dresselhaus, M.S., Ren, Z.F., and Chen, G. (2009) *Energy Environ. Sci.*, **2**, 466.
9 For a review see, e.g., Paschen, S. (2009) Chapter 4 (Thermoelectric materials), in *Complex Metallic Alloys – Properties and Applications of Complex Intermetallics*, vol. 2 (ed. Belin-Ferré Esther) World Scientific, London, ISBN 13 978-981-4261-63-0.
10 Sales, B.C., Chakoumakos, B.C., Jin, R., Thompson, J.R., and Mandrus, D. (2001) *Phys. Rev. B*, **63**, 245113.
11 Ogita, N., Kondo, T., Hasegawa, T., Takasu, Y., Udagawa, M., Takeda, N., Ishikawa, K., Sugawara, H., Kikuchi, D., Sato, H., Sekine, C., and Shirotani, I. (2006) *Physica B*, **383**, 128.
12 Hermann, R., Keppens, V., Bonville, P., Nolas, G., Grandjean, F., Long, G., Christen, H., Chakoumakos, B., Sales, B., and Mandrus, D. (2006) *Phys. Rev. Lett.*, **97**, 017401.
13 Zerec, I., Keppens, V., McGuire, M., Mandrus, D., Sales, B., and Thalmeier, P. (2004) *Phys. Rev. Lett.*, **92**, 185502.
14 Koza, M.M., Johnson, M.R., Viennois, R., Mutka, H., Girard, L., and Ravot, D. (2008) *Nature Mater.*, **7**, 805.

15 Christensen, M., Abrahamsen, A., Christensen, N., Juranyi, F., Andersen, N., Lefmann, K., Andreasson, J., Bahl, C., and Iversen, B. (2008) *Nature Mater.*, **7**, 811.

16 Williams, R.K., Butler, W.H., Graves, R.S., and Moore, J.P. (1983) *Phys. Rev. B*, **28**, 6316.

17 Urban, K. and Feuerbacher, M. (2004) *J. Non-Cryst. Solids*, **334–335**, 143.

18 Grin, Y. (2008) Chapter 12 (Chemical bonding and crystallographic features), in *Complex Metallic Alloys – Basics of Thermodynamics and Phase Transitions in Complex Intermetallics*, vol. **1** (ed. Belin-Ferré Esther) World Scientific, New Jersey, ISBN 13 978-981-279-058-3.

19 Jonson, M. and Mahan, G. (1980) *Phys. Rev. B*, **21**, 4223.

20 For a review see, e.g., Paschen, S. (2006) Chapter 15 (Thermoelectric aspects of strongly correlated electron systems), in *Thermoelectrics Handbook* (ed. D.M. Rowe) CRC Press, Boca Raton.

21 Buschinger, B., Geibel, C., Steglich, F., Mandrus, D., Young, D., Sarrao, J.L., and Fisk, Z. (1997) *Physica B*, **230–232**, 784.

22 Bentien, A., Johnsen, S., Madsen, G., Iversen, B., and Steglich, F. (2007) *Europhys. Lett.*, **80**, 17008.

23 Jaccard, D. and Sierro, J. (1982) Thermoelectric power of some intermediate valence compounds, in *Valence Instabilities* (eds P. Wachter and H. Boppart) North-Holland Publishing Company, Amsterdam, p. 409.

24 Kuroki, K. and Arita, R. (2007) *J. Phys. Soc. Jpn.*, **76**, 083707.

25 Held, K., Arita, R., Anisimov, V.I., and Kuroki, K. (2009) Chap. LDA + DMFT route to identify good thermoelectrics, in *Properties and Applications of Thermoelectric Materials* (eds V. Zlatic and A. C. Hewson) NATO ASJ Series B, Springer Dordrecht, Netherlands, p. 141.

26 Heremans, J., Jovovic, V., Toberer, E., Saramat, A., Kurosaki, K., Charoenphakdee, A., Yamanaka, S., and Snyder, G. (2008) *Science*, **321**, 554.

27 Rhyee, J., Lee, K.H., Lee, S.M., Cho, E., Kim, S.I., Lee, E., Kwon, Y.S., Shim, J.H., and Kotliar, G. (2009) *Nature*, **459**, 965.

28 Harman, T.C., Taylor, P.J., Walsh, M.P., and LaForge, B.E. (2002) *Science*, **297**, 2229.

29 Hsu, K.F., Loo, S., Guo, F., Chen, W., Dyck, J.S., Uher, C., Hogan, T., Polychroniadis, E.K., and Kanatzidis, M.G. (2004) *Science*, **303**, 818.

30 Poudel, B., Hao, Q., Ma, Y., Lan, Y., Minnich, A., Yu, B., Yan, X., Wang, D., Muto, A., Vashaee, D., Chen, X., Liu, J., Dresselhaus, M.S., Chen, G., and Ren, Z. (2008) *Science*, **320**, 634.

31 Kauzlarich, S.M., Brown, S.R., and Snyder, G.J. (2007) *Dalton Trans.*, **21**, 2099.

32 Brown, S.R., Kauzlarich, S.M., Gascoin, F., and Snyder, G.J. (2006) *Chem. Mater.*, **18**, 1873.

33 Zhang, H., Zhao, J.T., Grin, Y., Wang, X.J., Tang, M.B., Man, Z.Y., Chen, H.H., and Yang, X.X. (2008) *J. Chem. Phys.*, **129**, 164713.

34 Zhang, H., Fang, L., Tang, M., Chen, H., Yang, X., Guo, X.X., Zhao, J., and Grin, Y. (2010) *Intermetallics*, **18**, 193.

35 Zhang, H., Baitinger, M., Tang, M., Man, Z., Chen, H., Yang, X., Liu, Y., Chen, L., Grin, Y., and Zhao, J. (2010) *Dalton Trans.* doi: 10.1039/b916346h.

36 Bérardan, D., Godart, C., Alleno, E., Berger, S., and Bauer, E. (2003) *J. Alloys Compd.*, **351**, 18.

37 Leithe-Jasper, A., Schnelle, W., Rosner, H., Senthilkumaran, N., Rabis, A., Baenitz, M., Gippius, A., Morozova, E., Mydosh, J.A., and Grin, Y. (2003) *Phys. Rev. Lett.*, **91**, 037208.

38 Leithe-Jasper, A., Schnelle, W., Rosner, H., Baenitz, M., Rabis, A., Gippius, A.A., Morozova, E.N., Borrmann, H., Burkhardt, U., Ramlau, R., Schwarz, U., Mydosh, J.A., Grin, Y., Ksenofontov, V., and Reiman, S. (2004) *Phys. Rev. B*, **70**, 214418.

39 Gumeniuk, R., Schnelle, W., Rosner, H., Nicklas, M., Leithe-Jasper, A., and Grin, Y. (2008) *Phys. Rev. Lett.*, **100**, 017002.

40 Chen, L.D., Kawahara, T., Tang, X.F., Goto, T., Hirai, T., Dyck, J.S., Chen, W., and Uher, C. (2001) *J. Appl. Phys.*, **90**, 1864.

41 Nolas, G.S., Cohn, J.L., Slack, G.A., and Schujman, S.B. (1998) *Appl. Phys. Lett.*, **73**, 178.

42 Chen, B., Xu, J.H., Uher, C., Morelli, D.T., Meisner, G.P., Fleurial, J.P., Caillat, T., and Borshchevsky, A. (1997) *Phys. Rev. B*, **55**, 1476.

43 Berger, S., Paul, C., Bauer, E., Grytsiv, A., Rogl, P., Kaczorowski, D., Saccone, A., Ferro, R., and Godart, C. (2001) Proc. 20th Int. Conf. on Thermoelectrics, Beijing, China, IEEE, New York, USA, p. 77.

44 Anno, H., Nagamoto, Y., Ashida, K., Taniguchi, E., Koyanagi, T., and Matsubara, K. (2000) Proc. 19th Int. Conf. on Thermoelectrics, Babrow Press, Cardiff-Wales, UK, p. 90.

45 Sales, B.C., Chakoumakos, B.C., and Mandrus, D. (2000) *Phys. Rev. B*, **61**, 2475.

46 Sales, B.C. (1998) *MRS Bull.*, **23**, 15.

47 Chakoumakos, B.C., Sales, B.C., Mandrus, D., and Keppens, V. (1999) *Acta Crystallogr.*, **55**, 341.

48 Sales, B.C., Chakoumakos, B.C., Mandrus, D., and Sharp, J.W. (1999) *J. Solid State Chem.*, **146**, 528.

49 Sales, B.C., Chakoumakos, B.C., and Mandrus, D.G. (2000) Proc. Spring Meeting MRS 2000, San Francisco, Mater. Res. Soc., Warrendale, USA, Vol. 626, p. Z7.1.

50 Koza, M.M., Capogna, L., Leithe-Jasper, A., Rosner, H., Schnelle, W., Mutka, H., Johnson, M.R., Ritter, C., and Grin, Y. (2009) submitted.

51 Katsuyama, S., Kanayama, Y., Ito, M., Majima, K., and Nagai, H. (2000) *J. Appl. Phys.*, **88**, 3484.

52 Katsuyama, S., Watanabe, M., Kuroki, M., Maehala, T., and Ito, M. (2003) *J. Appl. Phys.*, **93**, 2758.

53 Katsuyama, S., Schichijo, Y., Ito, M., Majima, K., and Nagai, H. (1998) *J. Appl. Phys.*, **84**, 6708.

54 Wojciechowski, K.T., Malecki, A., Leszczynski, J., and Mickiewicza, A. (2001) Proc. 6th Europ. Workshop on Thermoelectrics, Fraunhofer-Institute of Physical Measurement Technique IPM, Freiburg im Breisgau, Germany.

55 Nagamoto, Y., Tanaka, K., and Koyanagi, T. (1998) Proc. 17th Int. Conf. on Thermoelectrics, Nagoya, Japan, IEEE, Piscataway, USA, p. 302.

56 Slack, G.A. and Tsoukala, V.G. (1994) *J. Appl. Phys.*, **76**, 1665.

57 Dyck, J.S., Chen, W., Uher, C., Chen, L., Tiang, X., and Hirai, T. (2002) *J. Appl. Phys.*, **91**, 3698.

58 Puyet, M., Dauscher, A., Lenoir, B., Dehmas, M., Stiewe, C., Müller, E., and Hejtmanek, J. (2005) *J. Appl. Phys.*, **97**, 083712.

59 Tang, X., Chen, L., Goto, T., and Hirai, T. (2001) *J. Mater. Res.*, **16**, 837.

60 Fleurial, J.P., Borshchevsky, A., Caillat, T., Morelli, D.T., and Meisner, G.P. (1996) Proc. 15th Int. Conf. on Thermoelectrics, Pasadena, CA, USA, p. 91.

61 Lamberton, G.A., Bhattacharya, S., Littleton, R.T., Kaeser, M., Tedstrom, R.H., Tritt, T.M., Yang, J., and Nolas, G.S. (2002) *Appl. Phys. Lett.*, **80**, 598.

62 He, T., Chen, J., Rosenfeld, H.D., and Subramanian, M.A. (2006) *Chem. Mater.*, **18**, 759.

63 Sales, B.C., Mandrus, D., and Williams, R.K. (1996) *Science*, **272**, 1325.

64 Kuznetsov, V.L., Kuznetsova, L.A., and Rowe, D.M. (2003) *J. Phys.: Condens. Matter*, **15**, 5035.

65 Anno, H., Nagao, J., and Matsubara, K. (2002) Proc. 21st Int. Conf. on Thermoelectrics, Long Beach, CA, USA, IEEE, New York, USA, p. 56.

66 Nolas, G.S., Kaeser, M., Littleton, R.T., and Tritt, T.M. (2000) *Appl. Phys. Lett.*, **77**, 1855.

67 Berardan, D., Alleno, E., Godart, C., Puyet, M., Lenoir, B., Lackner, R., Bauer, E., Girard, L., and Ravot, D. (2005) *J. Appl. Phys.*, **98**, 033710.

68 Kovnir, K.A. and Shevelkov, A.V. (2004) *Russ. Chem. Rev.*, **73**, 923.

69 Guloy, A.M., Ramlau, R., Tang, Z., Schnelle, W., Baitinger, M., and Grin, Y. (2006) *Nature*, **443**, 320.

70 Carrillo-Cabrera, W., Budnyk, S., Prots, Y., and Grin, Y. (2004) *Z. Anorg. Allg. Chem.*, **630**, 2267.

71 Aydemir, U., Gandolfi, C., Borrmann, H., Baitinger, M., Ormeci, A., Carrillo-Cabrera, W., Chubileau, C., Lenoir, B., Dauscher, A., Oeschler, N., Steglich, F., and Grin, Y. (2010) *Dalton Trans.* doi: 10.1039/b919726e.

72 Dubois, F., and Fässler, T.F. (2004) *Z. Anorg. Allg. Chem.*, **630**, 1718.

73 Carrillo-Cabrera, W., Borrmann, H., Paschen, S., Baenitz, M., Steglich, F., and

Grin, Y. (2005) *J. Solid State Chem.*, **178**, 715.

74 Fukuoka, H., Ueno, K., and Yamanaka, S. (2000) *J. Organomet. Chem.*, **611**, 543.

75 Zhao, J. and Corbett, J.D. (1994) *Inorg. Chem.*, **33**, 5721.

76 Fässler, T.F. and Kronseder, C. (1998) *Z. Anorg. Allg. Chem.*, **624**, 561.

77 Fässler, T.F. (2003) *Chem. Soc. Rev.*, **32**, 80.

78 Zaikina, J.V., Kovnir, K.A., Haarmann, F., Schnelle, W., Burkhardt, U., Borrmann, H., Schwarz, U., Grin, Y., and Shevelkov, A.V. (2008) *Chem. Eur. J.*, **14**, 5414.

79 von Schnering, H. and Menke, H. (1972) *Angew. Chem., Int. Ed. Engl.*, **11**, 43.

80 Atwood, J.L., Davies, J.E.D., and MacNicol, D.D. (eds) (1984) *Inclusion Compounds*, vol. **1** Academic Press, London, p. 2.

81 Beekman, M. and Nolas, G.S. (2008) *J. Mater. Chem.*, **18**, 842.

82 Zaikina, J.V., Schnelle, W., Kovnir, K.A., Olenev, A.V., Grin, Y., and Shevelkov, A.V. (2007) *Solid State Sci.*, **9**, 664.

83 Zaikina, J.V., Kovnir, K.A., Sobolev, A.V., Presniakov, I.A., Prots, Y., Baitinger, M., Schnelle, W., Olenev, A.V., Lebedev, O.I., van Tendeloo, G., Grin, Y., and Shevelkov, A.V. (2007) *Chem. Eur. J.*, **13**, 5090.

84 Rogl, P., Falmbigl, M., Bauer, E., Kriegisch, M., Mueller, H., and Paschen, S. (2009) Proc. Spring Meeting MRS 2009, San Francisco, Mater. Res. Soc., Warrendale, USA, Vol. 1166, N06-03.

85 Kasper, J.S., Hagenmüller, P., Pouchard, M., and Cros, C. (1965) *Science*, **150**, 1713.

86 Grin, Y., Melekhov, L.Z., Chutonov, K.A., and Yatsenko, S.P. (1987) *Sov. Phys. Crystallogr.*, **32**, 290; von Schnering, H.G., Kröner, R., Baitinger, M., Peters, K., Nesper, R., Grin, Y. (2000) *Z. Kristallogr. NCS*, **215**, 205.

87 Schäfer, H. (1985) *Ann. Rev. Mater. Sci.*, **15**, 1.

88 Pacheco, V., Bentien, A., Carrillo-Cabrera, W., Paschen, S., Steglich, F., and Grin, Y. (2005) *Phys. Rev. B*, **71**, 165205.

89 Bentien, A., Pacheco, V., Paschen, S., Grin, Y., and Steglich, F. (2005) *Phys. Rev. B*, **71**, 165206.

90 Cordier, G. and Woll, P. (1991) *J. Less-Common Met.*, **169**, 291.

91 Blake, N.P., Latturner, S., Bryan, J.D., Stucky, G.D., and Metiu, H. (2001) *J. Chem. Phys.*, **115**, 8060.

92 Mudryk, Y., Rogl, P., Paul, C., Berger, S., Bauer, E., Hilscher, G., Godart, C., and Noël, H. (2002) *J. Phys.: Condens. Matter*, **14**, 7991.

93 Anno, H., Hokazono, M., Kawamura, M., and Matsubara, K. (2003) Proc. 22nd Int. Conf. on Thermoelectrics, La Grande-Motte, France, IEEE, Piscataway, USA, p. 121.

94 Li, Y., Chi, J., Gou, W., Khandekar, S., and Ross, J.H.Jr. (2003) *J. Phys: Condens. Matter*, **15**, 5535.

95 Alleno, E., Maillet, G., Rouleau, O., Leroy, E., Godart, C., Carrillo-Cabrera, W., Simon, P., and Grin, Y. (2009) *Chem. Mater.*, **21**, 1485.

96 Nguyen, L.T.K., Aydermir, U., Baitinger, M., Bauer, E., Borrmann, H., Burkhardt, U., Custers, J., Haghighirad, A., Höfler, R., Luther, K.D., Ritter, F., Assmus, W., Grin, Y., and Paschen, S. (2010) *Dalton Tans.* **39**, 1071.

97 Bentien, A., Johnsen, S., and Iversen, B.B. (2006) *Phys. Rev. B*, **73**, 094301.

98 Chakoumakos, B.C., Sales, B.C., Mandrus, D.G., and Nolas, G.S. (2000) *J. Alloys Compd.*, **296**, 80.

99 Toberer, E.S., Christensen, M., Iversen, B.B., and Snyder, G.J. (2008) *Phys. Rev. B*, **77**, 075203.

100 Bentien, A., Christensen, M., Bryan, J.D., Sanchez, A., Paschen, S., Steglich, F., Stucky, G.D., and Iversen, B.B. (2004) *Phys. Rev. B*, **69**, 045107.

101 Anno, H., Hokazono, M., Kawamura, M., Nagao, J., and Matsubara, K. (2002) Proc. 21st Int. Conf. on Thermoelectrics, Long Beach, CA, IEEE, New York, USA, p. 77.

102 Bentien, A., Paschen, S., Pacheco, V., Grin, Y., and Steglich, F. (2003) Proc. 22nd Int. Conf. on Thermoelectrics, Vienna, Austria, IEEE, Piscataway, USA, ISBN 1-4244-0810-5, p. 131.

103 Saramat, A., Svensson, G., Palmqvist, A.E.C., Stiewe, C., Mueller, E., Platzek, D., Williams, S.G.K., Rowe, D.M., Bryan, J.D., and Stucky, G.D. (2006) *J. Appl. Phys.*, **99**, 023708.

104 Nylén, J., Lidin, S., Andersson, M., Iversen, B.B., Liu, H., Newman, N., and

Häussermann, U. (2007) *Chem. Mater.*, **19**, 834.
105 Snyder, G.J., Christensen, M., Nishibori, E., Caillat, T., and Iversen, B.B. (2004) *Nature Mater.*, **3**, 458.
106 Schweika, W., Hermann, R.P., Prager, M., Person, J., and Keppens, V. (2007) *Phys. Rev. Lett.*, **99**, 125501.
107 Ueno, K., Yamamoto, A., Noguchi, T., Inoue, T., Sodeoka, S., and Obara, H. (2005) *J. Alloys Compd.*, **388**, 118.
108 Yu, C., Zhu, T.J., Shi, R.Z., Zhang, Y., Zhao, X.B., and He, J. (2009) *Acta Mater.*, **57**, 2757.
109 Shutoh, N. and Sakurada, S. (2003) Proc. 22nd Int. Conf. on Thermoelectrics, La Grande Motte, France, IEEE, Piscataway, USA, p. 312.
110 Mastronardi, K., Young, D., Wang, C.C., Khalifah, P., Cava, R.J., and Ramirez, A.P. (1999) *Appl. Phys. Lett.*, **74**, 1415.
111 Oestreich, J., Probst, U., Richardt, F., and Bucher, E. (2003) *J. Phys.: Condens. Matter*, **15**, 635.
112 Gofryk, K., Kaczorowski, D., Plackowski, T., Mucha, J., Leithe-Jasper, A., Schnelle, W., and Grin, Y. (2007) *Phys. Rev. B*, **75**, 224426.
113 Kawano, K., Kurosaki, K., Sekimoto, T., Muta, H., and Yamanaka, S. (2007) *Appl. Phys. Lett.*, **91**, 062115.

10
Complex Metallic Phases in Catalysis
Marc Armbrüster, Kirill Kovnir, Yuri Grin, and Robert Schlögl

10.1
Introduction

Heterogeneous catalysis is the science and technology of modifying the course of chemical reactions by engineering the energy barriers between elementary steps of molecular transformations. Besides the traditional strategic relevance of catalysis for chemical industry and for environmental protection, the quest for sustainable energy supply systems poses new and additional challenges addressing transformations of small molecules with very high efficiency. Typical applications deal with water splitting and methane conversion. Catalysis further plays a pivotal role in efforts to optimize energy utilization in the production industry. In this arena, processes that are operated at large scale such as polymer synthesis need to be optimized with respect to selectivity of the reaction and to the velocity of transformations. Catalytic materials operate under enormous stress for their integrity as the function of an active surface involves continuous making and breaking of bonds between "corrosive" molecules and the intermetallic surface. Thus, only clean and single-phase materials will survive the chemical stress over any useful period of time. Finally, metallic catalysts need to offer substantial active surface areas in excess of $1\,m^2/g$ material, which is not easy to achieve with typical alloying techniques.

These challenges create a need for novel catalytic materials allowing rational development to replace the current empirical discovery processes. Intermetallic compounds and especially complex metallic phases are not among the first choice of novel catalytic materials as they are widely unknown to the catalysis experts and pose high hurdles in their availability due to unconventional synthesis pathways and due to their inherent structural complexity. For further considerations we understand hereafter an intermetallic compound as a chemical compound of two or more metallic elements adopting an – at least partly – ordered crystal structure that differs from those of the constituent metals. From the thermodynamic point of view intermetallic compounds represent single-phase materials and often possess more or less pronounced homogeneity ranges. On the other hand, an alloy is a – usually multiphase – mixture of elements or their solid solutions and intermetallic compounds.

Complex Metallic Alloys: Fundamentals and Applications
Edited by Jean-Marie Dubois and Esther Belin-Ferré
Copyright © 2011 WILEY-VCH Verlag GmbH & Co. KGaA, Weinheim
ISBN: 978-3-527-32523-8

The most well-known examples of metallic catalysts are the Raney-type catalysts, where an intermetallic compound or an alloy is leached before use, that is, decomposed, in order to produce a high-surface-area transition metal catalyst [1, 2]. In other cases where intermetallic compounds are applied as catalysts, they are decomposed during the catalytic process. A good and well-explored example is the hydrogenation of CO over intermetallic compounds of rare earth metals with nickel or copper, which are transformed *in situ* to elemental nickel or copper particles and the corresponding rare-earth oxide [3–7]. But intermetallic compounds can also be formed *in situ* by the chemical reaction between the active metal and the supporting material or between different supported metallic species. For example, during the selective hydrogenation of crotonaldehyde over Pt/ZnO [8, 9] and the steam reforming of methanol over Pd/ZnO [10, 11], the intermetallic compounds PtZn and PdZn are formed under reaction conditions, and the high selectivity and activity of the catalysts is attributed to their presence [12].

To circumvent these complications, the rational approach to apply intermetallic compounds with ordered crystal structures as stable and unsupported catalysts has been developed recently [13]. By selecting intermetallic compounds with suitable crystal structures, the geometric properties of the active sites can be preselected. Applying the active-site isolation concept [14], the intermetallic compound PdGa was successfully identified as catalyst for the selective acetylene hydrogenation reaction [13, 15–19]. The crystal structure of PdGa is crystallographically well ordered and the Pd atoms are surrounded solely by Ga atoms [20]. The Ga–Pd interactions were shown to have covalent nature [13] ensuring the thermodynamic stability of this phase under reaction conditions [16, 17].

Besides the geometric arrangement, the electronic structure of the compounds can be designed and verified both experimentally and theoretically as shown in other sections of this monograph. The achievable modification of the most relevant d-band structure for a given element can be considerable due to the combination of elements with dissimilar chemical properties (e.g., concept of early-late transition metals). The shifting of the d-band with respect to the Fermi level is the most powerful method to modify adsorption properties and reactivity of a solid surface. The substantial covalent contribution to chemical bonding in a wide range of complex metallic phases increases the chances of stability and integrity of the material even under chemically reactive environmental conditions. The structural features of complex metallic phases with motifs of transition-metal atoms or clusters exclusively surrounded by elements with different chemical properties allows realizing the concept of active-site isolation in a unique way. The high electron localization creates centers of strong Lewis basicity providing effective chemisorption of reactants without having to rely on typical metal–oxygen entities. The structural rigidity of many complex metallic phases prevents the dissolution of reactants underneath the surface forming subsurface compounds such as hydrides and carbides. These strongly alter the electronic properties of the surface in a feedback loop with the chemical potential of the reactants controlling the reaction properties of the catalyst. Embedding the centers of reactivity in a dense atomic matrix provides mechanical and structural stability and excellent thermal properties in comparison to conventional catalysts consisting of metal clusters deposited on typical support oxides.

This chapter describes in the form of a progress report the identification of generic combinations of complex metallic phases and catalytic reactions, which realize the advantages described above. The identification of such generic case studies is essential in developing strategies for overcoming the inherent hurdles towards a broader application of intermetallic compounds in catalysis. These hurdles are, in particular, the difficult synthesis of single-phase materials with structural integrity under reaction conditions and the nanostructuring of such systems. Only then are the active sites suitably dispersed without losing the intrinsic advantages of embedded and isolated sites due to excessive surface defects modifying the terminating intermetallic structure.

10.2
Why Use Intermetallic Compounds – The General Concept

Let's develop an ideal system for basic catalytic studies. What are the demands this system must fulfill? Trivial is the requirement of catalytic activity. Besides this, the system should be as simple as possible and be applicable in UHV as well as under real conditions to allow studies in both regimes. The simplicity allows concentrating the investigations on changes and properties of a small number of different catalytic species. In the ideal case only one, if we succeed in making the system as simple as possible. Using the same material in UHV and real conditions closes the materials gap, thus easing comparison of the obtained results. In contrast to industrial studies, basic studies are not in need of very high activity. This enables us to work with unsupported materials possessing rather low specific surface areas. Furthermore, our material needs to be homogeneous throughout and must be stable under reaction conditions. While the first requirement is supporting the "as simple as possible approach," the latter one allows us to connect the observed catalytic properties to the electronic and geometric structure of the material, which can be selected beforehand. The simplest realization of this concept is to use metallic elements in an unsupported state. The drawback is the very limited number of elements that are catalytically active for a specific reaction, restricting the data with which correlations between the catalytic properties and the electronic and structural properties can be revealed. In addition, this task would only achieve very coarse correlations because of the large property differences of the elements. Statistical alloys, used frequently in catalysis to overcome the gaps between the elements, are no alternative due to nonhomogeneity and lacking stability, resulting in frequently observed segregation.

Alternative materials are structurally ordered intermetallic compounds and among them complex metallic phases. These compounds very often comprise covalent bonding, making them more stable and reducing segregation. But to connect the observed catalytic properties with the structural and electronic properties of the compound itself, the structural stability under reaction conditions has to be proven in every combination of any intermetallic compound and a specific reaction.

The importance to prove the stability under *in situ* conditions becomes apparent when the pioneering work of Tsai *et al.* on the use of complex metallic phases and quasicrystals as catalysts for the steam reforming of methanol is taken into account [21–26]. Here, the compounds from the ternary Al-Cu-Fe system are decomposed to form a supported Cu catalyst [27]. Thus, the observed catalytic properties can not be ascribed to the complex metallic phases or quasicrystals but to the elemental copper particles.

On the other hand, correlating the structural and electronic circumstances of *in situ* stable complex metallic phases with observed catalytic properties enables a rational and knowledge-based approach to develop completely new catalysts that outperform conventional catalysts in terms of selectivity, activity and stability.

10.2.1
Chemical Bonding

The *in situ* stability of the intermetallic compounds is crucial in order to connect the catalytic properties to the well-defined crystal structure. This link may be made by investigation and understanding of atomic interactions, that is, chemical bonding. Since this is experimentally challenging and also time consuming, quantum chemical methods based on the concept of electron localizability, namely the electron localizability indicator (ELI), are employed beforehand to explore the chemical bonding in the intermetallic compounds. The electron localizability indicator in the ELI-D formulation [28–34] can be regarded as a quasicontinuous weighted electronic charge distribution of the given spin in position space, where the weighting factor (the so-called pair-volume function [32, 34]) represents a local measure for the volume that is needed to build a same-spin (with the given spin) electron pair [30–32]. The values of ELI-D are limited to the range of positive numbers. Typically, ELI values up to 2.5 can be found in the chemically relevant valence region of molecules (with the exception of H atoms). The topology of ELI-D (e.g., the location of local maxima) is well known to provide signatures of chemical bonding [32].

Because of computational difficulties caused by the structural complexity, complete investigations on the chemical bonding in real space in complex metallic phases are very rare. The most evaluated example is the binary compound o-Co_4Al_{13} that attracts special attention because of some basic structural features. Having a relatively small unit cell in comparison with other representatives of the family of complex metallic phases, the crystal structure possesses all typical characteristics: a large number of atoms in the unit cell, local atomic disorder and atomic arrangements with pseudopentagonal symmetry [35].

Caused by the (pseudo) penta- and decagonal arrangements of the atoms within these layers [36], o-Co_4Al_{13} is thought of as one of the simplest approximants to decagonal quasicrystals. To resolve disorder within the crystal structure of o-Co_4Al_{13}, high-quality single crystals of o-Co_4Al_{13} were prepared by Bridgman and Czochralski methods and structurally reinvestigated [37]. A high-resolution investigation of the

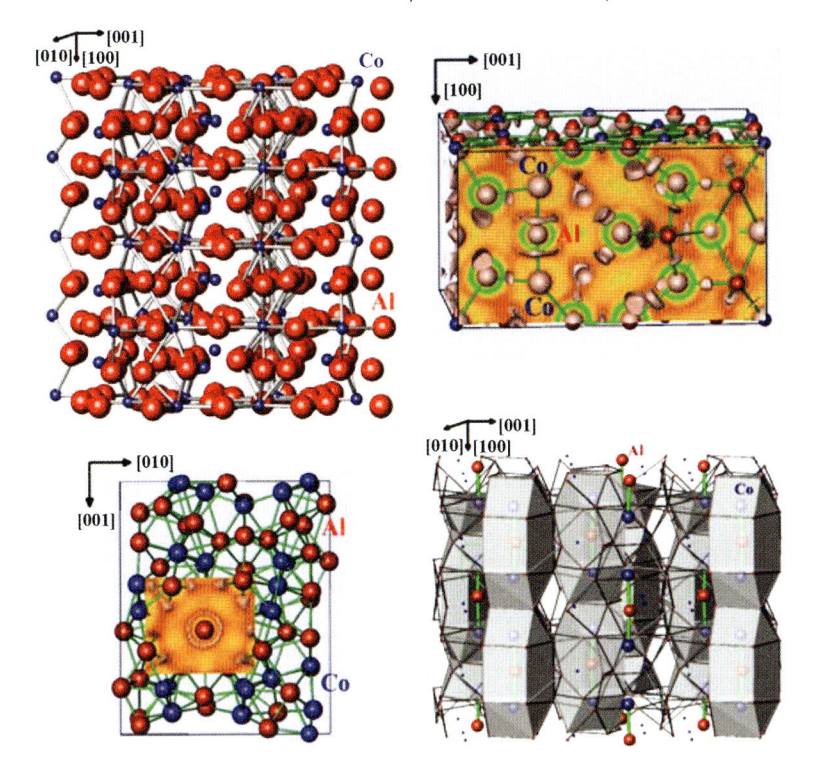

Figure 10.1 Crystal structure and chemical bonding o-Co$_4$Al$_{13}$: (top left) atomic layers along [100] and the shortest Co–Al contacts in one of the three structurally equivalent models; (bottom left) ELI distribution between the layers; (top right) ELI distribution in vicinity of the Co–Al–Co group; (bottom right) cage-like organization of the crystal structure from bonding analysis.

crystal structure was performed using X-ray diffraction data sets with $2\theta < 123°$ with AgKα radiation. While all cobalt positions have been found to be fully occupied, the aluminum sites in distinct regions of the crystal structure reveal positional disorder. Resolving of this disorder yielded three equivalent models, representing distortion variants of the pseudopentagonal columnar structural units characteristic of this group of complex metallic phases (Figure 10.1, top left [34, 38]).

Analysis of the chemical bonding in the ordered models by means of the electron localizability indicator (ELI-D) led to a highly unexpected result. Traditionally, the crystal structure is interpreted as consisting of atomic layers perpendicular to [100]. ELI-D revealed numerous directed (covalent) Co–Al and Al–Al bonds within the atomic layers perpendicular to [100], as well as in-between the layers (Figure 10.1 top right and bottom left). This indicates the formation of a 3D framework, contrary to the traditional consideration of atomic layers. In addition, in the elongated cavities parallel to [100] isolated three-atomic groups Co–Al–Co were found. From the topology of ELI-D, the atomic interactions within the group (directed, covalent) and between the group and the atoms of the framework (nondirected, ionic) are different (Figure 10.1, bottom left). This reveals a highly unexpected analogy between o-Co$_4$Al$_{13}$ and intermetallic clathrates. The latter represent a group of complex metallic phases whose crystal structures show covalently bonded 3D networks interacting ionically with the filler atoms in the cavities. Despite the complexity of the crystal structure, NMR spectroscopy confirmed the results of the bonding analysis and strongly supported the unique bonding situation of Al in the nearly linear Co–Al–Co groups [39]. In addition, the bonding model was supported by the results of the measurements of electrical conductivity and specific heat on oriented single crystals [40].

Further analysis of the ELI-D in the (100) and (400) planes revealed a larger amount of inplane covalent bonds in the puckered network parallel to (400) ($x = 0.22$–0.28) than in the flat network in (100). This indicates the possible terminating character of this plane along the [100] direction. Structure investigations of the (400) surface of the single crystal of o-Co$_4$Al$_{13}$ are in full agreement with the conclusions of the bonding analysis [37, 41]. The atomic decoration of this plane is not affected by the crystallographic disorder in the crystal structure. This results in distinct and ordered atomic configurations of cobalt atoms that are solely surrounded by Al on the surface (cf. geometric analogy to PdGa) – a promising feature for testing the compound in the active-site isolation concept.

Caused by similarities in atomic interactions, the electronic density of states of Co$_4$Al$_{13}$ and PdGa, despite huge structural differences, revealed one important common feature (Figure 10.2). In both cases the d-bands are shifted to lower energies with respect to the elemental metals, influencing the adsorption properties and, thus, the catalytic properties of the compounds.

10.2.2
Investigating the Stability *In Situ*

While heterogeneous catalysis is proceeding on the surface of the complex metallic phases, the underlying bulk is influencing the properties of the surface. Thus, the surface as well as the bulk have to be stable under reaction conditions in order to assign the observed catalytic properties to the electronic and geometric properties of the complex metallic phase in question.

10.2.2.1 Bulk Stability
One advantage of the complex metallic phases is their crystallinity, so powder diffraction under reaction conditions at different temperatures can be performed to detect changes in the bulk. The use of finely ground or filed material is advantageous. Surface areas of around $0.1 \, m^2/g$ result in detectable activity and the particles are small enough ($\sim 20 \, \mu m$) for the detection of changes that will start from the surface. For the measurements, the sample is placed in a cell allowing application of the reactive atmosphere and X-rays at the same time. To ensure good contact of the reactive atmosphere and the sample, the gas is sucked through the powder.

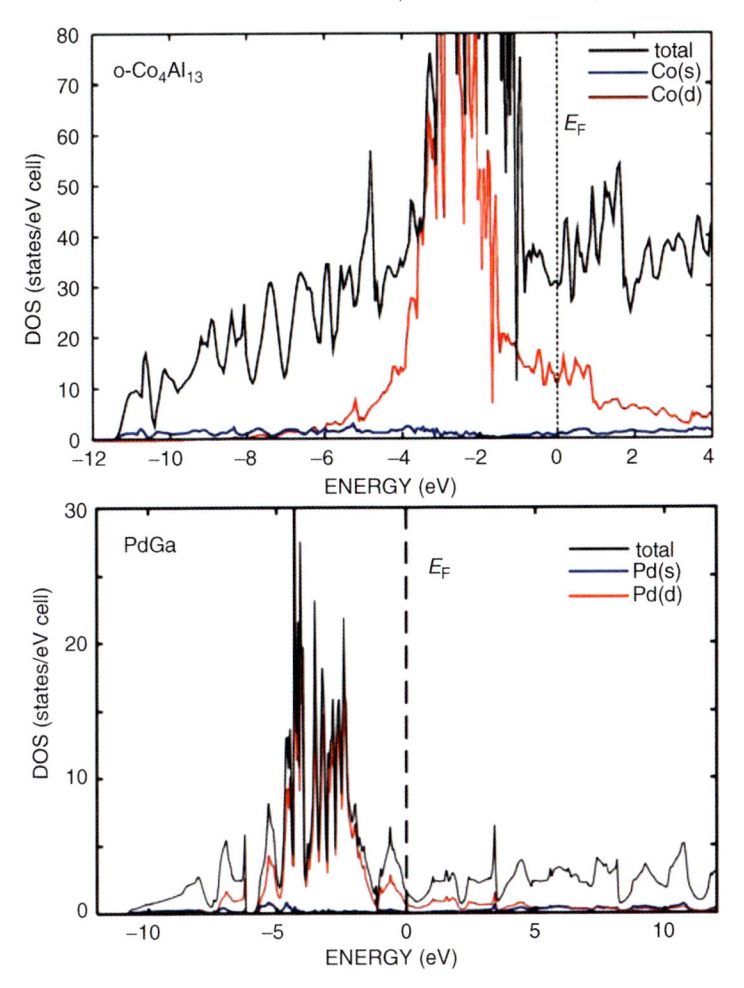

Figure 10.2 Electronic density of states of o-Co$_4$Al$_{13}$ (top) and PdGa (bottom). In both cases the d-bands are shifted to lower energy compared to the corresponding transition metal element.

In a typical supported catalyst, the metal particles are very small (<10 nm). This results in XRD patterns that are dominated by support signals. These hinder the detection of changes of the already broad metal reflections. The XRD pattern of a commercially available 5 wt.% Pd/Al$_2$O$_3$ catalyst is shown in Figure 10.3 as a typical example. Due to the small crystallite size the Pd signals are barely visible.

In contrast to supported catalysts, the obtained diffraction pattern of the unsupported complex metallic phase eases the detection of alterations in the material. Changes of the material can result in variations of the lattice parameters (e.g., hydride formation), decrease of signal intensity (e.g., formation of amorphous phases) or the appearance of additional components, for example, oxides in oxidizing environments. Testing the stability of compounds in reducing as well as oxidizing

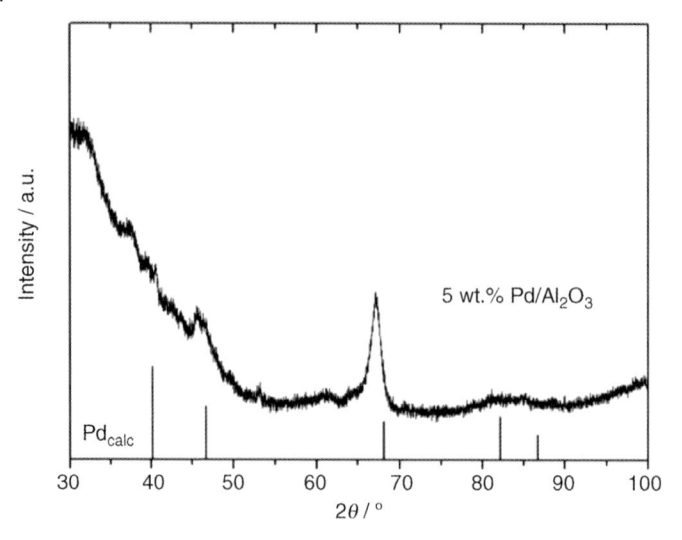

Figure 10.3 XRD pattern of 5 wt.% Pd/Al$_2$O$_3$ as an example for a typical supported catalyst.

atmosphere reveals the range of reactions in which they can be applied as catalysts. Furthermore, the range of conditions in which the catalyst could be reactivated, for example, by burning of carbonaceous deposits, is determined this way.

In the following, the focus lies on Co$_4$Al$_{13}$ (o-Co$_4$Al$_{13}$ type of structure, $Pmn2_1$, $a = 8.158(1)$ Å, $b = 12.342(1)$ Å, $c = 14.452(2)$ Å) [35] and Fe$_4$Al$_{13}$ (Fe$_4$Al$_{13}$ type of structure, $C2/m$, $a = 15.492(2)$ Å, $b = 8.078(2)$ Å, $c = 12.471(1)$ Å, $\beta = 107.69(1)°$) [36], which are excellent hydrogenation catalysts as will be shown below [42]. Due to the high fraction of covalent bonding, the complex metallic phase Fe$_4$Al$_{13}$ does not show changes in oxidizing or reducing environments up to temperatures of 450 °C (Figure 10.4).

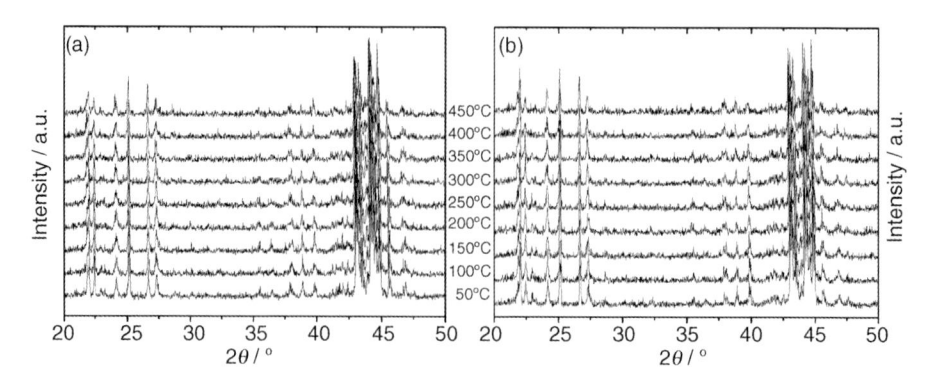

Figure 10.4 Temperature dependent X-ray powder diffraction of Fe$_4$Al$_{13}$ in (a) 25% H$_2$ and (b) 20% O$_2$ in helium.

The ability to form hydrides is widely spread among intermetallic compounds. The absence of hydride formation has to be proven carefully, because the formation of hydrides involves changes of the structural and electronic properties that directly influence the catalytic behavior. The observed catalytic properties thus would not be specific for the intermetallic compound that was originally placed in the reactor, but for the hydride formed *in situ*. This situation is even more complex due to the instability of some hydrides under normal conditions. The hydride decomposes as soon as the reactive atmosphere is removed, which makes the *ex situ* detection difficult or even impossible. To be able to assign the observed catalytic properties to a specific intermetallic compound, thus allowing a knowledge-based approach, hydride formation under reaction conditions has to be excluded.

A method that has only recently been extended to *in situ* investigations of reactions in the gas phase, is prompt gamma-activation analysis (PGAA) [43]. PGAA is based on the emission of gamma-rays of characteristic energy by the elements, when exposed to cold neutrons. In the first step, the neutrons are absorbed by the nucleus, which emits an element specific gamma-ray to release the gained energy. The sample is placed in a tubular reactor and while the reactive atmosphere and temperature are applied, the concentration of hydrogen in the sample is determined. The measurement has to be corrected for background contributions (mainly water in the path of the neutron beam) and results in an accurate measurement of the hydrogen concentration. As result, hydride formation can be detected at very low levels, allowing to exclude or to proof the presence of a hydride under reaction conditions.

To prove the feasibility of the method, measurements of the hydrogen concentration in elemental Pd in hydrogen atmosphere were performed and resulted in the formation of β-PdH$_{0.7}$, as expected. The complex metallic phases on the other hand showed practically no hydrogen uptake. Based on the PGAA and XRD results, the formation of hydrides in the complex metallic phases Co_4Al_{13} and Fe_4Al_{13} can clearly be ruled out.

10.2.2.2 Surface Stability

Heterogeneous catalysis takes place at the surface of the intermetallic compounds, so additionally surface sensitive methods have to be applied under reaction conditions to ensure the stability *in situ*.

Changes in the so-called near-surface region under reaction conditions can be detected by high pressure X-ray photoelectron spectroscopy (XPS) in the mbar range. Complementary to *in situ* XRD investigations, XPS does not give any structural information but is restricted to the electronic state of the elements within the material. To detect changes under reaction conditions, the XPS signals of core levels of the elements in the material as well as valence band spectra are first recorded under ultrahigh vacuum (UHV) conditions. In addition to the spectra of the metallic elements, the spectra of carbon and oxygen have to be also recorded to detect changes in these under reaction conditions. Using synchrotron radiation, the energy of the incoming photons can be tuned in such a way that information from different depths of the material can be obtained. This depth scan allows the detection of so-called subsurface chemistry that can strongly influence the catalytic properties. A recent example is the investigation of a subsurface palladium–carbon phase, when elemental palladium is used as catalyst for the hydrogenation of alkynes [44, 45]. Carbon, resulting from the decomposition of the alkynes, diffuses into the palladium and forms a subsurface Pd/C phase. The hydrogen in the feed causes the formation of β-PdH$_{0.7}$ as shown above. This catalytically highly active hydride usually leads to total hydrogenation of the respective alkynes. The subsurface Pd/C acts as a diffusion barrier that hinders the diffusion of the highly active hydridic hydrogen to the surface. This results in a complete change from the unwanted total hydrogenation of the alkyne to a selective semihydrogenation to the corresponding alkene, thus strongly altering the catalytic properties.

After recording UHV spectra and collecting the information for different depths, *in situ* measurements can be performed. The reactive atmosphere is introduced into the UHV chamber at pressures of around 1 mbar and then the sample is heated to the desired temperature. Mass spectrometry is applied to detect the catalytic activity and spectra are recorded after equilibration of the sample. Differences to the UHV spectra indicate alterations of the compounds, like decomposition or the involvement of subsurface species. Besides extra signals for an element, which result from the presence of an additional compound, also a shift of the signals relative to the UHV spectra is possible, for example, by hydride formation. Here, it becomes obvious why reference spectra have to be recorded beforehand. Furthermore, the elemental states of the metals constituting the complex metallic phase are significantly altered compared to the pure metals. This easily leads to shifts of more than 1 eV and the UHV spectra are necessary to assign the signals of the complex metallic phases unambiguously [15].

UHV and *in situ* XPS valence band spectra of Co$_4$Al$_{13}$ and Fe$_4$Al$_{13}$ are shown in Figure 10.5. Comparison of the spectra clearly shows the stability of the near-surface

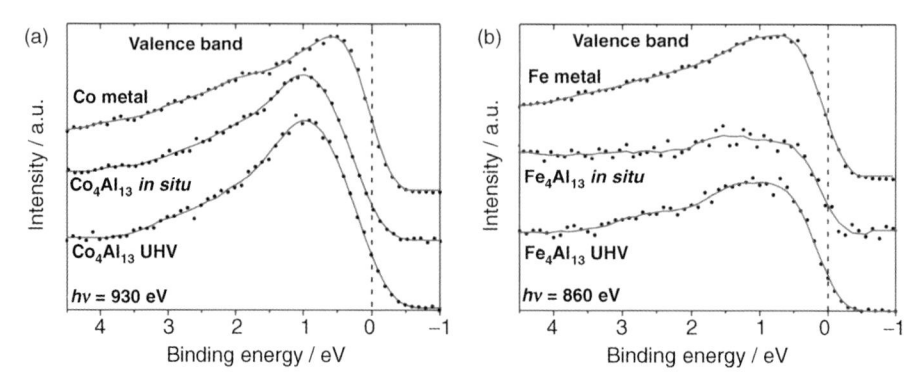

Figure 10.5 UHV XPS valence band spectra of (a) Co$_4$Al$_{13}$ and (b) Fe$_4$Al$_{13}$ and corresponding spectra under *in situ* conditions (1 mbar H$_2$, 0.1 mbar C$_2$H$_2$, recorded at beamline ISIS at BESSY). The spectra for elemental Co and Fe are shown for comparison.

region under *in situ* conditions. The influence of the covalent bonding on the electronic structure alters the band structure compared to cobalt and iron in the elemental states. Shifting the transition-metal d-bands below the Fermi energy leads to changes in the valence-band spectra expressed by the depletion of the density of states around the Fermi energy.

10.3
The Semihydrogenation of Acetylene

Ethylene for the production of polyethylene (2005: 50×10^6 t) is produced by steam-cracking of naphtha. Acetylene, which is present around 1% in the feed, is a poison for the polymerization catalyst [46]. To prevent detrimental effects during the polymerization, the acetylene concentration has to be reduced below 5 ppm. An elegant way to clean the ethylene feed is the selective semihydrogenation of acetylene to ethylene, thus turning the poison into valuable reactant. Selectivity is crucial in this process, since the hydrogenation of ethylene – which is present in a large excess – to ethane has to be prevented.

Selectivity can be enhanced by reducing the size of the active site, the so-called active-site isolation concept that was introduced in the 1970s [14]. Smaller active sites allow only a limited number of adsorption configurations, which possibly leads to a smaller number of by-products and higher selectivity towards ethylene. In addition, site-isolated catalysts possess higher stability since carbonaceous deposits, which present the dominant deactivation path in this reaction [47], are not formed on small active sites.

Palladium is a highly active catalyst for this reaction and to limit the size of the active sites, industrial catalysts are based on disordered palladium–silver alloys supported on Al_2O_3 [48]. One drawback of these catalysts is limited selectivity because, due to the random arrangement of the atoms, the active sites are not completely isolated from each other. Additionally, alloys are prone to segregation. This leads to larger active sites with time on stream resulting in lower selectivity and deactivation because of the formation of carbonaceous deposits.

First studies to realize the active-site isolation by intermetallic compounds were performed within the Pd–Ga system [13, 15–17, 49, 50]. For example, in the crystal structure of PdGa, the palladium atoms are surrounded by seven gallium atoms and no direct Pd–Pd bonds are present. After showing the feasibility of the concept, the focus changed to replacing the usually applied noble metals by cheaper transition metals. Despite the crystallographic disorder, complex metallic phases with suited crystal structures provide active-site isolation as an intrinsic property. In the crystal structure of Co_4Al_{13}, disorder is observed only for the aluminum atoms. Independently of that, the cobalt atoms are surrounded by nine to twelve aluminum atoms. Only for two out of ten crystallographic cobalt sites does a cobalt atom have one other Co atom at the periphery of the coordination sphere ($d_{Co-Co} = 2.89$–2.95 Å). Thus, the transition-metal atoms are completely isolated from one another by a shell of aluminum. Correspondingly, the number of transition-metal–transition-metal

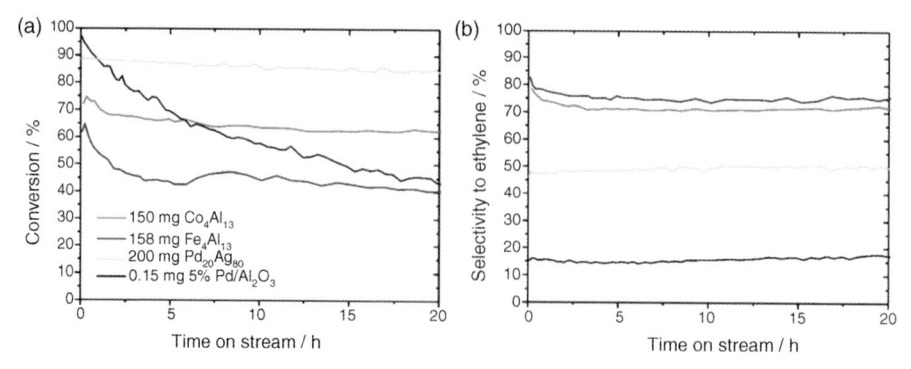

Figure 10.6 (a) Conversion and (b) selectivity to ethylene of unsupported Co_4Al_{13} and Fe_4Al_{13} in the semihydrogenation of acetylene. Pd/Al_2O_3 and an unsupported $Pd_{20}Ag_{80}$ alloy are shown for comparison.

contacts is much smaller than in the corresponding elements. This is in agreement with the active-site isolation concept.

In addition, segregation is drastically reduced by the covalent bonding present in the complex metallic phases. The resulting structural stability can preserve the geometric arrangement under reaction conditions and also prevents hydride formation, resulting in higher selectivity.

Figure 10.6 shows the catalytic properties of unsupported Co_4Al_{13} and Fe_4Al_{13} in the semihydrogenation of acetylene. The ratio of the reactants in the feed (0.5% acetylene, 5% hydrogen, 50% ethylene, rest He) resembles the feed composition under industrial conditions. The data represents isothermal runs at 200 °C.

Both compounds possess increased stability and selectivity compared to Pd/Al_2O_3. The selectivity is also higher than for an unsupported $Pd_{20}Ag_{80}$, where the active-site isolation is realized by random distribution of the Pd atoms in the Ag matrix. The detrimental effect of large active sites is obvious from the Pd/Al_2O_3 data. Here, side reactions lead to carbonaceous deposits that lead to deactivation of the catalyst. Deactivation is largely reduced in the complex metallic phases due to the isolated active sites that lead to ethylene with high selectivity.

The low activity compared to Pd/Al_2O_3 is due to the small surface area of the complex metallic phases, which have only been ground before use. That the intrinsic activity of intermetallic compounds should be of the same order of magnitude as Pd/Al_2O_3 has been shown in the case of nanoparticulate $PdGa/Al_2O_3$ and Pd_2Ga/Al_2O_3. Figure 10.7 shows the selectivity over the activity of different catalysts after 20 h time on stream under identical conditions.

The superior selectivity of the intermetallic compounds PdGa and Pd_2Ga as well as the complex metallic phases directly shows the advantageous influence of the isolated active sites. In the case of the Pd–Ga intermetallic compounds the nanoparticulate synthesis resulted in catalysts with higher activity than 5 wt.% Pd/Al_2O_3. Since the activity of ground Co_4Al_{13} and Fe_4Al_{13} is in the same regime as the ground Pd-Ga intermetallic compounds, nanoparticulate complex metallic phases should also be highly active and selective catalysts in this reaction. Due to the lower price of cobalt,

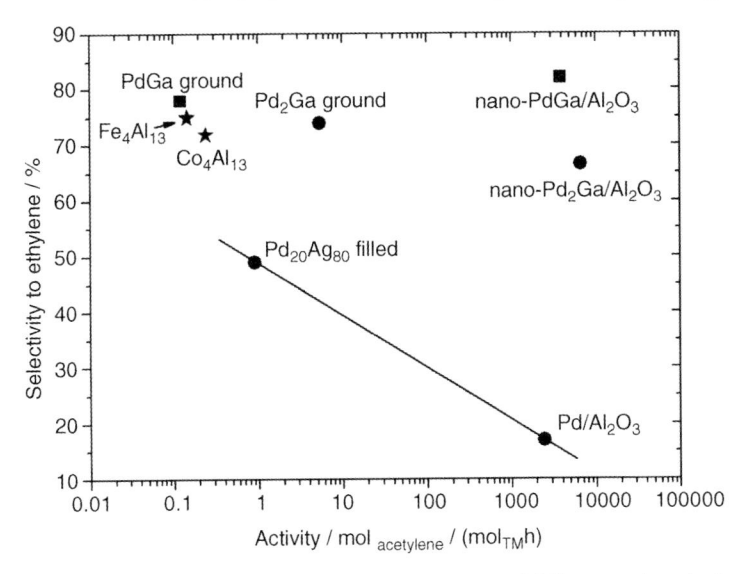

Figure 10.7 Selectivity towards ethylene versus activity of different catalysts in the semihydrogenation of acetylene (data after 20 h time on stream).

iron and aluminum, the development of a nanoparticulate synthesis route is a worthwhile target for future investigations.

10.4
Complex Metallic Phases as Platform Materials for Heterogeneous Catalysis

Using complex metallic phases and intermetallic compounds as *in situ* stable catalysts allows a knowledge-based improvement instead of a trial-and-error approach in heterogeneous catalysis. Since heterogeneous processes take place at the surface of the materials, exact knowledge of the nature of the surface, structurally as well as electronically, is required. As demonstrated in this chapter, our knowledge of the surface structures on selected complex metallic phases is comprehensive due to the effort of the European Network of Excellence "Complex Metallic Alloys" to focus different specialists on selected materials. The excellent knowledge about the surface of the compounds will enable a deep understanding of the ongoing processes in the future, making these compounds valuable platform materials not only for basic science, but also as industrial catalysts due to their low price compared to for example, palladium. Since identical materials can be investigated by UHV methods as well as by reactor studies and quantum chemical calculations, these materials have the potential to bridge the materials gap, thus allowing combined studies. The perspective of these studies is to identify and analyze the nature of the adsorption sites for the reactants and elucidate the mechanism of the semihydrogenation by kinetic measurements as well as quantum chemical calculations.

Acknowledgments

We thank D. Teschner for the PGAA measurements and for assisting in recording the XPS data and BESSY for providing beamtime. We are grateful to F. Girgsdies for the *in situ* XRD measurements. The European Network of Excellence on "Complex Metallic Alloys," contract No. NMP3-CT-2005-500145 is acknowledged for financial support.

References

1 Raney, M. (1925) US 1563587.
2 Raney, M. (1927) US 1628190.
3 Nix, R.M., Rayment, T., Lambert, R.M., Jennings, J.R., and Owen, G. (1987) *J. Catal.*, **106**, 216.
4 Hay, C.M., Jennings, J.R., Lambert, R.M., Nix, R.M., Owen, G., and Rayment, T. (1988) *Appl. Catal.*, **37**, 291.
5 Imamura, H. and Wallace, W.E. (1979) *Am. Chem. Soc. Div. Fuel Chem.*, **25**, 82.
6 France, J.E. and Wallace, W.E. (1988) *Lanthan. Actin. Res.*, **2**, 165.
7 Coon, V.T., Wallace, W.E., and Craig, R.S. (1978) *Rare Earths in Modern Science and Technology* (eds G.J. McCarthy and J.J. Rhyne) Plenum Press, New York, p. 93.
8 Consonni, M., Jokic, D., Murzin, D.Y., and Touroude, R. (1999) *J. Catal.*, **188**, 165.
9 Ammari, F., Lamotte, J., and Touroude, R. (2004) *J. Catal.*, **221**, 32.
10 Chin, Y.-H., Dagle, R., Hu, J., Dohnalkova, A.C., and Wang, Y. (2002) *Catal. Today*, **77**, 79.
11 Karim, A., Conant, T., and Datye, A. (2006) *J. Catal.*, **243**, 420.
12 Galloway, E., Armbrüster, M., Kovnir, K., Tikhov, M.S., and Lambert, R.M. (2009) *J. Catal.*, **261**, 60.
13 Kovnir, K., Armbrüster, M., Teschner, D., Venkov, T.V., Jentoft, F.C., Knop-Gericke, A., Grin, Yu., and Schlögl, R. (2007) *Sci. Technol. Adv. Mater.*, **8**, 420.
14 Sachtler, W.M.H. (1976) *Catal. Rev. Sci. Engin.*, **14**, 193.
15 Kovnir, K., Teschner, D., Armbrüster, M., Schnörch, P., Hävecker, M., Knop-Gericke, A., Grin, Yu., and Schlögl, R. (2008) BESSY Highlights 2007, 22.
16 Osswald, J., Giedigkeit, R., Jentoft, R.E., Armbrüster, M., Girgsdies, F., Kovnir, K., Grin, Yu., Schlögl, R., and Ressler, T. (2008) *J. Catal.*, **258**, 210.
17 Osswald, J., Kovnir, K., Armbrüster, M., Giedigkeit, R., Jentoft, R.E., Wild, U., Grin, Yu., Schlögl, R., and Ressler, T. (2008) *J. Catal.*, **258**, 219.
18 Kovnir, K., Osswald, J., Armbrüster, M., Giedigkeit, R., Ressler, T., Grin, Yu., and Schlögl, R. (2006) *Stud. Surf. Sci. Catal.*, **162**, 481.
19 Osswald, J., Giedigkeit, R., Armbrüster, M., Kovnir, K., Jentoft, R.E., Ressler, T., Grin, Yu., and Schlögl, R. (2006) European patent pending, EP1834939A1.
20 Giedigkeit, R., Borrmann, H., and Armbrüster, M., Borrmann, A., Wedel, M., Prots, Y., Giedigkeit, R., and Gille, P. (2010) *Z. Kristallogr.*, NCS, in print.
21 Kameoka, S., Tanabe, T., and Tsai, A.P. (2004) *Catal. Today*, **93–95**, 23.
22 Kameoka, S. and Tsai, A.P. (2008) *Catal. Today*, **132**, 88.
23 Kameoka, S. and Tsai, A.P. (2008) *Catal. Lett.*, **121**, 337.
24 Tanabe, T., Kameoka, S., and Tsai, A.P. (2006) *Catal. Today*, **111**, 153.
25 Tsai, A.P. and Yoshimura, M. (2001) *Appl. Catal. A*, **214**, 237.
26 Yoshimura, M. and Tsai, A.P. (2002) *J. Alloys Compd.*, **342**, 451.
27 Belin-Ferré, E., Fontaine, M.-F., Thirion, J., Kameoka, S., Tsai, A.P., and Dubois, J.M. (2006) *Philos. Mag.*, **86**, 687.
28 Kohout, M. (2004) *Int. J. Quantum Chem.*, **97**, 651.
29 Kohout, M. (2007) *Faraday Discuss.*, **135**, 43.

30 Kohout, M., Pernal, K., Wagner, F.R., and Grin, Yu. (2004) *Theor. Chem. Acc.*, **112**, 453.
31 Kohout, M., Wagner, F.R., and Grin, Yu. (2006) *Int. J. Quantum Chem.*, **106**, 1499.
32 Wagner, F.R., Bezugly, V., Kohout, M., and Grin, Yu. (2007) *Chem. Eur. J.*, **13**, 5724.
33 Kohout, M., Wagner, F.R., and Grin, Yu. (2008) *Theor. Chem. Acc.*, **119**, 413.
34 Wagner, F.R., Kohout, M., and Grin, Yu. (2008) *J. Phys. Chem. A*, **112**, 9814.
35 Grin, J., Burkhardt, U., Ellner, M., and Peters, K. (1994) *J. Alloys Compd.*, **206**, 243.
36 Grin, J., Burkhardt, U., and Ellner, M. (1994) *Z. Kristallogr.*, **209**, 479.
37 Fournée. V., Ledieu, J., and Park, J.Y. (2010) *Complex Metallic Alloys Fundamentals and Applications* (eds. J. M. Dubois and E. Belin-Ferré) Wiley-VCH, Weinheim, p. 155.
38 Grin, Yu., Bauer, B., Burkhardt, U., Cardoso-Gil, R., Dolinšek, J., Feuerbacher, M., Gille, P., Haarmann, F., Heggen, M., Jeglič, P., Müller, M., Paschen, S., Schnelle, W., and Vrtnik, S. (2007) EUROMAT 2007: European Congress on Advanced Materials and Processes, Book of Abstracts, Nürnberg, Germany, p. 30.
39 Jeglič, P., Heggen, M., Feuerbacher, M., Bauer, B., Gille, P., and Haarmann, F. (2009) *J. Alloys Compd.*, **480**, 141.
40 Dolinšek, J., Komelj, M., Jeglič, P., Vrtnik, S., Stanić, D., Popčević, P., Ivkov, J., Smontara, A., Jagličić, Z., Gille, P., and Grin, Yu. (2009) *Phys. Rev. B*, **79**, 184201.
41 Addou, R., Gaudry, E., Deniozou, T., Heggen, M., Feuerbacher, M., Gille, P., Widmer, R., Gröning, O., Grin, Yu., Fournée, V., Dubois, J.-M., and Ledieu, J. (2009) *Phys. Rev. B*, **80**, 014203.
42 Armbrüster, M., Kovnir, K., Grin, Yu., Schlögl, R., Gille, P., Heggen, M., and Feuerbacher, M. (2009) European patent pending EP09157875.7.
43 Révay, Z., Belgya, T., Szentmiklósi, L., Kis, Z., Wootsch, A., Teschner, D., Swoboda, M., Schlögl, R., Borsodi, J., and Zepernick, R. (2008) *Anal. Chem.*, **80**, 6066.
44 Teschner, D., Vass, E., Havácker, M., Zafeiratos, S., Schnörch, P., Sauer, H., Knop-Gericke, A., Schlögl, R., Chamam, M., Wootsch, A., Canning, A.S., Gamman, J.J., Jackson, S.D., McGregor, J., and Gladden, L.F. (2006) *J. Catal.*, **242**, 26.
45 Teschner, D., Borsodi, J., Wootsch, A., Révay, Z., Hävecker, M., Knop-Gericke, A., Jackson, S.D., and Schlögl, R. (2008) *Science*, **320**, 86.
46 Borodzinski, A. and Bond, G.C. (2006) *Chem. Rev.*, **48**, 91.
47 Ahn, I.Y., Lee, J.H., Kim, S.K., and Moon, S.H. (2009) *Appl. Catal. A*, **360**, 38.
48 Johnson, M.M., Walker, D.W., and Nowack, G.P. (1982) EP 0 064 301.
49 Kovnir, K., Armbrüster, M., Teschner, D., Venkov, T., Szentmiklósi, L., Jentoft, F.C., Knop-Gericke, A., Grin, Yu., and Schlögl, R. (2009) *Surf. Sci.*, **603**, 1784.
50 Kovnir, K., Osswald, J., Armbrüster, M., Teschner, D., Weinberg, G., Wild, U., Knop-Gericke, A., Ressler, T., Grin, Yu., and Schlögl, R. (2009) *J. Catal.*, **264**, 93.

Index

a

acoustic dispersion 82, 103
activation enthalpy 29
activation volume 29
ADF detector 352
adhesion
– on air-oxidized quasicrystal surfaces 193 f
– on clean quasicrystal surfaces 191 f
– on in-situ oxidized quasicrystal surfaces 191 f
adhesion energy of CMAs 254
adhesion force 195
– under fretting of CMAs 301 ff
adhesion measurement using AFM 191 f
adiabatic demagnetization 317
adiabatic temperature change 317
adsorption on quasicrystal surfaces 181 ff
– growth of a copper pseudomorphic multilayer film 183
– growth of a xenon thin film 182
– growth of lead monolayer 181
– molecular adsorption onto d-Al-Ni-Co 249
– molecular adsorption onto i-Al-Pd-Mn 249
adsorption sites on a quasicrystal surface 179 f
AFM. see atomic force microscopy
AFM/FFM measurement 195 f
i-Ag-In-Yb QC 50
o-Al$_{13}$Co$_4$ 9, 26 ff
– ab initio calculation of the electronic band structure of 142
– anisotropic physical properties of 117 ff, 148 ff
– behaviour in catalytic hydrogenation 392 ff, 396
– comparison with PdGa 390 f, 396
– ELI-D of 388 f
– Fermi surface of 142 f
– mechanical properties of 275
– single-crystal growth of 215 f
– structure of 389
Al$_4$(Cr,Fe) phase
– ab initio calculation of the electronic band structure of 142 ff
– anisotropic physical properties of 119 f
– Fermi surface of 142 f
– localized corrosion model for 265 f
Al-Cr-Fe phase system 3, 22, 48, 166
– abrasion resistance of quasicrystalline Al-Cu-Fe coatings 306 f
– air-oxidation of 252 f
– Al-Cu-Fe QCs as reinforcement in Al-based matrix composites 291 ff
– Al-Cu-Fe-B QC 47, 296, 309
– Al-Cu-Fe-Cr QC 30
– θ-Al$_2$Cu 3, 14, 229, 236, 277
– ω-Al$_7$Cu$_2$Fe 15, 248, 293
– φ-Al$_{10}$Cu$_{10}$Fe 14
– Al$_{63}$Cu$_{25}$Fe$_{12}$ 45, 57, 71, 227
– corrosion experiments 255 ff
– corrosion experiments 255 ff, 267
– electrical conductivity 54
– i-Al-Cu-(Fe,Ru,Os) QC 49, 55, 59, 66
– mechanical properties of compounds of the 277
– MOCVD processing of Al-Cu-Fe thin films 227 ff
– phase diagram 3, 228
– photoelectrochemical behaviour of surface-passivating oxide layers 261
– surface energy of Al-Cu-Fe QCs 24
– surface oxidation of 247 ff, 267, 269
– surface oxidation under UHV environment of 248
– Vickers microhardness of Al-Cu-Fe QCs 274 ff

– Vickers microhardness of *i*-Al-Cu-Fe
  reinforced composites 296
Al-Cu-Ru phase system 15
alloying 185 f
Al-Mg alloy 15 ff, 278 ff
– β-Al$_3$Mg$_2$ 16, 26 ff, 268, 275
Al-Mn-Si approximant phase 66, 68 ff
Al-Pd-Fe phase system 8
– C$_1$-Al-Pd-Fe 12
– C$_2$-Al-Pd-Fe 11
– ε$_{22}$-Al-Pd-Fe 8
– ξ-Al-Pd-Fe 8
Al-Pd-Mn phase system 7, 15, 22, 69
– ε$_{28}$-Al-Pd-Mn 7
– ε$_6$-Al-Pd-Mn 6, 26 ff
– *i*-Al-Pd-Mn 22, 47, 86 f, 158, 245
– ψ-Al-Pd-Mn 221
Al-Pd-TM phase system
– *i*-Al-Pd-(Mn,Re) QC 49 f, 55, 59, 66
– electrical conductivity of Al-Pd-Mn
  CMAs 69
Al-Pd-Mn phase system
– molecular adsorption onto *i*-Al-Pd-Mn 249
– oxidation characteristics of fivefold
  i-Al$_{70}$Pd$_{21}$Mn$_9$ surface 244
– single-crystal growth of ξ'-Al-Pd-Mn
  221 f
– surface oxidation of 244 ff
– surface-structure determination
  of *i*-Al-Mn-Pd 160 ff, 165, 168
– surface-structure determination of
  ξ'-Al-Mn-Pd approximant phase 164
– thermal conductivity of Al-Pd-Mn
  CMAs 105
Al$_{13}$TM$_4$-structured CMA 117 ff
aluminum-based CMA
– Al$_{56}$Li$_{33}$Cu$_{11}$ 56
– Al$_{70}$Pd$_{20}$(V,Cr,Mn,W)$_5$(Co,Fe,Ru,Os)$_5$ 55
– in the Al-Co-Cu-Si system 48
aluminum matrix composite reinforced with
  CMAs
– friction properties of 309
– mechanical properties of 294 f
– preparation of 291 ff, 309
– preservation of complex phases in 292 f
– thermal stability of 292 f
aluminum oxide layer 266 ff
analytical coefficient 67 f
anharmonic interaction, effect on phonon
  spectrum 79
anisotropic physical property 117 ff
– electrical resistivity 124 ff, 129, 148 f
– Hall coefficient 132 ff, 146 ff
– magnetic properties 120 ff

– of Al$_4$(Cr,Fe) CMAs 122 f, 126 f, 129,
  132, 135, 138
– of o-Al$_{13}$Co$_4$ 120 ff, 125 f, 129, 131,
  134 ff, 148 ff
– of the Y-Al-Ni-Co phase 120, 124 f, 129,
  130, 132 f, 136, 147
– thermal conductivity 136 ff
– thermoelectric power 130 ff
annealing treatment in CMA surface
  preparation 158 f
anticrossing effect 98 f, 101
antiphase boundary 6
antistick properties of QCs 30
aperiodic system 61 ff
application of QCs
– related to heat insulation 34
– related to light absorption 34
– related to magnetocaloric effect 33
– related to particle dispersion 35
– related to surface energy 30 f
Arrhenius plot 233, 235, 238 f
as-cast quasicrystalline composite 281 f,
  288 ff, 295, 308, 326
asymmetry parameter 174 f
atmospheric aging 251
atomic force microscopy (AFM) 191 ff
atomic long-range order 41
atomic-scale adhesion property of CMAs 190

b
Ba$_8$Ga$_{16}$Ge$_{30}$ 99 f, 370 f
BDTT. *see* brittle-to-ductile transition
  temperature
Bergman cluster 162
Bergman phase 12, 18, 50, 82
BF detector 352 f
BF image. *see* bright-field image
Bi$_2$Te$_3$/Sb$_2$Te$_3$ superlattice 369 f
binary intermetallic magnetocaloric
  compound 334 ff
Bloch wave 71
BMG. *see* bulk metallic glass
Bragg diffraction 351
Bragg peak 76 f, 84, 86 ff, 99 f, 165
γ-brass phase 14
Bravai rule, modified for CMAs 162
Bridgman technique 210 ff, 218 ff, 221 ff, 224
bright-field (BF) image 11, 349 ff
Brillouin zone 13 f, 71 ff, 82, 103, 140 ff, 175
– Brillouin-Jones zone 56
– Brillouin-zone boundary 13, 72 ff, 79, 82
Brinell hardness 296 f, 310
brittle-to-ductile transition temperature
  (BDTT) 276

bulk metallic glass (BMG)   284 ff, 285, 335 f
bulk structure of quasicrystal samples   160, 167
Burger vector   6, 8

c
cage compound   95 f
capillarity   208
casting   219, 280 f, 288, 291 f, 296
i-Cd(Yb,Ca) QC   57
Cd-Yb(rare earth) QC   49 f
centered dark-field (CDF) image   349
charge-density wave instability   372
chemical bonding of intermetallic compounds   388
chemical vapor deposition (CVD)   225 ff. see also molecular chemical vapor deposition
– codeposition of two metals   229
– precursor selection   229 f, 240
– process steps   226
clathrate   50, 96 ff, 370
– structural classification of   377 f
– thermoelectric properties of   370, 377 f
CMA coating   227 ff, 235 ff, 306 f. see also thin film deposition
CMA. see complex metallic alloy
CMA single crystal   208
CMAs surface
– atmospheric aging of   251 ff
– electrochemical characterization of oxidized   256 f, 262 f
– hardness of   23
– interaction with oxygen under UHV environment   244 ff
– localized degradation reactions of surface oxides   262 ff
– molecular adsorption onto   248 ff
– of d-Al-Ni-Co   163 f
– pentagonal surface of i-Al-Pd-Mn   244 ff, 250
– preparation of   20
– properties of   21 f, 254 f
– thermodynamic stability of   255 ff
– treatment of   21 f
cold crucible   210
cold welding of CMAs   299 f
complex metallic alloy (CMA)
– aluminum-based   4 ff
– anisotropic properties of   20
– as hydrogenation catalyst   388 ff
– atomic-scale adhesion properties of   190
– characterization of   2
– coating application of   31 ff
– complexity index   4 f

– definition of   1
– electrical resistivity of   19
– electronic structure of   13 f
– heat insulation application of   34
– magnetocaloric properties of   324 ff
– mechanical engineering properties of   273 ff
– phonon modes of   71 ff
– plasticity of   25
– reactions in aqueous solutions   255 ff
– stabilization of   13 f, 17 f
– surface chemistry of   243 ff
– surface energy of   22 f
– surface properties of   20 ff
– TEM investigation of   353 ff
– transport properties of   18 f, 33 f
– vibrational properties of   71 f
– wetting properties of   188 ff
complexity index   4 f
concept of "active-site isolation"   386, 395
concept of "electron engineering"   371 f
concept of "phonon engineering"   370
concept of "phonon glass electron crystal (PGEC)"   368 ff
concept of "phonon-blocking electron-transmitting"   369
concept of "rattling atoms"   102 f, 369 f
congruently melting phase   217
contact angle of water on quasicrystal surfaces   188 f
continuum-mechanics models for atomic-scale adhesion   190 f
Conway's theorem   61
copper cyclopentadienyl triethyl phosphine (CpCuPEt$_3$)   231
corrosion
– high-temperature oxidation of Al-Cu-Fe thin films   268
– high-temperature oxidation of Al-rich bulk samples   266 f
– localized   264 ff
– of CMA powders   269
– resistance of QCs   30
CpCuPEt$_3$. see copper cyclopentadienyl triethyl phosphine
Cr-Cr$_2$Nb two-phase alloy   280 ff
critical electronic state   60 ff
critical normal mode   63
critical wavefunction   61 f
crystal interface   209
crystal/quasicrystal interface   185
current-voltage curve   44 f
CVD. see chemical vapor deposition
Czochralski technique   211 ff, 215 ff, 220

d

dark-field (DF) image   349 ff
"dark star" motif   162, 168, 179 f
Debye approximation   74
Debye-Waller factor   105
decagonal quasicrystalline approximant   118
defect   6 ff
– line   6
– one-dimensional   6
– planar   6, 9 f
– zero-dimensional   6
deformation of CMAs   26 ff, 192 f, 274 f, 284
density functional theory (DFT)   167, 320 ff
density of states (DOS)   13 f, 44, 56
– *ab initio* calculation of   14, 140 ff
– distribution in Al_8V_5 16
– distribution in β-Al_3Mg_2 17
– effect of the electronic DOS on the thermopower   371
– effect on electrical conductivity   127
– fine spectral features   57 f
– modification of the phonon DOS   370
– of a free-electron system   13
– of a Hume-Rothery alloy   13, 172
– of thallium-substituted lead telluride   373
– relation to adhesion energy   254 f
Derjaguin-Muller-Toporov (DMT) model   190 f
DF detector   352
DF image. *see* dark-field image
DFT. *see* density functional theory
diffusion barrier in reinforced MMCs   293
diffusion length   209
dimethyl-ethyl-amine alane (DMEAA)   231
dislocation   6 ff
dispersion relation   72
DMEAA. *see* dimethyl-ethyl-amine alane
Doniach-Sunjic line shape analysis   174
DOS. *see* density of states
Drude model for optical conductivity   45 ff
Drude peak   46, 48
dynamical matrix   73, 78, 83 f, 93

e

EAM potential calculation   84
eigenmode   71
elastic limit   300
electrical conductivity   66
– model for calculation of   127 ff
electrical resistivity   19, 34. *see also* anisotropic electrical resistivity
– *ab initio* calculation of   148 f
– Boltzmann-type PTC   125 ff

electrochemical polarization measurement   256 ff, 263 ff
electrocodeposited quasicrystalline composite   309
electron confinement in thin films   186 f
electron density of states. *see* density of states (DOS)
electron localizability indicator (ELI)   388
electronic band structure   140, 175, 371
electronic state   175 f
electron-phonon scattering   125
ELI. *see* electron localizability indicator
ELI-D formulation   388
entropy-temperature relation of ferromagnetic material   381
Erlich-Schoebel barrier   179
$EuZn_2Sb_2$ 374

f

Fe_4Al_{13} as hydrogenation catalyst   392 ff, 396
Fe-B-Nb bulk metallic glass   284 ff
Fe-Nb alloy system   280 ff
Fermi energy   13 f, 56, 371
Fermi sphere pseudo-Brillouin zone interaction   56
Fermi surface   140 ff, 322
Fermi-Dirac distribution function   64
Fermi-sphere Bragg-plane interaction   171
Fermi-surface Brillouin-zone diffraction effect   56 f
Fe-Zr alloy system   283
FFM. *see* friction force microscopy
Fibonacci lattice   61 ff
filled skutterudite   375
floating zone melting technique   212
flux growth technique   214 f, 218
force-distance curve   192
fretting test   299 ff
Friauf polyhedron   82
friction   191
– adhesive part of   23, 25 f
– electronic   199
– on atomically clean quasicrystal surfaces   196 f
– on *in-situ* oxidized quasicrystal surfaces   196 f
– phononic   199
– properties of CMA matrix composites   306 ff
– relation of low friction with wetting and adhesion   200
friction anisotropy   197 f
– after surface modification   200
– of clean two-fold Al-Ni-Co surface   198 f

friction coefficient 22 ff, 197, 310
friction force microscopy (FFM) 195 f
friction measurement 194 f

g
gadolinium 319 f, 320 f
$Gd_5(Si_{1-x}Ge_x)_4$ alloy 324 f, 327, 330, 337 f
– structure of 346 f
$Gd_5Si_2Ge_2$ 317, 321, 324 f, 338
– alloys with iron substitutions 339 ff
– magnetic measurements on iron substituted phases of 345 f
– phase formation with iron 339 ff
– $\alpha \rightarrow \beta$ structural transformation 348 f
– structure of 348
– TEM analysis of 353 ff
– TEM analysis of iron substituted phases of 354 f
– X-ray diffraction measurements of iron substituted phases of 344
generalized vibrational density of states (GVDOS) 78, 97, 101
grain-boundary phase 325, 339 ff, 345, 357 f
GVDOS. see generalized vibrational density of states

h
half-Heusler phase 379 f
Hall coefficient 20, 132 ff. see also anisotropic physical property
– ab initio calculation of 146 ff, 150 f
Hall-Petch equation 277 f
hardness of CMAs 275 f, 295 f
harmonic approximation 73, 78, 83 f, 93 f
heterogeneous catalyst 386
Heusler phase 328 f, 379
horizontal gradient freeze technique 212
Hume-Rothery (H-R) phase 13, 171
Hume-Rothery mechanism 13 f, 44, 59, 68, 171
Hume-Rothery rules 13 f, 56, 170 f
Hurwitz Zeta function 65
hybridization mechanism 14 f, 44, 57, 59
hysteresis loss of MC, reduction of 324 f, 339 ff, 346 f

i
identification of trap sites 179 f
IEM transition. see itinerant-electron metamagnetic transition
incongruent melting phase 212, 214, 216
inelastic X-ray scattering 75, 89
in-situ analysis of gas phase reactions 393
in-situ analysis of the near-surface region of CMA catalyts 393
in-situ stability of intermetallic catalysts 390
interband transition 46
interlayer relaxation 169
intermetallic catalyst 386
– selectivity to ethylene versus activity in hydrogenation 397
– system requirements for the design of 387 f
intermixing 185 f
interstitial 6
intraband transition 45
intrinsic surface 156
inverse Matthiessen rule 43 ff
iron bis(N,N′-di-tert-butylacetamidinate) 231
island formation during thin-film growth 178
isothermal magnetization 317
itinerant-electron metamagnetic (IEM) transition 330 ff

j
Johnson-Kendall-Roberts (JKR) model 190 f

k
Kondo lattice system 372
Kondo resonance 372
Kramers-Krönig transformation 45
Kubo-Greenwood formalism 63 f
Kyropoulus equipment 215
Kyropoulus technique, liquid-encapsulated 224 f

l
$La(Fe_{13-x}M_x)$-based magnetocaloric compounds 330 ff
$La(Fe_{13-x}Si_x)$ alloy 330 ff
$La_{1-x}Ca_xMnO_3$ alloy 324 f
$LaFe_{13-x}Si_x$ 321 f
laser-cladded quasicrystalline composite 309
lattice dynamic
– of the i-Al-Pd-Mn icosahedral QC 87
– of xenon clathrate hydrates 96
Laves phase 2, 18, 81, 224 f
– in two-phase intermetallic compounds 280 ff
– magnetocaloric properties of 323 f
– mechanical properties of 274 f
LEED. see low-energy electron diffraction
LEIS. see low-energy ion scattering
Lifshitz-Van der Waals term 30 f
linear response theory 63
local-spin-density approximation (LSDA) 321
Lorenz function 50 ff, 64
Lorenz number 49, 66, 368

low-energy electron diffraction (LEED) 156, 161, 165, 181 f
low-energy ion scattering (LEIS) 161
LSDA. see local-spin-density approximation
LSDA + U method 321

m

magic island formation 187
magnetic entropy change 318 ff, 323, 328 ff, 337 f, 344 f
magnetic phase transformation
– first-order (FOMT) 319 ff
– FOMT coupled with a structural transformation 326
– of a ferromagnet 318 f
– second-order (SOMT) 319 ff
magnetic refrigeration 317
magnetic susceptibility 20, 117, 121 ff
magnetocaloric (MC) material 317 ff
– binary intermetallic magnetocaloric compounds 334 ff
– comparison of 336 f
– effect of interstitial hydrogen 332
– gadolinium 322 f, 337
– $Gd_5(Si_{1-x}Ge_x)_4$ alloys 324 f, 327, 330, 337 f
– $Gd_5Si_2Ge_2$ 321, 324 f
– $Gd_5Si_2Ge_2$ phases with iron substituents 338 ff
– $La(Fe_{13-x}Si_x)$ alloys 330 ff
– $La(Fe_{13-x}Si_x)H_y$ alloys 322 f
– $LaFe_{13-x}Si_x$ 321 f, 337
– Laves phases 323 f
– $Mn(As_{1-x}Sb_x)$ alloys 326 f
– $MnFe(P_{1-x}As_x)$ alloys 327 f
– $Mn_{1-x}Fe_xAs$ alloys 327, 331
– Mn_3GaC 329 f
– Mn_5Ge_3 329
– $Mn_5Ge_{3-x}Si_x$ alloys 330
– modelling of MC behaviour 320 ff
– nanocomposites 336
– $Nd_5(Si_{1-x}Ge_x)_4$ alloys 325
– Ni_2MnX Heusler alloys 328 f
– rare-earth manganites $R_{1-x}M_xMnO_3$ 332 f
– substituted $La(Fe_{13-x}M_x)$-based compounds 330 ff
– $Tb_5(Si_{1-x}Ge_x)_4$ alloys 325
magnetocaloric effect (MCE) 317 ff
– inverse 329
mass effect 81
mass-thickness contrast 350 f
material diffusion 209
matrix phase 273, 339 ff, 355, 357 f
MCE. see magnetocaloric effect
mean-field approximation 321

mechanical properties of quasicrystals 188 ff
melting behaviour 209 f
metadislocation 6 f, 9
metal deposition on a CMA surface 185
metal matrix composite (MMC) 35
metal matrix composite reinforced with CMAs 290 ff
– friction properties of 309 f
metal-oxide-solution capacitance 259 ff
microcell technique 262 ff
microhardness (Vickers) 229, 276 f, 296, 308 f
microhardness indentation test 274
microstructure 268, 278 ff, 340 ff, 357 f
– multiphase 283 ff, 312
– refinement of 278 ff
MMC. see metal matrix composite
$Mn(As_{1-x}Sb_x)$ alloy 326 f
$MnFe(P_{1-x}As_x)$ alloy 327 f
$Mn_{1-x}Fe_xAs$ alloy 327, 331
Mn_3GaC 329 f
Mn_5Ge_3 329
$Mn_5Ge_{3-x}Si_x$ alloy 330
MOCVD 225 ff. see molecular chemical vapor deposition
modelling
– in the 1/1 Zn-Sc approximant of i-Zn-Mg-Sc 88 ff
molecular chemical vapor deposition (MOCVD) 226 ff, 239 f
– of a copper coating processed from $CpCuPEt_3$ 233 ff
– of Al-Cu-Fe thin films 227 ff
– of an aluminum coating processed from DMEAA 232
– of an iron film from Fe2 237 f
– of the Al_4Cu_9 approximant phase 235 f
– precursors for codeposition of Al-Cu-Fe films 230 f
– process strategy 230
Mott-Schottky analysis 260
Mott's formula 67, 371
multi-phase intermetallic compound 280 ff, 312. see also bulk metallic glass

n

nanocomposite 336
nanostructured CMA 277 ff, 284 ff, 289
– preparation of 278 f
$Nd_5(Si_{1-x}Ge_x)_4$ alloy 325
neutron scattering
– coherent inelastic 75 ff
– incoherent inelastic 78, 100 f, 104
Ni_2MnX Heusler alloy 328 f
nucleation mechanism 178 f

o

optical conductivity   45 ff
optical phonon mode   46
oxidation of CMAs surface   244 ff

p

pair-volume function   388
passivating oxide layer   247, 253, 257 f
– electronic properties of   259 ff
– formed on aluminum-based CMAs at high temperatures   266 f
– oxide growth model   253
PBZ. see pseudo-Brillouin zone
PdGa as hydrogenation catalyst   386, 390, 395 f
Peierls transition   372
Penrose tiling. see pentagonal tiling
pentagonal hollow site   179 ff
pentagonal tiling of a CMA surface   160 ff
peritectic reaction   216, 219, 222, 292
PGEC. see concept of "phonon glass electron crystal"
ω-phase   292 ff, 310
phase diagram
– of the Al-Cu-Fe system   3
phase-transition temperature   321
phason line   6
phason plane   7, 9, 221
phonon   71 ff
– in quasicrystals and approximants   81 ff
– measuring of   75 f
phonon backscattering   369
phonon dispersion   74, 82, 101, 371
phonon-glass electron-crystal (PGEC)   96, 101 f
phonon lifetime   100
phonon mode
– effect of disorder on   81
– in the 1/1 Zn-Sc approximant of i-Zn-Mg-Sc   88 ff
– of skutterudites   101 ff
– of the i-Al-Pd-Mn icosahedral quasicrystal   86 ff
– of the i-Zn-Mg-Sc quasicrystal   88 ff
– of the Zn_2Mg laves phase   81 ff
– of xenon clathrate hydrates   95 ff
– of Zn_4Sb_3 alloy   103 f
phonon-phonon interaction   77, 79
photoemission spectra of quasicrystals   173, 177, 187
pin-on-disk friction experiment   22, 24, 194 f
plasma edge   46
plasma frequency value   46
plasma-arc sprayed quasicrystalline composite   306 f, 310
plasticity   25 ff, 274 f, 284

PMI. see pseudo-Mackay cluster
positive temperature coefficient (PTC)   125
prompt gamma-activation analysis (PGAA)   393
pseudo-Brillouin zone (PBZ)   14, 56, 171
pseudo-Brillouin zone boundary   87, 91
pseudogap   13 ff, 17, 44, 56, 94, 105, 168, 170 ff
– detection of, at the Fermi level   173 f
pseudo-Mackay cluster (PMI)   161, 221
pseudomorphic layer   181 f
PTC. see positive temperature coefficient

q

QC. see quasicrystalline compound
quantum size effect (QSE)   177, 187
quantum-well state (QWS)   186
quasicrystal   1 ff
– chemical constitution of thermodynamically stable   55
– decagonal systems   43, 48, 117 f
– electrical properties of   43 ff
– electrical resistivity of   19
– electronic-structure-related properties of   43 ff
– icosahedral systems   43, 46, 51, 68, 86 f
– optical conductivity of   45 ff
– photoemission spectrum of a Al-TM quasicrystal   172 f
– physical properties of, compared with metallic systems   53
– rare-earth-bearing   49
– stable binary   2, 57
– thermal conductivity of   105 ff
quasicrystalline approximant   43 ff, 48, 68, 81 ff. see also quasicrystal
– 1/1 Al-Mn-Si   48
– to decagonal quasicrystals   117 f
quasicrystalline composite coating   306 ff
– preparation of   30
quasicrystalline compound (QC)   13 f
quasiperiodic optical conductivity   48
QWS. see quantum-well state

r

rare-earth manganites $R_{1-x}M_xMnO_3$ 332 f
"rattling". see thermal oscillation
"rattling" mode   42, 96
real-space method   160 ff
reduction of DOS. see pseudogap
refrigerant capacity (RC)   338
"resonant scattering" of mobile charge carriers   372
RHEED pattern   181, 185
Riemann Zeta function   65

s

Samson phase 12, 16 ff
scanning transmission electron microscopy (STEM) 349 f
scanning tunneling microscopy (STM) 160 f, 164
SCAP. *see* structurally complex alloy phase
scattering rate 106
secondary electron imaging (SEI) 161
Seebeck coefficient 49, 66, 131, 366. *see also* thermoelectric power
SEI. *see* secondary electron imaging
self hybridization 15 ff
self-flux growth. *see* flux growth technique
semihydrogenation of acetylene 395 f
SEXS. *see* soft X-ray emission spectroscopy
single-crystal growth
– basic concepts of 208 ff
– Bridgman equipment 211
– crucible material 210
– growth velocity 209
– of Al-Pd-Mn approximants 221 f
– of $Al_{13}Co_4$ 215 f
– of $Al_{13}Fe_4$ 215 f
– of β-Al_3Mg_2 217
– of $MgZn_2$ 224 f
– of $Mg_{32}(Al,Zn)_{49}$ 219 f
– of Yb-Cu superstructural phases 222 f
– techniques for 210 ff
single-phase intermetallic compound 274 ff, 311 f. *see also* Laves phase
skutterudite 101 ff, 375 f
$SmRu_4P_{12}$ 101
soft X-ray emission spectroscopy (SEXS) 14
Sommerfeld's value 50 f, 368
SPA. *see* spot profile analysis
SPA-LEED technique 166 f
spectral conductivity function 59 f, 63, 67, 70
spectral conductivity model 59 f
spot profile analysis (SPA) 166
sputtered surface of CMA 157 f
stabilisation of alloy surfaces 169 f
stacking fault 6
STEM. *see* scanning transmission electron microscopy
STEM diffraction contrast 352 f
STEM image 351
STM. *see* scanning tunneling microscopy
stress-strain curve 27 f, 297, 299
structurally complex alloy phase (SCAP) 1 ff. *see also* CMA structure
– of $Al_4(Cr,Fe)$ CMAs 119 f
– of clathrates 377 f
– of decagonal quasicrystals 160

– of $EuZn_2Sb_2$ 367, 374
– of half Heusler MgAgAs-type phase 379
– of Heusler $MgCu_2Al$-type phase 379
– of *i*-CdYb 89
– of icosahedral quasicrystals 160, 167
– of o-$Al_{13}Co_4$ 119
– of PdGa 395
– of the Y-Al-Ni-Co phase 118 f
– of $Yb_{14}MnSb_4$ 367
– of Zn_4Sb_3 378 f
supercooled liquid (SCL) region of metallic glasses 287, 290
superstructure 7, 91, 222
– Moiré 174 f
– of Al-Pd-Mn approximants 221 f
– type-I 160, 163, 166
surface chemistry of CMAs 243
– under UHV environment 244 ff
surface energy 22 ff, 169
– applications related to 30 f
– broken-band approximation 169
– minimization of 170
surface kinetics 208 ff
surface orientation 169
surface preparation of CMA 156 ff
– of intrinsic surfaces in UHV starting from monocrystal 157
– sputter-annealing technique 157 f
surface reconstruction 169
surface relaxation 169
surface segregation in alloys 170
surface-structure determination 156 ff
– using *ab initio* methods 167 ff
– using real-space methods 160 ff
– using reciprocal-space methods 163 ff
symmetry of CMAs 160 ff
– icosahedral 35, 56, 91, 95, 161
– of decagonal QCs 160
– pentagonal symmetry elements 161
– pseudodecagonal 183
– pseudopentagonal 388
– ten-fold symmetry axis in decagonal QCs 160, 176

t

Ta-HfV_2 two-phase alloy 280 ff
Taylor phase 9
$Tb_5(Si_{1-x}Ge_x)_4$ alloy 325
TEM. *see* transmission electron microscopy
TEM diffraction contrast 351 f
TEM image 350 f
temperature coefficient 20
thermal conductivity 20, 49 f, 105 ff
– anisotropic 136 ff

– electron contribution to 64, 105, 149 f, 368
– enhancement parameter 51
– lattice contribution to 368 ff
– of silicium nanowires 369
– phonon contribution to 49 ff, 79
– phononic temperature dependence of 106 f
thermal diffusion 209
thermal diffusivity 105
thermal oscillation 109, 376
thermal surface vacancy 159
thermoelectric figure of merit 106 f, 365 ff, 376 ff
thermoelectric material 99, 101, 103, 106 f, 365 ff, 372 f
– clathrates 377 f
– design concepts for 368 ff
– filled skutterudites 375 f
– half-Heusler phases 379
– Zintl phases 374 f
– Zn_4Sb_3 378 f
thermoelectric power 20, 49, 64, 67, 95 ff. see also Seebeck coefficient
thermoelectric power factor 366, 369
thermoelectricity 105 ff, 365
thermopower. see thermoelectric power
thin-film deposition 225 ff. see also molecular chemical vapor deposition
thin-film growth on a CMA surface 177 ff. see also nucleation mechanism
transition-metal-containing metalloid glas 285
transmission electron microscopy (TEM) 349 ff
– analysis of $Gd_5Si_2Ge_2$ 353 f
– analysis of low hysteresis-loss CMA sample 354 f
transport coefficient 63 f
– for different quasicrystalline families 108
– theoretical ab initio calculation of 144 ff
transport process 208
transport properties of QC 43 ff, 60 f
transverse acoustic (TA) mode 82 f, 87, 92 ff
trap site 179
tribometer 194 f
twin boundary 355
twinning 160, 184 ff, 275, 348
two-beam CDF image 352
two-phase intermetallic compound 280 ff, 312. see also Laves phase

u

ultrahigh vacuum (UHV) environment 156, 244 ff, 393 f

Umklapp process 95, 100, 106

v

vacancy 6
VASP. see Vienna ab initio simulation package
VDOS. see vibrational density of states
vibrational density of states (VDOS) 74, 95
Vienna ab initio simulation package (VASP) 167

w

wetting property of CMA surfaces 188 f
WFL. see Wiedemann-Franz law
Wiedemann-Franz law (WFL) 49 ff, 105, 136 f, 368
"white flower" motif 162, 168, 180
work term 29

x

XPS. see X-ray photoelectron spectroscopy
X-ray photoelectron spectroscopy (XPS) 393 f

y

Y-Al-Ni-Co phase 118 ff
– ab initio calculation of the electronic band structure of 140 ff
– anisotropic magnetic properties of 120
– Fermi surface of 141
– friction anisotropy on two-fold clean surface of 198 f
– molecular adsorption onto d-Al-Ni-Co 249
– structure of 160
– surface-structure determination of 162 ff, 166, 168
Young's equation 188
ytterbium-based CMA
– $Yb_{14}MnSb_{11}$ 374
ytterbium-copper superstructural phase system 222 f

z

zinc-based CMA
– i-Zn-Mg(Y,Tb,Ho,Er) system 49 f
– Zn-Al-Mg phase 82
– Zn_2Mg 81 ff
– Zn_4Sb_3 103 f, 378 f
– $Zn_{43}Mg_{37}Ga_{20}$ 56
– $Zn_{60}Mg(rare earth)_{10}$ 56
– $Zn_{80}Sc_{15}Mg_5$ 56
Zintl phase 374 f
zirconium-based bulk glass-forming alloy 287 f
zone melting 212